AF352103

Horizons in World Physics
Volume 224

Dynamics of Transition Metals and Alloys

Horizons in World Physics

Horizons in World Physics
Volume 224

DYNAMICS OF TRANSITION METALS AND ALLOYS

S. Prakash

NOVA SCIENCE PUBLISHERS, INC.
Commack, New York
1998

Editorial Production: Susan Boriotti
Assistant Vice President/Art Director: Maria Ester Hawrys
Office Manager: Annette Hellinger
Graphics: Frank Grucci and John T'Lustachowski
Information Editor: Tatiana Shohov
Book Production: Christine Mathosian, Tammy Sauter and Diane Sharp
Circulation: Maryanne Schmidt
Marketing/Sales: Cathy DeGregory

Library of Congress Cataloging-in-Publication Data

Prakash, S.
Dynamics of transition metals and alloys / by S. Prakash
      p.   cm. – (Horizons in world physics; vol. 224)
    Includes index.
    ISBN 1-56072-574-5
    1. Solid State Physics. 2. Lattice dynamics. 3. Transition metals. 4. Transition metal
    alloys. 5. Electronic structure. I. Title. II. Series: Horizons in world physics; v. 224
QC176.P668   1998   98-21169
530.4'16—dc21   CIP

Copyright 1998 by   Nova Science Publishers, Inc.
                      6080 Jericho Turnpike, Suite 207
                      Commack, New York 11725
                      Tele. 516-499-3103   Fax 516-499-3146
                      E-Mail: Novascience@earthlink.net
                      Web Site: http://www.nexusworld.com/nova

Dedicated to

*the sacred and loving memory of my parents Shri S. L. Tiwari and
Smt. Kalawati Tiwari and my brother Shri R. S. Tiwari*

Contents

PREFACE

The complete theoretical basis of wave propagation in the perfect crystals was laid down by 1940 and by the same time the crucial and concerning experiments were carried out. In 1942, Max Born introduced the term "phonon". These principles along with their applications were summarized in the classic book of Born and Huang. A few years later revolutionary experimental developments of neutron scattering, Mösbaur effect and lasers techniques generated enormous amount of data on the dynamics of crystalline and non-crystalline solids. Dr. Brockhouse's first communication in 1956 caused a deep sensation and the later decades produced an abundant crop of phonon dispersion curves by neutron scattering method for the main reference materials. Phenomenological theories were developed to analyze the experimental data and to understand the nature of interatomic forces in these materials. A general account of these developments, with particular emphasis on ionic solids, is carefully scripted by Maradudin, Montroll, Weiss and Ipatova and Venkataraman, Feldkamp and Sahni in their classic volumes.

The Cauchy relations for elastic constants were not satisfied in the metallic crystals. Fuchs attributed this failure to the presence of conduction electrons in the metals. Since then it continue to be a puzzle to understand the effect of conduction electrons on the dynamics of atoms in metals. The gravity of this problem was more deeply realized in 1962 when neutron scattering revealed the anomalous phonon dispersion relation of Pb and later many materials with transition and rare earth elements showed these characteristics. Many of these materials were found superconductors. Therefore the understanding of interatomic interactions in these metallic materials became a challenge. Various phenomenological models were developed to account for the contribution of delocalized and partially localized conduction electrons towards the interatomic forces and phonon frequencies.

The microscopic analysis of interatomic forces and phonon frequencies of metals was started as early as in 1958 by T. Toya. With the development of fast computers, the calculations of electron energy band structure of metals became available. At the same time a huge wealth of phonon measurements in alkali metals, noble metals, transition and rare earth metals and their alloys got accumulated. Correspondingly, the first principle theories were formulated to understand the long range interatomic interactions and the anomalous behavior of phonon dispersion relations. Few reviews were also written on this subject. However, there does not exist an unified presentation of these developments of lattice dynamics of transition metals and their alloys. The present book is planned to fill up this gap.

It is not possible to sum up all the aspects of first principle description of dynamics of atoms in the solids in one go. Therefore and also due to personal limitations, I have restricted myself to transition metals and their alloys. The dynamics of non-transition metals and their alloys can be achieved just by dropping the contribution of partially localized conduction electrons. The book coherently presents the microscopic theories which have been developed in the last four decades. The available experimental data are presented with an emphasis on their anomalous features. The subjects of anharmonicity and surface phonons which need separate volumes, are excluded.

A text book on the "Dynamics of Transition Metals and Alloy" which covers advanced topics, is the need of a post graduate student and a researcher. A graduate student of physics and material science goes through the basic courses of quantum mechanics, mathematical physics and solid state physics. With this prerequisite, the book is designed for post graduate students, researchers and scientists working on the frontiers of material science and metal physics in particular. The intricate mathematical derivations are presented within the reach of a student and the final results are compared with the experimental observations. The book is self-contained with a long list of references and diagrams. I believe, researchers will find it as a useful desk book.

I was introduced and inspired to take up this project by my teachers Professor S.K. Joshi and Professor Rajender Singh and my collegue Dr. Narender Singh. I had very fruitful discussions on the subject of transition metals with Professor J. Friedel, Professor P.Lucasson, Dr. S.K.Sinha and Dr.P.K. Iyengar and continuous encouragement from my collegues Professor K.N. Pathak, Professor H.S. Hans, Professor U.S. Kushawaha and Professor Natthi Singh. The firm support to complete this project was given by my wife Mira Tiwari and children Deepankar and Hema. The credit to put this volume in the LaTeX language goes to my student Sanjeev Gautam. I am grateful to all of them and to those near and dears who suppoprted in silence.

S.Prakash

January 24, 1998

Chapter 1

INTRODUCTION

Contents

Chapter 1

INTRODUCTION

In the crystalline solids the atoms are arranged in a regular periodic structure. These atoms can be static only at absolute zero temperature. Otherwise, at all the times these are vibrating about their equilibrium positions. These motions constitute thermal energy of the solid. Historically, the study of dynamics of the solids was developed to understand their thermal properties [1]. Later on, the interest shifted to know the nature of interatomic forces which govern the atomic motions and binding of the solids [2]. In recent years, neutron inelastic scattering measurements revealed the phonon anomalies in the form of dips, peaks and wiggles in the frequency versus wave vector relations of simple metals [2], transition and rare earth metals and their compounds and alloys [3], [4]. These features indicate an intricate radial and angular dependence of interatomic forces [5]. The phonons and consequently the interatomic forces in the simple metals are well understood [5], [6]. However, the interatomic forces in the majority of the materials which consist of partially filled d and f-bands are only partially understood [3], [4].

The existence of Ni, Pd and Pt at the end of long periods in the periodic table corresponding to the filling of 3d, 4d and 5d shells suggest that the peculiar properties of these metals are due to strong d-character in their valence states. Similarly the properties of rare earth metals are governed by f-character of their valence states. These quasi-localized d and f-states continue to affect the elastic and electronic properties of the compounds and alloys of these metals too. The study of lattice dynamics of these materials makes it possible to understand various properties of these solids such as specific heat, optical and dielectric properties, certain aspects of interaction with radiation and their thermal and electrical conductivities in a unified way [7] [8].

It is not possible to consider the dynamical properties of a solid in complete isolation from other structural, electronic or magnetic properties. In fact, the dynamical properties of solids strongly depend on their structural, electronic and magnetic properties. With this view point, we have presented the microscopic analysis of phonon spectra of transition metals (TMS), rare-earth metals (REMs), transition metal compounds (TMCs) and their alloys in this monograph. In these materials the conduction electrons have both the free (s or p) and quasi localized (d or f) characters. Some of these materials show the magnetic ordering effects and some are even superconductors with sufficiently high superconducting transition tamperature T_c (upto 23^0 K for Nb_3 Ge).

In the study of dynamics of solids one has to consider three entities: (i) The phonon frequencies which are mostly determined through the experiment (ii) The polarization vectors whose mathematical structure is determined by the crystal symmetries but their phonon wave-vector dependence is known from the measurements. (iii) The dynamical matrix which depends upon the physical model adopted for a particular solid. It is hard to decide which one of these entities should be regarded as the fundamental and remaining two derived from it in the formulation of a theory of the dynamics of the solid. At first, the polarization vectors are calculated from the group theoretical analysis [7], [8]. Then the components of these vectors are related to the elements of dynamical matrix which is derived using a physical model for the system in question. This allows one to interpret the observed phonon spectra and to test the validity of the physical model. A discussion can also be made as to what new measurements are required. The calculations of polarization vectors are presented in the classic reviews of Born and Huang [7] and Maradudin et al [8]. The dynamics of perfect solids is excellently described by Venkataraman et al [9] and Brüech [10]. Böttger [11] has given the detailed description of phonons in disordered solids.

Theoretical analysis of lattice vibrations is undertaken on two distinct levels. At the first level, the Born-von-Kármán (BvK) force constant models are used. The most successful models have been the shell model [12], charge density distortion model [13] and the valence force field model [14]. These models are generalized to include the local and non-local deformations in the symmetries of the crystal structure [15]. The difficulties lie in the quantum mechanical interpretation of these models. Tolpygo [16] showed that the shell model is equivalent to the forces derived from the distortion of electronic bonding orbitals. However, many parameters are required to fit the phonon dispersion curves. Therefore, in practice, it is difficult to understand the deviations from the homology among the phonon dispersion curves. These models have been successful in explaining the phonon dispersion relations of insulating and semiconducting materials. While applying to transition metals and their alloys, appropriate modifications are to be incorporated.

At the second level, the dynamical properties are derived from the electronic structure of undistorted crystal. This is known as the microscopic approach which has two aims: First to understand in detail, how the dynamical spectrum of a solid is related to its crystal structure and electronic properties (i.e. atomic bonding, energy band structure, Fermi surface effects etc.) and secondly to use the same microscopic theory to explain the related properties such as superconductivity, crystallographic phase transitions, optical and transport properties and others. In this sense, the microscopic approach occupies a central place in the study of dynamics of crystalline as well as non-crystalline solids.

1.1   The Central Questions

The phonon spectra of transition metals, rare earth metals and their compounds and alloys possess a complex structure which reflects a variety of interesting and intricate properties which are not found in the simple metals and their alloys. The presence of phonon anomalies, discontinuties, lattice instabilities and superconductivity found in these materials pose some of the difficult questions to microscopic theory. In these systems, the conduction electrons consist of free as well as localized characters. The d and f-electrons in the outer atomic shells are not rigidly bound to the core. These are partially localized. Therefore these must be included in the calculation of response of the conduction electrons to the ionic motion. This response, known as dielectric screening function (or in more general "dielectric matrix"), will contain a free electron part and a part which is due to quasi localized d or f-electrons.

There is no clear cut separation between the core and conduction electrons in these materials. The s and d or f-electrons hybridize with one another and give rise to additional s-d and s-f hybridization interactions respectively. This further reduces the total energy of the crystal. In this process the s and d or f-electrons loose their individual characteristics [17]. The calculation of dynamical matrix is quite sensitive to the number and the characteristics of conduction electrons per atom. Therefore, the first question arises, how should the partially localised d or f- electrons be represented and how the contribution of the core and the conduction electrons to the dynamical matrix be separated in a theoretical description ?

The unusual behaviour in the phonon dispersion curves of superconducting Nb was first observed by Nakagawa and Woods [18]. Since then the numerous experimental data on TMs, REMs and TMCs show that high T_c elements and their compounds and alloys have anomalous dips in their phonon spectra whereas the neighbouring elements with similar crystal structure are non-superconducting and exhibit no anomalous features in the phonon spectra. For example, the phonon spectra of Nb [18], [19], V [20] and Cr [21] have anomalous features which are absent in the phonon spectra of Mo [22], Ta [23] and W [24]. Although hexagonal closed packed structure is least favourable for superconductivity but technetium (Tc) is a superconductor with superconducting transition temperature $T_c = 8^0$ K. The phonon spectrum of technetium [25] has strong anomalies in the longitudinal optical (LO) and transverse optical (TO) modes. Ruthenium (Ru), neighbouring to technetium has low superconducting transition temperature and the phonon spectrum shows only a small anomaly in the (LO) mode along c-direction. Zr is also a low temperature superconductor and shows some softening in the longitudinal mode along c-direction.

The phonon spectra of TiC, ZrC and HfC are free from anomalies and are non-superconducting ($T_c < 0.05^0$ K) [3], [4]. Hydrides, deutrides and nitrides of transition metals like PdH, PdD, NbD, TaH, TaD, TiN, ZrN and HfN [3], [4], [26], [27] are superconducting and exihibit anomalous behaviour in their phonon spectra. Similar trend is also observed in the transition metal alloys. For example, the $Nb_{0.25} Mo_{0.75}$ alloy does not have any anomaly in the phonon spectrum and $T_c = 0.04^0$ K. For higher and lower values of electron per atom (e/a) ratios in NbMo alloys, T_c increases rapidly and phonon anomalies appear in the phonon spectrum [27]. Pd is a non-superconductor but when it is alloyed with hydrogen and deutrium, it becomes a superconductor [28].

Phonon anomalies signify incipient lattice instabilities. For instance, addition of Zr to Nb, makes the phonon anomaly deeper in the longitudinal mode and ultimately at about $Zr_{0.80} Nb_{0.20}$ structural phase transition appears [29]. Although the structural instability or anomalous phonon softening in many materials appear to be correlated with the occurence of high superconducting transition temperature but there also exist counter examples [30]. Here a number of questions arise: What is the mechanism of the origin of phonon anomalies? Are the anomalies related to high T_c causing significant increase in it as argued by Testardi [30]? Is there any relationship between the phonon anomalies and the structural phase transition? Are the anomalies and high T_c both are the manifestations of the same strong electron-phonon (e-p) coupling? Do the magnetic interactions also contribute significantly to these anomalous characteristics ? At least some of the questions, if not all, can be answered by an appropriate microscopic theory of lattice vibrations of these materials [31]. Therefore, the central question is to discover a microscopic theory for the phonon spectra of these matrials and to understand some of the intricasies produced by partially localized conduction electrons.

1.2   Outlines of The Monograph

In this monograph we have examined the central questions mentioned above and their possible solutions with particular reference to transition metals, rare earth metals, their compunds and alloys. We confess our inability to discuss all the related topics, in particular the literature related to superconducting transition tamperature [32]. However, we have tried to present the developments of dynamics of these materials in the recent past in a unified way which can be easily adopted by a graduate student. The plan of the monograph is as follows:

The essentials of Born-von-Kármán (BvK) theory are presented in Chapter 2. Solution of dynamical equations in the harmonic approximation and the generalized tensor force models which are used to explain the experimental data of phonon dispersion relations are discussed in detail. The central force model, axially symmetric and modified axially symmetric force models which are based on abinito interatomic interactions are also described. The validity of the proposed interatomic interactions in reference to phonons in these materials is examined.

Owing to the partially localized character of conduction electrons in transition metals and their compounds, a similarity is drawn between these metallic compounds and insulating compounds. Therefore, in the frame work of BvK theory, the shell model, screened shell model, double shell model and their other extensions are used frequently to understand the interatomic interactions in transition metal compounds. These phenomenological multipole models and their applications are presented in Sec. 2.11 of Chapter-2.

The survey of experimental results on d and f-band metals, their compounds and alloys is presented in Chapter 3. The experimental procedures and the underlying physical concepts of inelastic coherent and incoherent neutron scattering, diffuse X- ray scattering, low energy diffuse electron scattering and He atom inelastic scattering are discribed in Secs. 3.1 to 3.4. The experimental results are tabulated in Sec. 3.5. The detailed information about the measured phonon frequencies, the models used to explain the experimental data, the observed phonon anomalies and appropriate references are compiled together in Sec. 3.5. The unique characteristics of the observed phonons and their relative comparision with other similar systems are discussed with the help of diagrams in Sec. 3.6.

Chapter 4 is uniquely devoted to develop a unified microscopic theory of lattice dynamics of these metallic systems. The derivation of longitudinal dielectric matrix in the Hartree approximation and further improvements due to exchange-correlation corrections are given in Sec. 4.1 and Appendix-A. The dynamical matrix using dielectric screening theory and second order perturbation theory is worked out in Sec.4.2 and Appendix C. The Coulomb part of the dynamical matrix is derived in Appendix-B. The abinitio derivation of phenomenological multipole models described in Sec. 2.11 is also presented there.

The quasi-localized conduction electrons are well described in the Wannier representation. Therefore, the electron-ion interaction in these materials is calculated in this representation. To make the calculations tractable, the linear combination of atomic orbitals (LACO) and the mixed band schemes are used. For further simplifications, the non-interacting band scheme is used in the numerical calculations. The observed phonon frequencies of transition metals and their compounds are explained. All these details are given in Chapter 4 [33], [34].

Harrison [35] extended the simple metal pseudopotential theory and Animalu [36] extended the Heine-Avarenkov model potential theory for transition metals. These model potentials in conjunction with second order perturbation theory are used to calculate the change in energy of a vibrating lattice, electron-ion matrix elements, dynamical matrix and finally the phonon frequencies. These aspects are discussed in Sec. 4.5. Recently there have been attempts to generate the central interatomic potentials for transition metals and rare-earth metals [37]. These potentials are used to explain the lattice dynamical properties of these metals. These details are given in Sec.4.6.

In practice it is difficult to calculate the lattice vibrational properties of transition metals and their alloys using microscopic theories. Therefore, there have been various attempts to develop the microscopic models to understand the interatomic interactions in these metallic systems. One such model is charge fluctuation model. Its general theory, applications and its phenomenological equivalence is given in Sec.5.1. Charge density distortion model and its applications are given in Sec. 5.2. Many of the transition metal properties are well described in the tight-binding approximation. As the tight-binding wave-functions are not orthogonal, full non-orthogonal tight-binding theory is developed for electron-phonon interaction in transition metals. These electron-phonon matrix elements along with the appropriate dielectric screening functions are used to calculate the dynamical matrix and the phonon frequencies of transition metals, their compounds and alloys. Attempts are also made to understand the anomalous behaviour of phonon frequencies of these materials. All these details are given in Secs. 5.3 and 5.4.

If it is believed that the anomalous behaviour of phonon frequencies is due to electronic structure of these materials, the electronic response function must exhibit the singular behaviour at the appropriate momentum transfer vectors. This aspect is examined in Chapter 6. The generalized susceptibility function and its abinitio calculations are described in detail. The procedure for numerical integration over the Brillouin zone is illustrated through diagrams. The results are given for transition metals, rare earth metals and their alloys. The possible correlations between the phonon anomalies and the susceptibility function are discussed.

The last Chapter of the monograph is devoted to the recent developments of lattice dynamics of transition metal alloys. With an introduction to Green's function for the perfect crystal, the dynamics of a substitutional defect crystal is described. The atomic displacements, phonon frequencies, density of states and other experimentally observed quantities are expressed in terms of Green's function of the perfect crystal [8]. The formalism is further generalized to calculate the various dynamical aspects of disordered alloys in Sec. 7.3. The averaged t-matrix approximation (ATA), virtual crystal approximation (VCA) and coherent potential approximation (CPA) are discussed in detail in the context of calculations of phonon frequencies and phonon density of states. The available calculations of transition metal alloys are reviewed [11], [38]. Recently developed shell model and double shell model for disordered alloys and their limited applications are discussed in Sec. 7.5 [39]. An epilogue is presented at the end.

Bibliography

[1] P.Debye, Ann. Phys. 43, 49 (1914).

[2] S.C. Ng and B.N. Brockhouse, in "Neutron Inelastic Scattering " (IAEA, Vienna, 1968) Vol.1.

[3] H.G.Smith, in "Superconductivity in d and f-band Metals" Ed, D.H. Douglass, (AIP, New York, 1972).

[4] H.G.Smith, N.Wakabayashi, and M.Möstoller, in "Superconductivity in d and f-band Metals" Ed. D.H. Douglass (Plenum Press, New York, 1976).

[5] S.K.Joshi and A.K.Rajagopal, Solid State Physics, Vol.22, (Academic Press, New York, 1969).

[6] L.J.Sham, in "Dynamical Properties of Solids" Vol.1, Eds. G.K.Horton and A.A. Maradudin, (North Holland Pub.Co., 1974).

[7] M.Born and K.Huang, "Dynamical Theory of Crystal Lattices" (Oxford Univ. Press, Oxford, 1954).

[8] A.A.Maradudin, E.W.Montroll, G.H.Weiss and I.P.Ipatova, Solid State Physics, Suppl. 3, (2nd Edition, Academic Press, New York, 1971).

[9] G.Venkataraman, L.A.Feldkamp and V.C.Sahni, "Dynamics of Perfect Crystals" (MIT Press, London, 1975).

[10] P.Brüch, "Phonons: Theory and Experiment I, II" (Springer-Verlag, Heidelberg, 1982-83).

[11] H.Böttgar, "Principles of the Theory of Lattice Dynamics" (Physik-Verlag, Weinheim, 1986).

[12] W. Cochran, Crit. Rev. in Solid State Sciences, 2, 1 (1971).

[13] J.R.Hardy, in "Dynamical Properties of Solids" Vol.1, Eds. G.K.Horton and A.A.Maradudin, (North Holland, Amsterdam, 1974).

[14] H.Bilz, B. Gliss and W.Hanke, in "Dynamical Properties of Solids" Vol.1, Eds. G.K.Horton and A.A.Maradudin (North Holland, Amsterdam, 1974).

[15] B.Dorner, W.Van der Osten and W.Buhrar, J.Phys. C9, 723 (1976).

[16] K.B.Tolpygo, Sovt. Phys.-Solid State 3, 685 (1961).

[17] J.Friedel, in "The Physics of Metals" Ed. J.M.Ziman, (Cambridge Univ. Press. 1969).

[18] Y.Nakagawa and A.D.B. Woods, Phys. Rev. Letters, 11, 271 (1963).

[19] R.I.Sharp, J.Phys. C2, 421, 432 (1969).

[20] R.Collela and B.W.Batterman, Phys. Rev. B1, 3913 (1970).

[21] W.M.Shaw and L.D.Mühlestein, Phys. Rev. B4, 969 (1971).

[22] C.B.Walker and P.A.Egelstaff, Phys. Rev. 177, 1111 (1969).

[23] A.D.B.Woods, Phys. Rev. 136, A781 (1964).

[24] S.H.Chen and B.N.Brockhouse, Solid State Comm. 2, 73 (1964).

[25] H. G. Smith, N. Wakabyashi, R. M. Nicklow and S. Mihailovich, in "Proc. of Low Temperature Physics" LT-13, Eds. K. D. Timmerhause, W. J. O'Sullivan, and E.F.Hammel (Plenum Press, N.Y. 1974).

[26] H.G.Smith and W.Gläser, in "Proc. of International Conference on Phonons", Rennes, France, Ed. M. A. Nusivovici (Flammarion, Paris, 1971).

[27] R. L. Cappelletti, N. Wakabayashi, W. A. Kamitakahara, J. G. Taylor and A. J. Bevolo, Phys. Rev. B 25, 6096 (1982).

[28] J.M.Rowe, J.J.Rush, M.Möstoller, H.G.Smith and H.E.Flotow, Phys. Rev. Letters, 33, 1297 (1974).

[29] B. M. Powell, P. Martel and A. D. B. Woods, Phys. Rev. 171, 727 (1968).

[30] L.R.Testardi, Comments in Solid State Physics, 6, 131 (1975).

[31] P. B. Allen and B. Mitrovic, Solid State Physics Vol.37, (Academic New York, 1982); P. B. Allen, in "Dynamical Properties of Solids" Vol.3, Eds. G. K. Horton and A. A. Maradudin (North Holland, Amsterdam, 1980).

[32] J. P. Carbotte, Rev. Mod. Phys. 62, 1027 (1990).

[33] S. Prakash and S. K. Joshi, Phys. Rev. B2, 915 (1970); B4, 1770 (1971).

[34] S. Prakash, in "Current Trends in Lattice Dynamics" Ed. K. R. Rao (IPA Publication, Bombay, 1979); S.Prakash in "Lattice Dynamics" Ed. M. Balkanski (Flammarion Sciences, Paris, 1977).

[35] W. A. Harrison, Phys. Rev. 181, 1036 (1969).

[36] A. O.E. Animalu, Phys. Rev. B8, 3542, 3555 (1973).

[37] W. A.Harrison and G.K.Straub, Phys.Rev. B36, 2695 (1987).

[38] R. J. Elliot, J.A.Krumhansl and P.L.Leath, Rev. Mod.Phys. 46, 465 (1974).

[39] G. Grünewald, J. Phys. C14, 595 (1981).

Chapter 2

BORN-VON-KÁRMÁN THEORY

Contents

List of Figures

List of Tables

21

Chapter 2

BORN-VON-KÁRMÁN THEORY

A. Basic Concepts

A metal in the perfect crystalline state consists of the array of positive ions embedded in a more or less uniform sea of negative charge of conduction electrons. The equations of motion for thermal vibrations of the electron-ion system involve many-body interactions. The exact solutions of these equations are formidable. Therefore following approximations are adopted [1]-[5]

2.1   Rigid Ion Approximation

The core electrons are assumed to move rigidly with its nuclei and cannot be excited with available thermal energies. However, the questions arise: What is the configuration of rigid ion core and how many conduction electrons per atom are there in a metal? These questions are answered to some extent for simple metals but are unanswered for d and f-band metals. For example, in Al (atomic number 13) the rigid core has $1s^2\ 2s^2\ 2p^6$ configuration and remaining three electrons per atom are regarded as the conduction electrons. In a transition metal Ni (atomic number 28), the preassumption that the rigid core configuration is $1s^2\, 2s^2\, 2p^6\, 3s^2\, 3p^6\, 3d^8$ and there are two electrons per atom as the conduction electrons does not explain the electrical conductivity and the intrinsic magnetic moment 0.606 Bohr magnaton [6]

For the moment, it is assumed that the rigid core is upto a particular electronic configuration and the remaining electrons are the conduction electrons per atom in a metal. The Hamiltonian for such a thermally perturbed metallic solid is

$$H = H_I + H_e \tag{2.1}$$

where H_I and H_e are the ionic and electronic parts of Hamiltonian which are defined as,

$$
\begin{aligned}
H_I &= -\sum_{l\kappa} \frac{\hbar^2}{2M_{l\kappa}} \frac{\partial^2}{\partial R^2(l\kappa)} + \sum_{l'\kappa'>lk} V_I\left(\vec{R}(l\kappa) - \vec{R}(l'\kappa')\right) \\
&= T_I + V_{II} \tag{2.2}
\end{aligned}
$$

$$
\begin{aligned}
H_e &= -\sum_{i} \frac{\hbar^2}{2m} \frac{\partial^2}{\partial r_i^2} + \sum_{l\kappa,i} V_b\left(\vec{r}_i - \vec{R}(l\kappa)\right) + \sum_{i>j} \frac{e^2}{|\vec{r}_i - \vec{r}_j|} \\
&= T_e + V_b(\vec{r},\vec{R}) + V_{ee}(\vec{r}) \tag{2.3}
\end{aligned}
$$

Here $M_{l\kappa}$ and $\vec{R}(l\kappa)$ are the mass and instantaneous position coordinates of the κth atom in the lth unit cell, while m and $\vec{r}_i$ are the corresponding quantities for the electron. In Eq. (2.2), the first term is kinetic energy operator and the second term is potential energy of the ions which includes both the Coulomb and non-Coulomb (core-core exchange overlap) interactions [7]. In a similar description to Eq. (2.3), the first term is kinetic energy operator of conduction electrons, the second term is bare electron-ion interaction and the third term is electron-electron Coulomb interaction. The bare ion potential $V_b(\vec{r}_i - \vec{R}(l\kappa))$ includes both the Coulomb and non-Coulomb (core-conduction exchange overlap) interactions. In Eqs. (2.2) and (2.3) the sum is over all the ionic and electronic coordinates.

2.2   Adiabatic Approximation

The adiabatic approximation is also known as Born- Oppenheimer approximation [8]. It simplifies the problem of determining the eigenvalues of H by separating the dynamical aspects of ionic and electronic motions. The essential idea is as follows :

As the rigid ion cores move, the conduction electron wave-functions do not remain unaffected. The ions which are $2 \times (10^3$ to $10^5)$ times heavier than the electrons, move much slower than the electrons. Therefore at any instant, the electrons see the ions in some fixed configuration $\{\vec{R}(l\kappa)\} = \vec{R}$. In other words, one can say that the electrons follow ionic motion adiabatically i.e. the electrons do not make abrupt transitions from one state to another. Instead, an electronic state is deformed progressively by ionic displacements.

For any fixed configuration $\vec{R}$ of ions, the electrons in the configuration $\vec{r}(\equiv \{\vec{r}_i\})$ follow the wave equation

$$H_e(\vec{r}, \vec{R})\psi_n(\vec{r}, \vec{R}) = E_n(\vec{R})\psi_n(\vec{r}, \vec{R}), \qquad (2.4)$$

where $\psi_n(\vec{r}, \vec{R})$ and $E_n(\vec{R})$ are the electronic eigenfunction and eigen value of the nth state and depend parametrically on $\vec{R}$. $\psi_n(\vec{r}, \vec{R})$ make a complete set. Therefore the eigenfunctions of H satisfy the equation

$$H\phi_\lambda(\vec{r}, \vec{R}) = E_\lambda\phi_\lambda(\vec{r}, \vec{R}), \qquad (2.5)$$

where E_λ is the energy of the entire system in the state λ. To determine the wave function ϕ_λ, it is expanded in terms of complete orthonormal set of functions $\psi_n(\vec{r}, \vec{R})$ i.e.

$$\phi_\lambda(\vec{r}, \vec{R}) = \sum_n \chi_{n\lambda}(\vec{R})\psi_n(\vec{r}, \vec{R}). \qquad (2.6)$$

The coefficients $\chi_{n\lambda}(\vec{R})$ are essentially the eigenfunctions of the effective ionic Hamiltonian and these are determined as discussed below.

We assume that the lattice is monotonic. Then the use of Eqs. (2.6) and (2.1) in Eq. (2.5) gives

$$\left[-\sum_l \frac{\hbar^2}{2M}\nabla_l^2 + \sum_{l>l'} V_I\left(\vec{R}(l) - \vec{R}(l')\right) + H_e \right] \sum_n \chi_{n\lambda}(\vec{R})\psi_n(\vec{r}, \vec{R})$$

$$= E_\lambda \sum_n \chi_{n\lambda}(\vec{R})\psi_n(\vec{r}, \vec{R}). \qquad (2.7)$$

Operation of ∇_l^2 and use of Eq. (2.4) simplifies Eq. (2.7) as

$$\sum_n \psi_n(\vec{r}, \vec{R})\left[H_I + E_n\right]\chi_{n\lambda}(\vec{R}) \;-\; \frac{\hbar^2}{2M}\sum_n\sum_l\left[\chi_{n\lambda}(\vec{R})\nabla_l^2\psi_n(\vec{r}, \vec{R})\right.$$

$$\left. +2\vec{\nabla}_l\psi_n(\vec{r}, \vec{R})\cdot\vec{\nabla}_l\chi_{n\lambda}(\vec{R})\right] \;=\; E_\lambda\sum_n\chi_{n\lambda}(\vec{R})\psi_n(\vec{r}, \vec{R}). \qquad (2.8)$$

Premultiplying by $\psi_{n'}^*(\vec{r}, \vec{R})$ on both the sides and integrating over $\vec{r}$ one gets

$$\left(H_I(\vec{R}) + E_n(\vec{R}) + C_{nn}(\vec{R})\right)\chi_{n\lambda}(\vec{R}) + \sum_{n'\neq n} C_{nn'}\chi_{n'\lambda}(\vec{R}) = E_\lambda\chi_{n\lambda}(\vec{R})$$

$$(2.9)$$

where

$$C_{nn'} = A_{nn'} + B_{nn'}, \qquad (2.10)$$

$$A_{nn'} = -\frac{\hbar^2}{2M}\sum_l\int d\vec{r}\,\psi_n^*(\vec{r}, \vec{R})\nabla_l^2\psi_{n'}(\vec{r}, \vec{R}), \qquad (2.11)$$

$$B_{nn'} = -\frac{\hbar^2}{M}\sum_l\int d\vec{r}\,\psi_n^*(\vec{r}, \vec{R})\vec{\nabla}_l\psi_{n'}(\vec{r}, \vec{R})\cdot\vec{\nabla}_l. \qquad (2.12)$$

E_n and C_{nn} are inversely proportional to electronic and ionic masses respectively, therefore $C_{nn} \ll E_n$. Similarly on the left side of Eq. (2.9), the second term is very small as compared to the first term. Therefore $\chi_{n\lambda}(\vec{R})$ is adequately determined by solving the equation.

$$\left[H_I(\vec{R}) + E_n(\vec{R})\right]\chi_{n\lambda}(\vec{R}) = E_\lambda\chi_{n\lambda}(\vec{R}). \qquad (2.13)$$

This shows that the electronic contribution towards ionic motion is simply given by an effective potential $E_n(\vec{R})$ which is the *nth* state electronic energy and it depends parametrically upon $\vec{R}$. Substitution of $\phi_\lambda \approx \chi_{n\lambda}(\vec{R})\psi_n(\vec{r}, \vec{R})$ in Eq. (2.5) also gives Eq. (2.13). Consequently adiabatic approximation leads to separable electronic and ionic motions. In practice $E_n(\vec{R})$ is calculated for the electronic ground state i.e. $E_n(\vec{R}) \approx E_0(\vec{R})$.

The question arises: What are the conditions when adiabatic approximation holds? The answer is in Eq. (2.13). As the ions move, the electronic configuration continuously readjusts itself according to

Eq. (2.4) and keeps itself in the same *nth* state. This is possible only when the ions move very slowly as compared to electrons so that the latter make continuous readjustment to remain in the same state. In other words the frequencies of ionic motion must be much smaller than the characteristic electronic transition frequencies. This may always be true for insulators because the energy gap between the filled and unfilled states leads to higher electronic transitions frequencies. Therefore, one may expect that this approximation is always valid for insulators and large gap semiconductors while it may be poor for metals as there is no corresponding energy gap. However, Chester [8] has shown that the coupling coefficients C_{nn} which govern nonradiative transitions are negligible for all the states except for those few lying within the thermal layer of typical phonon energy $\hbar\omega$ near Fermi energy E_F. The number of these electrons is very small as compared to the total number of electrons in the conduction band. The exclusion principle also disallows most of these nonradiative transitions. Thus the adiabatic approximation is found to be valid for all those physical properties of a metal which involve entire electronic distribution. This is applicable to phonon frequencies of metals even with strong electron-phonon interaction leading to Kohn anomalies. This approximation breaks down for phenomenon where the dynamics of electrons is directly involved such as resistance due to scattering of electrons by phonons and superconductivity [9] to [11] .

2.3 Dynamical Equations in the Harmonic Approximation

Consider a perfect crystal of N unit cells where N is typically of the order of the number of unit cells in the laboratory specimen. If there are r atoms in each cell, there will be rN atoms in the crystal. The equilibrium position of κth atom in the *lth* cell is (Fig. 2.1)

$$\vec{R}^0(l\kappa) = \vec{R}^0_l + \vec{R}^0_\kappa, \tag{2.14}$$

where

$$\vec{R}^0_l = l_1\vec{a}_1 + l_2\vec{a}_2 + l_3\vec{a}_3. \tag{2.15}$$

l_1, l_2, l_3 (collectively referred as l) are integers and $\vec{a}_1, \vec{a}_2, \vec{a}_3$ are primitive translation vectors which generate the lattice. $\vec{R}_\kappa^0$ is the position vector of κth atom ($\kappa = 1, 2, \ldots, r$) and defines the basis of the lth cell. Each $\vec{R}_\kappa^0$ do generate its own sublattice which may have the same symmetry as that of the crystal lattice or lower than that. Thus the crystal lattice consists of r sublattices each of which is a Bravias lattice. In some of the crystals these sublattices belong to similar atoms such as in Zn and Y the two interpenetrating hexagonal sublattices are occupied by similar atoms. These sublattices may also be occupied by different atoms such as in NbC and ZrN the two interpenetrating fcc sublattices are occupied by dissimilar atoms. The details of the crystal structure are given elsewhere [3].

The reciprocal lattice of the crystal lattice defined by Eq. (2.15) is generated by the translation vectors

$$\vec{G}(h) = h_1\vec{b}_1 + h_2\vec{b}_2 + h_3\vec{b}_3, \tag{2.16}$$

where h_1, h_2, h_3 are integers and $\vec{b}_1, \vec{b}_2, \vec{b}_3$ are primitive translation vectors of reciprocal lattice. The vectors $\vec{b}_1, \vec{b}_2, \vec{b}_3$ are related to $\vec{a}_1, \vec{a}_2, \vec{a}_3$ by the relation,

$$\vec{a}_i \cdot \vec{b}_j = 2\pi\delta_{ij}, \tag{2.17}$$

where δ_{ij} is Kronecker delta and $i, j = 1, 2, 3$. Evidently

$$\vec{G}(h) \cdot \vec{R}_l^0 = 2\pi \times \text{integer}. \tag{2.18}$$

The volume of unit cell in the crystal lattice is

$$v = \mid \vec{a}_1 \cdot (\vec{a}_2 \times \vec{a}_3) \mid, \tag{2.19}$$

and that of the corresponding cell of reciprocal lattice is

$$\mid \vec{b}_1 \cdot (\vec{b}_2 \times \vec{b}_3) \mid = (2\pi)^3/v. \tag{2.20}$$

The thermally displaced position $\vec{R}(lk)$ of each atom is

$$\vec{R}(\vec{l}k) = \vec{R}^0(lk) + \vec{u}(lk, t), \tag{2.21}$$

where $\vec{u}$ is the small displacement of the $l\kappa th$ atom from the equilibrium position $\vec{R}^0(lk)$. We drop the time argument of $\vec{u}$ in the following discussion for simplicity till Eq. (2.52). In the adiabatic approximation, the kinetic energy part of the Hamiltonian given in Eq. (2.13), is

$$T_I = -\sum_{l\kappa\alpha} \frac{\hbar^2}{2M_{l\kappa}} \frac{\partial^2}{\partial u_\alpha^2(l\kappa)} = \frac{1}{2} \sum_{l\kappa\alpha} M_{l\kappa} \dot{u}_\alpha^2(l\kappa), \qquad (2.22)$$

where $\alpha(= x, y, z)$ is the Cartesian component index and dot over $\vec{u}$ denotes its time derivative. The potential energy part is

$$\Phi_n(\vec{R}) = \sum_{l'\kappa'>l\kappa} V_I\left(\vec{R}(l\kappa) - \vec{R}(l'\kappa')\right) + E_n(\vec{R}). \qquad (2.23)$$

In the further discussion, it is assumed that $E_n(\vec{R})$ is the ground state energy and therefore the subscript n is dropped. The Taylor series expansion of $\Phi\left(\{\vec{R}^0(l\kappa) + \vec{u}(l\kappa)\}\right)$ in powers of $\vec{u}(l\kappa)$ is

$$\Phi = \Phi^{(0)} + \Phi^{(1)} + \Phi^{(2)} + \Phi^{(3)} + \ldots, \qquad (2.24)$$

where

$$\Phi^{(0)} = \Phi\left(\vec{R}^0(l\kappa)\right)$$

$$\begin{aligned}
\Phi^{(1)} &= \sum_{l\kappa\alpha} \frac{\partial\Phi}{\partial u_\alpha(l\kappa)}\bigg)_0 u_\alpha(l\kappa) \\
&= \sum_{l\kappa\alpha} \Phi_\alpha(l\kappa) u_\alpha(l\kappa)
\end{aligned}$$

$$(2.25a)$$

$$\begin{aligned}
\Phi^{(2)} &= \frac{1}{2!} \sum_{l\kappa\alpha} \sum_{l'\kappa'\beta} \frac{\partial^2\Phi}{\partial u_\alpha(l\kappa)\partial u_\beta(l'\kappa')}\bigg)_0 u_\alpha(l\kappa) u_\beta(l'\kappa') \\
&= \frac{1}{2!} \sum_{l\kappa\alpha} \sum_{l'\kappa'\beta} \Phi_{\alpha\beta}(l\kappa, l'\kappa') u_\alpha(l\kappa) u_\beta(l'\kappa')
\end{aligned}$$

$$(2.25b)$$

$$
\begin{aligned}
\Phi^{(3)} &= \frac{1}{3!}\sum_{l\kappa\alpha}\sum_{l'\kappa'\beta}\sum_{l''\kappa''\gamma}\left.\frac{\partial^3\Phi}{\partial u_\alpha(l\kappa)\partial u_\beta(l'\kappa')\partial u_\gamma(l''\kappa'')}\right)_0 \\
&\quad \times u_\alpha(l\kappa)u_\beta(l'\kappa')u_\gamma(l''\kappa'') \\
&= \frac{1}{3!}\sum_{l\kappa\alpha}\sum_{l'\kappa'\beta}\sum_{l''\kappa''\gamma}\Phi_{\alpha\beta\gamma}(l\kappa,l'\kappa',l''\kappa'')u_\alpha(l\kappa)u_\beta(l'\kappa')u_\gamma(l''\kappa'').
\end{aligned}
$$

$$(2.25c)$$

$$(2.25)$$

The series is expected to converge rapidly if $\vec{u}(lk)$ is small as compared to interatomic spacing. For most of the solids this condition is satisfied at the temperatures below the melting point and if the zero-point displacements are negligible as compared to interatomic spacing. The subscript $)_0$ denotes that the derivatives are evaluated for the equilibrium configuration.

For the vibrational properties of solids, it is sufficient to retain the terms upto $\Phi^{(2)}$. This leads to harmonic vibrations of atoms in the crystal and therefore referred as "Harmonic Approximation". $\Phi^{(3)}$ and other anharmonic terms are important for temperature dependent properties such as thermal expansion, thermal conductivity etc. We shall not deal with these vital properties of the solids at this stage [12].

In Eq. (2.24), $\Phi^{(0)}$ is the static part of potential energy and does not contribute towards the dynamics. Therefore $\Phi^{(0)}$ is dropped altogether unless total energy of the crystal is required. The force acting on an atom at $\vec{R}(l\kappa)$ in the α-direction due to the displacement of all other atoms at $\vec{R}(l'\kappa')$ is given as (using Eqs.(2.25a) and (2.25b))

$$
\begin{aligned}
F_\alpha(lk) &= -\frac{\partial\Phi}{\partial u_\alpha(l\kappa)} &\qquad (2.26)\\
&= -\Phi_\alpha(l\kappa) - \sum_{l'\kappa'\beta}\Phi_{\alpha\beta}(l\kappa,l'\kappa')u_\beta(l'\kappa'). &\qquad (2.27)
\end{aligned}
$$

If all the atoms are in their equilibrium positions, $u_\beta(l'\kappa') = 0$, therefore $F_\alpha(l\kappa)|_0 = -\Phi_\alpha(l\kappa)$. But the force on each atom vanishes in the equilibrium position, therefore

$$\Phi_\alpha(l\kappa) = 0, \qquad \text{for each}(l\kappa\alpha), \qquad (2.28)$$

which leads to $\Phi^{(1)} = 0$. Thus in the harmonic approximation the potential energy

$$\Phi = \frac{1}{2} \sum_{l\kappa\alpha} \sum_{l'\kappa'\beta} \Phi_{\alpha\beta}(l\kappa, l'\kappa') u_\alpha(l\kappa) u_\beta(l'\kappa'), \qquad (2.29)$$

and hence the crystal Hamiltonian in the harmonic approximation is

$$H = \frac{1}{2} \sum_{l\kappa\alpha} M_{l\kappa} \dot{u}_\alpha^2(l\kappa) + \frac{1}{2} \sum_{l\kappa\alpha} \sum_{l'\kappa'\beta} \Phi_{\alpha\beta}(l\kappa, l'\kappa') u_\alpha(l\kappa) u_\beta(l'\kappa'). \qquad (2.30)$$

The force on the $(l\kappa)th$ atom in the α-direction is

$$F_\alpha(l\kappa) = - \sum_{l'\kappa'\beta} \Phi_{\alpha\beta}(l\kappa, l'\kappa') u_\beta(l'\kappa'). \qquad (2.31)$$

The equations of motion of the crystal lattice with the help of Eqs. (2.28), (2.30) and (2.31) can be written as

$$M_{l\kappa} \ddot{u}_\alpha(l\kappa) = - \sum_{l'\kappa'\beta} \Phi_{\alpha\beta}(l\kappa, l'\kappa') u_\beta(l'\kappa'). \qquad (2.32)$$

Thus for a crystal of N unit cells there are $3rN$ equations of motion. The coefficients $\Phi_{\alpha\beta}(l\kappa, l'\kappa')$ are atomic force constants and are interpreted as follows : Suppose all the atoms are in their equilibrium positions except a particular atom $(l'\kappa')$ which is displaced a distance d_β in the β-direction, then from Eq. (2.31)

$$F_\alpha(l\kappa) = -\Phi_{\alpha\beta}(l\kappa, l'\kappa') d_\beta. \qquad (2.33)$$

Thus the coefficient $\Phi_{\alpha\beta}(l\kappa, l'\kappa')$ is the negative of the force exerted in the α- direction on the atom $(l\kappa)$ when the atom $(l'\kappa')$ is displaced a unit distance $(d_\beta = 1)$ in the β-direction while all other atoms are in their equilibrium positions. Schematically it is shown in the Fig. 2.2. These are the force constants which are introduced in the dynamics of Hook's spring [1]-[3].

2.4 Restrictions on the Atomic Force Constants

There are number of restrictions on the atomic force constants due to translation and rotation symmetries. These restrictions are mentioned here briefly [2].

2.4.1 Permutation Symmetry

In the Eq.(2.25b), if the order of differentiation is changed, the atomic force constant does not change, therefore

$$\Phi_{\alpha\beta}(l\kappa, l'\kappa') = \Phi_{\beta\alpha}(l'\kappa', l\kappa). \qquad (2.34)$$

2.4.2 Rigid Body Translation and Rotation Sum Rules

The invariance of numerical value of the potential energy against rigid body translation and rotation of the crystal leads to the following restrictions.

1. For rigid body translation we replace $u_\alpha(l\kappa)$ and $u_\beta(l\kappa)$ by v_α and v_β independent of $(l\kappa)$ in Eqs. (2.29) and (2.30) and use the condition $\Phi = \Phi^{(0)}$ and $F_\alpha(l\kappa) = 0$ since v_α and v_β are arbitrary. These simplifications lead to the force constant sum rules

$$\sum_{l\kappa}\sum_{l'\kappa'} \Phi_{\alpha\beta}(l\kappa, l'\kappa') = 0, \qquad (2.35a)$$

   and

$$\sum_{l'\kappa'} \Phi_{\alpha\beta}(l\kappa, l'\kappa') = 0. \qquad (2.35b)$$

   Thus the sum of atomic force constants vanishes. The sum rule (2.35b) implies (2.35a) but not the vice versa. Therefore, Eq. (2.35b) is more restrictive than the Eq. (2.35a).

It is important to note that the self-force constant $\Phi_{\alpha\beta}(l\kappa, l\kappa)$ is not the second derivative of Φ with respect to $u_\alpha(l\kappa)$ but it is given by the sum rule (2.35b) i.e.

$$\Phi_{\alpha\beta}(l\kappa, l\kappa) = -\sum_{l'\kappa' \neq l\kappa} \Phi_{\alpha\beta}(l\kappa, l'\kappa'). \qquad (2.35c)$$

$$(2.35)$$

This relation is useful for many applications.

2. For an infinitesimal rigid body rotation of the crystal, the displacement is written as

$$u_\alpha(l\kappa) = \sum_\beta \omega_{\alpha\beta} R_\beta^0(l\kappa), \qquad (2.36)$$

where $\omega_{\alpha\beta}$ are the elements of an infinitesimal antisymmetric matrix i.e. $\omega_{\alpha\beta} = -\omega_{\beta\alpha}$ [13]. Using Eq. (2.36) in (2.27) one gets

$$-F_\alpha(l\kappa) = \Phi_\alpha(l\kappa) + \sum_{l'\kappa'} \sum_{\beta,\gamma} \Phi_{\alpha\beta}(l\kappa, l'\kappa')\omega_{\beta\gamma}(l\kappa, l'\kappa')R_\gamma^0(l'\kappa').$$

$$(2.37)$$

However, the $F_\alpha(l\kappa)$ component of the force on $(l\kappa)$ atom must transform as the α-component of another vector (force) when lattice is rigidly rotated i.e. upto the first order in $\omega_{\alpha\beta}$.

$$\begin{aligned}
F_\alpha(l\kappa) &= \sum_\beta (\delta_{\alpha\beta} + \omega_{\alpha\beta}) F_\beta(l\kappa) \\
&= -\sum_\beta (\delta_{\alpha\beta} + \omega_{\alpha\beta}) \Phi_\beta(l\kappa). \qquad (2.38)
\end{aligned}$$

Comparing Eqs. (2.37) and (2.38)

$$\sum_{l'\kappa'} \sum_{\beta,\gamma} \Phi_{\alpha\beta}(l\kappa, l'\kappa')\omega_{\beta\gamma}R_\gamma^0(l'\kappa')$$

$$= \sum_{\beta,\gamma} \omega_{\alpha\beta}\Phi_\beta(l\kappa)\delta_{\gamma\alpha}. \qquad (2.39)$$

Summing over β and γ explicitly on both the sides and collecting the coefficients of $\omega_{\alpha\beta}, \omega_{\beta\alpha}$ and $\omega_{\gamma\alpha}$ and equating the coefficients of $\omega_{\alpha\beta}$ one gets

$$\sum_{l'\kappa'} \left[\Phi_{\alpha\beta}(l\kappa, l'\kappa') R^0_\gamma(l'\kappa') - \Phi_{\alpha\gamma}(l\kappa, l'\kappa') R^0_\beta(l'\kappa') \right]$$

$$= \delta_{\alpha\beta}\Phi_\gamma(l\kappa) - \delta_{\alpha\gamma}\Phi_\beta(l\kappa). \qquad (2.40)$$

Since $\Phi_\gamma = \Phi_\beta = 0$ for a system in equilibrium, therefore

$$\sum_{l'\kappa'} \Phi_{\alpha\beta}(l\kappa, l'\kappa') R^0_\gamma(l'\kappa')$$

$$= \sum_{l'\kappa'} \Phi_{\alpha\gamma}(l\kappa, l'\kappa') R^0_\beta(l'\kappa'). \qquad (2.41a)$$

From the cyclic order of $(\alpha\beta\gamma)$ other relations can also be obtained. Similarly one can also derive the relation

$$\sum_{l\kappa} \left[\Phi_\beta(l\kappa) R^0_\alpha(l\kappa) - \Phi_\alpha(l\kappa) R^0_\beta(l\kappa) \right] = 0. \qquad (2.41b)$$

$$(2.41)$$

The restrictions (2.35) to (2.41) are independent of the geometry of the atomic arrangements, therefore these are valid for any collection of finite number of atoms whose potential energy is well defined and depends on $\vec{R}(l\kappa)$.

2.4.3   Crystal Symmetry

The most general symmetry operation which brings the crystal into own coincidence is a combination of rotation S of the crystal about an axis passing through some point plus a translation $\left[\vec{V}(S) + \vec{R}^0(m) \right]$. In the Seitz notation, such a symmetry operation is written as

$$G = \left[S \mid \vec{V}(S) + \vec{R}^0(m) \right], \qquad (2.42)$$

where S is a 3×3 real orthogonal matrix representation of one of the proper or improper rotation (pure rotation or a combination of pure rotation and an inversion or reflection) of the point group of the space group, $\vec{V}(S)$ is a fractional primitive translation vector associated with

S and $\vec{R}^0(m)$ is the translation vector of the lattice. The nonzero vector $\vec{V}(S)$ are associated with the glide planes and screw axes symmetry elements. The space groups which do not have any essential fractional translation $\vec{V}(S)$ are called symmorphic and those with nonzero $\vec{V}(S)$ are called nonsymmorphic.

If G is applied to $\vec{R}(l\kappa)$ given in Eq. (2.21), we obtain

$$GR(l\kappa) = S\vec{R}(l\kappa) + \vec{V}(S) + \vec{R}^0(m).$$ (2.43)

Use of Eq. (2.21) for $\vec{R}(l\kappa)$ gives

$$G\vec{R}(l\kappa) = \frac{S\vec{R}^0(l\kappa) + \vec{V}(S) + \vec{R}^0(m)}{G\vec{R}^0(l\kappa) + S\vec{u}(l\kappa)} + S\vec{u}(l\kappa).$$

Here G is an element which transforms the undistorted crystal into itself. Therefore the point $(l\kappa)$ is transformed into an equivalent point $(GlG\kappa)$ and we can write

$$G\vec{R}^0(l\kappa) = \vec{R}^0(GlG\kappa) = \vec{R}^0(LK),$$

which gives

$$G\vec{R}(l\kappa) = \vec{R}^0(LK) + S\vec{u}(l\kappa).$$ (2.44)

The potential energy Φ is a function of positions $\{\vec{R}(l\kappa)\}$ and its invariance with respect to G can be expressed as

$$\begin{aligned}
\Phi(\vec{R}(l\kappa)) &= \Phi(G\vec{R}(l\kappa)) \\
&= \Phi\left(\vec{R}^0(LK) + S\vec{u}(l\kappa)\right).
\end{aligned}$$ (2.45)

Equation (2.45) shows that Φ can be expressed in Taylor series about the transformed equilibrium position $\vec{R}^0(LK)$ as well, but now the new variables are the transformed displacements $S\vec{u}(l\kappa)$. Confining upto harmonic terms

$$\begin{aligned}
\Phi = {}& \Phi^{(0)}(\vec{R}^0(LK)) + \sum_{l\kappa\alpha} \Phi_\alpha(LK)\left[Su(l\kappa)\right]_\alpha \\
& + \frac{1}{2}\sum_{l\kappa\alpha}\sum_{l'\kappa'\beta} \Phi_{\alpha\beta}(LK, L'K')\left[S\vec{u}(l\kappa)\right]_\alpha\left[S\vec{u}(l'\kappa')\right]_\beta,
\end{aligned}$$ (2.46)

where

$$[S\vec{u}(l\kappa)]_\alpha = \sum_{\alpha'} S_{\alpha\alpha'} u_{\alpha'}(l\kappa). \tag{2.47}$$

Using Eq. (2.47) in Eq. (2.46) and comparing the coefficients of $u_\alpha(l\kappa)$ etc. term by term with Eq. (2.24), one gets

$$\Phi(\vec{R}^0(l\kappa)) = \Phi(\vec{R}^0(LK)), \tag{2.48a}$$

$$\Phi_\mu(l\kappa) = \sum_\alpha \Phi_\alpha(LK) S_{\alpha\mu}, \tag{2.48b}$$

$$\Phi_{\mu\nu}(l\kappa, l'\kappa') = \sum_{\alpha\beta} \Phi_{\alpha\beta}(LK, L'K') S_{\alpha\mu} S_{\beta\nu}$$

$$= \sum_{\alpha\beta} S_{\mu\alpha}^T \Phi_{\alpha\beta}(LK, L'K') S_{\beta\nu}. \tag{2.48c}$$

Eqations (2.48b) and (2.48c) in the matrix notation are

$$\mathbf{\Phi}(l\kappa) = \mathbf{\Phi}(LK)S = S^T\mathbf{\Phi}(LK)$$

$$\mathbf{\Phi}(l\kappa, L'K') = S^T\mathbf{\Phi}(LK, L'K')S$$

respectively. Using $S^{-1} = S^T$ (S being real orthogonal matrix) one gets

$$\Phi_\alpha(LK) = \sum_\mu S_{\alpha\mu} S_{\alpha\mu} \Phi_\mu(l\kappa)$$

or

$$\mathbf{\Phi}(LK) = S\mathbf{\Phi}(l\kappa)$$

$$\Phi_{\alpha\beta}(LK, L'K') = \sum_{\mu\nu} S_{\alpha\mu} S_{\beta\nu} \Phi_{\mu\nu}(l\kappa, l'\kappa') \tag{2.48d}$$

or

$$\mathbf{\Phi}(LK, L'K') = S\mathbf{\Phi}(l\kappa, l'\kappa')S^T \tag{2.48e}$$

$$\tag{2.48}$$

Equation (2.48a) expresses the invariance of potential energy of static lattice against a symmetry operation which leads the crystal into itself while the Eqs. (2.48b), (2.48c), (2.48d) and (2.48e) describe the transformation of force constants under such symmetry operations. If the symmetry operations leave a given pair of sites $(l\kappa)$ and $(l'\kappa')$ fixed or interchange them i.e., $(LK) = (l\kappa)$ and $(L'K') = (l'\kappa')$ or $(LK) = (l'\kappa')$ and $(L'K') = (l\kappa)$, then Eq. (2.48e) becomes

$$\mathbf{\Phi}(l\kappa, l'\kappa') = \mathbf{S}\mathbf{\Phi}(l\kappa, l'\kappa')\mathbf{S}^{\mathbf{T}}. \tag{2.49}$$

This relation is very useful in determining, which of the matrix elements will vanish and what will be the inter-relation between non-vanishing matrix elements (force constants).

A few important cases of (2.48e) are as follows :
(i) When $G = \{E \mid \vec{R}^0(m)\}$ where E is a 3×3 unit matrix. This is an operation of displacing the crystal through the lattice translation $\vec{R}^0(m)$. In this case $L = l + m$ and $K = \kappa$. Then from Eqs. (2.48d) and (2.48e)

$$\Phi_\alpha(l + m\kappa) = \Phi_\alpha(l\kappa) \tag{2.50a}$$

$$\Phi_{\alpha\beta}(l + m\kappa, l' + m'\kappa') = \Phi_{\alpha\beta}(l\kappa, l'\kappa') \tag{2.50b}$$

$$\tag{2.50}$$

Setting $m = -l$ in (2.50a) and $m = -l$ and $m' = -l'$ in (2.50b), one finds that $\Phi_\alpha(l\kappa)$ is independent of l and $\Phi_{\alpha\beta}(l\kappa, l'\kappa')$ depends on l and l' only through their differences i.e.

$$\Phi_\alpha(l\kappa) = \Phi_\alpha(0\kappa) \equiv \Phi_\alpha(\kappa), \tag{2.51a}$$

$$\Phi_{\alpha\beta}(l\kappa, l'\kappa') = \Phi_{\alpha\beta}(l - l'\kappa, 0\kappa') \equiv \Phi_{\alpha\beta}(0\kappa, l' - l\kappa'). \tag{2.51b}$$

$$\tag{2.51}$$

We shall stop here exploring the additional relations among force constants due to other symmetry operations of space group [3], [14] due to paucity of space. However, these will be recalled as and when needed.

2.5   Solution of Dynamical Equations

2.5.1   Dynamical Matrix and Eigenvectors

The equations of motion (2.32) are a set of $3rN$ (infinite) simultaneous linear equations. Their solutions are simplified by the periodicity of lattice which suggests that the solutions must be such that the displacement of the corresponding atoms in different cells is equivalent, apart from a possible phase factor. Accordingly, we choose the wave like solution of the form

$$u_\alpha(l\kappa, t) = \frac{1}{\sqrt{M_\kappa}} u_\alpha(\vec{q} \mid \kappa) \exp\left[i\left\{\vec{q} \cdot \vec{R}^0(l\kappa) - \omega(\vec{q})t\right\}\right], \qquad (2.52)$$

where $\vec{q}$ is the wave vector and $\omega(\vec{q})$ is the angular frequency associated with the wave. The index l of $M_{l\kappa}$ is dropped due to cell periodicity of the lattice. The amplitude $u_\alpha(\vec{q} \mid \kappa)$ of the wave which is in general a complex number is independent of cell index l and depends on $\vec{q}$ and κ while the exponential depends only on $\vec{R}^0(l\kappa)$. In trying the complex solutions (2.52), it is implied that when all the independent solutions for each $\vec{q}$ are superimposed, the displacement will be real.

Substitution of (2.52) in (2.32) gives the following $3r$ simultaneous equations

$$\omega^2(\vec{q})u_\alpha(\vec{q} \mid \kappa) = \sum_{\kappa'\beta} D_{\alpha\beta}(\vec{q}, \kappa\kappa')u_\beta(\vec{q} \mid \kappa'), \qquad (2.53a)$$

$$(2.53)$$

where the elements of dynamical matrix $\mathbf{D}(\vec{q})$, also called the Fourier transformed dynamical matrix, are

$$D_{\alpha\beta}(\vec{q}, \kappa\kappa') = \frac{1}{\sqrt{M_\kappa M_{\kappa'}}} \sum_{l'} \Phi_{\alpha\beta}(lk, l'\kappa')$$

$$\times \exp\left[-i\vec{q} \cdot \left\{\vec{R}^0(l\kappa) - \vec{R}^0(l'\kappa')\right\}\right]. \qquad (2.54)$$

Thus $D_{\alpha\beta}(\vec{q}, \kappa\kappa')$ depends upon $(l-l')$. Equation (2.53a) in the matrix notation is

$$\omega^2(\vec{q})\mathbf{u}(\vec{q}) = \mathbf{D}(\vec{q})\mathbf{u}(\vec{q}) \qquad (2.53b)$$

where $\mathbf{D}(\vec{q})$ is a $3r$ dimensional square matrix and $\mathbf{u}(\vec{q})$ is a $3r$ component column matrix i.e.

$$\mathbf{D}(\vec{q}) = \left(\begin{array}{c} \kappa\alpha \end{array} \right) \begin{pmatrix} \leftarrow & & \kappa'\beta & & \rightarrow \\ & & \vdots & & \\ & & \vdots & & \\ \cdots & \cdots & D_{\alpha\beta}(\vec{q}, \kappa\kappa') & \cdots & \cdots \\ & & \vdots & & \\ & & \vdots & & \end{pmatrix} \tag{2.55}$$

$$\mathbf{u}(\vec{q}) = \left(\begin{array}{c} \kappa\alpha \end{array} \right) \begin{pmatrix} u_x(\vec{q} \mid \kappa = 1) \\ u_y(\vec{q} \mid \kappa = 1) \\ u_z(\vec{q} \mid \kappa = 1) \\ u_x(\vec{q} \mid \kappa = 2) \\ \vdots \\ \vdots \\ u_z(\vec{q} \mid \kappa = r) \end{pmatrix} \tag{2.56}$$

Equations (2.53) define the eigenvalue problem where the eigenvalues $\omega^2(\vec{q})$ and eigenvectors $\vec{u}(\vec{q} \mid \kappa)$ are to be determined for each $\vec{q}$. As usual, the non-trivial eigenvalues are calculaed by solving the characterstic equation

$$\mid D_{\alpha\beta}(\vec{q}, \kappa\kappa') - \omega^2(\vec{q})\delta_{\alpha\beta}\delta_{\kappa\kappa'} \mid = 0, \tag{2.57}$$

which is obtained by rearranging Eq. (2.53a) as

$$\sum_{\kappa',\beta} \left[D_{\alpha\beta}(\vec{q}, \kappa\kappa') - \omega^2(\vec{q})\delta_{\alpha\beta}\delta_{\kappa\kappa'} \right] u_\beta(\vec{q} \mid \kappa') = 0. \tag{2.58}$$

Equation (2.57) is of $3r$ degree in $\omega^2(\vec{q})$ and the $3r$ solutions for each $\vec{q}$ are denoted as $\omega_j^2(\vec{q})$ where $j = 1, 2, \ldots, 3r$. All of these solutions need not to be distinct, depending upon the symmetry of the crystal and the value of $\vec{q}$ the solutions may be degenerate.

Few important properties of $D_{\alpha\beta}(\vec{q}, \kappa\kappa')$ are as follows :

1. Use of Eqs. (2.14) and (2.51b) in Eq. (2.54) gives

$$
\begin{aligned}
D_{\alpha\beta}(\vec{q}, \kappa\kappa') \;=\; & \frac{1}{\sqrt{M_\kappa M_{\kappa'}}} \exp\left\{ -i\vec{q}\cdot(\vec{R}_\kappa^0 - \vec{R}_{\kappa'}^0) \right\} \\
& \times \sum_{l'} \Phi_{\alpha\beta}(0\kappa, l' - l\kappa') \exp\left\{ i\vec{q}\cdot(\vec{R}_{l'}^0 - \vec{R}_l^0) \right\}.
\end{aligned}
\tag{2.59}
$$

Thus $D_{\alpha\beta}(\vec{q}, \kappa\kappa')$ depends upon the difference $(l - l')$ or $(l' - l)$.

2. Using Eq. (2.34) in Eq. (2.54) and then replacing $\vec{R}_{l'}^0$, by $(\vec{R}_l^0 - \vec{R}_{l'}^0)$ one gets

$$
\begin{aligned}
D_{\alpha\beta}(\vec{q}, \kappa\kappa') \;=\; & \frac{1}{\sqrt{M_\kappa M_{\kappa'}}} \sum_{l'} \Phi_{\beta\alpha}(l'\kappa', l\kappa) \times \\
& \exp\left[i\vec{q}\cdot\left\{ \vec{R}^0(l'\kappa') - \vec{R}^0(lk) \right\} \right] \\
\;=\; & \frac{1}{\sqrt{M_\kappa M_{\kappa'}}} \sum_{l'} \Phi_{\beta\alpha}(l\kappa', l'\kappa) \times \\
& \exp\left[i\vec{q}\cdot\left\{ \vec{R}^0(l\kappa') - \vec{R}^0(l'\kappa) \right\} \right] \\
\;=\; & D_{\beta\alpha}^*(\vec{q}, \kappa'\kappa).
\end{aligned}
\tag{2.60}
$$

Thus $\mathbf{D}(\vec{q})$ is Hermitian, its eigenvalues $\omega_j^2(\vec{q})$ are real and $\omega_j(\vec{q})$ are either real or imaginary. An imaginary $\omega_j(\vec{q})$ would imply that the displacement (2.52) increases indefinitely with time leading to the distruction of the solid. For stable solid, only positive $\omega_j(\vec{q})$ are admitted. The condition that this be satisfied is that the principal minors of $\mathbf{D}(\vec{q})$ must be positive [1], hence it imposes further restrictions on the force constants. As one is concerned with the stable crystals, it is assumed that these additional restrictions are satisfied.

3. Including self-force constants in Eq. (2.54), it is rewritten as

$$
\begin{aligned}
D_{\alpha\beta}(\vec{q}, \kappa\kappa') \;=\; & \frac{1}{M_\kappa} \delta_{\kappa\kappa'} \delta_{ll'} \Phi_{\alpha\beta}(l\kappa, l\kappa) \\
& + \frac{1}{\sqrt{M_\kappa M_{\kappa'}}} \sum_{l',l(l'\neq l\, if\, \kappa'=\kappa)} \Phi_{\alpha\beta}(lk, l'\kappa') \times \\
& \exp\left[-i\vec{q}\cdot\left\{ \vec{R}^0(l\kappa) - \vec{R}^0(l'\kappa') \right\} \right].
\end{aligned}
\tag{2.61}
$$

Further use of Eq. (2.35c) and keeping the origin at $\vec{R}_l^0 = 0$ one gets,

$$D_{\alpha\beta}(\vec{q}, \kappa\kappa') = \bar{D}_{\alpha\beta}(\vec{q}, \kappa\kappa') - \delta_{\kappa\kappa'} \sum_{\kappa''} \bar{D}_{\alpha\beta}(0, \kappa\kappa''), \qquad (2.62)$$

where

$$\bar{D}_{\alpha\beta}(\vec{q}, \kappa\kappa') = \frac{1}{\sqrt{M_\kappa M_{\kappa'}}} \sum_{l'(l' \neq 0 \, if \, \kappa = \kappa')} \Phi_{\alpha\beta}(0\kappa, l'\kappa') \times$$
$$\exp\left[-i\vec{q} \cdot \left\{\vec{R}_\kappa^0 - \vec{R}^0(l'\kappa')\right\}\right]. \qquad (2.63)$$

These equations are extremely useful in the evaluation of dynamical matrix and force constants.

4. It is left to the reader to prove that the dynamical matrix and eigenvalues are periodic with the periodicity of reciprocal lattice i.e.

$$\mathbf{D}(\vec{q}) = \mathbf{D}(\vec{q} + \vec{G}) \qquad (2.64a)$$
$$\omega_j^2(\vec{q}) = \omega_j^2(\vec{q} + \vec{G}). \qquad (2.64b)$$

Consequently

$$\omega_j^2(\vec{q}) = \omega_j^2(S\vec{q}). \qquad (2.64c)$$

$$(2.64)$$

Using known eigenvalues $\omega_j^2(\vec{q})$ in Eq. (2.53a), the corresponding components of eigenvectors $u_\alpha(\vec{q}j \mid \kappa)$ can be calculated. If $\omega_j^2(\vec{q})$ are nondegenrate, $u_\alpha(\vec{q}j \mid \kappa)$ are determined by choosing one component as arbitrary and then solving $(3r - 1)$ homogeneous equations. For example, if there is two fold degeneracy, it is possible to construct two eigenvectors $u_\alpha(\vec{q}j_1 \mid \kappa)$ and $u_\alpha(\vec{q}j_2 \mid \kappa)$ which are orthonormal. These two eigenvectors are arbitrary, therefore one can take linear combinations of $u_\alpha(\vec{q}j_1 \mid \kappa)$ and $u_\alpha(\vec{q}j_2 \mid \kappa)$ and thus one can construct an infinite number of pairs of vectors which are equally acceptable as the eigenvectors. But the presence of double degeneracy permits only $3r - 2$ components of each eigenvector to be determined and so on. A group theoretic solution to this problem is given in Chapter 3 of Ref [3].

While solving for eigenvectors a suitable normalizing condition is required. It is interesting to apply such constraints not on $\vec{u}(\vec{q}j \mid \kappa)$ but on a dimensionless quantity $\vec{e}(\vec{q}j \mid \kappa)$ with components $e_\alpha(\vec{q}j \mid \kappa)$ related to $u_\alpha(\vec{q}j \mid \kappa)$ as

$$u_\alpha(\vec{q}j \mid \kappa) = A'(\vec{q}j \mid \kappa)e_\alpha(\vec{q}j \mid \kappa). \tag{2.65}$$

Here $A'(\vec{q}j)$ is independent of α, κ and time and has the same dimensions as $\vec{u}$ and

$$e_\alpha(\vec{q}j \mid \kappa) = \mid e_\alpha(\vec{q}j \mid \kappa) \mid e^{i\phi(\vec{q}j)}. \tag{2.66}$$

The phase factor $e^{i\phi(\vec{q}j)}$ is arbitrary. Using Eq. (2.53a) for (jth) eigenvalue, one finds

$$\omega_j^2(\vec{q})e_\alpha(\vec{q}j \mid \kappa) = \sum_{\kappa',\beta} D_{\alpha\beta}(\vec{q}, \kappa\kappa')e_\beta(\vec{q}j \mid \kappa') \tag{2.67a}$$

or

$$\Lambda(\vec{q})e(\vec{q}) = D(\vec{q})e(\vec{q}) \tag{2.67b}$$

$$\tag{2.67}$$

where the eigenvector matrix

$$e(\vec{q}) = \begin{pmatrix} \kappa\alpha \end{pmatrix} \begin{bmatrix} & & j & & \\ & & \vdots & & \\ \cdots & \cdots & e_\alpha(\vec{q}j) \mid \kappa) & \cdots & \cdots \\ & & \vdots & & \\ & & \vdots & & \end{bmatrix} \tag{2.68}$$

and eigenvalue matrix

$$\Lambda(\vec{q}) = \begin{bmatrix} \omega_1^2(\vec{q}) & \cdots & 0 & \cdots & \cdots \\ 0 & \omega_2^2(\vec{q}) & \cdots & 0 & \cdots \\ & & & & \\ \cdots & 0 & \cdots & \cdots & \omega_{3r}^2(\vec{q}) \end{bmatrix}. \tag{2.69}$$

Thus $\vec{e}(\vec{q}j \mid \kappa)$ is as acceptable an eigenvector of $D(\vec{q})$ as $\vec{u}(\vec{q}j \mid \kappa)$. However, the homogeneous equations (2.67a) and (2.67b) define the matrix $\mathbf{e}(\vec{q})$ too within a constant factor only. On the other hand $\mathbf{e}(\vec{q})$ diagonalizes the Hermitian matrix $\mathbf{D}(\vec{q})$ i.e. $\mathbf{e}^{-1}(\vec{q})\mathbf{D}(\vec{q})\mathbf{e}(\vec{q}) = \mathbf{\Lambda}(\vec{q})$ i.e. the principal axis transformation of the present problem. Therefore $\mathbf{e}(\vec{q})$ can be chosen to be a unitary matrix which fixes the undetermined factor. Thus one requires

$$\mathbf{e}^{-1}(\vec{q}) = \mathbf{e}^{\dagger}(\vec{q}) = \mathbf{e}^{*\mathbf{T}}(\vec{q}), \quad \mathbf{e}^{\dagger}(\vec{q})\mathbf{e}(\vec{q}) = \mathbf{I}_{3r} = \mathbf{e}(\vec{q})\mathbf{e}^{\dagger}(\vec{q}), \qquad (2.70)$$

where $\mathbf{I}_{3r}$ is a unit matrix of order 3r. The rows and columns of a unitary matrix are orthogonal. Thus the following orthogonality and closure (completeness) relations exist [15],

$$\sum_{\kappa} \vec{e}^{*}(\vec{q}j \mid \kappa)\vec{e}(\vec{q}j' \mid \kappa) = \delta_{jj'}, \qquad (2.71a)$$

$$\sum_{j} e_{\beta}^{*}(\vec{q}j \mid \kappa')e_{\alpha}(\vec{q}j \mid \kappa) = \delta_{\alpha\beta}\delta_{\kappa\kappa'}. \qquad (2.71b)$$

$$(2.71)$$

The significance of (2.71a) and (2.71b) is that the $3r$ vectors $\vec{e}(\vec{q}j)$ given by Eq.(2.68) form a complete set and therefore these are suitable basis vectors in the $3r$-dimensional space.

From Eq. (2.54),

$$D_{\alpha\beta}(\vec{q}, \kappa\kappa') = D_{\alpha\beta}^{*}(-\vec{q}, \kappa\kappa'). \qquad (2.72)$$

We replace $\vec{q}$ by $-\vec{q}$ in Eq. (2.67a), take the complex conjugate of the resulting equation, use Eq. (2.71) and put $\omega_{j}^{2*}(\vec{q}) = \omega_{j}^{2}(\vec{q})$. This gives

$$\sum_{\kappa'\beta} D_{\alpha\beta}(\vec{q}, \kappa\kappa')e_{\beta}^{*}(-\vec{q}j \mid \kappa') = \omega_{j}^{2}(-\vec{q})e_{\alpha}^{*}(-\vec{q}j \mid \kappa). \qquad (2.73)$$

Thus $\omega_{j}^{2}(-\vec{q})$ and $\omega_{j}^{2}(\vec{q})$ are the eigenvalues of the same matrix $\mathbf{D}(\vec{q})$ and hence for non-degenerate eigenvalues

$$\omega_{j}^{2}(-\vec{q}) = \omega_{j}^{2}(\vec{q}). \qquad (2.74)$$

At the points of degeneracy, Eq. (2.74) allows to label the modes at $-\vec{q}$ in terms of $\vec{q}$. Using Eq. (2.74) in Eq. (2.73) and comparing with

Eq. (2.67a) one finds that $e_\alpha^*(-\vec{q}j \mid \kappa)$ or $e_\beta^*(-\vec{q}j \mid \kappa)$ satisfy the same equation as $e_\alpha(\vec{q}j \mid \kappa)$. Therefore the two vectors can differ within a factor of modulus unity i.e.

$$\vec{e}^*(-\vec{q}j \mid \kappa) = e^{-i\phi}\vec{e}(\vec{q}j \mid \kappa).$$

Since the physical properties do not depend upon the choice of phase factor, therefore one chooses $\exp(-i\phi) = 1$ [1] and gets

$$\vec{e}^*(-\vec{q}j \mid \kappa) = \vec{e}(\vec{q}j \mid \kappa). \tag{2.75}$$

Using Eq. (2.65) in Eq. (2.52), one finds

$$u_\alpha(l\kappa, t) = \frac{1}{\sqrt{M_\kappa}} A'(\vec{q}j) e_\alpha(\vec{q}j \mid \kappa) exp\left[i\left\{\vec{q}\cdot\vec{R}^0(l\kappa) - \omega_j(\vec{q})t\right\}\right]. \tag{2.76}$$

Here $e_\alpha(\vec{q}j \mid \kappa)$, as given in Eq. (2.66), determine the direction and phase and $A'(\vec{q}j)$ the amplitude of vibration of atoms in the wave $(\vec{q}, \omega_j(\vec{q}))$. The quantities $\mid e_\alpha(\vec{q}j \mid \kappa) \mid$ are the direction cosines of the displacement. Thus the components of the vector $\vec{e}(\vec{q}j \mid \kappa)$ essentially give the pattern of atomoc displacements in a particular lattice wave. For this reason $\vec{e}(\vec{q}j \mid \kappa)$ are also referred as polarization vectors. In special situations, the polarization vectors of all the atoms are either parallel or perpendicular to $\vec{q}$, the mode is then said to be longitudinal (L) or transverse (T) respectively. Summing over all the allowed values of $\vec{q}$ and j one gets the general solution which is the superposition of independent solutions, i.e.

$$u_\alpha(l\kappa, t) = \frac{1}{\sqrt{M_\kappa}} \sum_{\vec{q}j} A'(\vec{q}j) e_\alpha(\vec{q}j \mid \kappa) \exp\left[i\left\{\vec{q}\cdot\vec{R}^0(l\kappa) - \omega_j(\vec{q})t\right\}\right], \tag{2.77}$$

which is the real displacement of the $(l\kappa)$ atom in the α-direction at time t.

2.5.2 Cyclic Boundary Conditions, Dispersion Relations and Acoustic and Optical Modes

Once the vibrational frequency $\omega_j(\vec{q})$ and the polarization vectors $\vec{e}(\vec{q}j, \kappa)$ are determined, one needs to know allowed values of wave vectors $\vec{q}$, introduced in Eq. (2.52), to complete the solution. These are determined by the boundary conditions imposed on $\vec{u}(l\kappa, t)$. It is customary to apply periodic or cyclic boundary conditions for the perfect crystal which can be thought of as describing a situation in which n-dimensional lattice is wrapped on the n-dimensional tours. For example, one dimensional lattice closed on itself to form a ring, two dimensional lattice first rolled into a cylinder and then into a doughnut and so on [16].

Technically we consider an infinite crystal to be subdivided into large "macrocells" each containing a very large number N of unit cells. For convenience one chooses the macrocell of edge $L\vec{a}_1, L\vec{a}_2$ and $L\vec{a}_3$ and of the same shape as the unit cell and therefore it follows that $N = L^3$. The cyclic conditions are now applied to macrocell and for this reason it is frequently referred to as a "cyclic" crystal.

The cyclic boundary conditions require that the atomic displacements of equivalent atoms in different macrocells are identical i.e.

$$\vec{u}(l\kappa) = \vec{u}(l + L, \kappa)$$
$$u_\alpha(l_1 l_2 l_3, \kappa) = u_\alpha(l_1 + L, l_2 + L, l_3 + L, \kappa). \qquad (2.78)$$

Here the time index is dropped for neatness. Use of Eq. (2.78) in (2.52) in conjuction with (2.15) gives

$$e^{i\vec{q}\cdot(L\vec{a}_1 + L\vec{a}_2 + L\vec{a}_3)} = 1,$$

or

$$e^{i\vec{q}\cdot L\vec{a}_1} = e^{i\vec{q}\cdot L\vec{a}_2} = e^{i\vec{q}\cdot L\vec{a}_3} = 1. \qquad (2.79)$$

From Eqs. (2.16) and (2.17), these conditions are satisfied if [17],

$$\vec{q} = \xi_1 \vec{b}_1 + \xi_2 \vec{b}_2 + \xi_3 \vec{b}_3 = (\xi\vec{b}), \qquad (2.80a)$$

where

$$\xi_i = h_i/L; h_1, h_2, h_3 = 1, 2, \ldots, L. \qquad (2.80b)$$

$$(2.80)$$

Thus there are $L^3 (= N;$ number of unit cells) allowd values of $\vec{q}$. The values of $\vec{q}$ defined in Eqs. (2.80) form a fine mesh of points in the reciprocal space. The single value of $\vec{q}$ occupies the volume $(\vec{b}_1.(\vec{b}_2 \times \vec{b}_3)/L^3)$ in the reciprocal space (using Eqs. 2.20). The density of $\vec{q}$ points $(= \Omega/8\pi^3)$ where $\Omega = Nv$ is the macrocell volume. An important consequence of dense and uniform distribution of $\vec{q}$ vectors is that whenever needed, $\vec{q}$ is treated a continuous variable and summation over $\vec{q}$ is replaced by integration according to the rule

$$\sum_{\vec{q}} \rightarrow \frac{\Omega}{(2\pi)^3} \int d\vec{q}. \tag{2.81}$$

If $\vec{q}$ is replaced by $(\vec{q} + \vec{G})$ in Eq. (2.79), the boundary conditions are still satisfied, therefore $\vec{q}$ and $\vec{q} + \vec{G}$ are equivalent allowed wavevectors, $\vec{q}$ is in the reduced zone scheme and $\vec{q} + \vec{G}$ in the extended zone scheme. All the distinct values of $\vec{q}$ are obtained by restricting $\vec{q}$ in the first Brillouin zone (BZ) which is primitive cell of reciprocal lattice. The BZ explicitly displays the symmetry of the crystal lattice and its volume is $(8\pi^3/v)$. Details are referred to any text book [18].

Raman and his co-workers raised the question regarding the applicability of the above results to finite crystals as these are deduced for the infinite crystals on the basis of cyclic boundary conditions [19]. In a crystal obeying cyclic boundary conditions, the elements $D_{\alpha\beta}(l\kappa, l'\kappa')$ depend upon the cell indices through $(l - l')$. This remains no more true for the cells lying in the range of interatomic force near the surface in a finite crystal as these cells are subjected to external pressure. Thus the dynamical matrix for a lattice obeying cyclic boundary conditions differs from that for a lattice satisfying natural boundary conditions for the rows and columns which correspond to the atoms lying in the surface shell. However, if the ratio of number of unit cells in the surface shell to the total number of unit cells, is small, the cyclic boundary conditions will remain practically valid. Even for long range Coulomb interaction in ionic crystals, Hardy [20] has shown that the use of cyclic boundary conditions lead to negligible error in the calculation of frequency spectrum.

Since the allowed values of $\vec{q}$ are closely spaced, it is possible to represent 3rN frequencies graphically by a continuous curve as a function of $\vec{q}$. In each direction of $\vec{q}$ space, there will be $3r$

$$\omega = \omega_j(\vec{q}) \tag{2.82}$$

curves. These curves are called dispersion curves or dispersion relations. Each of 3r curves is a branch. Thus there are 3r branches of dispersion curves. These branches, in general, are distinct but may be degenerate also for crystal symmetry reasons.

From Eqs.(2.54) and (2.67a), we have for $\vec{q} = 0$,

$$\omega_j^2(0)\frac{e_\alpha(0j \mid \kappa)}{\sqrt{M_\kappa}} = \frac{1}{M_\kappa}\sum_\beta \left[\sum_{l'\kappa'}\frac{\Phi_{\alpha\beta}(l\kappa, l'\kappa')}{\sqrt{M_{\kappa'}}}\right]e_\beta(0j \mid \kappa'). \qquad (2.83)$$

If we choose

$$\frac{\vec{e}(0j \mid \kappa)}{\sqrt{M_\kappa}} = \frac{\vec{e}(0j \mid \kappa')}{\sqrt{M_{\kappa'}}} = \vec{u}(l\kappa \mid 0j) = \vec{u}(l\kappa' \mid 0j) \qquad (2.84)$$

i.e. all the r particles in each unit cell move in parallel with equal amplitudes then the right side of (2.83) vanishes by sum rule (2.35b). Thus $\omega_j^2(0) = 0$ for each α and hence as $\vec{q} \to 0$, $\omega_j^2(0) = 0$ for 3 branches. These are called acoustic modes because of characteristics of atomic displacements. The remaining $(3r - 3)$ modes whose frequencies do not vanish at $\vec{q} = 0$, are called optical modes for historical reasons [1],[2].

2.6 Normal Coordinate Transformations

Redefining the time dependent amplitude of atomic displacements in Eq.(2.77) we write

$$u_\alpha(l\kappa, t) = \frac{1}{\sqrt{NM_\kappa}}\sum_{\vec{q}j}Q(\vec{q}j, t)e_\alpha(\vec{q}j \mid \kappa)\exp\left[i\vec{q}\cdot\vec{R}^0(l\kappa)\right], \qquad (2.85)$$

where the factor $(1/\sqrt{N})$ is introduced for convenience and the complex time dependent amplitude

$$Q(\vec{q}j, t) = A'(\vec{q}j)\exp\left[-i\omega_j(\vec{q})t\right]. \qquad (2.86)$$

Equation (2.85) amounts to a Fourier expansion of displacements and generates the principal axis transformation of Hamiltonion (2.30). The two terms of Eq. (2.30) are of quadratic form, one in the momentum and other in the displacement coordinates. From matrix algebra it follows [15] that if the quadratic forms of kinetic and potential energies are postive definite, it is possible to find a transformation essentially unitary, which diagonalizes, both kinetic and potential energies simultaneously. Since $u_\alpha(l\kappa, t)$ is real therefore equating $u_\alpha^*(l\kappa, t)$ and $u_\alpha(l\kappa, t)$ and using Eq.(2.75) one finds

$$Q^*(\vec{q}j, t) = Q(-\vec{q}j, t). \tag{2.87}$$

Substituting (2.85), the kinetic energy part of Hamiltonian (2.30) becomes

$$T_I = \frac{1}{2} \sum_{\kappa\alpha} \sum_{\vec{q}\vec{q}',jj'} \frac{1}{N} e_\alpha(\vec{q}j \mid \kappa) e_\alpha(\vec{q'}j' \mid \kappa) \dot{Q}(\vec{q}j, t) \dot{Q}(\vec{q'}j', t)$$

$$\times e^{i(\vec{q}+\vec{q'})\cdot\vec{R}_\kappa^0} \sum_l e^{i(\vec{q}+\vec{q'})\cdot\vec{R}_l^0} \tag{2.88a}$$

$$= \frac{1}{2} \sum_{\vec{q}jj'} \left[\sum_{\kappa,\alpha} e_\alpha(\vec{q}j, \kappa) e_\alpha^*(\vec{q}j' \mid \kappa) \right] \dot{Q}(\vec{q}j, t) \dot{Q}(-\vec{q}j', t) \tag{2.88b}$$

$$= \frac{1}{2} \sum_{\vec{q}j} \dot{Q}(\vec{q}j, t) \dot{Q}^*(\vec{q}j, t). \tag{2.88c}$$

$$\tag{2.88}$$

Equation (2.88b) is obtained from (2.88a) using Eqs. (2.75), (2.87) and the relation

$$\sum_l e^{i(\vec{q}+\vec{q'})\vec{R}_l^0} = N\delta(\vec{q}+\vec{q'}), \tag{2.89}$$

which is a consequence of the periodicity of lattice. The vectors $\vec{q}$ and $\vec{q'}$ are in the reduced zone scheme. The orthonormality condition (2.71a) is used to get (2.88c). The dot over Q denotes its time derivative.

Similarly, the potential energy part of Eq. (2.30) is

$$\Phi(\vec{R}) = \frac{1}{2}N \sum_{l\kappa\alpha,l'\kappa'\beta} \frac{1}{\sqrt{M_\kappa M_{\kappa'}}} \Phi_{\alpha\beta}(l\kappa, l'\kappa') \sum_{\vec{q}j,\vec{q'}j'} e_\alpha(\vec{q}j \mid \kappa) e_\beta(\vec{q'}j' \mid \kappa')$$

$$\times Q(\vec{q}j, t) Q(\vec{q'}j', t) \exp\left[i\left\{ \vec{q}\cdot\vec{R}^0(l\kappa) + \vec{q'}\cdot\vec{R}^0(l'\kappa') \right\} \right]. \tag{2.90}$$

Now we combine the terms containing indices l and l', separate $\exp[i\vec{q} \cdot \vec{R}^0_\kappa + i\vec{q'} \cdot \vec{R}^0_{\kappa'}]$, multiply and divide by $\exp[i\vec{q} \cdot \vec{R}^0_l]$, rearrange and use Eqs. (2.54) and (2.89). These simplifications give

$$\Phi(\vec{R}) = \frac{1}{2} \sum_{\kappa\alpha} \sum_{\vec{q}jj'} e_\alpha(\vec{q}j \mid \kappa) \left[\sum_{\kappa'\beta} D_{\alpha\beta}(-q, \kappa\kappa') e_\beta(-\vec{q}j' \mid \kappa') \right]$$
$$\times Q(\vec{q}j, t) Q(-\vec{q}j, t).$$

Further use of Eqs. (2.67a), (2.74), (2.71a) and (2.87) yields

$$\Phi(\vec{R}) = \frac{1}{2} \sum_{\vec{q}j} \omega_j^2(\vec{q}) Q(\vec{q}j, t) Q^*(\vec{q}j, t). \tag{2.91}$$

Thus the lattice Hamiltonian in the new coordinates is

$$\begin{aligned} H &= T_I + \Phi \\ &= \frac{1}{2} \sum_{\vec{q}j} \left[\dot{Q}(\vec{q}j, t) \dot{Q}^*(\vec{q}j, t) + \omega_j^2(\vec{q}) Q(\vec{q}j, t) Q^*(\vec{q}j, t) \right]. \end{aligned} \tag{2.92}$$

From Lagrangian $L = T_I - \Phi$, the momentum, conjugate to $Q^*(\vec{q}j, t)$, is

$$P(\vec{q}j, t) = \frac{\partial L}{\partial \dot{Q}^*(\vec{q}j, t)} = \dot{Q}(\vec{q}j, t) \tag{2.93}$$

which simplifies Eq. (2.92) as

$$H = \frac{1}{2} \sum_{\vec{q}j} \left[P(\vec{q}j, t) P^*(\vec{q}j, t) + \omega_j^2(\vec{q}) Q(\vec{q}j, t) Q^*(\vec{q}j, t) \right]. \tag{2.94}$$

From Hamilton's equations

$$\begin{aligned} \dot{P}(\vec{q}j, t) &= -\frac{\partial H}{\partial Q^*(\vec{q}j, t)} \\ &= -\omega_j^2(\vec{q}) Q(\vec{q}j, t). \end{aligned} \tag{2.95}$$

Combining Eqs. (2.93) and (2.95) one gets

$$\ddot{Q}(\vec{q}j, t) + \omega_j^2(\vec{q}) Q(\vec{q}j, t) = 0. \tag{2.96}$$

This is the equation of motion of an uncoupled harmonic oscillator whose solution can be written as

$$Q(\vec{q}j, t) = Q_+(\vec{q}j)e^{i\omega_j(\vec{q})t} + Q_-(\vec{q}j)e^{-i\omega_j(\vec{q})t}. \tag{2.97}$$

Also from Eq. (2.85) one can write

$$Q(\vec{q}j, t) = \frac{1}{\sqrt{N}} \sum_{l\kappa\alpha} \sqrt{M_\kappa} e_\alpha^*(\vec{q}j \mid \kappa) \exp\left[-i\vec{q}\cdot\vec{R}^0(l\kappa)\right] u_\alpha(l\kappa, t). \tag{2.98}$$

$Q(\vec{q}j, t)$ are collective coordinates as the sum extends over all the local atomic displacements. These are complex in nature and are called "Normal Coordinates". Each of the coordinates describes an independent mode of vibration of the crystal with only one frequency $\omega_j(\vec{q})$. Such vibrational modes are called normal modes. In each mode, all the atoms vibrate with the same frequency and in the same phase. The number of modes is equal to the degrees of freedom in the crystal i.e. $3rN$.

In studying the quantum aspects of lattice vibrations one needs real normal coordinates. This can be achieved in many ways [1], [2], [4]. One of the methods is to rewrite

$$Q(\vec{q}j, t) = \frac{1}{\sqrt{2}}\left[Q_1(\vec{q}j, t) + iQ_2(\vec{q}j, t)\right],$$

and

$$Q(-\vec{q}j, t) = \frac{1}{\sqrt{2}}\left[Q_1(\vec{q}j, t) - iQ_2(\vec{q}j, t)\right], \tag{2.99}$$

where $Q_i(\vec{q}j, t)$ are real. Evidently $Q_1(\vec{q}j, t) = Q_1(-\vec{q}j, t)$ and $Q_2(\vec{q}j, t) = -Q_2(-\vec{q}j, t)$, therefore only one half of the quantities $Q_1(\vec{q}j, t)$ and $Q_2(\vec{q}j, t)$ are independent. These independent coordinates may be chosen by taking $\vec{q}$ vectors lying only on one side of a plane passing through the center of the Brillouin zone. The Hamiltonian now becomes

$$H = \frac{1}{2} \sum_{q>0j} \sum_{i=1,2} \left[\dot{Q}_i^2(\vec{q}j, t) + \omega_{ji}^2(\vec{q})Q_i^2(\vec{q}j, t)\right]. \tag{2.100}$$

Here $Q_i(\vec{q}j, t)$ and $\dot{Q}_i(\vec{q}j, t)\, (= P_i(\vec{q}j, t))$ are conjugate variables. Writing $\dot{Q}_i(\vec{q}j, t)$ in the operator form, the lattice wave equation becomes

$$\frac{1}{2} \sum_{q>0} \sum_{j} \sum_{i} \left[-\hbar^2 \frac{\partial^2}{\partial Q_i^2(\vec{q}j, t)} + \omega_{ji}^2(\vec{q})Q_i^2(\vec{q}j, t)\right]\chi_\lambda = E_\lambda\chi_\lambda. \tag{2.101}$$

Since the Hamiltonian is the sum of independent single oscillator Hamiltonians, the lattice wavefunction χ_λ is the product of single oscillator wavefunction ϕ_{nq} i.e.,

$$\chi_{\{n\}} = \Pi_{\vec{q}j,i}\phi_{nq}\left(Q_i(\vec{q}j)\right), \quad \lambda \equiv \{n\} \tag{2.102}$$

where $nq = n_{ji}(\vec{q})$ is the quantum number associated with the mode $(\vec{q}j, i)$ and describes the state of excitation of the $(\vec{q}j, i)$ oscillation. The $\phi_{nq}(Q_i(\vec{q}j))$ is the corresponding wavefunction which satisfies the single oscillator wave equation

$$\left[-\frac{\hbar^2}{2}\frac{\partial^2}{\partial Q_i^2(\vec{q}j,t)} + \frac{1}{2}\omega_{ji}^2(\vec{q})Q_i^2(\vec{q}j,t)\right]\phi_{nq}\left(Q_i(\vec{q}j)\right) = E_{nq}\phi_{nq}\left(Q_i(\vec{q}j)\right)$$

$$\tag{2.103}$$

where

$$\phi_{nq}\left(Q_i(\vec{q}j)\right) = \left(\frac{\alpha_{ji}}{\pi^{1/2}2^n n!}\right)^{1/2}\exp\left(-\frac{1}{2}\alpha_{ji}Q_i^2\right)H_n\left(\alpha_{ji}Q_i\right)$$

and

$$\alpha_{ji}^2 = \omega_{ji}(\vec{q})/\hbar. \tag{2.104}$$

$H_n(x)$ is the nth order Hermite polynomial. The associated energy levels are

$$E_{nq} = \left[n_{ji}(\vec{q}) + \frac{1}{2}\right]\hbar\omega_{ji}(\vec{q}) \tag{2.105}$$

with $n_{ji}(\vec{q}) = 0, 1, 2, \ldots$, all integer values. The total energy of the crystal described by $3rN$ quantum numbers $\{n_{ji}(\vec{q})\}$ is

$$E_{\{n\}} = \sum_{\vec{q}j,i} E_{nq}. \tag{2.106}$$

Another transformation of Hamiltonian (2.94) which is very convenient for quantum mechanical formulation is as follows : Q and P are operators and can be expressed in terms of another set of creation operators $a^\dagger_{\vec{q}j}$ and annihilation operators $a_{\vec{q}j}$ as

$$Q(\vec{q}j,t) = \left[\frac{\hbar}{2\omega_j(\vec{q})}\right]^{1/2}\left(a_{\vec{q}j} + a^\dagger_{-\vec{q}j}\right)$$

and

$$P(\vec{q}j,t) = -i\left[\frac{\hbar\omega_j(\vec{q})}{2}\right]^{1/2}\left(a_{\vec{q}j} - a^{\dagger}_{-\vec{q}j}\right).\tag{2.107}$$

Here the condition $Q(\vec{q}j,t) = Q^*(-\vec{q}j,t)$ is automatically satisfied. Q^* is Hermitian conjugate of Q. Equations (2.107) can be used to express $a_{\vec{q}j}$ and $a^{\dagger}_{-\vec{q}j}$ in terms of $Q(\vec{q}j,t)$ and $P(\vec{q}j,t)$. With the help of commutation relations

$$\begin{aligned}
[u_\alpha(l\kappa), u_\beta(l'\kappa')] &= [p_\alpha(l\kappa), p_\beta(l'\kappa')] = 0,\\
[u_\alpha(l\kappa), p_\beta(l'\kappa')] &= i\hbar\delta_{\alpha\beta}\delta_{ll'}\delta_{\kappa\kappa'},
\end{aligned}\tag{2.108}$$

one finds

$$\begin{aligned}
\left[Q^*(\vec{q}j,t), Q(\vec{q}'j',t)\right] &= \left[P^*(\vec{q}j,t), P(\vec{q}'j',t)\right] = 0,\\
[Q^*(\vec{q}j,t), P(\vec{q}j,t)] &= [Q(\vec{q}j,t), P^*(\vec{q}j,t)] = i\hbar\delta_{\vec{q}\vec{q}'}\delta_{jj'}.
\end{aligned}\tag{2.109}$$

Here

$$p_\alpha(l\kappa) = M_\kappa \dot{u}_\alpha(l\kappa).$$

From Eqs. (2.107) and (2.109) one can write

$$\begin{aligned}
\left[a_{\vec{q}j}, a^{\dagger}_{\vec{q}'j'}\right] &= \delta_{\vec{q}\vec{q}'}\delta_{jj'},\\
\left[a_{\vec{q}j}, a_{\vec{q}'j'}\right] &= \left[a^{\dagger}_{\vec{q}j}, a^{\dagger}_{\vec{q}'j'}\right] = 0.
\end{aligned}\tag{2.110}$$

Thus $a_{\vec{q}j}$ and $a^{\dagger}_{\vec{q}j}$ are boson operators.

Using Eqs. (2.107) in (2.94) the Hamiltonian becomes

$$H = \sum_{\vec{q}j} \hbar\omega_j(\vec{q})\left[a^{\dagger}_{\vec{q}j}a_{\vec{q}j} + \frac{1}{2}\right],\tag{2.111}$$

which is the standard form of the Hamiltonian for a collection of independent harmonic oscillators. From Eqs. (2.110), one finds that the number operators $(a^{\dagger}_{\vec{q}j}a_{\vec{q}j})$ commute among themselves i.e.

$$\left[a^{\dagger}_{\vec{q}j}a_{\vec{q}j}, a^{\dagger}_{\vec{q}'j'}a_{\vec{q}'j'}\right] = 0.\tag{2.112}$$

Consequently, each sub-Hamiltonian of (2.111)

$$H_{\vec{q}j} = \hbar\omega_j(\vec{q}) \left[a^\dagger_{\vec{q}j} a_{\vec{q}j} + \frac{1}{2} \right] \qquad (2.113)$$

can be diagonalized simultaneously. Thus if the eigenstates $\mid n_j(\vec{q}) >$ and eigenenrgies E_{nq} of $H_{\vec{q}j}$ are known, the eigenstate $\mid n >$ and eigenenergy $E_{\{n\}}$ of total Hamiltonian H are given as

$$\mid n >=\mid \{n_j(\vec{q})\} >, \qquad (2.114)$$

$$E_{\{n\}} = \sum_{\vec{q}j} \hbar\omega_j(\vec{q}) \left[n_j(\vec{q}) + \frac{1}{2} \right], \qquad (2.115)$$

where

$$\begin{aligned}
a_{\vec{q}j} \mid n > &= [n_j(\vec{q})]^{1/2} \mid n' >, \\
a^\dagger_{\vec{q}j} \mid n > &= [1 + n_j(\vec{q})]^{1/2} \mid n'' >, \\
a^\dagger_{\vec{q}j} a_{\vec{q}j} \mid n > &= n_j(\vec{q}) \mid n > .
\end{aligned} \qquad (2.116)$$

Here $\mid n >$ denotes the crystal state in which the oscillator with frequency $\omega_{j_1}(\vec{q_1})$ is excited to the state $\mid n_{j_1}(\vec{q_1}) >$, oscillator with frequency $\omega_{j_2}(\vec{q_2}) >$ to state $\mid n_{j_2}(\vec{q_2}) >$ and so on. $\mid n' >=\mid n-1 >$ denotes a state in which all the oscillator quantum numbers are the same as in $\mid n >$ except that pertaining to frequency $\omega_j(\vec{q})$ with quantum number $(n_j(\vec{q}) - 1)$. Similarly $\mid n'' >=\mid n+1 >$ is similar to $\mid n >$ except that pertaining to frequency $\omega_j(\vec{q})$ with quantum number $(n_j(\vec{q}) + 1)$. Thus $a_{\vec{q}j}$ and $a^\dagger_{\vec{q}j}$ refer to annhilation and creation operators respectively.

Thermal average of number operator is Bose-Einstein distribution function

$$< a^\dagger_{\vec{q}j} a_{\vec{q}j} >=< n_j(\vec{q}) >= \frac{1}{\exp(\hbar\omega_j(\vec{q})/k_B T) - 1} \qquad (2.117)$$

where $n_j(\vec{q})$ is eigenvalue of the number operator and takes any integer value, k_B is the Boltzmann constant and T is temperature. Therefore one says that the energy is quantized and the quantum of excitation energy of lattice vibrations is called a "phonon".

Thus the quantum treatment gives the following two equivalent and interesting interpretations. (i) The crystal may be regarded as a collection of $3rN$ independent distinguishable oscillators with frequencies $\omega_j(\vec{q})$. Each oscillator is capable of being in various stationary states defined by quantum numbers $n_{ji}(\vec{q}) = 0, 1, 2, \ldots$, and having energies $E_{n_{ji}q} = \left[n_{ji}(\vec{q}) + \frac{1}{2} \right] \hbar \omega_{ji}(\vec{q})$. (ii) The system may also be considered as a gas of indistinguishable quasiparticles called phonons. A phonon can be in varous stationery states defined by quantum numbers $(\vec{q}j)$. The number of phonons in the state $\mid \vec{q}j >$ is $n_j(\vec{q})$ and operators $a^{\dagger}_{\vec{q}j}$ and $a_{\vec{q}j}$ respectively create and annhilate a phonon in the state $\mid \vec{q}j >$ In the harmonic theory phonons are non-interacting, therefore, the crystal may be as non-interacting boson gas. Anharmonicity brings in the phonon- phonon interaction,consequently the phonon-width and phonon life time [2].

B. Analysis of Experimental Data

The theme of lattice dynamical studies has been to understand the the interatomic potential and hence the interatomic force constants. Historically, in the beginning, the measurements for lattice specific heat and elastic constants were available. Therefore, a lattice dynamical model involving an interatomic potential was proposed. The phonon density of states was calculated for this model which was further used to calculate the lattice specific heat [18]. The elastic constants were calculated by taking the long wave-length limit of dynamical Eq.(2.67). From the varification of the measured specific heat and elastic constants, the validity of the proposed interatomic potential (or force constants) for dynamical equations was checked. Some times the interatomic potential (or force constants) was predicted with the help of measured specific heat, elastic constants and lattice parameter. However, the measurements of phonon dispersion relations by X-ray and neutron scattering techniques [3] have brought a revolution. These measurements have made it possible to carry out the above exercise of confirming the proposed interatomic potential or predicting a new one (and hence the related properties) directly by analysing the phonon dispersion relations. Therefore in this monograph we confine ourselves within the analysis of phonon dispersion relations to understand the interatomic potential.

There are two aspects of above analysis: (1) An interatomic potential with its character and range is proposed and its parameters are determined by fitting the phonon dispersion relations. One goes on changing the character and range of potential, until one gets the best fit and hence the final interatomic potential. This is called the phenomenological modelling of dispersion relations. (2) One can also write the interatomic potential from "abinitio" as the sum of all basic interactions as defined in Eqs. (2.2) and (2.3) for a metallic system and can compute phonon frequencies as discussed in Sec. A. Any discrepancy with the experimental measurements may lead to add new interactions or to question the validity of adiabatic and harmonic approximations in deriving the dynamical equations of motion. In this section, the phenomenological models for d and f-band metals are described. The abinito calculations will be presented in Chapter 4.

Before getting into actual calculations, it is important to answer why most of the measurements of phonon dispersion relations and their calculations are carried out only along the principal symmetry directions of a given crystal structure. Post multiplication of $e_\alpha^*(\vec{q}j \mid \kappa)$ in Eq. (2.67a) and sum over $(\kappa\alpha)$ gives

$$\omega_j^2(\vec{q}) = \sum_{\kappa\alpha,\kappa'\beta} e_\alpha^*(\vec{q}j \mid \kappa) D_{\alpha\beta}(\vec{q}, \kappa\kappa') e_\beta(\vec{q}j \mid \kappa') \qquad (2.118a)$$

$$= \sum_{\kappa\alpha,\kappa'\beta} e_\alpha^*(\vec{q}j \mid \kappa) \left[\frac{1}{\sqrt{M_\kappa M_{\kappa'}}} \sum_{l'} \Phi_{\alpha\beta}(l\kappa, l'\kappa') \right.$$

$$\times \exp(-i\vec{q} \cdot \{\vec{R}^0(l\kappa) - \vec{R}^0(l'\kappa')\}) \left] e_\beta(\vec{q}j \mid \kappa'). \qquad (2.118b) \right.$$

$$(2.118)$$

Thus $\omega_j^2(\vec{q})$ are completely determined if all the polarization vectors $e_\alpha(\vec{q}j \mid \kappa)$ and force constants $\Phi_{\alpha\beta}(l\kappa, l'\kappa')$ are known. Further use of Eqs. (2.67a), (2.71), (2.72) and the relation

$$\sum_{\vec{q}} \exp\left[-i\vec{q} \cdot (\vec{R}_l^0 - \vec{R}_{l'}) \right] = N\delta_{ll'}$$

gives

$$\Phi_{\alpha\alpha}(l\kappa, l'\kappa) = \frac{M_\kappa}{N} \sum_{\vec{q}j} e_\alpha^*(\vec{q}j \mid \kappa) \omega_j^2(\vec{q}) e_\alpha(\vec{q}j \mid \kappa)$$

$$\times \exp\left[i\vec{q} \cdot \{\vec{R}^0(l\kappa) - \vec{R}^0(l'\kappa)\} \right], \qquad (2.119a)$$

or

$$Tr\left[\Phi_{\alpha\alpha}(l\kappa, l'\kappa) \right] = \frac{M_\kappa}{N} \sum_{\vec{q}} \left(\sum_j \omega_j^2(\vec{q}) \right) \exp\left[i\vec{q} \cdot \{\vec{R}^0(l\kappa) - \vec{R}^0(l'\kappa)\} \right].$$

$$(2.119b)$$
$$(2.119)$$

Thus the knowledge of both $\omega_j^2(\vec{q})$ and $\vec{e}(\vec{q}j \mid \kappa)$ is required to know the harmonic force constants even for monatomic lattice [2]. However, form Eq.(2.119b) the trace of harmonic tensor can be calculated from the knowledge of $\omega_j^2(\vec{q})$ only. This trace may contain useful informations about the nature of interatomic forces but may not reflect all their characteristics. For example, if $\Phi_n(\vec{R})$ in Eq.(2.23) is the sum of two body interatomic potentials $V\left(\vec{R}(l\kappa) - \vec{R}(l'\kappa)\right)$ which satisfies Laplace's equation, $Tr\left[\Phi_{\alpha\alpha}(l\kappa, l'\kappa)\right]$ will vanish identically and nothing definite can be said about interatomic forces. If $V(|\vec{R}|)$ is screened Coulomb or Lennard-Jones potential, the decrease in $Tr\left[\Phi_{\alpha\alpha}(l\kappa, l'\kappa)\right]$ with increasing distance may correctly predict the decrease in the magnitude of force constants with increasing distance between atoms. Thus we conclude that atleast two out of the three quantities $\omega_j^2(\vec{q})$, $e_\alpha(\vec{q}j \mid \kappa)$ and $\Phi(\vec{R})$ or $\Phi_{\alpha\beta}(l\kappa, l'\kappa')$ must be known for the complete analysis of dispersion relations.

One can determine the number of distinct eigenvalues, their degeneracies and the symmetry properties of the corresponding eigenvectors [2],[3] through group theoretical calculations. The problem need not be solved even numerically. In principle, the eigenvalues are determined by diagonalizing $\mathbf{D}(\vec{q})$ and eigenvectors by substituting eigenvalues in Eq. (2.67b). But these eigenvevtors, although linearly independent, may not be necessarily orthogonal. The orthogonal eigenvectors can be generated using Schmidt process but then the eigenfrequencies corresponding to these new eigenvectors may no more remain normal mode frequencies, rather these become mixed mode frequencies and the mixing ratio of different modes is not unique. Hence normal mode frequencies cannot be determined uniquely if eigenvectors are not orthogonal.

Again the expressions for scattered X-ray intensity and neutron flux [2] simplify only when the polarization vector is either parallel or perpendicular to phonon wavevector $\vec{q}$. Therefore, the analysis and interpretation of the experimental results along all the $\vec{q}$ directions in the irreducible part of BZ is not possible. However, it turns out, as will be seen later, that along the principal axes for a given crystal structure, the polarization vectors are linearly independent and orthogonal and eigenvalues are normal mode frequencies. The modes do not get mixed up as $\vec{q}$ increases along the principal axes. Therefore, along the principal axes, the eigenvectors and eigenvalues are determined uniquely and the measured results can also be analyzed. This does not mean that the measurements cannot be carried out in the off symmetry directions. In fact, it has been done but their interpretation is no more unique.

In the preview of the BvK theory for d and f- band metals and for other metallic and non-metallic solids too the experimental results are interpreted by the following phenomenological models.

2.7 Generalized Tensor Force Model

In the generalized tensor force (GTF) model, the elements of dynamical matrix $D_{\alpha\beta}(\vec{q}, \kappa\kappa')$ are expressed in terms of the force constants $\Phi_{\alpha\beta}(l\kappa, l'\kappa')$ upto the desired nth nearest neighbours (nNNs) including self-force constants (for $l = l'$). The symmetry properties of the lattice are used to determine the irreducible number of force constants which further simplify the expression for $D_{\alpha\beta}(\vec{q}, \kappa\kappa')$. Once all the elements of $3r \times 3r$ matrix $\mathbf{D}(\vec{q})$ are known, the matrix is diagonalized to obtain the eigenvalues and eigenvectors. The diagonalization of $\mathbf{D}(\vec{q})$ is not always trivial. However, the diagonalization is simplified for $\vec{q}$ along the principal symmetry directions. Standard "computer programs" are available to generate the dynamical matrices for different crystal structures upto desired nearest neighbours (NNs) and to compute the eigenvalues and eigenvectors. The above procedure is illustrated here by presenting the calculations for monatomic fcc, bcc and hcp lattices upto $2NNs$.

2.7.1 fcc lattice

The cubic unit cell with $1NNs$ and $2NNs$ and the rhombohedral primitive cell for fcc lattice are shown in Fig. 2.3. The BZ and irreducible prism are shown in Fig. 2.4. For monatomic lattice $\kappa = \kappa' = 1$ and hence the index κ is dropped. Substituting $M_\kappa = M_{\kappa'} = M$ and choosing the origin at $\vec{R}_l^0$, Eq. (2.54) simplifies as

$$D_{\alpha\beta}(\vec{q}) = \frac{1}{M} \sum_l \Phi_{\alpha\beta}(0l) \exp(i\vec{q} \cdot \vec{R}_l^0)$$

$$= \frac{1}{M}\left[\Phi_{\alpha\beta}(0,0) + \Phi_{\alpha\beta}(0,l^{(1)})\exp(i\vec{q}\cdot\vec{R}^0_{l(1)})\right.$$
$$\left. + \Phi_{\alpha\beta}(0,l^{(2)})\exp(i\vec{q}\cdot\vec{R}^0_{l(2)}) + \ldots\ldots\right], \qquad (2.120)$$

where $l^{(1)}$ and $l^{(2)}$ correspond to $1NNs$ and $2NNs$ shells.

Consider the $1NN$ force matrix $\Phi_{\alpha\beta}(0l^{(1)}) = \Phi_{\alpha\beta}(000,1\bar{1}0)$ for atom number 1 with $l^{(1)} = (1\bar{1}0)$. We can write

$$\mathbf{\Phi}(000,1\bar{1}0) = -\begin{pmatrix} \Phi_{xx} & \Phi_{xy} & \Phi_{xz} \\ \Phi_{yx} & \Phi_{yy} & \Phi_{yz} \\ \Phi_{zx} & \Phi_{zy} & \Phi_{zz} \end{pmatrix}, \qquad (2.121)$$

where the negative sign is introduced for convenience.

1. The mirror reflection $\sigma(1\bar{1}0)$ in the $(1\bar{1}0)$ plane can be written in the matrix form as

$$\sigma(1\bar{1}0) = \begin{pmatrix} 0 & 1 & 0 \\ 1 & 0 & 0 \\ 0 & 0 & 1 \end{pmatrix}. \qquad (2.122)$$

Noting that $\sigma(1\bar{1}0)$ leaves the $(0-1)$ bond invariant and using Eq.(2.49) one can write

$$\mathbf{\Phi}(000,1\bar{1}0) = \sigma(1\bar{1}0)\mathbf{\Phi}(000,1\bar{1}0)\sigma^{\mathrm{T}}(1\bar{1}0). \qquad (2.123)$$

Further simplifications and the comparison of right sides of Eqs. (2.121) and (2.123) leads to inter-relations

$$\Phi_{xx} = \Phi_{yy}, \quad \Phi_{xy} = \Phi_{yx}, \quad \Phi_{yz} = \Phi_{xz}, \quad \Phi_{zx} = \Phi_{zy}. \qquad (2.124)$$

2. The reflection in the plane (001) is represented by the matrix

$$\sigma(001) = \begin{pmatrix} 1 & 0 & 0 \\ 0 & 1 & 0 \\ 0 & 0 & -1 \end{pmatrix}, \qquad (2.125)$$

and it leaves the bond $(0-1)$ invariant. A calculation similar to (2.123) and its comparison with (2.121) yields

$$\begin{aligned} \Phi_{xz} = -\Phi_{xz} = 0, \quad \Phi_{zx} = -\Phi_{zx} = 0, \\ \Phi_{yz} = -\Phi_{yz} = 0, \quad \Phi_{zy} = -\Phi_{zy} = 0. \end{aligned} \qquad (2.126)$$

3. With further symmetry considerations along these lines, no more simplification of the tensor is possible.

Thus $\Phi(000, 1\bar{1}0)$ is left with only three independent components $\Phi_{xx} = \alpha_1, \Phi_{xy} = \gamma_1$ and $\Phi_{zz} = \beta_1$ i.e.

$$\Phi(000, 1\bar{1}0) = - \begin{pmatrix} \alpha_1 & \gamma_1 & 0 \\ \gamma_1 & \alpha_1 & 0 \\ 0 & 0 & \beta_1 \end{pmatrix}. \tag{2.127}$$

The force constants for the remaining eleven nearest neighbours are obtained by using the appropriate operators in Eq. (2.48c) which transform the bond $(0-1)$. For example, the bond $(0-6)$ is obtained from $(0-1)$ by an anti-clockwise rotation by 90^0 about [010] axis which is represented by the matrix

$$C_4[010] = \begin{pmatrix} 0 & 0 & 1 \\ 0 & 1 & 0 \\ 1 & 0 & 0 \end{pmatrix}.$$

From Eq. (2.49) the rotation $C_4[010]$ gives

$$\begin{aligned} \Phi(000, 0\bar{1}1) &= C_4[010]\Phi(000, 1\bar{1}0)C_4^T[010] \\ &= - \begin{pmatrix} \beta_1 & 0 & 0 \\ 0 & \alpha_1 & -\gamma_1 \\ 0 & -\gamma_1 & \alpha_1 \end{pmatrix}. \end{aligned}$$

Similarly the remaining force constant matrices are obtained by continued application of symmetry operations. These force constant tensors will also be checked against rotaional sum rules, [see Eq. (2.41)]. This may remove the redundancies among the force constants [23]. Similar calculations are carried out for the $2NNs$. Table 2.1 summarizes the matrices appropriate to the $1NNs$ and $2NNs$ counted in Fig. (2.3a).

The self-force constant matrix is determined by the sum rule (2.35c) i.e.

$$\Phi(0,0) = -\sum_{l=1}^{18} \Phi(0,l)$$

$$= (8\alpha_1 + 4\beta_1 + 2\alpha_2 + 4\beta_2)\begin{pmatrix} 1 & 0 & 0 \\ 0 & 1 & 0 \\ 0 & 0 & 1 \end{pmatrix}. \qquad (2.128)$$

Thus $\boldsymbol{\Phi}(0,0)$ depends upon the number of NNs and need not to show the permutation symmetry always [24], [see Eq. (2.34)]. Using Eq. (2.128), $\Phi_{\alpha\beta}(0l)$ from Table 2.1 and the corresponding values of $\vec{R}_l^0$, in the Cartesian coordinates $\left[\vec{R}_l^0 = l_1\vec{a}_1 + l_2\vec{a}_2 + l_3\vec{a}_3, \vec{a}_1 = \frac{a}{2}(\hat{i}+\hat{j}), \vec{a}_2 = \frac{a}{2}(\hat{i}+\hat{k}), \vec{a}_3 = \frac{a}{2}(\hat{j}+\hat{k})\right]$ in Eq. (2.120), one finds

$$\begin{aligned} D_{xx}(\vec{q}) = \ & \frac{1}{M}\left[8\alpha_1 + 4\beta_1 + 2\alpha_2 + 4\beta_2 - 4\alpha_1\cos\left(\frac{q_x a}{2}\right) \right.\\ & \left\{\cos\left(\frac{q_y a}{2}\right) + \cos\left(\frac{q_z a}{2}\right)\right\} - 4\beta_1\cos\left(\frac{q_y a}{2}\right)\cos\left(\frac{q_z a}{2}\right) \\ & \left. -2\alpha_2\cos(q_x a) - 2\beta_2\{\cos(q_y a) + \cos(q_z a)\}\right], \qquad (2.129) \end{aligned}$$

where $\vec{q}$ is in units of $(2\pi/a)$ throughout. $D_{yy}(\vec{q})$ and $D_{zz}(\vec{q})$ can be written from cyclic permutations of (xyz). Similarly

$$D_{xy}(\vec{q}) = \frac{4\gamma_1}{M}\sin\left(\frac{q_x a}{2}\right)\sin\left(\frac{q_y a}{2}\right), \qquad (2.130)$$

and $D_{yz}(\vec{q})$ and $D_{zx}(\vec{q})$ can be obtained from cyclic permutations of (xyz).

In cubic crystals, pure longitudinal and transverse modes are along the principal symmetry directions [100], [110] and [111].

1. $\underline{\omega^2(\vec{q}) \text{ along [100] direction}}$: For the wave propagating along [100] direction with wave vector $\vec{q} = [q_x, 0, 0] = [q, 0, 0]$, the elements of dynamical matrix are

$$\begin{aligned} D_{xx}(\vec{q}) &= \frac{1}{M}\left[8\alpha_1\left(1 - \cos\frac{qa}{2}\right) + 2\alpha_2(1 - \cos qa)\right], \\ D_{yy}(\vec{q}) &= D_{zz}(\vec{q}) = \frac{1}{M}\left[(4\alpha_1 + 4\beta_1)\left(1 - \cos\frac{qa}{2}\right)\right.\\ &\qquad \left. +2\beta_2(1 - \cos qa)\right], \\ D_{xy}(\vec{q}) &= D_{yx}(\vec{q}) = D_{xz}(\vec{q}) = D_{zx}(\vec{q}) = 0 \\ &= D_{yz}(\vec{q}) = D_{zy}(\vec{q}) = 0. \qquad (2.131) \end{aligned}$$

Substituting Eqs. (2.131) in Eq. (2.57), one gets

$$\begin{pmatrix} D_{xx}(\vec{q}) - \omega_j^2(\vec{q}) & 0 & 0 \\ 0 & D_{yy}(\vec{q}) - \omega_j^2(\vec{q}) & 0 \\ 0 & 0 & D_{zz}(\vec{q}) - \omega_j^2(\vec{q}) \end{pmatrix} = 0.$$

(2.132)

The solutions are

$$\omega_L^2(\vec{q}) = D_{xx}(\vec{q}) \tag{2.133}$$

$$\omega_T^2(\vec{q}) = D_{yy}(\vec{q}) = D_{zz}(\vec{q}). \tag{2.134}$$

Thus there are three modes of vibrations, one of frequency $\omega_L(\vec{q})$ and other two are doubly degenerate with frequency $\omega_T(\vec{q})$. The polarization vectors associated with these modes are obtained by rewriting Eqs. (2.67a) and (2.67b) as

$$\begin{pmatrix} D_{xx}(\vec{q}) & 0 & 0 \\ 0 & D_{yy}(\vec{q}) & 0 \\ 0 & 0 & D_{zz}(\vec{q}) \end{pmatrix} \begin{pmatrix} e_x(\vec{q}j) \\ e_y(\vec{q}j) \\ e_z(\vec{q}j) \end{pmatrix} = \omega_j^2(\vec{q}) \begin{pmatrix} e_x(\vec{q}j) \\ e_y(\vec{q}j) \\ e_z(\vec{q}j) \end{pmatrix}.$$

(2.135)

If $\vec{e}(\vec{q}L) = \begin{pmatrix} 1 \\ 0 \\ 0 \end{pmatrix}$ i.e.parallel to $\vec{q}$, one gets Eq. (2.133), therefore $\omega_L(\vec{q})$ are the frequencies of longitudinal modes. If $\vec{e}(\vec{q}T_1) = \begin{pmatrix} 0 \\ 1 \\ 0 \end{pmatrix}$ or $\vec{e}(\vec{q}T_2) = \begin{pmatrix} 0 \\ 0 \\ 1 \end{pmatrix}$ i.e. perpendicular to $\vec{q}$, Eq. (2.134) is obtained. Therefore $\omega_T(\vec{q})$ are the frequencies of doubly degenerate transverse modes.

The displacement of the *lth* atom in the particular mode $(\vec{q}j)$ from Eq. (2.85) is

$$\vec{u}(l,t) = \frac{1}{\sqrt{NM}} \sum_{\vec{q}j} Q(\vec{q}j,t)\vec{e}(\vec{q}j)\exp(i\vec{q}\cdot\vec{R}_l^0). \tag{2.136}$$

For $\vec{q} = (2\pi/a)[100]$ i.e. the X-point in the [100] direction in Fig.2.4, the values of $\vec{R}_l^0 = (0,0,0), \frac{a}{2}(1,1,0), \frac{a}{2}(1,-1,0)$ etc. are used and the sign and magnitude of $\vec{u}(l,t)$ are determined. The relative orientations of $\vec{u}(l,t)$ for all the atoms for polarization vector $\vec{e}(\vec{q}L), \vec{e}(\vec{q}T_1)$ and $\vec{q}(\vec{q}T_2)$ are shown in Fig. 2.5 . The displacement pattern will remain the same for all the values of $\vec{q}$ in the [100] direction. The point to be noted is that the displacement of all the atoms is along the same direction in one {100} atomic plane but the displacements in the alternate {100} planes are in the opposite directions. The vibrational modes do not mix up.

2. $\underline{\omega^2(\vec{q}) \text{ along } [110] \text{ direction}}$: For the waves propagating along [100] direction with $\vec{q} = (q_x, q_y, 0) = \frac{1}{\sqrt{2}}(q, q, 0)$, the elements of the dynamical matrix are

$$
\begin{aligned}
D_{xx}(\vec{q}) \;=\; D_{yy}(\vec{q}) &= \frac{1}{M}\left[4(\alpha_1 + \beta_2)\left(1 - \cos\frac{qa}{2\sqrt{2}}\right)\right. \\
&\quad \left. +2(\alpha_1 + \alpha_2 + \beta_2)\left(1 - \cos\frac{qa}{\sqrt{2}}\right)\right], \\
D_{zz}(\vec{q}) &= \frac{1}{M}\left[8\alpha_1\left(1 - \cos\frac{qa}{2\sqrt{2}}\right) + 2(\beta_1 + 2\beta_2)\left(1 - \cos\frac{qa}{\sqrt{2}}\right)\right], \\
D_{xy}(\vec{q}) \;=\; D_{yx}(\vec{q}) &= \frac{1}{M}\left[2\gamma_1\left(1 - \cos\frac{qa}{\sqrt{2}}\right)\right], \\
D_{xz}(\vec{q}) \;=\; D_{zx}(\vec{q}) &= D_{yz}(\vec{q}) = D_{zy}(\vec{q}) = 0. \qquad (2.137)
\end{aligned}
$$

Using these values in Eq. (2.57), one gets

$$
\begin{aligned}
\omega_L^2(\vec{q}) &= D_{xx}(\vec{q}) + D_{xy}(\vec{q}) \\
\omega_{T_1}^2(\vec{q}) &= D_{xx}(\vec{q}) - D_{xy}(\vec{q}) \\
\omega_{T_2}^2(\vec{q}) &= D_{zz}(\vec{q}). \qquad (2.138)
\end{aligned}
$$

Using Eq. (2.138) in Eq. (2.67),

$$
\vec{e}(\vec{q}_L) = \frac{1}{\sqrt{2}}\begin{pmatrix} 1 \\ 1 \\ 0 \end{pmatrix},
$$

$$
\vec{e}(\vec{q}T_1) = \frac{1}{\sqrt{2}}\begin{pmatrix} 1 \\ -1 \\ 0 \end{pmatrix},
$$

and

$$\vec{e}(\vec{q}T_2) \;=\; \begin{pmatrix} 0 \\ 0 \\ 1 \end{pmatrix}. \tag{2.139}$$

Thus $\vec{e}(\vec{q}_L)$ is $\parallel$ to $\vec{q}$ and $\vec{e}(\vec{q}T_1), \vec{e}(\vec{q}T_2)$ are $\perp$ to $\vec{q}$. Hence $\omega_L^2(\vec{q})$ are the frequencies of longitudinal modes and $\omega_{T_1}^2(\vec{q})$ and $\omega_{T_2}^2(\vec{q})$ are the frequencies of transverse modes . The relative orientation of atomic displacements for a particular $(\vec{q}j)$ can be determined with the help of Eq. (2.136). The atomic displacements of all the atoms in any $\{110\}$ plane will be along the same direction. However, the atomic displacements in alternate $\{110\}$ planes will be in the opposite direction.

3. $\underline{\omega^2(\vec{q}) \text{ along } [111] \text{ direction}}$: For the waves propagating along [111] direction, the expressions for $D_{\alpha\beta}(\vec{q})$ are calculated by substituting $\vec{q} = (q_x, q_y, q_z) = \frac{1}{\sqrt{3}}(q, q, q)$. The final results for eigenvalues and eigenvectors are as follows :

$$\begin{aligned}
\omega_L^2(\vec{q}) &= D_{xx}(\vec{q}) + 2D_{xy}(\vec{q}) \\
\omega_{T_1}^2(\vec{q}) &= \omega_{T_2}^2(\vec{q}) \\
&= D_{xx}(\vec{q}) - D_{xy}(\vec{q}). \tag{2.140}
\end{aligned}$$

$$\vec{e}(\vec{q}L) \;=\; \frac{1}{\sqrt{3}} \begin{pmatrix} 1 \\ 1 \\ 1 \end{pmatrix},$$

$$\vec{e}(\vec{q}T_1) \;=\; \frac{1}{\sqrt{2}} \begin{pmatrix} 1 \\ -1 \\ 0 \end{pmatrix},$$

$$\vec{e}(\vec{q}T_2) \;=\; \frac{1}{\sqrt{6}} \begin{pmatrix} 1 \\ 1 \\ -2 \end{pmatrix}. \tag{2.141}$$

The atomic displacements are such that the atoms are displaced in the same phase in a particular $\{111\}$ plane and in opposite phases in the alternate $\{111\}$ planes. Thus, in the fcc lattice the vibrational modes are longitudinal or transverse along [100], [110] and [111] directions. For other directions of $\vec{q}$, the eigenvectors will depend on both the direction and the magnitude of $\vec{q}$ and hence on the force constants. The vibrational modes will be mixed in these directions.

The above analysis can further be extended upto $nNNs$. In transition metals the forces are long ranged. Therefore the data analysis involves large number of NNs. A typical example of $5NN$ model is given as follows [25]:

[100] direction, longitudinal mode

$$
\begin{aligned}
M\omega_L^2(\vec{q}) \; = \; & (8\alpha_1 + 16\beta_3 + 8\beta_5)\left\{1 - \cos(\frac{qa}{2})\right\} \\
& + (2\alpha_2 + 8\alpha_3 + 8\alpha_4)\left\{1 - \cos(qa)\right\} \\
& + 8\alpha_5\left\{1 - \cos(\frac{3qa}{2})\right\}.
\end{aligned}
\tag{2.142}
$$

[100] direction, transverse mode

$$
\begin{aligned}
M\omega_T^2(\vec{q}) \; = \; & (4\alpha_1 + 4\beta_1 + 8\alpha_3 + 8\beta_3 + 4\alpha_5 + 4\gamma_5)\left(1 - \cos\frac{qa}{2}\right) \\
& + (2\beta_2 + 8\beta_3 + 4\alpha_4 + 4\beta_4)(1 - \cos qa) \\
& + (4\beta_5 + 4\gamma_5)\left(1 - \cos\frac{3qa}{2}\right),
\end{aligned}
\tag{2.143}
$$

and so on. A typical fit of phonon frequencies for Ni upto $4NN$ is shown in the Fig. 2.6 and the values of force constants are tabulated in Table 2.2 [26].

2.7.2   bcc Lattice

The unit cell, nearest neighbours, Brillouin zone and its irreducible part for bcc lattice are shown in Figs. (2.7) and (2.8). The force constants matrix $\boldsymbol{\Phi}(\mathbf{000}, \mathbf{111})$ is written in the form of Eq. (2.121) and complete set of symmetry operations is carried out [27],[28] such that the bond $(000 - 111) \equiv (0 - 1)$ remains invarient. The mirror reflections $\sigma(1\bar{1}1), \sigma(11\bar{1}), \sigma(\bar{1}11)$ etc. leave the bond $(0 - 1)$ invarient. Using the matrix representation

$$\sigma(1\bar{1}1) = \begin{pmatrix} 0 & 1 & 0 \\ 1 & 0 & 0 \\ 0 & 0 & 1 \end{pmatrix}, \qquad \sigma(11\bar{1}) = \begin{pmatrix} 1 & 0 & 0 \\ 0 & 0 & 1 \\ 0 & 1 & 0 \end{pmatrix}, \qquad \text{etc}$$

in Eq. (2.49) one gets

$$\boldsymbol{\Phi}(000,111) = - \begin{pmatrix} \alpha_1 & \beta_1 & \beta_1 \\ \beta_1 & \alpha_1 & \beta_1 \\ \beta_1 & \beta_1 & \alpha_1 \end{pmatrix}, \tag{2.144}$$

with $\alpha_1 = \Phi_{xx}$ and $\beta_1 = \Phi_{xy}$. The force constant matrix for bond $(0-2)\left[\vec{R}^0_{l(2)} = (a/2)(\bar{1}00)\right]$ is obtained using reflection symmetry about yz plane, represented by the matrix

$$\sigma(\bar{1}11) = \begin{pmatrix} -1 & 0 & 0 \\ 0 & 1 & 0 \\ 0 & 0 & 1 \end{pmatrix},$$

and Eq.(2.49). This gives

$$\boldsymbol{\Phi}(000,\bar{1}11) = - \begin{pmatrix} \alpha_1 & -\beta_1 & -\beta_1 \\ -\beta_1 & \alpha_1 & \beta_1 \\ -\beta_1 & \beta_1 & \alpha_1 \end{pmatrix}. \tag{2.145}$$

All other force constant matrices for the $1NNs$ and $2NNs$ are derived in the same manner. These results are tabulated in Table 2.3. The self-force constant (upto 2NNs) is

$$\Phi(0,0) \;=\; -\sum_{l=1}^{14} \Phi(0,l)$$

$$= \; (8\alpha_1 + 2\alpha_2 + 4\beta_2) \begin{pmatrix} 1 & 0 & 0 \\ 0 & 1 & 0 \\ 0 & 0 & 1 \end{pmatrix}. \tag{2.146}$$

Using Eq. (2.120) the elements of dynamical matrix are

$$D_{xx}(\vec{q}) \;=\; \frac{1}{M}\left[8\alpha_1\left(1 - \cos\frac{q_x a}{2}\cos\frac{q_y a}{2}\cos\frac{q_z a}{2}\right)\right.$$
$$+ 2\alpha_2(1 - \cos q_x a) + 2\beta_2(1 - \cos q_y a)$$
$$\left. + 2\beta_2(1 - \cos q_z a)\right]. \tag{2.147}$$

$D_{yy}(\vec{q})$ and $D_{zz}(\vec{q})$ are obtained by interchanging x by y and z respectively.

$$D_{xy}(\vec{q}) = D_{yx}(\vec{q}) = \frac{8\beta_1}{M}\left[\sin\frac{q_x a}{2}\cos\frac{q_y a}{2}\cos\frac{q_z a}{2}\right].\qquad(2.148)$$

The second nearest neighbours do not contribute to $D_{xy}(\vec{q}).D_{yz} = D_{zy}$ and $D_{xz} = D_{zx}$ can be written by cyclic permutation of (xyz).

(1) $\omega^2(\vec{q})$ along [100] direction

Using $\vec{q} = (q_x, 0, 0) = (q, 0, 0)$ in Eqs. (2.147) and (2.148), the elements $D_{\alpha\beta}(\vec{q})$ can readily be obtained. Putting these $D_{\alpha\beta}(\vec{q})$ in Eq. (2.57) one finds

$$\begin{pmatrix} D_{xx}(\vec{q}) - \omega^2(\vec{q}) & 0 & 0 \\ 0 & D_{yy}(\vec{q}) - \omega^2(\vec{q}) & 0 \\ 0 & 0 & D_{zz}(\vec{q}) - \omega^2(\vec{q}) \end{pmatrix} = 0.$$

Therefore

$$\omega_L^2(\vec{q}) = D_{xx}(\vec{q})$$

and

$$\omega_{T_1}^2(\vec{q}) = \omega_{T_2}^2(\vec{q}) = D_{yy}(\vec{q}).\qquad(2.149)$$

The transverse mode is doubly degenerate. The eigenvectors are calculated by the method which has been discusssed earlier. These are given as

$$\vec{e}(\vec{q}L)\begin{pmatrix} 1 \\ 0 \\ 0 \end{pmatrix},\quad \vec{e}(\vec{q}T_1) = \begin{pmatrix} 0 \\ 1 \\ 0 \end{pmatrix},\quad \vec{e}(\vec{q}T_2) = \begin{pmatrix} 0 \\ 0 \\ 1 \end{pmatrix}.\qquad(2.150)$$

(2) $\omega^2(\vec{q})$ along [110] direction

Here $\vec{q} = \frac{1}{\sqrt{2}}(q, q, 0)$ is used to calculate $D_{\alpha\beta}(\vec{q})$ from Eqs. (2.147) and (2.148). The determinental equation (2.57) gives the eigenvalues

$$\begin{aligned} \omega_L^2(\vec{q}) &= D_{xx}(\vec{q}) + D_{xy}(\vec{q}) \\ \omega_{T_1}^2(\vec{q}) &= D_{xx}(\vec{q}) - D_{xy}(\vec{q}) \\ \omega_{T_2}^2(\vec{q}) &= D_{zz}(\vec{q}). \end{aligned}\qquad(2.151)$$

The corresponding eigenvectors are:

$$\vec{e}(\vec{q}L) = \frac{1}{\sqrt{2}}\begin{pmatrix} 1 \\ 1 \\ 0 \end{pmatrix}, \quad \vec{e}(\vec{q}T_1) = \frac{1}{\sqrt{2}}\begin{pmatrix} 1 \\ -1 \\ 0 \end{pmatrix}, \quad \vec{e}(\vec{q}T_2) = \begin{pmatrix} 0 \\ 0 \\ 1 \end{pmatrix}. \tag{2.152}$$

(3) $\omega^2(\vec{q})$ along [111] direction

Here $\vec{q} = (1/\sqrt{3})(q, q, q)$ is used to calculate $D_{\alpha\beta}(\vec{q})$ from Eqs. (2.147) and (2.148). The solutions of Eq. (2.57) are

$$\begin{aligned} \omega_L^2(\vec{q}) &= D_{xx}(\vec{q}) + 2D_{xy}(\vec{q}) \\ \omega_{T_1}^2(\vec{q}) &= \omega_{T_2}^2(\vec{q}) = D_{xx}(\vec{q}) - D_{xy}(\vec{q}). \end{aligned} \tag{2.153}$$

The transverse modes are degenerate. The corresponding eigenvectors are:

$$\vec{e}(\vec{q}L) = \frac{1}{\sqrt{3}}\begin{pmatrix} 1 \\ 1 \\ 1 \end{pmatrix}, \quad \vec{e}(\vec{q}T_1) = \frac{1}{\sqrt{2}}\begin{pmatrix} 1 \\ -1 \\ 0 \end{pmatrix}, \quad \vec{e}(\vec{q}T_2)\frac{1}{\sqrt{6}}\begin{pmatrix} 1 \\ 1 \\ -2 \end{pmatrix}. \tag{2.154}$$

The relevant $D_{\alpha\beta}(\vec{q})$ are tabulated in Table 2.4.

Similar calculations in the lower symmetry directions lead to mixed modes. For example for $\vec{q}$ along $[\frac{1}{2}, \frac{1}{2}, q]$ and $[q, q, 1]$ directions, the waves are neither longitudinal nor transverse although their polarization vectors are completely determined from the symmetry considerations [29].

The direction of atomic displacements for $\vec{q} = (2\pi/a)(100)$ and $\vec{q} = (\pi/a)(110)$ in the [100] and [110] directions are calculated using Eq. (2.136). These are shown in Figs. (2.9a) and (2.9b), respectively. Again a pattern of planer vibrations exists.

The general treatment for determining the force constants for the cubic systems is given by Squires [30]. A particular example upto $5NNs$ is as follows [29]:

[001] direction, longitudinal mode

$$\begin{aligned} M\omega_L^2(\vec{q}) &= (8\alpha_1 + 16\beta_4)\{1 - \cos(qa/2)\} \\ &\quad +(2\alpha_2 + 8\alpha_3 + 8\alpha_5)\{1 - \cos(qa)\} \\ &\quad +8\alpha_4\{1 - \cos(3qa/2)\}. \end{aligned} \tag{2.155}$$

[001] direction, transverse mode

$$\begin{aligned}
M\omega_T^2(\vec{q}) &= (8\alpha_1 + 8\alpha_4 + 8\beta_4)\{1 - \cos(qa/2)\} \\
&\quad + (2\beta_2 + 4\alpha_3 + 4\beta_3 + 8\alpha_5)\{1 - \cos(qa)\} \\
&\quad + 8\beta_4\{1 - \cos(3qa/2)\}.
\end{aligned} \tag{2.156}$$

A typical least square fit of the phonon frequencies of Cr [31] is given in Fig.2.10 and resulting force constants are given in Table 2.5.

2.7.3   hcp Lattice

The hcp unit cell, the $1NNs$ of two basis atoms, and the BZ are shown in Fig. (2.11). The basis atoms of equal mass M are distinguished by $\kappa = 1, 2$. The cell vectors $\vec{a}_1, \vec{a}_2, \vec{a}_3$ are chosen such that the angle between $\vec{a}_1$ and $\vec{a}_2$ is $(2\pi/3)$ and $\vec{a}_3$ is perpendicular to the plane of $\vec{a}_1$ and $\vec{a}_2$. Here $|\vec{a}_1| = |\vec{a}_2| = a$ and $|\vec{a}_3| = c$.

To describe the force constant analysis Begbie and Born [22] procedure is followed. Let us consider the force constant tensor $\mathbf{\Phi}(1\kappa, 1'\kappa')$ related to $1NNs$ of two atoms at O and O' shown in Fig. (2.11b). Each atom has twelve $1NNs$, six in the same plane, three in each upper and lower planes at the distance $c/2$. For non-ideal hcp lattice, these atoms may be a set of the $1NNs$ and $2NNs$. Let p denotes the bonds of O and O' to their neighbours and $\mathbf{\Phi}^\mathbf{p}$ the corresponding force tensors. The indices p for the neighbours of O and O' are $(1, 2, 3, 4, 5, 6, \bar{1}, \bar{2}, \bar{3}, \bar{4}, \bar{5}, \bar{6})$ and $(1', 2', 3', 4', 5', 6', \bar{1}', \bar{2}', \bar{3}', \bar{4}', \bar{5}', \bar{6}')$ respectively. The six atoms which lie in a plane containing an atom at O are of type $\kappa = 1$, the six atoms which lie in the plane containing an atom at O' are of type $\kappa = 2$ and so on, such that the atoms of one type lie on one plane and of other type on the next plane. The connection between symbol p and $(o\kappa, l'\kappa')$ is given in Table 2.6.

The lattice symmetries are as follows:

1. A six fold rotation about z-axis together with an inversion in the

origin. This operation is represented by the matrix.

$$C_6^I(001) = \begin{pmatrix} -\frac{1}{2} & -\frac{\sqrt{3}}{2} & 0 \\ \frac{\sqrt{3}}{2} & -\frac{1}{2} & 0 \\ 0 & 0 & -1 \end{pmatrix}. \qquad (2.157a)$$

2. A two fold axis of rotation about y-axis which is represented by the matrix

$$C_2(010) = \begin{pmatrix} -1 & 0 & 0 \\ 0 & 1 & 0 \\ 0 & 0 & -1 \end{pmatrix}. \qquad (2.157b)$$

3. The glide plane symmetry containing a reflection in the $y = 0$ plane and a translation with vector $\vec{R}^0_{\kappa\kappa'} = \pm\frac{a}{\sqrt{3}}\hat{j} \mp \frac{1}{2}c\hat{k}$. This reflection symmetry is represented by the matrix

$$\sigma(101) = -C_2(010) = \begin{pmatrix} 1 & 0 & 0 \\ 0 & -1 & 0 \\ 0 & 0 & 1 \end{pmatrix} \qquad (2.157c)$$

$$(2.157)$$

and the translation leads to the atoms in the next plane.

The following symmetry operations are used to reduce the number of force matrices and force constants:

1. the permutation symmetry (2.34) leads to

$$\begin{aligned}
\Phi^1 &= \{\Phi^{\bar{1}}\}^{\mathrm{T}}, & \Phi^6 &= \{\Phi^{\bar{6}'}\}^{\mathrm{T}}, & \Phi^{5'} &= \{\Phi^{\bar{5}}\}^{\mathrm{T}}, \\
\Phi^2 &= \{\Phi^{\bar{3}}\}^{\mathrm{T}}, & \Phi^{1'} &= \{\Phi^{\bar{1}'}\}^{\mathrm{T}}, & \Phi^{6'} &= \{\Phi^{\bar{6}}\}^{\mathrm{T}}. \\
\Phi^3 &= \{\Phi^{\bar{2}}\}^{\mathrm{T}}, & \Phi^{2'} &= \{\Phi^{\bar{3}'}\}^{\mathrm{T}}, \\
\Phi^4 &= \{\Phi^{\bar{4}'}\}^{\mathrm{T}}, & \Phi^{3'} &= \{\Phi^{\bar{2}'}\}^{\mathrm{T}}, \\
\Phi^5 &= \{\Phi^{\bar{5}'}\}^{\mathrm{T}}, & \Phi^{4'} &= \{\Phi^{\bar{4}'}\}^{\mathrm{T}},
\end{aligned} \qquad (2.158)$$

Thus only 12 matrices out of 24 Φ^P or $\Phi^{p'}$ matrices are to be calculated.

2. The reflection symmetry in the $z = 0$ plane leads to

$$\begin{pmatrix} \Phi^1_{11} & \Phi^1_{12} & \Phi^1_{13} \\ \Phi^1_{21} & \Phi^1_{22} & \Phi^1_{23} \\ \Phi^1_{31} & \Phi^1_{32} & \Phi^1_{33} \end{pmatrix} = \begin{pmatrix} 1 & 0 & 0 \\ 0 & 1 & 0 \\ 0 & 0 & -1 \end{pmatrix} \left(\Phi^1 \right) \begin{pmatrix} 1 & 0 & 0 \\ 0 & 1 & 0 \\ 0 & 0 & -1 \end{pmatrix}$$

$$= \begin{pmatrix} \Phi^1_{11} & \Phi^1_{12} & -\Phi^1_{13} \\ \Phi^1_{21} & \Phi^1_{22} & -\Phi^1_{23} \\ -\Phi^1_{31} & -\Phi^1_{32} & \Phi^1_{33} \end{pmatrix}, \qquad (2.159)$$

and hence $\Phi^1_{13} = \Phi^1_{31} = \Phi^1_{23} = \Phi^1_{32} = 0$. The matrix Φ^1 is written as

$$\Phi^1 = - \begin{pmatrix} \alpha_1 + 2\beta_1 & \delta_1 & 0 \\ \epsilon_1 & \alpha_1 - 2\beta_1 & 0 \\ 0 & 0 & \gamma_1 \end{pmatrix}, \qquad (2.160)$$

where $\Phi^1_{11}(01, l1) = \alpha_1 + 2\beta_1$ etc. are written for notational convenience. Since

$$\Phi^{\bar{1}} = C_2(010)\Phi^1 C_2^T(010) = - \begin{pmatrix} \alpha_1 + 2\beta_1 & -\delta_1 & 0 \\ -\epsilon_1 & \alpha_1 - 2\beta_1 & 0 \\ 0 & 0 & \gamma_1 \end{pmatrix},$$

$$(2.161)$$

and from (2.158)

$$\Phi^1 = \{\Phi^{\bar{1}}\}^T = - \begin{pmatrix} \alpha_1 + 2\beta_1 & -\epsilon_1 & 0 \\ -\delta_1 & \alpha_1 - 2\beta_1 & 0 \\ 0 & 0 & \gamma_1 \end{pmatrix}. \qquad (2.162)$$

Comparison of Eqs. (2.160) and (2.162) gives $\epsilon_1 = -\delta_1 . \Phi^2, \Phi^{\bar{2}}, \Phi^3$ and $\Phi^{\bar{3}}$ are calculated using the relation $\Phi^2 = C_6^I(001)\Phi^1 \{C_6^I(001)\}^T$, $\Phi^3 = C_6^I(001)\Phi^2 \{C_6^I(001)\}^T$ and so on, where in general

$$C_6^I(001)\Phi\{C_6^I(001)\}^T = \begin{pmatrix} A_{11} & A_{12} & A_{13} \\ A_{21} & A_{22} & A_{23} \\ A_{31} & A_{32} & A_{33} \end{pmatrix} \qquad (2.163)$$

where

$$A_{11} = \frac{1}{4}\left[\Phi_{11} + \sqrt{3}\Phi_{21} + \sqrt{3}\Phi_{12} + 3\Phi_{22}\right]$$

$$A_{12} = \frac{1}{4}\left[-\sqrt{3}\Phi_{11} - 3\Phi_{21} + \Phi_{12} + \sqrt{3}\Phi_{22}\right]$$

$$A_{13} = \frac{1}{2}\left[\Phi_{13} + \sqrt{3}\Phi_{23}\right]$$

$$A_{21} = \frac{1}{4}\left[\Phi_{11} + \Phi_{21} - 3\Phi_{12} + \sqrt{3}\Phi_{22}\right]$$

$$A_{22} = \frac{1}{4}\left[3\Phi_{11} - \sqrt{3}\Phi_{21} - \sqrt{3}\Phi_{12} + \Phi_{22}\right]$$

$$A_{23} = \frac{1}{2}\left[-\sqrt{3}\Phi_{13} + \Phi_{23}\right]$$

$$A_{31} = \frac{1}{2}\left[\Phi_{31} + \sqrt{3}\Phi_{32}\right]$$

$$A_{32} = \frac{1}{2}\left[-\sqrt{3}\Phi_{31} + \Phi_{22}\right]$$

$$A_{33} = \Phi_{33}.$$

3. $\boldsymbol{\Phi}^4$ is simplified using the reflection symmetry in the $x = 0$ plane (yz plane) i.e.

$$\begin{pmatrix} \Phi^4_{11} & \Phi^4_{12} & \Phi^4_{13} \\ \Phi^4_{21} & \Phi^4_{22} & \Phi^4_{23} \\ \Phi^4_{31} & \Phi^4_{32} & \Phi^4_{33} \end{pmatrix} = \begin{pmatrix} -1 & 0 & 0 \\ 0 & 1 & 0 \\ 0 & 0 & 1 \end{pmatrix} (\boldsymbol{\Phi}^4) \begin{pmatrix} -1 & 0 & 0 \\ 0 & 1 & 0 \\ 0 & 0 & 1 \end{pmatrix},$$

$$(2.164)$$

which leads $\Phi^4_{12} = \Phi^4_{21} = \Phi^4_{13} = \Phi^4_{31} = 0$. The resulting matrix $\boldsymbol{\Phi}^4$ is written as

$$\boldsymbol{\Phi}^4 = -\begin{pmatrix} \lambda_1 + 2\mu_1 & 0 & 0 \\ 0 & \lambda_1 - 2\mu_1 & 2\sigma_1 \\ 0 & 2\sigma_1 & \nu_1 \end{pmatrix} \qquad (2.165)$$

where $\Phi^4_{11} = \lambda_1 + 2\mu_1$ etc. The remaining matrices $\boldsymbol{\Phi}^{\bar{4}}, \boldsymbol{\Phi}^6, \boldsymbol{\Phi}^{\bar{6}}, \boldsymbol{\Phi}^5, \boldsymbol{\Phi}^{\bar{5}}$ are calculated in the same way as $\boldsymbol{\Phi}^2, \boldsymbol{\Phi}^3$ etc. using (2.163). The final results are collected in Table 2.7.

The self-force constant matrix from Eq. (2.35c) is

$$\left.\begin{matrix} \boldsymbol{\Phi}^0 \\ \boldsymbol{\Phi}^{0'} \end{matrix}\right\} = \begin{pmatrix} 6\alpha_1 + 6\lambda_1 & 0 & 0 \\ 0 & 6\alpha_1 + 6\lambda_1 & 0 \\ 0 & 0 & 6\alpha_1 + 6\lambda_1 \end{pmatrix}. \qquad (2.166)$$

The elements of dynamical matrix are evaluated using Eq. (2.54). These are given as

$$\mathbf{D}(\vec{q}, 11) = \frac{1}{M} \sum_{\mathbf{p}} \mathbf{\Phi^p} \exp\left[-i\vec{q} \cdot \mathbf{R}^0_\rho\right],$$

where $\vec{R}^0_\rho = \vec{R}^0_l$ and $p = 0, 1, 2, 3, \bar{1}, \bar{2}, \bar{3}$ and

$$\mathbf{D}(\vec{q}, 21) = \frac{1}{M} \sum_{\mathbf{p}} \mathbf{\Phi^p} \exp\left[-i\vec{q} \cdot \mathbf{R}^0_\rho\right] \tag{2.167}$$

where $\vec{R}^0_\rho = \vec{R}^0_l + \vec{R}^0_{21}$ and $p = 4, 5, 6, \bar{4}, \bar{5}, \bar{6}$. The values of $\vec{R}^0_\rho$ in Cartesian coordinates are tabulated in Table 2.8. Using (2.167) and the symmetry properties of $\Phi_{\alpha\beta}(\vec{q}, \kappa\kappa')$, one finds

$$
\begin{aligned}
D_{\alpha\beta}(\vec{q}, 11) &= D^*_{\beta\alpha}(\vec{q}, 11) = D^*_{\alpha\beta}(\vec{q}, 22), \\
D_{\alpha\beta}(\vec{q}, 12) &= D^*_{\beta\alpha}(\vec{q}, 12) = D^*_{\alpha\beta}(\vec{q}, 21).
\end{aligned}
\tag{2.168}
$$

Thus the final expressions for elements of $\mathbf{D}$ upto $1NNs$ (or $2NNs$) are

$$
\begin{aligned}
D_{xx}(\vec{q}, 11) &= \frac{1}{M}\left[6\lambda_1 + 4(\alpha_1 + 2\beta_1)\left\{1 - \cos^2\left(\frac{q_x a}{2}\right)\right\}\right. \\
&\quad \left. +4(\alpha_1 - \beta_1)\left\{1 - \cos\left(\frac{q_x a}{2}\right)\cos\left(\frac{\sqrt{3}}{2}q_y a\right)\right\}\right], \\
D_{yy}(\vec{q}, 11) &= \frac{1}{M}\left[6\lambda_1 + 4(\alpha_1 - 2\beta_1)\left\{1 - \cos^2\left(\frac{q_x a}{2}\right)\right\}\right. \\
&\quad \left. +4(\alpha_1 + \beta_1)\left\{1 - \cos\left(\frac{q_x a}{2}\right)\cos\left(\frac{\sqrt{3}}{2}q_y a\right)\right\}\right], \\
D_{zz}(\vec{q}, 11) &= \frac{1}{M}\left[6\nu_1 + 4\gamma_1\left\{2 - \cos^2\left(\frac{q_x a}{2}\right) - \cos\left(\frac{q_x a}{2}\right)\right.\right. \\
&\quad \left.\left. \times \cos\left(\frac{\sqrt{3}}{2}q_y a\right)\right\}\right], \\
D_{xy}(\vec{q}, 11) &= \frac{1}{M}\left[4\sqrt{3}\beta_1 \sin\left(\frac{q_x a}{2}\right)\sin\left(\frac{\sqrt{3}}{2}q_y a\right) + 4i\delta_1 \sin\left(\frac{q_x a}{2}\right)\right. \\
&\quad \left. \times\left\{\cos\left(\frac{q_x a}{2}\right) - \cos\left(\frac{\sqrt{3}}{2}q_y a\right)\right\}\right],
\end{aligned}
$$

$$D_{yx}(\vec{q},11) = \frac{1}{M}\left[4\sqrt{3}\beta_1\sin\left(\frac{q_x a}{2}\right)\sin\left(\frac{\sqrt{3}}{2}q_y a\right) - 4i\delta_1\sin\left(\frac{q_x a}{2}\right)\right.$$
$$\left.\times\left\{\cos\left(\frac{q_x a}{2}\right) - \cos\left(\frac{\sqrt{3}}{2}q_y a\right)\right\}\right],$$

$$D_{yz}(\vec{q},11) = D_{zy}(\vec{q},11) = D_{zx}(\vec{q},11) = D_{xz}(\vec{q},11) = 0. \qquad (2.169)$$

$$D_{xx}(\vec{q},21) = \frac{1}{M}\left[-2(\lambda_1+2\mu_1)\cos\left(\frac{q_z c}{2}\right)\exp\left(\frac{icq_z}{2}\right) - 4(\lambda_1-\mu_1)\right.$$
$$\left.\times\cos\left(\frac{cq_z}{2}\right)\cos\left(\frac{q_x a}{2}\right)\exp\left\{-\frac{1}{2}i\left(\sqrt{3}q_y a - q_z c\right)\right\}\right],$$

$$D_{yy}(\vec{q},21) = \frac{1}{M}\left[-2(\lambda_1-2\mu_1)\cos\left(\frac{q_z c}{2}\right)\exp\left(\frac{icq_z}{2}\right) - 4(\lambda_1+\mu_1)\right.$$
$$\left.\times\cos\left(\frac{cq_z}{2}\right)\cos\left(\frac{q_x a}{2}\right)\exp\left\{-\frac{1}{2}i\left(\sqrt{3}q_y a - q_z c\right)\right\}\right],$$

$$D_{zz}(\vec{q},21) = -\frac{1}{M}2\nu_1\cos\left(\frac{cq_z}{2}\right)\left[\exp\left(\frac{icq_z}{2}\right) + 2\cos\left(\frac{\pi q_x}{2}\right)\right.$$
$$\left.\times\exp\left\{-\frac{1}{2}i(\sqrt{3}aq_y - cq_z)\right\}\right],$$

$$D_{yz}(\vec{q},21) = D_{zy}(\vec{q},21)$$
$$= \frac{4i}{M}\sigma_1\sin\left(\frac{cq_z}{2}\right)\left[\cos\left(\frac{q_x a}{2}\right)\right.$$
$$\left.\times\exp\left\{-\frac{1}{2}i(\sqrt{3}aq_y - cq_z)\right\} - \exp\left(\frac{icq_z}{2}\right)\right],$$

$$D_{zx}(\vec{q},21) = D_{xz}(\vec{q},21) = \frac{4\sqrt{3}}{M}\sigma_1\sin\left(\frac{cq_z}{2}\right)\sin\left(\frac{q_x a}{2}\right)$$
$$\times\exp\left\{-\frac{1}{2}i(\sqrt{3}q_y a - cq_z)\right\},$$

$$D_{xy}(\vec{q},21) = D_{yx}(\vec{q},21) = \frac{-4\sqrt{3}}{M}i\mu\cos\left(\frac{cq_z}{2}\right)\sin\left(\frac{q_x a}{2}\right)$$
$$\times\exp\left\{-\frac{1}{2}i(\sqrt{3}aq_y - cq_z)\right\}. \qquad (2.170)$$

Using Eqs. (2.169) and (2.170) in Eq. (2.57) one gets

$$\begin{vmatrix} \mathbf{A}(\vec{q},11) - \omega^2\delta_{\alpha\beta} & \mathbf{B}(\vec{q},12) \\ \mathbf{B}^*(\vec{q},12) & \mathbf{A}^*(\vec{q},11) - \omega^2\delta_{\alpha\beta} \end{vmatrix} = 0, \qquad (2.171)$$

where

$$\mathbf{A}(\vec{q}, 11) - \omega^2 \delta_{\alpha\beta} = \begin{pmatrix} D_{xx}(\vec{q}, 11) - \omega^2 & D_{xy}(\vec{q}, 11) & D_{xz}(\vec{q}, 11) \\ D_{yx}(\vec{q}, 11) & D_{yy}(\vec{q}, 11) - \omega^2 & D_{yz}(\vec{q}, 11) \\ D_{zx}(\vec{q}, 11) & D_{zy}(\vec{q}, 11) & D_{zz}(\vec{q}, 11) - \omega^2 \end{pmatrix}$$

$$(2.172a)$$

$$\mathbf{B}(\vec{q}, 12) = \begin{pmatrix} D_{xx}(\vec{q}, 12) & D_{xy}(\vec{q}, 12) & D_{xz}(\vec{q}, 12) \\ D_{yx}(\vec{q}, 12) & D_{yy}(\vec{q}, 12) & D_{yz}(\vec{q}, 12) \\ D_{zx}(\vec{q}, 12) & D_{zy}(\vec{q}, 12) & D_{zz}(\vec{q}, 12) \end{pmatrix}.$$

$$(2.172b)$$

$$(2.172)$$

It may not be possible to diagonalize the matrix (2.171) for ω^2 at each point in the Brillouin zone. Few simplifications are as follows :

1. $\underline{\vec{q} \text{ along } [0001] \text{ direction}}$
   Here $\vec{q} = (0, 0, q)$ with $q_x = q_y = 0$ and $q_z = q$ in the q_z -direction. The determinant (2.171) factorizes into three 2×2 determinants. The eigenvalues and eigenvectors are as follows :

   (a) <u>Longitudinal optical modes</u>

   $$\omega_{LO}^2(\vec{q}) = D_{zz}(\vec{q}, 11) + |D_{zz}(\vec{q}, 12)|,$$

   $$\vec{e}(\vec{q}L \mid 1) = \begin{pmatrix} 0 \\ 0 \\ 1 \end{pmatrix}, \vec{e}(\vec{q}L \mid 2) = \begin{pmatrix} 0 \\ 0 \\ -1 \end{pmatrix}. \qquad (2.173a)$$

   (b) <u>Longitudinal acoustic modes</u>

   $$\omega_{LA}^2(\vec{q}) = D_{zz}(\vec{q}, 11) - \mid D_{zz}(\vec{q}, 12) \mid,$$

   $$\vec{e}(qL \mid 1) = \begin{pmatrix} 0 \\ 0 \\ 1 \end{pmatrix}, \vec{e}(\vec{q}L \mid 2) = \begin{pmatrix} 0 \\ 0 \\ 1 \end{pmatrix}. \qquad (2.173b)$$

   (c) <u>Degenerate transverse optical modes</u>

   $$\omega_{T_1O}^2(\vec{q}) = \omega_{T_2O}^2(\vec{q}) = D_{xx}(\vec{q}, 11) + \mid D_{xx}(\vec{q}, 12), \mid$$

$$\vec{e}(\vec{q}T_1 \mid 1) = \begin{pmatrix} 1 \\ 0 \\ 0 \end{pmatrix}, \vec{e}(\vec{q}T_1 \mid 2) = \begin{pmatrix} -1 \\ 0 \\ 0 \end{pmatrix},$$

$$\vec{e}(\vec{q}T_2 \mid 1) = \begin{pmatrix} 0 \\ 1 \\ 0 \end{pmatrix}, \vec{e}(\vec{q}T_2 \mid 2) = \begin{pmatrix} 0 \\ -1 \\ 0 \end{pmatrix}. \tag{2.173c}$$

(d) <u>Degenerate transverse acoustic modes</u>

$$\omega_{T_1A}^2(\vec{q}) = \omega_{T_2A}^2(\vec{q}) = D_{xx}(\vec{q},11) - \mid D_{xx}(\vec{q},12), \mid$$

$$\vec{e}(\vec{q}T_1 \mid 1) = \begin{pmatrix} 1 \\ 0 \\ 0 \end{pmatrix}, \vec{e}(\vec{q}T_1 \mid 2) = \begin{pmatrix} 1 \\ 0 \\ 0 \end{pmatrix},$$

$$\vec{e}(\vec{q}T_2 \mid 1) = \begin{pmatrix} 0 \\ 1 \\ 0 \end{pmatrix}, \vec{e}(\vec{q}T_2 \mid 2) = \begin{pmatrix} 0 \\ 1 \\ 0 \end{pmatrix}. \tag{2.173d}$$

$$\tag{2.173}$$

2.   <u>$\vec{q}$ along $[01\bar{1}0]$ direction</u>

$\vec{q} = (0, q, 0)$ with $q_x = q_z = 0$ and $q_y = q$ in the q_y direction. The determinant (2.171) factorizes into three 2×2 determinants. The eigenvalues and eigenvactors are as follows :

(a) <u>Longitudinal optical modes</u>

$$\omega_{LO}^2(\vec{q}) = D_{yy}(\vec{q},11) + \mid D_{yy}(\vec{q},12) \mid,$$

$$\vec{e}(qL \mid 1) = \begin{pmatrix} 0 \\ 1 \\ 0 \end{pmatrix}, \vec{e}(\vec{q}L \mid 2) = \begin{pmatrix} 0 \\ -1 \\ 0 \end{pmatrix}. \tag{2.174a}$$

(b) <u>Longitudinal acoustic modes</u>

$$\omega_{LA}^2(\vec{q}) = D_{yy}(\vec{q},11) - \mid D_{yy}(\vec{q},12) \mid,$$

$$\vec{e}(qL \mid 1) = \begin{pmatrix} 0 \\ 1 \\ 0 \end{pmatrix}, \vec{e}(\vec{q}L \mid 2) = \begin{pmatrix} 0 \\ 1 \\ 0 \end{pmatrix}. \tag{2.174b}$$

(c) <u>Transverse optical modes</u>

$$\omega_{T_1O}^2(\vec{q}) = D_{xx}(\vec{q},11) + \mid D_{xx}(\vec{q},12)\mid,$$

$$\vec{e}(\vec{q}T_1\mid 1) = \begin{pmatrix} 1 \\ 0 \\ 0 \end{pmatrix}, \vec{e}(\vec{q}T_1\mid 2) = \begin{pmatrix} -1 \\ 0 \\ 0 \end{pmatrix}. \qquad (2.174c)$$

(d) <u>Transverse acoustic modes</u>

$$\omega_{T_1A}^2(\vec{q}) = D_{xx}(\vec{q},11) - \mid D_{xx}(\vec{q},12)\mid,$$

$$\vec{e}(\vec{q}T_1\mid 1) = \begin{pmatrix} 1 \\ 0 \\ 0 \end{pmatrix}, \vec{e}(\vec{q}T_1\mid 2) = \begin{pmatrix} 1 \\ 0 \\ 0 \end{pmatrix}. \qquad (2.174d)$$

(e) <u>Transverse optical modes</u>

$$\omega_{T_2O}^2(\vec{q}) = D_{zz}(\vec{q},11) - \mid D_{zz}(\vec{q},12)\mid,$$

$$\vec{e}(\vec{q}T_2\mid 1) \begin{pmatrix} 0 \\ 0 \\ 1 \end{pmatrix}, \vec{e}(\vec{q}T_2\mid 2) = \begin{pmatrix} 0 \\ 0 \\ -1 \end{pmatrix}. \qquad (2.174e)$$

(f) <u>Transverse acoustic modes</u>

$$\omega_{T_2A}^2(\vec{q}) = D_{zz}(\vec{q},11) - \mid D_{zz}(\vec{q},12)\mid,$$

$$\vec{e}(\vec{q}T_2\mid 1) = \begin{pmatrix} 0 \\ 0 \\ 1 \end{pmatrix}, \vec{e}(\vec{q}T_2\mid 2) = \begin{pmatrix} 0 \\ 0 \\ 1 \end{pmatrix}. \qquad (2.174f)$$

$$(2.174)$$

3. <u>$\vec{q}$ along [1000] direction</u>

$\vec{q} = (q,0,0)$ with $q_y = q_z = 0$ and $q_x = q$ in the q_x direction. The determinant (2.171) factorizes into 4×4 and 2×2 determinants. The 4×4 determinant gives the mixed modes. The 2×2 determinant gives pure optical and acoustic modes with the following eigenvalues and eigenvactors.

(a) <u>Transverse optical modes</u>

$$\omega_{TO}^2(\vec{q}) = D_{zz}(\vec{q}, 11) + \mid D_{zz}(\vec{q}, 12) \mid,$$

$$\vec{e}(\vec{q}T \mid 1) = \begin{pmatrix} 0 \\ 0 \\ 1 \end{pmatrix}, \vec{e}(\vec{q}T \mid 2) = \begin{pmatrix} 0 \\ 0 \\ -1 \end{pmatrix}. \qquad (2.175a)$$

(b) <u>Transverse acoustic modes</u>

$$\omega_{TA}^2(\vec{q}) = D_{zz}(\vec{q}, 11) - \mid D_{zz}(\vec{q}, 12) \mid,$$

$$\vec{e}(\vec{q}T \mid 1) = \begin{pmatrix} 0 \\ 0 \\ 1 \end{pmatrix}, \vec{e}(\vec{q}T \mid 2) = \begin{pmatrix} 0 \\ 0 \\ 1 \end{pmatrix}. \qquad (2.175b)$$

$$(2.175)$$

As an example, the expressions along [0001] direction (given in Eqs.(2.173)) simplify as

$$M\omega_{LO(A)}^2\vec{q} = 6\nu_1 \left[1 \pm \cos\left(\frac{q_z c}{2}\right) \right],$$

and

$$M\omega_{TO(A)}^2(\vec{q}) = 6\lambda_1 \left[1 \pm \cos\left(\frac{q_z c}{2}\right) \right]. \qquad (2.176)$$

Similarly the Eqs.(2.174) for $[01\bar{1}0]$ direction simplify as

$$\begin{aligned} M\omega_{LO(A)}^2(\vec{q}) \;=\; & 6\lambda_1 + 4(\alpha_1 + \beta_1)\left\{ 1 - \cos\left(\frac{\sqrt{3}q_y a}{2}\right) \right. \\ \pm \;& 2\left\{ (\lambda_1 - 2\mu_1)^2 + 4(\lambda_1 + \mu_1)^2 \right. \\ + \;& 16(\lambda_1 - 2\mu_1)(\lambda_1 + \mu_1)\cos\left(\frac{\sqrt{3}q_y a}{2}\right) \right\}^{1/2} \end{aligned}$$

and

$$M\omega^2_{TO(A)}(\vec{q}) = 6\lambda_1 + 4(\alpha_1 - \beta_1)\left\{1 - \cos\left(\frac{\sqrt{3}q_y a}{2}\right)\right\}$$
$$\pm\ 2\left\{(\lambda_1 + 2\mu_1)^2\right.$$
$$+\ 4(\lambda_1 - \mu_1)^2 + 4(\lambda_1 + 2\mu_1)$$
$$\times\ (\lambda_1 - \mu_1)cos\left(\frac{\sqrt{3}q_y a}{2}\right)\left.\right\}^{1/2}. \qquad (2.177)$$

The plus sign is for the optical modes and minus sign for acoustic modes. It is not always possible to get the closed expressions for $\omega^2_j(\vec{q})$ in directions other than y and z. Czachor [32] generalized the GTF model for hcp lattice upto all NNs. Mostly the least square fit of the experimental data along [0001] and [01$\bar{1}$0] directions in the expressions similar to (2.176) and (2.177) upto higher NNs is carried out and the force constants are obtained [33], [34], [35]. The atomic displacements for $\vec{q} = (2\pi/c)(001)$ obtained by Eq. (2.136) are shown in Fig. 2.12. It is noted that the atomic planes perpendicular to c-axis vibrate as a whole.

2.7.4 Planer Force Constants

Figures 2.5 and 2.9 indicate that in a cubic lattice when $\vec{q}$ is along a principal symmetry direction, the atoms in the planes perpendicular to the principal axis get displaced simultaneously and the relative displacements are independent of one another for each plane in the longitudinal and transverse modes. Thus the propagation of a lattice wave along a particular symmetry direction involves the vibration of complete plane of atoms in the same phase. Thus the mathematical problem is reduced to that of a linear chain of particles each connected to its neighbour by harmonic forces where each particle represents a plane of atoms. Therefore, each longitudinal and transverse mode of vibration may be treated separately as a linear chain problem [21]. The Eqs. (2.142), (2.143), (2.155), (2.156) and (2.177) etc. can be expressed as

$$Mw^2 = \sum_{n=1}^{\infty} \Phi_n \left[1 - \cos\left(\frac{nq}{2q_m} \right) \right], \qquad (2.178)$$

where Φ_n is the harmonic force constant between the nNN particles and q_m is the maximum value of q. Φ_n are known as planer force constants. For $n = 1, 2, 3$, one can write from Eq. (2.155),

$$\begin{aligned}
\Phi_1 &= (8\alpha_1 + 16\beta_4), \\
\Phi_2 &= 2\alpha_2 + 8\alpha_3 + 8\alpha_5, \\
\Phi_3 &= 8\alpha_4, \qquad (2.179)
\end{aligned}$$

and so on. Equations (2.179) gives the relation between palner force constants and atomic force constants. By Fourier analyzing the longitudinal and transverse mode phonon frequencies the interplaner force constants can be deduced. The convergence of fitting will give an indication about the range of interplaner force constants and hence the range of atomic force constants. If sufficient number of Φ_n are known, interatomic force constants may be determined by solving either simultaneous equations or by the least square fit. The accuracy of this analysis will depend upon the number of experimental points and separation between them [25], [28], [29]. For hcp lattice [32] the same systematics exist for planes perpendicular to c-axis i.e. $c = 0, \pm(1/2), \pm1, \pm(3/2)$, if all the atoms are assumed of the same mass. The planer vibrations can be represented by Eq. (2.178) and the results can be interpreted in terms of interplaner force constants.

In the lower symmetry directions, in the cubic and hcp lattices, the measurements can be Fourier analyzed using equations similar to Eq. (2.178) but these Fourier Coefficients will not be interpreted as planer force constants [27], [34].

2.7.5   Metallic Compound

The structural symmetry of lattice in metallic compounds is used to fit the GTF model. The mass normalized sum of the squares of optical and acoustic frequencies is fitted with the expression

$$\omega_O^2 + \omega_A^2 = \Phi_0 + \sum_{n=1}^{N} \Phi_n \left[1 - \cos\left(\frac{nq}{2q_m} \right) \right], \qquad (2.180)$$

where Φ_n are interpreted as metal-metal, metal-nonmetal and nonmetal-nonmetal force constants. Ten planes were needed to have a reasonable fit of phonon frequencies of TaC [36] and HfC [37] which indicated the presence of long range forces in these systems.

2.7.6 Alloys

In the strict sense the BvK theory is not applicable to structurally disordered alloys. However, to interpret the measured phonon frequencies, the alloy is assigned an averaged crystal structure of averaged atomic masses and the phonon frequencies are fitted in the expressions (2.178) and (2.180). Standard computer programmes are available for the least square fit of phonon frequencies. Φ_n are regarded as the averaged force constants and not the interplaner force constants as for the perfect systems. A fit of the phonon frequencies of $Ni_{0.3}Fe_{0.7}$ of average fcc structure is given in Fig.2.13. The resulting force constants are tabulated in Table 2.9 [38]. The subject of phonons in alloys will be discussed in detail in chapter 7.

2.8　Axially Symmetric Force Model

In GTF model, crystal symmetry is used to reduce the number of force constants. But the force constants also depend upon the characterstics of interatomic potential. For example, if one assumes that the force between a given pair of atoms $(l\kappa, l'\kappa')$ is derived from a two body potential $V(|\ \vec{R}(l\kappa) - \vec{R}(l'\kappa')\ |)$ which depends only on the magnitude of separation between the atoms, one can write

$$\Phi_{\alpha\beta}(lk, l'\kappa') = \left[-\frac{1}{R}\frac{\partial V(r)}{\partial R}\left\{ \delta_{\alpha\beta} - \frac{R_\alpha R_\beta}{R^2} \right\} - \frac{R_\alpha R_\beta}{R^2}\frac{\partial^2 V(R)}{\partial R^2} \right]_{\vec{R}=R^0}$$

$$(2.181)$$

Thus the force constants $\Phi_{\alpha\beta}$ are characterized by two parameters namely the first and second derivatives of pair potential although the symmetry considerations given in Eq. (2.49) may allow more parameters. Here the first term is associated with the tangential force and the second term with radial force between a given pair of atoms. If the potential is known, the two parameters and hence the force constants are completely known. If the potential is not known, these two derivatives are treated as parameters and determined by fitting the experimental data. Lehmann et al. [39] used this concept and proposed the axially symmetric (AS) model for lattice dynamic of metals.

It is assumed that all atomic pairs in the crystal interact via two body potential. Thus

$$\Phi(\{R\}) \;=\; \frac{1}{2}\sum_{l\kappa}\sum_{l'\kappa'}{}' V(\mid \vec{R}(l'\kappa') - \vec{R}(l\kappa) \mid)$$

$$=\; \frac{N}{2}\sum_{\kappa}\sum_{l'\kappa'}{}' V(R), \qquad (2.182)$$

where the dash over summation sign means $1\kappa \neq l'\kappa'$. Writing $|\vec{R}^0| = |\vec{R}^0(l'\kappa') - \vec{R}^0(l\kappa)|$, and $|\vec{u}| = |\vec{u}(l'\kappa') - \vec{u}(l\kappa)|$, where $|\vec{u}|$ is relative displacement, the Taylor expansion of potential energy becomes

$$\Phi\left(\{|\vec{R}^0 + \vec{u}|\}\right) \;=\; \Phi^{(0)} + \frac{\Phi'(R^0)}{R_0}(\vec{u}\cdot\vec{R}^0) + \frac{1}{2}\frac{\Phi'(R^0)}{R^{0^3}}(\vec{u}\times\vec{R}^0)^2$$

$$+\frac{1}{2}\frac{\Phi''(R^0)}{R^{0^4}}(\vec{u}\cdot\vec{R}^0)^3 + \ldots O(\vec{u}^4), \qquad (2.183)$$

where the dashes denote the space derivatives of potential $\Phi(R^0)$. The first two terms are unimportant. Third term which involves the first derivative of potential, corresponds to the relative displacement perpendicular to $\vec{R}^0$ and causes a tangential or bond bending force. Fourth term which involves second derivative of potential gives rise to radial or bond stretching force. If it is assumed that all the directions perpendicular to $\vec{R}^0$ are equivalent, the interaction potential and corresponding forces are axailly symmetric. Therefore, this model will require only two parameters to determine the force constants for each group of neighbours. For example, in fcc lattice this model will require four parameters upto $2NNs$ while GTF model requires five parameters. Thus the number of parameters is reduced.

In Eq. (2.183) the quadratic term associated with the κth atom in the zeroth unit cell in analogy with Eq. (2.181) is

$$\Phi^{(2)}(0\kappa) \;=\; \frac{1}{2}\sum_{l'\kappa'} \mid \vec{R}^0(0\kappa, l'\kappa') \mid^{-2} \Big[C_B(0\kappa, l'\kappa') \big\{ \vec{R}_0(0\kappa, l'\kappa')$$

$$\times \vec{u}(0\kappa, l'\kappa') \big\}^2 + C_I(0\kappa, l'\kappa') \big\{ \vec{R}^0(0\kappa, l'\kappa') \cdot \vec{u}(0\kappa, l'\kappa') \big\}^2 \Big].$$

$$(2.184)$$

$C_B(0\kappa, l'\kappa')$ is the effective bond bending force constant for the interaction of κ' atom in the l' cell with κth atom in the cell at the origin. Similarly $C_I(0\kappa, l'\kappa')$ represents the effective "bond stretching" force constant. For monatomic cubic lattices (sc, bcc, fcc), use of Eq. (2.184) in Eqs. (2.54) and (2.61) gives

$$D_{\alpha\beta}(\vec{q}) = \bar{D}_{\alpha\beta}(\vec{q}) - \bar{D}_{\alpha\beta}(0), \qquad (2.185)$$

with

$$\bar{D}_{\alpha\beta}(\vec{q}) = \frac{1}{M}\sum_{l'}\left[\frac{K_1(l')}{|\vec{R}^0_{l'}|^2}\frac{\partial^2}{\partial q_\alpha \partial q_\beta} - C_B(l')\delta_{\alpha\beta}\right]\exp(-i\vec{q}\cdot\vec{R}^0_{l'}), \quad (2.186)$$

where

$$K_1(l') = C_I(l') - C_B(l'). \qquad (2.187)$$

The sum over l' is for different NN shells of the atom at the origin.

Varshni and Shukla [40] calculated the elements of dynamical matrix for AS model extending interactions to any NN shells for bcc and fcc lattices. A typical expression for bcc lattice upto $2NN$ shells is as follows.

$$\begin{aligned} MD_{\alpha\alpha} \;=\;& 4C_B(1)\left[3 - C_\alpha C_\beta - C_\beta C_\gamma - C_\gamma C_\alpha\right] \\ &+\; 2C_B(2)\left[3 - C_{2\alpha} - C_{2\beta} - C_{2\gamma}\right] + 2K_1(1)\left[2 - C_\alpha(C_\beta + C_\gamma)\right] \\ &+\; 2K_1(2)\left[1 - C_{2\alpha}\right], \end{aligned} \qquad (2.188)$$

and

$$MD_{\alpha\beta} = 2K_1(1)S_\alpha S_\beta, \qquad (2.189)$$

where

$$\begin{aligned} C_{n\alpha} \;&=\; \cos(nq_\alpha a/2) \\ S_{n\alpha} \;&=\; \sin(nq_\alpha a/2). \end{aligned}$$

A comparision of above equations with Eqs. (2.129) and (2.130) gives the relation between force constants of AS and GTF models. These relations are

$$
\begin{aligned}
\alpha_1 &= C_B(1) + \frac{1}{2}K_1(1), \beta_1 = C_B(1), \\
\gamma_1 &= \alpha_1 - \beta_1 = \frac{1}{2}K_1(1), \alpha_2 = C_B(2) + K_1(2), \\
\beta_2 &= C_B(2).
\end{aligned}
\tag{2.190}
$$

Similar relations are obtained for bcc lattice [38]. In the bcc lattice there are four force constants upto $2NNs$ in both the AS and GTF models. However the number of force constants is much less for higher order NNs in AS model than in GTF model. The force constants obtained by fitting the observed phonon frequencies in the AS model are found to decrease rapidly for higher NNs for simple metals. However, these are found significant upto $10NNs$ or beyond to explain the anomalous features of the phonon dispersion curves of transition metals [35], [41].

The general characterstics of phonon spectra of hexagonal metals are discussed in AS model by Lehman et al [42]. For metals which are elastically isotropic such as Mg, the AS model gave a satisfactory explanation. An additional degree of freedom in the AS model was required to explain the phonon dispersion curves of less isotropic metals such as Be and Zn. This lead DeWames et al [43] to propose a modified axially symmetric (MAS) model which is frequently used to explain the phonon spectra of hcp transition metals.

2.9   Modified Axially Symmetric (MAS) Force Model

Lehman et al [42] showed that the AS model is a special form of a more general lattice dynamical model based on phenomenological electron gas theory earlier developed by Bhatia [44], deLaunay [45], Sharma and Joshi [46] and Krebs [47]. In general, the dynamical forces in a system of ions and electrons in a metal lead to the following form of total potential energy.

$$\Phi_{\Gamma} = \sum_{l\kappa}\sum_{l'\kappa'}{}' V_{\Gamma}(\vec{R}, a) + NE_0(a),\qquad(2.191)$$

where E_0 is electron gas self-energy and the parameter a in both the terms denotes the explicit volume dependence of V_{Γ} and E_0. The subscript Γ denotes that V_{Γ} should transform according to the point group symmetry of the centre of the BZ. Equations (2.191) is a particular case of Eq. (2.23) where ionic potential is a two body potential and electronic energy depends upon the volume of the crystal. The AS model is achieved replacing V_{Γ} by V which is volume dependent spherically symmetric potential.

For cubic crystals V_{Γ} is expanded as,

$$\begin{aligned}
V_{\Gamma}(R, a) &= V(R, a) + (X^4 + Y^4 + Z^4)V_1(R, a)\\
&\quad + (X^2Y^2 + Y^2Z^2 + Z^2X^2)V_2(R, a) + \dots
\end{aligned}\qquad(2.192)$$

where X, Y, Z are the components of $\vec{R}$ and V_1, V_2 are the expansion coefficients. Here the fourth and higher order harmonics provide a cubic anisotropic correction to V of AS model and these corrections lead to GTF model.

For hexagonal crystal

$$V_{\Gamma}(R, a) = V(R, a) + Z^2V_1(R, a) + Z^4V_2(R, a) + \dots\qquad(2.193)$$

where Z is along the c-axis of the crystal. The second and third terms in Eq. (2.193) introdeuce c-axis anisotropy in the model. Therefore, in the MAS model, the bond-bending and bond-stretching force constants in the basal plane are taken different from those perpendicular to the basal plane i.e. Z-direction. The elements of dynamical matrix, associated with MAS model are derived by DeWames et al [43]. For Example, $D_{xx}(\vec{q}, 11)$ upto $2NN$ shells is

$$\begin{aligned}
MD_{xx}(\vec{q}, 11) &= (12\delta_1 + 6\epsilon_{1y}) + (3\alpha_2 + 4\beta_{2y})\left\{1 - \cos\left(\frac{\sqrt{3}}{2}q_ya\right)\right.\\
&\quad \left.\times \cos\left(\frac{q_xa}{2}\right)\right\} + 4\beta_{2x}\left\{1 - \cos(q_xa)\right\},\qquad(2.194)
\end{aligned}$$

and so on. Here the constants δ_1, ϵ_{1y} etc. can be expressed in terms of C_I and C_B. A relation with force constants in the GTF model can be obtained by comparing (2.194) with (2.169). We notice that upto $1NNs$, MAS model involves four constants while GTF involves eight constants. The MAS is frequently used to fit the phonon frequencies of hcp transition metals [34], [48]. A typical fit of phonon frequencies of Y is given in Fig. 2.14 and the force constants are tabulated in Table 2.10.

2.10   Central Force Constant Models (An Over View)

Born and Huang [49] pointed out that the equilibrium lattice configuration is characterized not only by translational and rotational invarience of potential energy but also by zero stress or pressure on the crystal. This is specified by minimum of potential energy $\Phi(\{\vec{R}\})$ with respect to structural constants and may be determined as follows.

The instantaneous distance $|\vec{R}(0\kappa, l'\kappa')|$ depends upon three basis vectors, 3 angles between them and 3r displacement components $u_x(l\kappa), u_y(l\kappa)$ and $u_z(l\kappa)$ with $\kappa = 1, 2, \ldots, r$. Therefore the variational parameters are $(3r + 6)$. Since the displacements $u_x(l0), u_y(l0)$ and $u_z(l0)$ etc. depend upon the choice of origin which is arbitrary, therefore, the total number of independent parameters is $(3r + 3)$. Now the equilibrium configuration is specified by $(3r + 3)$ equations

$$\left(\frac{\partial \Phi}{\partial p_i}\right)_0 = \sum_\kappa \sum_{l'\kappa'} \left(\frac{\partial V(R)}{\partial R} \frac{\partial R}{\partial p_i}\right)_0 = 0 \qquad (2.195)$$

where p_i denotes ith variational parameter. There are $(3r + 3)$ constraints in the form of sum rules involving first derivative of Φ. For example, for the fcc lattice with $a/\sqrt{2}, a$ and $a\sqrt{6}/2 \ldots$, distances of the first three nearest neighbours, Eq. (2.195) simplifies as

$$6\sqrt{2}\left(\frac{\partial V}{\partial R}\right)_1 + 6\left(\frac{\partial V}{\partial R}\right)_2 + 12\sqrt{6}\left(\frac{\partial V}{\partial R}\right)_3 + \ldots = 0, \qquad (2.196)$$

where the subscripts indicate the neighbours at which the derivatives are evaluated. Similar relations can be derived for other lattices also.

If the two body potential $V(R)$ is known for all the values of R and for all the pairs of atoms, Eq. (2.196) is used to determine equilibrium configuration. The force constants are determined by evaluating appropriate second derivatives of V and no parameters are involved. If the geometry of the lattice is known, but V is not known, then Eq. (2.196) leads to useful constraints on force constant parameters. These constraints may further reduce the number of independent force constants. If the crystal is supposed to be held together entirely by two body interaction and the equilibrium condition (2.195) is used, the calculated force constants are called "central force constants" as per convention in the literature. If Eq.(2.195) is ignored, the force constants are AS or MAS in nature.

Similar to AS model, deLaunay [50] and Clark-Gazis-Wallis (CGW) [51] developed angular force models to explain the phonon spectra of metals. If these models are extended to all NNs and translational and rotational invariance of potential energy along with the equilibrium condition (2.195) are used, these models become identical [52]. Slutsky and Garland [53] and Metzbower [54] proposed angular force models for hexagonal metals. These phenomenological models are reviewed by Prakash [55].

2.11 Phenomenological Screened Multipole Models

Carbides, nitrides and hydrides of transition metals of groups IV, V and VI show the unusual combination of physical properties [56]. Their melting points and hardness are very high like homopolar diamond, while these are very good metals and many of them are even superconductors with high transition temperature [57]. These compounds possess NaCl, hexagonal and other complex structures, while many of them consist of wide range of non-stoichiometricity in which the lattice structure remains unchanged. These facts are tabulated in Chapter 3.

There are two models for chemical bonding of these compounds; the interstitial alloy model and covalent bond model [58]. In the interstitial alloy model, it is assumed that the small non-metal atoms fill the empty space of metal host lattice and the physical properties are dominated by metallic bonding. The fact that the conductivity, hardness and melting point exceed of those of host metals, is attributed to electron transfer from non-metal atoms to metal conduction band [59]. Covalent model postulates that the covalent metal-nonmetal interaction is essential for the stability of compounds. Bilz [60] used this concept to calculate the band structure of these compounds using linear combination of atomic orbital (LCAO) method and obtained the low lying bands of bonding character which are formed by the non-metal p-states and metal d-states of E_g symmetry. This feature is connected with an electron transfer from metal to non-metal atoms. Metallic d-bands of T_{2g} symmetry lie above the p-bands and partially overlap. The elctron density of states at the Fermi energy is found minimal for 8VE (valence electron)compounds and found to increase for higher VE compounds. This concept was further supported by detailed band structure calculations [61] and measured magnetic susceptibility and electronic specific heat [62]. The superconducting transition temperature T_c is also found to increase rapidly for compound of $8VE$ to $10VE$ [57]. A schematic crystal structure, band structure and density of states for a particular material (NbN) is shown in Fig.2.15.

The measured phonon dispersion relations of these compounds [62] show strong anomalies (Fig.2.17). The high values of elastic constants show steep rise of the acoustic modes. The optical modes are separated by a very large gap from acoustic modes and are splitted largely at the shorter wave lengths which indicates a certain amount of ionic interation. The longitudinal and transverse optical modes become degenerate at $\vec{q} = 0$, as all the compounds are conductors. These characterstics of phonon spectrum are not only due to large mass difference of atoms but also due to much stronger force constants between nearest neighbour metal and nonmetal ions. This supports the idea that the binding in these compounds is strongly covalent with some ionic contribution. In view of this, the shell model [63] and its extensions are used to explain the phonon spectrum of these compounds. A systematic development of these models is presented here.

2.11.1 Simple Shell Model

In the Kellermann model of the solids [64], the short range overlap interaction between the atoms is limited to $1NNs$ only while the Coulomb interaction among the ions extends throughout the crystal. The polarizability of ions is neglected and potential between the NNs is taken to be spherically symmetric. Thus the total energy per unit cell for a cubic lattice is $-\alpha_M(Z^2 e^2/q) + 6V(a)$ where α_M is Madelung constant and 2a is lattice parameter. In the dynamics of crystals, the radial and tangential force constants defined in Eq. (2.184) for Coulomb interaction are completely known while these force constants for $V(R)$ are not known. The above description of an ionic solid is used by Dick and Overhauser, Havilon and Lawson and Woods et al. [63] to develop the shell model (SM) for ionic and covalent solids to explain their dielectric and lattice dynamical properties.

In the simplest form of the SM it is assumed that as ion consists of a rigid core of mass M_κ and charge $X_\kappa e$ and and a rigid shell of zero mass and charge $Y_\kappa e$. The charge neutrality condition of the primitive cell is expressed as [65],

$$\sum_\kappa (X_\kappa + Y_\kappa)\, e = \sum_\kappa Z_\kappa e = 0. \qquad (2.197)$$

$Z_\kappa e$ is ionic charge. If the center of the shell is displaced relative to its core, the electronic polarizability can be brought in. Let $\vec{u}^C(l\kappa)$ and $\vec{u}^S(l\kappa)$ are the core and shell displacements relative to ionic equilibrium position $\vec{R}^0(l\kappa)$, then the electronic displacement of the $l\kappa$th ion is

$$\vec{P}_e(l\kappa) = Y_\kappa e \left[\vec{u}^S(l\kappa) - \vec{u}^C(l\kappa) \right]. \qquad (2.198)$$

The two contributions, one from long range electrostatic interaction and other from short range polarization machanism [66] will contribute to $\vec{P}_e(l\kappa)$. Cochran and co-workers generalized this concept to study the lattice dynamics of ionic crystals in the frame-work of BvK theory [67], [68]. A schematic diagram for this simple SM is given in Fig.2.16.

In analogy with Eq. (2.25b) the harmonic potential energy of vibrating core-shell system is

$$\Phi^{(2)} = \frac{1}{2} \sum_{i,j=C,S} \sum_{l\kappa\alpha,l'\kappa'\beta} \Phi^{ij}_{\alpha\beta}(l\kappa)\, u^i_\alpha(l\kappa)\, u^j_\beta(l'\kappa'), \qquad (2.199)$$

where the indices C and S are for the core and shell respectively. $\vec{u}^i$ are the core and shell displacements from the equilibrium positions and $\Phi^{ij}_{\alpha\beta}$ are core-core, core-shell and shell-shell force constants which include both the long range (Coulomb) and short range (non-Coulomb) components. It is further assumed that the core-shell self terms for the long range interactions vanish i.e.

$$\Phi^{CS}_{\alpha\beta}(l\kappa, l\kappa) \;=\; \Phi^{SC}_{\alpha\beta}(l\kappa, l\kappa) = 0 \qquad (2.200)$$

for each (α, β) and $(l\kappa)$. The short range core-shell (or shell-core) interaction energy of the $(l\kappa)$ ion is $K_{\alpha\beta}(\kappa)\left[u^C_\alpha(l\kappa) - u^S_\alpha(l\kappa)\right]\left[u^C_\beta(l\kappa) - u^S_\beta(l\kappa)\right]$ where $K_{\alpha\beta}(\kappa)$ is force constant similar to that as defined in Eq.(2.187). Thus Eq.(2.199) is explicitly written as:

$$
\begin{aligned}
\Phi^{(2)} \;=\;& \frac{1}{2}\sum_{l\kappa\alpha}\sum_{l'\kappa'\beta}\left[\Phi^{CC}_{\alpha\beta}(l\kappa, l'\kappa')u^C_\alpha(l\kappa)u^C_\beta(l'\kappa')\right. \\
&+\; \left.\Phi^{SS}_{\alpha\beta}(l\kappa, l'\kappa')u^S_\alpha(l\kappa)u^S_\beta(l'\kappa')\right] \\
&+\; \frac{1}{2}\sum_{l\kappa\alpha}\sum_{l'\kappa'\beta(l\kappa\neq l'\kappa')}\left[\Phi^{CS}_{\alpha\beta}(l\kappa, l'\kappa')u^C_\alpha(l\kappa)u^S_\beta(l'\kappa')\right. \\
&+\; \left.\Phi^{SC}_{\alpha\beta}(l\kappa, l'\kappa')u^S_\alpha(l\kappa)u^C_\beta(l'\kappa')\right] \\
&+\; \frac{1}{2}\sum_{l\kappa,\alpha\beta}K_{\alpha\beta}(\kappa)\left[u^C_\alpha(l\kappa) - u^S_\alpha(l\kappa)\right]\left[u^C_\beta(l\kappa) - u^S_\beta(l\kappa)\right].
\end{aligned}
$$

$$(2.201)$$

If one writes $\vec{u}(l\kappa) = \vec{u}^C(l\kappa)$ and $\vec{w}(l\kappa) = \vec{u}^S(l\kappa) - \vec{u}^C(l\kappa)$ and adds Eqs.(2.200) and (2.201) [This will include short range part of $\Phi^{CS}_{\alpha\beta}(l\kappa, l\kappa)$ and $\Phi^{SC}_{\alpha\beta}(l\kappa, l\kappa)$ once more. However, the final results are not affected as all these interactions are taken as parameters], it gives

$$
\begin{aligned}
\Phi^{(2)} \;=\;& \frac{1}{2}\sum_{l\kappa\alpha}\sum_{l'\kappa'\beta}\left[\Phi^{AA}_{\alpha\beta}(l\kappa, l'\kappa')u_\alpha(l\kappa)u_\beta(l'\kappa')\right. \\
&+\; \Phi^{AD}_{\alpha\beta}(l\kappa, l'\kappa')u_\alpha(l\kappa)w_\beta(l'\kappa') \\
&+\; \Phi^{DA}_{\alpha\beta}(l\kappa, l'\kappa')w_\alpha(l\kappa)u_\beta(l'\kappa') \\
&+\; \left.\Phi^{DD}_{\alpha\beta}(l\kappa, l'\kappa')w_\alpha(l\kappa)w_\beta(l'\kappa')\right.
\end{aligned}
$$

$$+ \; \frac{1}{2} \sum_{l\kappa\alpha;l'\kappa'\beta} K_{\alpha\beta}(\kappa)w_\alpha(l\kappa)w_\beta(l\kappa). \tag{2.202}$$

where

$$\Phi_{\alpha\beta}^{AA} \;=\; \sum_{i=C,S}\sum_{j=C,S} \Phi_{\alpha\beta}^{ij},$$

$$\Phi_{\alpha\beta}^{AD} \;=\; \sum_{i=C,S} \Phi_{\alpha\beta}^{iS},$$

$$\Phi_{\alpha\beta}^{DA} \;=\; \sum_{i=C,S} \Phi_{\alpha\beta}^{Si},$$

$$\Phi_{\alpha\beta}^{DD} \;=\; \Phi_{\alpha\beta}^{SS}. \tag{2.203}$$

Equation (2.202) can be interpreted as the harmonic potential energy of a crystal composed of atoms (index A) and induced dipoles (indexD). The equations of motion for the core and shell dipoles are:

$$
\begin{aligned}
M_\kappa \ddot{u}_\alpha(l\kappa) \;&=\; -\frac{\partial \Phi^{(2)}}{\partial u_\alpha(l\kappa)} \\
&=\; -\sum_{l'\kappa'\beta}\left[\Phi_{\alpha\beta}^{AA}(l\kappa,l'\kappa')u_\beta(l'\kappa') + \Phi_{\alpha\beta}^{AD}(l\kappa,l'\kappa')w_\beta(l'\kappa')\right],
\end{aligned}
$$
$$\tag{2.204}$$

and

$$
\begin{aligned}
M_S \ddot{w}_\alpha(l\kappa) \;&=\; 0 \\
&=\; -\frac{\partial \Phi^{(2)}}{\partial w_\alpha(l\kappa)} \\
&=\; -\sum_{l'\kappa'\beta}\left[\Phi_{\alpha\beta}^{DA}(l\kappa,l'\kappa')u_\beta(l'\kappa') + \Phi_{\alpha\beta}^{DD}(l\kappa,l'\kappa')w_\beta(l'\kappa')\right] \\
&\quad -\sum_{\beta} K_{\alpha\beta}(\kappa)w_\beta(l\kappa).
\end{aligned}
$$
$$\tag{2.205}$$

In Eq.(2.205) the shell mass M_S is considered negligible which is equivalent to adiabatic ápproximation as the electrons in the shell are assumed to follow the core motion faithfully. Equation (2.205) may be regarded as minimization condition on deformation energy $\Phi^{(2)}$ with respect to induced dipole moments.

Use of Eq.(2.205) in Eq.(2.198) gives

$$
\begin{aligned}
\vec{P}_e(l\kappa) &= Y_\kappa e \vec{w}(l\kappa) \\
&= Y_\kappa e \left[-\mathbf{K}(\kappa) - \mathbf{\Phi}^{DD}(l\kappa, l\kappa) \right]^{-1} \left[\sum_{l'\kappa'} \mathbf{\Phi}^{DA}(l\kappa, l'\kappa')\vec{u}(l'\kappa') \right. \\
&\qquad\left. + \sum_{l'\kappa'}' \mathbf{\Phi}^{DD}(l\kappa, l'\kappa')\vec{w}(l'\kappa') \right].
\end{aligned}
\tag{2.206}
$$

Thus the electronic dipole moment induced at the site $(l\kappa)$ depends on displacement at all other sites and the electronic moment at the site $(l\kappa)$. In consonance with $\mathbf{\Phi}^{DA}$ and $\mathbf{\Phi}^{DD}$, $\vec{P}_e(l\kappa)$ also consists of both the long range and short range forces. Thus Eq.(2.205) is the self-consistency requirement on the induced dipoles.

Introducing the wave like solution(2.52), Eqs.(2.204) and (2.205) simplify as

$$
\omega^2(\vec{q})M_\kappa u_\alpha(\vec{q}|\kappa) = \sum_{\kappa'\beta} \left[H^{AA}_{\alpha\beta}(\vec{q}, \kappa\kappa')u_\beta(\vec{q}|\kappa') + H^{AD}_{\alpha\beta}(\vec{q}, \kappa\kappa')w_\beta(\vec{q}|\kappa') \right],
\tag{2.207}
$$

$$
\begin{aligned}
0 = \sum_{\kappa'\beta} \Big[& H^{DA}_{\alpha\beta}(\vec{q}, \kappa\kappa')u_\beta(\vec{q}|\kappa') \\
& + \left\{ H^{DD}_{\alpha\beta}(\vec{q}, \kappa\kappa') + \delta_{\kappa\kappa'}K_{\alpha\beta}(\kappa) \right\} w_\beta(\vec{q}|\kappa') \Big],
\end{aligned}
\tag{2.208}
$$

where $\vec{u}(\vec{q}|\kappa)$ and $\vec{w}(\vec{q}|\kappa)$ are the amplitudes of core and shell dipole displacements respectively in the qth mode and

$$
\begin{aligned}
H^{ij}_{\alpha\beta}(\vec{q}, \kappa\kappa') &= \sum_{l'} \Phi^{ij}_{\alpha\beta}(0\kappa, l'\kappa') \exp\left[-i\vec{q}\cdot\vec{R}^0(0\kappa, l'\kappa') \right]. \tag{2.209} \\
&= H^{ijC}_{\alpha\beta} + H^{ijS}_{\alpha\beta}, \tag{2.210}
\end{aligned}
$$

with $i, j = A, D$. Here $H^{ijC}_{\alpha\beta}$ and $H^{ijS}_{\alpha\beta}$ are the Coulomb and short range parts of $H^{ij}_{\alpha\beta}$. In analogy of Eqs.(2.61) and (2.63), the elements of Coulomb matrix are written as

$$
C_{\alpha\beta}(\vec{q}, \kappa\kappa') = \bar{C}_{\alpha\beta}(\vec{q}, \kappa\kappa') - \delta_{\kappa\kappa'} \sum_{\kappa''} \frac{Z_{\kappa''}}{Z_\kappa} \bar{C}_{\alpha\beta}(0, \kappa\kappa''),
\tag{2.211}
$$

where

$$\bar{C}_{\alpha\beta}(\vec{q},\kappa\kappa') = -\sum_{l'(l'\neq 0 \, if \, \kappa'=\kappa)} \frac{\partial^2}{\partial R_\alpha \partial R_\beta}\left(\frac{1}{R}\right)\Big|_{\vec{R}=\vec{R}^0(0\kappa,l'\kappa')}$$
$$\times \ \exp\left[-i\vec{q}\cdot\vec{R}^0(0\kappa,l'\kappa')\right]. \tag{2.212}$$

If it is assumed that the charge neutrality condition of primitive cell is satisfied, the matrix elements $H_{\alpha\beta}^{ijC}$ become

$$H_{\alpha\beta}^{AAC}(\vec{q},\kappa\kappa') = Z_\kappa e C_{\alpha\beta}(\vec{q},\kappa\kappa')Z_{\kappa'}e, \tag{2.213a}$$

$$H_{\alpha\beta}^{ADC}(\vec{q},\kappa\kappa') = Z_\kappa e C_{\alpha\beta}(\vec{q},\kappa\kappa')Y_{\kappa'}e, \tag{2.213b}$$

$$H_{\alpha\beta}^{DAC}(\vec{q},\kappa\kappa') = Y_\kappa e C_{\alpha\beta}(\vec{q},\kappa\kappa')Z_{\kappa'}e, \tag{2.213c}$$

$$H_{\alpha\beta}^{DDC}(\vec{q},\kappa\kappa') = Y_\kappa e C_{\alpha\beta}(\vec{q},\kappa\kappa')Y_{\kappa'}e. \tag{2.213d}$$

$$\tag{2.213}$$

For convenience, the notations for $H_{\alpha\beta}^{ijS}$ are introduced as

$$H_{\alpha\beta}^{AAS}(\vec{q},\kappa\kappa') = R_{\alpha\beta}(\vec{q},\kappa\kappa'),, \tag{2.214a}$$

$$H_{\alpha\beta}^{ADS}(\vec{q},\kappa\kappa') = T_{\alpha\beta}(\vec{q},\kappa\kappa'), \tag{2.214b}$$

$$H_{\alpha\beta}^{DAS}(\vec{q},\kappa\kappa') = T_{\alpha\beta}^{\dagger}(\vec{q},\kappa\kappa'), \tag{2.214c}$$

$$H_{\alpha\beta}^{DDS}(\vec{q},\kappa\kappa') = S_{\alpha\beta}(\vec{q},\kappa\kappa'). \tag{2.214d}$$

$$\tag{2.214}$$

Using Eqs.(2.210), (2.213) and (2.214) in Eqs.(2.207) and (2.208), the final results for the equation of motion of the core and shell dipoles are written in the matrix form as:

$$\omega^2(\vec{q})\mathbf{Mu} = [\mathbf{R} + \mathbf{Z_A C Z_A}]\,\mathbf{u}$$
$$+ \ [\mathbf{T} + \mathbf{Z_A C Y_D}]\,\mathbf{w}, \tag{2.215}$$
$$\mathbf{0} = \left[\mathbf{T^\dagger} + \mathbf{Y_A C Z_A}\right]\mathbf{u}$$
$$+ \ [\mathbf{S} + \mathbf{K} + \mathbf{Y_D C Y_D}]\,\mathbf{w}. \tag{2.216}$$

Here $\mathbf{Z_A}$, $\mathbf{Y_D}$ and $\mathbf{M}$ are $3r \times 3r$ real matrices with elements

$$
\begin{aligned}
(Z_A)_{\alpha\beta}(\kappa, \kappa') &= Z_\kappa e \delta_{\alpha\beta} \delta_{\kappa,\kappa'}, \\
(Y_D)_{\alpha\beta}(\kappa, \kappa') &= Y_\kappa e \delta_{\alpha\beta} \delta_{\kappa,\kappa'}, \\
(M)_{\alpha\beta}(\kappa, \kappa') &= M_\kappa \delta_{\alpha\beta} \delta_{\kappa,\kappa'}.
\end{aligned}
\tag{2.217}
$$

$\mathbf{C}$ is a $3r \times 3r$ matrix with elements $C_{\alpha\beta}(\vec{q}, \kappa\kappa')$ etc. Both $\mathbf{C}$ and $\bar{\mathbf{C}}$ are Hermitian and $\mathbf{K}$ is a diagonal matrix. A more detailed discussion of Eqs.(2.215) and(2.216) is given in Refs.[3], [67] and [69]. The quantum mechanical justification of these equations is given by Sinha [70] and the applications of SM to ionic and covalent solids are reviewed by Singh [71].

Eliminating $\mathbf{w}$ from Eqs.(2.215) and (2.216) one gets

$$
\begin{aligned}
\omega^2(\vec{q})\mathbf{Mu} &= [\mathbf{R} + \mathbf{Z_A C Z_A}]\,\mathbf{u} - [\mathbf{T} + \mathbf{Z_A C Y_D}] \\
&\times\ [\mathbf{S} + \mathbf{K} + \mathbf{Y_D C Y_D}]^{-1} \left[\mathbf{T}^\dagger + \mathbf{Y_D C Z_D}\right]\mathbf{u}.
\end{aligned}
\tag{2.218}
$$

In this equation if the first term is identified with the Born model, the second term is entirely a contribution from electronic polarization. The long range contribution associated with both the terms have regular and nonregular parts. The regular part arises from the contribution of the form $q_\alpha q_\beta/|\vec{q}|^2$ while the nonregular part of the first term is of the form $Z_\kappa Z_{\kappa'}(q_\alpha q_\beta/|\vec{q}|^2)$ and the form of the nonregular part of the second term is more complicated. These two nonregular terms lead to apparent charge matrix $\mathbf{Z}(\vec{q})$ as introduced by Cochran and Cowley [72]. This matrix with elements $Z_{\alpha\beta}(\vec{q}|\kappa)$ is regarded as representing the charge interacting with macroscopic field with wave vector $\vec{q}$ in the κth sublattice.

In fact the shell charges $Y_\kappa e$ and shell displacements $\vec{w}$ are ficticious, the physically meaningful quantities are dipole moments $\vec{P}_e(l\kappa)$. Therefore, these ficticious quantities are eliminated from Eqs.(2.215) and (2.216) by defining new short range interaction coefficients $\mathbf{T'}, \mathbf{S'}$ and $\mathbf{K'}$ related to $\mathbf{T}, \mathbf{S}$ and $\mathbf{K}$ as:

$$
\begin{aligned}
\mathbf{T} &= \mathbf{T'Y_D}, \\
\mathbf{S} &= \mathbf{Y_D S' Y_D}, \\
\mathbf{K} &= \mathbf{Y_D K' Y_D}.
\end{aligned}
\tag{2.219}
$$

Equations (2.206) and (2.219) transform Eqs. (2.215) and (2.216) as:

$$\omega^2(\vec{q})\mathbf{Mu} = [\mathbf{R} + \mathbf{Z_A C Z_A}]\,\mathbf{u} + [\mathbf{T'} + \mathbf{Z_A C}]\,\mathbf{P}_e, \qquad (2.220)$$

$$\mathbf{O} = \left[\mathbf{T'}^\dagger + \mathbf{C Z_A}\right]\mathbf{u} + [\mathbf{S'} + \mathbf{K'} + \mathbf{C}]\,\mathbf{P}_e. \qquad (2.221)$$

If $\mathbf{P}_e$ is eliminated in Eqs.(2.220) and (2.221) to get phonon frequencies, two types of nonuniquenesses crop up in the SM bonding coefficients if these are determined from the given dispersion relations. In the first place if $\mathbf{F'}$ is any $3r \times 3r$ matrix, which is periodic function of $\vec{q}$ and satisfies the same symmetry requirements as $\mathbf{R}$, $\mathbf{T'}$, $\mathbf{S'}$ and $\mathbf{K'}$, then the transformations

$$[\mathbf{T'} + \mathbf{Z_A C}] \rightarrow \mathbf{F'}[\mathbf{T'} + \mathbf{Z_A C}]\,\mathbf{F'},$$

$$[\mathbf{S'} + \mathbf{K'} + \mathbf{C}] \rightarrow \mathbf{F'}[\mathbf{S'} + \mathbf{K'} + \mathbf{C}]\,\mathbf{F'}, \qquad (2.222)$$

will not affect the phonon frequencies or eigenvectors. Further if $\mathbf{F'} = 1$ for $\vec{q} \rightarrow 0$, the transformation (2.222), will not affect $\vec{P}_e$ determined from (2.220) and (2.221) of the atom either. Hence the SM fitted to these polarizabilities and phonon frequencies will not be unique relative to such a transformation.

In general the transformations (2.222) may change the values of both the atom-dipole and dipole-dipole bonding coefficients. To eliminate the lack of uniqueness in the conventional shell model, it is assumed that

$$\mathbf{R} = \mathbf{S} = \mathbf{T} = \mathbf{T}^\dagger \qquad (2.223)$$

for all the values of $\vec{q}$. Some times $\mathbf{S} + \mathbf{K}$ is written as $\mathbf{S}$ and equated in Eq.(2.223). Here we keep $\mathbf{K}$ separate with reference to its applications to the transition metal compounds. Sinha [70] evaluated the interaction matrices $\mathbf{R}, \mathbf{S}, \mathbf{T}$ and $\mathbf{T}^\dagger$ and found them similar but not equal. Use of Eq.(2.223) in Eq.(2.218) gives,

$$\omega^2(\vec{q})\mathbf{Mu} = [\mathbf{R} + \mathbf{Z_A C Z_A}]\,\mathbf{u} - [\mathbf{R} + \mathbf{Z_A C Y_D}]$$
$$\times\; [\mathbf{R} + \mathbf{K} + \mathbf{Y_D C Y_D}]^{-1}\,[\mathbf{R} + \mathbf{Y_D C Z_D}]. \qquad (2.224)$$

This equation is used to fit the phonon frequencies of ionic and covalent solids [71]. For $\vec{q}$ along the principal symmetry directions, the Coulomb coefficients are determined by Ewald's Θ-function transformation [64], [3]. For most of the crystal structures, these are available in the tabulated form [71] and can directly be adopted for any $Z_\kappa, Z_{\kappa'}, Y_\kappa$ and $Y_{\kappa'}$.

For a central short range potential between any two NNs, the matrix elements $R_{\alpha\beta}(\vec{q}, \kappa\kappa')$, defined in Eqs. (2.210) and (2.214a), are written in terms of two parameters $C_I(0\kappa, l'\kappa') = A_{l'}(\kappa\kappa')$ and $C_B(0\kappa, l'\kappa') = B_{l'}(\kappa\kappa')$ for each NN shell. Here $A_{l'}(\kappa\kappa')$ and $B_{l'}(\kappa\kappa')$ are radial and tangential force constant parameters defined in Eq.(2.184). For a crystal of NaCl structure, $R_{\alpha\beta}(\vec{q}, \kappa\kappa')$ including $1NN$ and $2NN$ interactions are as follows [71],[76]:

$$
\begin{aligned}
R_{xx}(\vec{q}, \kappa\kappa') &= R_{xx}(\vec{q}, \kappa'\kappa) \\
&= [A_1(\kappa\kappa')\cos(q_x a/2) + B_1(\kappa\kappa')\left(\cos(q_y a/2) + \cos(q_z a/2)\right)], \\
R_{xx}(\vec{q}, \kappa\kappa) &= [A_1(\kappa\kappa') + 2B_1(\kappa\kappa')] + [A_2(\kappa\kappa) + B_2(\kappa\kappa)] \\
&\quad \times \ [2 - \cos(q_x a/2)\cos(q_y a/2) - \cos(q_x a/2)\cos(q_z a/2)] \\
&\quad + \ B_2(\kappa\kappa)\left[1 - \cos(q_y a/2)\cos(q_z a/2)\right], \\
R_{xy}(\vec{q}, \kappa\kappa') &= 0, \\
R_{xy}(\vec{q}, \kappa\kappa) &= [A_2(\kappa\kappa) - B_2(\kappa\kappa)]\sin(q_x a/2)\sin(q_y a/2).
\end{aligned}
\tag{2.225}
$$

Other elements can be obtained by cyclic permutation of $(q_x q_y q_z)$. Similarly the diagonal matrix $\mathbf{K}$ is represented by two parameters K_1 and K_2. In the actual calculations of phonon frequencies of transition metal compounds, the ion and the shell charges $Z_\kappa e$ and $Y_\kappa e$ are also treated as parameters. Thus in a $2NN$ shell model, there are ten parameters $A_1(12), B_1(12), A_2(22), B_2(22), Z_1, Z_2, Y_1, Y_2, K_1$ and K_2. Here $\kappa = 1$ is for nonmetal atom and $\kappa' = 2$ is for metal atom. For simplicity some of the parameters are put to zero. For example, for an unpolarizable nonmetallic atom $Z_1 = 0$. The number of these parameters is further increased by extending the interactions upto higher order NNs. The calculated results for a typical compound TaC are shown in Fig. 2.17 [73]. The simple SM overestimates the phonon frequencies.

2.11.2 Screened and Extended Shell Models

To improve upon simple SM, Weber et al [73] proposed that the covalent bonds between metal-nonmetal atoms are stronger and metal-metal bonding is less important. This is accounted for by assuming that the free electrons (metallic character) produce screening of Coulomb interaction between ion cores. Using Hartree dielectric function $\epsilon(\vec{q})$, these authors modified Eq. (2.213a) as

$$H_{\alpha\beta}^{AAC}(\vec{q},\kappa\kappa') = Z_\kappa e C_{\alpha\beta}(\vec{q},\kappa\kappa')Z_{\kappa'}e + D_{\alpha\beta}^{SC}(\vec{q},\kappa\kappa'), \qquad (2.226)$$

where

$$D_{\alpha\beta}^{SC}(\vec{q},\kappa\kappa') = 4\pi Z_\kappa e Z_{\kappa'}e\frac{q_\alpha q_\beta}{|\vec{q}|^2}\left[\epsilon^{-1}(\vec{q}) - 1\right], \qquad (2.227)$$

$$\epsilon(\vec{q}) = 1 + \frac{k_s^2}{q^2}\left[\frac{1}{2} + \frac{4k_F^2 - q^2}{8k_F q}\ln\left|\frac{2k_F + q}{2k_F - q}\right|\right], \qquad (2.228)$$

$$k_s = 4k_F\left(m^*/m\right)/(2\pi a).$$

k_F is Fermi wavevector, k_s is Thomas-Fermi screening length and m^* is the effective mass of the electron. Z_κ are determined by the condition $\omega_{LO} = \omega_{TO}$ at $\vec{q} = 0$ and k_s is treated as a parameter. This augments Eq. (2.224) as

$$\omega^2(\vec{q})\mathbf{Mu} = \left[\mathbf{R} + \mathbf{Z_A C Z_A} + \mathbf{D}^{SC}\right]\mathbf{u} - \left[\mathbf{R} + \mathbf{Z_A C Y_D}\right]$$
$$\times \left[\mathbf{R} + \mathbf{K} + \mathbf{Y_D C Y_D}\right]^{-1}\left[\mathbf{R} + \mathbf{Y_D C Z_A}\right]. \qquad (2.229)$$

This equation is used to fit the phonon frequencies and the parameters are determined by the method of least square fit. This so called "screened shell model (SSM)" explained overall shape of dispersion curves of some of the transition metal compounds.

In NbC and TaC compounds where dynamics is dominated by heavy metal ions, the phonon anomolies in the acoustic branches are found at $q_\alpha \sim \pm(1/2)q_m, \alpha = x, y, z$. This defines the surface of a cube which touches the first BZ of fcc lattice at eight L points. In real space this means that there exist a strong coupling in each simple cubic metal sublattice. This coupling corresponds to very long range interatomic forces [36] and cannot be described by using even ten additional parameters in the SM or SSM. Weber et al [73] proposed that the origin of these long range forces might be a strong short range correlation in the electronic charge densities. This lead to $\vec{q}$-dependent polarizability of metal ions which becomes resonant nesr $q_\alpha \sim (1/2)q_m$ and could result in a singularity in the dielectric function. Similarly the anomalous behavior in the optical branches for HfC may be associated with the short range metal-nonmetal coupling.

To describe the above resonant behaviour, the SSM is further modified by introducing a coupling of induced electronic polarization $\mathbf{W}$ at different ions via a short range correlation field W_1 (an additional degree of freedom). Thus the modified equations of motion of the core-shell system become

$$\omega^2(\vec{q})\mathbf{Mu} = \left[\mathbf{R} + \mathbf{Z_A C Z_A} + \mathbf{D^{SC}}\right]\mathbf{u} + \left[\mathbf{R} + \mathbf{Z_A C Y_D}\right]\mathbf{W} \qquad (2.230a)$$

$$0 = \left[\mathbf{R} + \mathbf{Y_D C Z_A}\right]\mathbf{u} + \left[\mathbf{K} + \mathbf{R} + \mathbf{Y_D C Y_D}\right]\mathbf{W} + \mathbf{R_1 W_1} \qquad (2.230b)$$

$$0 = \mathbf{R_1^\dagger W} + \mathbf{K_1 W_1} \qquad (2.230c)$$

$$(2.230)$$

where the interionic and intraionic correlation matrices R_1 and K_1 respectively, have the same structure as $\mathbf{R}$ and $\mathbf{K}$. Eliminating $\mathbf{W}$ and W_1 in the above equations

$$\omega^2(\vec{q})\mathbf{Mu} = \left[\mathbf{R} + \mathbf{Z_A C Z_A} + \mathbf{D^{SC}}\right]\mathbf{u} - \left[\mathbf{R} + \mathbf{Z_A C Y_D}\right]$$
$$\times \left[\mathbf{K}(\vec{q}) + \mathbf{R} + \mathbf{Y_D C Y_D}\right]^{-1}\left[\mathbf{R} + \mathbf{Y_D C Z_A}\right]\mathbf{u}.$$

$$(2.231)$$

where

$$\mathbf{K}(\vec{q}) = \mathbf{K} - \mathbf{R_1}(\mathbf{K_1})^{-1}R_1^\dagger. \qquad (2.232)$$

Here $\vec{q}$ dependendce in the polarizability matrix arises from $\mathbf{R_1}$. Evidently the additional coupling does not affect the elastic behaviour as the elastic constants are determined by the first term of the Eq. (2.231) while the anomalous behaviour of phonon frequencies is explained substantially. This model is usually called extended shell model(ESM) in literature.

In Fig. 2.17 the fitting of phonon frequencies of TaC by using SM, SSM and ESM is compared [73]. The model parameters are given in Table 2.11. In the ESM for NbC the additional coupling parameter R_1 describes correlations in the metal sublattice only. The main features of anomalous behaviour are described by metal-metal correlation parameter A_2. The NN coupling constants A_1 and B_1 influence the overall shape and the shift of the positions of minima to some higher values of $\vec{q}$. A similar explanation is also given for the phonon dispersion relation of HfC.

In the ESM the anomalies in the phonon spectra are related to resonance like behaviour of electric polarization contribution given by the second term of Eq.(2.231). In HfC which is (8VE) compound, the anomalous behaviour in the vicinity of $\vec{q}$=0 in the optical modes is related to few uncoupled p-d bonding states. This mechanism fades away in the compounds with filled p-d bands such as TaC, NbC (9VE) and in non-stiochiometric $NbC_{0.75}$. In TaC and NbC, the anomalous behaviour at $q_\alpha \approx (1/2)q_m$ in the acoustic modes is associated with the filling of d-bands. If one assumes that the strength of resonance is proportional to n_d which is the number of additional d-electrons, one obtains phonon instabilities at $q_\alpha \approx (1/2)q_m$ for TaC for a critical value of n_d=0.5 VE. Such a phase transition from fcc to hcp structure occurs in $TaCN_{1-x}$ near x=0.4 which corresponds to 9.6 VE [75]. This resonance like behaviour in the acoustic phonons can be thought due to overlap of d-shells while the metallic atoms in the cubic lattice are vibrating in opposite pahses as shown in Fig.2.18. The arrows show the pattern of motion of metallic atoms for the acoustic phonons with $q_\alpha \approx (1/2)q_m$. The distribution of charge density is indicated by the loops for metallic d-electrons with T_{2g} symmetry. This explanation for the correlation parameters in the ESM leads to the conclusion that the additional degree of freedom is also associated with polarization of d- or f-electrons which led Weber to propose double shell model [76].

2.11.3 Double Shell Model

It was realized that the long range forces between the ions in transition metal compounds are of electronic origin. This effect could not be included adding more parameters in the simple SM. The additional degree of freedom due to short range correlations in the electron charge density was added in the ESM to explain anomalies in the phonon spectra. Further an additional degree of freedom was identified which should represent the d-electron charge density. Thus an extra shell of electrons (usually called a supershell) on metallic ions will also move in the potential energy due to metal ion core and the inner shell. This supershell will also interact with the neighbouring supeshells as shown in Fig. 2.19. The supershell interaction with core plus inner shell in represented by coupling constant K^s and supershell-supershell interaction by force constant R^s. These short range interactions in the electronic system between first and second neighbours are found to cause very long range ion-ion forces.

The equations of motion for double shell model (DSM) are derived by extending Eqs.(2.207) and (2.208) for simple SM. These equations are modified for the presence of supershell interactions (denoted by s) and one more equation of motion for supershell is added. Equations (2.197) to (2.206) are modified by adding supershell-core plus innershell and supershell-supershell interactions. The sum over i,j are extended for core, shell and supershell and so on. Finally Eq.(2.207) by separating long and short range parts become

$$\omega^2(\vec{q})M_\alpha u_\alpha(\vec{q}|\kappa) = \sum_{\kappa'\beta}\left[\left(H^{AAS}_{\alpha\beta} + H^{AAL}_{\alpha\beta}\right)u_\beta(\vec{q}|\kappa') + \left(H^{ADS}_{\alpha\beta} + H^{ADL}_{\alpha\beta}\right)\right.$$

$$\left. \times\ w_\beta(\vec{q}|\kappa') + \left(H^{ASS}_{\alpha\beta} + H^{ASL}_{\alpha\beta}\right)w^S_\beta(\vec{q}|\kappa')\right]. \qquad (2.233)$$

The last term is added for the atom-supershell interaction. $H^{ASS}_{\alpha\beta}$ and $H^{ASL}_{\alpha\beta}$ are atom-supershell short and long range interactions respectively and

$$w^S(\vec{q}\mid\kappa) = u^{S'}_\beta(\vec{q}\mid\kappa') - u^C_\beta(\vec{q}\mid\kappa'), \qquad (2.234)$$

where $u^{s'}(\vec{q}\mid\kappa')$ is supershell displacement from the equilibrium position. Further

$$\begin{aligned}
H^{AAS}_{\alpha\beta} &= R_{\alpha\beta} + R^S_{\alpha\beta}, \\
H^{ASS}_{\alpha\beta} &= T^S_{\alpha\beta} + R^S_{\alpha\beta}, \\
H^{ASL}_{\alpha\beta} &= 0.
\end{aligned} \qquad (2.235)$$

Thus the atom-supershell short range interactions are represented by matrix $\mathbf{R_S}$ and atom-supershell long range interactions are neglected. Including all these terms and simplifying in the same way as in the shell model, Eq.(2.233) is reduced in the matrix form as

$$\omega^2(\vec{q})\mathbf{Mu} = \left[\mathbf{R} + \mathbf{R}^S + \mathbf{Z_A}\mathbf{C}\mathbf{Z_A}\right]\mathbf{u}$$
$$+ \left[\mathbf{R} + \mathbf{Z_A}\mathbf{C}\mathbf{Y_D}\right]\mathbf{w} + \mathbf{R^S}\mathbf{w^S} \qquad (2.236)$$

Adding dipole-supershell interaction $H_{\alpha\beta}^{DS}$, Eq.(2.208) becomes

$$0 = \sum_{\kappa'\beta}\left[\left(H_{\alpha\beta}^{DAL} + H_{\alpha\beta}^{DAS}\right)u_\beta(\vec{q}\mid\kappa') + \left(H_{\alpha\beta}^{DDL} + H_{\alpha\beta}^{DDS} + \delta_{\kappa\kappa'}K_{\alpha\beta}\right)\right.$$
$$\left. \times\ w_\beta(\vec{q}\mid\kappa') + \left(H_{\alpha\beta}^{DSL} + H_{\alpha\beta}^{DSS}\right)\left(w_\beta(\vec{q}\mid\kappa') - w_\beta^S(\vec{q}\mid\kappa')\right]\right). \qquad (2.237)$$

Here $\left(w_\beta(\vec{q}\mid\kappa') - w_\beta^S(\vec{q}\mid\kappa')\right)$ is diploe-supershell relative displacement. Further adopting $H_{\alpha\beta}^{DSL} = 0$ and $H_{\alpha\beta}^{DSS} = K_{\alpha\beta}$, Eq.(2.237) in the matrix form becomes

$$0 = \left(\mathbf{Y_D}\mathbf{C}\mathbf{Z_A} + \mathbf{R}\right)\mathbf{u} + \left(\mathbf{Y_D}\mathbf{C}\mathbf{Y_D} + \mathbf{R} + \mathbf{K}\right)\mathbf{w}$$
$$+ \mathbf{K}^S\left(\mathbf{w} - \mathbf{w}^S\right). \qquad (2.238)$$

From the analogy of Eq.(2.237), the equation of motion of the supershell is

$$0 = \sum_{\kappa'\beta}\left[\left(H_{\alpha\beta}^{SAL} + H_{\alpha\beta}^{SAS}\right)u_\beta(\vec{q}\mid\kappa') + \left(H_{\alpha\beta}^{SSL} + H_{\alpha\beta}^{SSS}\right)w_\beta^S(\vec{q}\mid\kappa')\right.$$
$$\left. + \left(H_{\alpha\beta}^{SDL} + H_{\alpha\beta}^{SDS}\right)\left(w_\beta^S(\vec{q}\mid\kappa') - w_\beta(\vec{q}\mid\kappa')\right)\right]. \qquad (2.239)$$

The first, second and third terms represent suprershell-atom, supershell-supershell and supershell-dipole interactions respectively. All the long range interaciotns are neglected and short range interactions are defined as

$$H_{\alpha\beta}^{SAS} = H_{\alpha\beta}^{SSS} = R_{\alpha\beta}^S, H_{\alpha\beta}^{SDS} = K_{\alpha\beta}^S. \qquad (2.240)$$

Using Eq.(2.240) in Eq.(2.239) and simplifying, it becomes in the matrix form as

$$0 = \mathbf{R^S}\mathbf{u} + \left(\mathbf{R^S} + \mathbf{K^S}\right)\mathbf{w}^S - \mathbf{K^S}\mathbf{w}. \qquad (2.241)$$

Eliminating $\mathbf{w}, \mathbf{w}^S$ from Eqs.(2.236),(2.238) and (2.241) and adding the screening matrix $\mathbf{D}^{SC}$ to $\mathbf{Z_A CZ_A}$ as in Eq.(2.229), one gets

$$\omega^2(\vec{q})\mathbf{Mu} \;=\; \left(\bar{\mathbf{R}} + \mathbf{Z_A CZ_A} + \mathbf{D}^{SC}\right)\mathbf{u} - \left(\bar{\mathbf{R}} + \mathbf{Z_A CY_D}\right)$$
$$\times\; \left(\mathbf{K} + \bar{\mathbf{R}} + \mathbf{Y_D CY_D}\right\}^{-1}\left(\bar{\mathbf{R}} + \mathbf{Y_D CZ_A}\right), \quad (2.242)$$

where $\bar{\mathbf{R}}$ is function of $\vec{q}$ through $\mathbf{R}(\vec{q})$ i.e.

$$\begin{aligned}
\bar{\mathbf{R}}(\vec{q}) \;&=\; \mathbf{R}(\vec{q}) + \mathbf{R}^S(\vec{q}) - \mathbf{R}^S(\vec{q})\left[\mathbf{R}^S(\vec{q}) + \mathbf{K}^S\right]^{-1}\mathbf{R}^S(\vec{q}) \\
&=\; \mathbf{R} + \mathbf{K}^S\left[\mathbf{R}^S + \mathbf{K}^S\right]^{-1}\mathbf{R}^S \\
&=\; \mathbf{R} + \mathbf{R}^S\left[\mathbf{R}^S\mathbf{K}^S\right]^{-1}\mathbf{K}^S \\
&=\; \mathbf{R} + \mathbf{K}^S - \mathbf{K}^S\left[\mathbf{R}^S + \mathbf{K}^S\right]^{-1}\mathbf{K}^S. \quad\quad (2.243)
\end{aligned}$$

These are the dynamical equations derived by Weber [76] for DSM. The matrices $\mathbf{R}^S$ and $\mathbf{K}^S$ have the same structure as $\mathbf{R}$ and $\mathbf{K}$ except that the radial and tangential force constants A_1^S and B_1^S are considered in the metallic sublattice. The supershell on nonmetallic atoms is not introduced. The results for the simple SM are achieved by putting $\mathbf{K}^S = \mathbf{R}^S = \mathbf{0}$.

In the limit $q \to 0$, $\mathbf{R}^S\left(\mathbf{K}^S + \mathbf{R}^S\right)\mathbf{R}^S + \mathbf{D}^{SC}$ become independent of q. Therefore the elastic constants are determined by the expression $\mathbf{R} + \mathbf{Z_A CZ_A} + \mathbf{R}^S$ and hence depend on supershell interactions. However, in actual calculations it is found that elastic constants remain practically unchanged while going from TaC to HfC. Therefore $\left[\|\mathbf{R}^S\|/\|\mathbf{R} + \mathbf{Z_A CZ_A}\|\right] << 1$ has to be valid. To get the instabilities in certain regions of q, the condition

$$\bar{\mathbf{R}}(\vec{q}) = \mathbf{R}(\vec{q}) + \mathbf{K}^S\left(\mathbf{K}^S + \mathbf{R}^S(\vec{q})\right)^{-1}\mathbf{R}^S(\vec{q}) \approx \mathbf{0}. \quad\quad (2.244)$$

must hold. Therfore one expects $\mathbf{R}^S$ to be small but on the other hand $\mathbf{K}^S(\mathbf{K}^S + \mathbf{R}^S)^{-1}\mathbf{R}^S$ must be very large, negative and rapidly varying with $\vec{q}$ to cancel the slowly varying term $\mathbf{R}(\vec{q})$ in the soft mode region. To fulfil this condition $\left(\mathbf{K}^S + \mathbf{R}^S(\vec{q})\right)$ must behave like a resonance denominator which becomes very small in the critical $\vec{q}$ region i.e.

$$\left(\mathbf{K}^S + \mathbf{R}^S(\vec{q})\right) \geq \mathbf{0}. \quad\quad (2.245)$$

Equation (2.245) holds if the force constants entering the matrix $\mathbf{R}^S(\vec{q})$ are negative i.e. if the forces between the supershells are attractive. Thus the new electronic degree of freedom represents a charge density which is rather weakly bound at the metal ions and is in the unstable position with respect to the displacement of the neighbouring charge densities. Equation (2.245) holds for certain values of q where the crystal tends to show a resonance like increase in the polarizability and a sudden softening of the phonon frequencies. Thus the DSM shows that the attractive short range interactions between the first and second nearest neighbour supershells cause very long range forces which are connected with the soft modes.

The DSM is used to explain the anomolous behaviour of phonon spectra of HfC, ZrC, NbC, TaC and UC in the [110] LA and [110] TA branches. The typical calculations for NbC are shown in Fig. (2.17) and additional parameters for DSM are included in Table 2.11. There is no one-to-one correspondence between additional parameters for ESM and DSM. The DSM reproduces the resonance cube of NbC and TaC. The position of soft modes is found sensitive to supershell-supershell coupling constant while the width of the soft mode is mainly determined by the core-shell coupling constant $\mathbf{K}$. The covalent model postulates that the metallic d-states of T_{2g} symmetry become filled as one goes from 8VE to 10 VE compounds. This fact is also supported by DSM calculations. However, few questions remain unanswered such as: (1) The carbon ions are taken negatively charged, however there is no evidence in UC. (2) The brittleness and hardness etc. are related to strong covalent bonding but the addition of metallic bonding explains the phonon spectra.

The DSM has also been used to calculate the phonon spectra of NbMo binary alloy [77], TiN, ZrN and NbN [78] to [80]. Some of the features of these calculations are as follows:

1. The contributions to the dynamical matrix due to supershell interactions are negligible outside a narrow region of $\vec{q}$-space and these contributions give rise to anomalies and dips in the phonon spectra of nitrides.

2. It is the shell-supershell coupling constant which determines the

localization (or width) of the dips and anomalies. Smaller the value of shell-supershell coupling constant more localized are the anomalies. The DSM results show that the width of dips in the the phonon spectra of TiN is about (1/3) of those found in the phonon spectra of NbC. This fact shows that the anomalies in TiN are more sharply defined than in NbC which is consistent with the experimental information.

3. The positions of anomalies are found to depend upon the ratio of $1NN$ and $2NN$ interactions in the metal sublattice. Further, the magnitude of anomalies is also related to DSM force constants.

4. A point to be noted is that the $2NN$ coupling constants in 10VE and 9VE materials are of apposite sign (negative in TiN and ZrN and positive in δ-NbN)[80].

5. There are some qualitative diffrences in the contributions of the supershell interactions to the dynamical matrix for TM carbides and nitrides. For example in TiN, the LA branch in the vicinity of X-symmetry point is well described by simple SM contribution where as in NbC the supershell contribution lowers the frequency by 20%. A similar situation also exists for acoustic modes in Λ direction.

The above information obtained from the DSM calculations further incourages the investigations of phonon spectra of other d and f-band materials. The resonant polarization discussed above might be related to high superconducting transition temperature T_c in TaC, NbC etc. The possible effect of short wavelength polarizability of d-electrons on T_c has been discussed by Matthias et al [81].

2.11.4   Extensions of Shell Model

One of the worth mentioning extensions is the inclusion of many body (especially three body) interactions in the shell model. In the Hitler-London model of an ionic solid if $Z_\kappa e$ is the charge on the κth free ion and $\Delta\rho(\vec{r})$ is overlap electron charge density, than the electron density in the solid is

$$\rho(\vec{r}) = \sum_{l\kappa} Z_\kappa e \delta \left(\vec{r} - \vec{R}^0(l\kappa) \right) + \Delta\rho(\vec{r}). \qquad (2.246)$$

Localizing the electron wave functions on ionic sites, the overlap charge density is expanded in the spherical harmonics as

$$\Delta\rho(\vec{r}) = \sum_{\mu\nu} \rho_{\mu\nu} \left(| \vec{R}(l\kappa) - \vec{R}(l'\kappa') | \right) \psi_\mu^*(l\kappa)\psi_\nu(l'\kappa'). \qquad (2.247)$$

Here μ, ν refer to all quantum numbers including spin and $\psi_\mu(\vec{r})$ are normalized one electron wavefunctions. $\rho_{\mu\nu}$ is a function of $| \vec{R}(l\kappa) - \vec{R}(l'\kappa') |$ and it gives rise to monopole, dipole etc. The potential energy of such a distribution of electron-ion system is expressed as [82].

$$\begin{aligned}
\Phi = {} & \frac{1}{2} \sum_{l\kappa>l'\kappa'} \sum_{l'\kappa'} \frac{Z_\kappa Z_{\kappa'} e^2}{| \vec{R}(l\kappa, l'\kappa') |} + \frac{1}{2} \sum_{l\kappa} \sum_{l'\kappa'} V^S \left(| \vec{R}(l\kappa, l'\kappa') | \right) \\
& + \frac{1}{2} \sum_{l\kappa>} \sum_{l'\kappa'>} \sum_{l''\kappa''} e f_\kappa \left(| \vec{R}(l\kappa, l'\kappa') | \right) \frac{Z_{\kappa''} e}{| \vec{R}(l\kappa, l''\kappa'') |}. \qquad (2.248)
\end{aligned}$$

The first two terms are the usual Coulomb and two body short range potentials. In the third term $e f_\kappa \left(| \vec{R}(l\kappa, l'\kappa') | \right)$ is the depletion charge caused by the overlap of ionic wavefunctions at $(l\kappa)$ and $(l'\kappa')$ sites. Hence this term is the electrostatic interaction between depletion charge and ionic charge of ions which do not overlap. It is called as three-body interaction term. Similarly four-body term, five-body term etc. and in genral many-body terms can be added in Eq. (2.248). However these terms are ingored here.

This model implies charge transfer in the sense that a charge $e f_\kappa(\vec{R})$ is transferred from the ion $(l'\kappa')$ to the ion $(l\kappa)$ or vice-versa which then interacts with an ion at $(l''\kappa'')$ via the Coulomb law. This charge transfer is a function of relative separation of overlapping ions and therefore, it is a function of their respective displacements.

If one works out the expressions for force constants for $\kappa \neq \kappa'$ from Eq. (2.248) considering only $1NN$ interactions for NaCl crystal structure, one finds

$$\Phi_{\alpha\beta}(l\kappa, l'\kappa') = -Z_\kappa e \left[1 + f_\kappa(d_0) \right] \left\{ \frac{\partial^2}{\partial R_\alpha \partial R_\beta} \left(\frac{1}{R} \right) \right\} Z_{\kappa'} e \left[1 + f_{\kappa'}(d_0) \right]$$

$$
-\frac{\partial^2}{\partial R_\alpha \partial R_\beta}\left[V_1(R) + f_{\kappa'}(R)\frac{Z_{\kappa'}\alpha_M e^2}{d_0} + f_\kappa(R)\frac{Z_\kappa \alpha_M e^2}{d_0}\right]
$$

$$
-\sum_{l''\kappa''}\left\{Z_\kappa e^2\left[\frac{\partial}{\partial R_\alpha}\left(\frac{1}{R}\right)\right]\left[\frac{\partial}{\partial R_\beta}f_{\kappa''}(R)\right]\right.
$$

$$
\left. + Z_{\kappa'}e^2\left[\frac{\partial}{\partial R_\alpha}f_{\kappa''}(R)\right]\right\}\cdot\left[\frac{\partial}{\partial R_\beta}\left(\frac{1}{R}\right)\right]. \tag{2.249}
$$

Here all the derivatives are evaluated in the equilibrium position, d_0 denotes the $1NN$ distance, α_M is Madelung constant and

$$
V_1(R(l\kappa, l'\kappa')) = V^S(R(l\kappa, l'\kappa')) - 2f_\kappa(R(l\kappa, l'\kappa'))\frac{Z_{\kappa'}e}{R(l\kappa, l'\kappa')}. \tag{2.250}
$$

Comparing the Born force constant

$$
\Phi_{\alpha\beta}(l\kappa, l'\kappa') = -Z_\kappa e\left[\frac{\partial^2}{\partial R_a \partial R_\beta}\left(\frac{1}{R}\right)\right]_{R=R^0}Z_{\kappa'}e - \frac{\partial^2 V^S(R)}{\partial R_a \partial R_\beta}\Big|_{R=R^0(l\kappa.l'\kappa')}.
$$
$$
\tag{2.251}
$$

with Eq.(2.249), one finds that the three-body forces modify the Coulomb and short range parts and also contribute explicitly via third term of Eq.(2.249).

Adopting the concept of overlap of outer shells, Feldkamp [83] proposed that the shell charge during lattice vibrations can vary as

$$
Y(l\kappa) = Y_\kappa e - \frac{Y_\kappa e}{d_0}\frac{\sqrt{3}}{2}\frac{Z_\kappa}{|Z_\kappa|}\sum_{l'\kappa'}\left[u^S(l'\kappa') - u^S(l\kappa)\right]R^0(l\kappa, l'\kappa'). \tag{2.252}
$$

Here the second term on the right side denotes the fluctuation in the shell charge. The physical content of Eq. (2.252) is that when the shells of neighbouring ions are displaced, an extra charge which depends on the variation in the overlap integral appears on the shell. In the lowest order, the characterstic of this extra charge should be s-s overlap charge. This fluctuating charge is allowed to interact electrostatically with all the other charges in the system which lead to three-body and four body forces etc.

Verma and Singh [84] evaluated three-body term explicitly.The detailed calculations give the dynamical equations in the following form:

$$
\omega^2(\vec{q})\mathbf{Mu} = \mathbf{R'} + \mathbf{Z_A}\left\{1 + 6\mathbf{f(d_0)}\right\}\mathbf{CZ_A}\left\{1 + 6\mathbf{f(d_0)}\right\}
$$
$$
+\mathbf{Z}_d(\mu\lambda + \bar{\lambda}\bar{\mu})]\,\mathbf{u} \tag{2.253}
$$

where $\mathbf{R}'$ represents the short range contribution to the dynamical matrix and includes all the modifications to the two-body potential $V^s(R)$ given in Eqs. (2.248) and (2.250). Here $f(d_0)$, μ and λ are defined as

$$f(d_0) = f_\kappa(d_0) Z_\kappa / |Z_\kappa|, \qquad (2.254a)$$

$$\mu_\alpha(\vec{q}|\kappa\kappa') = \sum_{l'}{}' \frac{R_\alpha^0(l\kappa, l'\kappa')}{\left|\vec{R}^0(l\kappa, l'\kappa')\right|^3} \exp\left\{-i\vec{q}\cdot\vec{R}^0(l\kappa, l'\kappa')\right\}, \qquad (2.254b)$$

and

$$\lambda_\beta(\vec{q}|\kappa\kappa') = \sum_{l'}{}' \frac{R_\beta^0(l\kappa, l'\kappa')}{|\vec{R}^0(l\kappa, l'\kappa')|^2} \exp\left\{-i\vec{q}\cdot\vec{R}^0(l\kappa, l'\kappa')\right\}. \qquad (2.254c)$$

$$(2.254)$$

The prime implies that if $\kappa = \kappa'$ then $l' = 0$ term must be ommited. Noting that $[1 + 6f(d)]^2 \approx 1 + 12f(d)$, Eq. (2.253) can be written as

$$\omega^2(\vec{q})\mathbf{Mu} = [\mathbf{R}' + \mathbf{Z_A C' Z_A}]\,\mathbf{u}, \qquad (2.255)$$

where

$$\mathbf{C}' = \left[\mathbf{C}\left\{1 + 12\mathbf{f(d_0)}\right\} + \mu\lambda + \bar{\lambda}\bar{\mu}\right]. \qquad (2.256)$$

Verma and Singh [84] instead of solving Eq. (2.255) replaced $\mathbf{C}$ by $\mathbf{C}'$ in simple SM equations (2.215) and (2.216). Finally other simplifications of SM were adoped and Eq. (2.224) was solved replacing $\mathbf{R}$ by $\mathbf{R}'$ and $\mathbf{C}$ by $\mathbf{C}'$. Although this modification of SM equations is phenomenological in nature and does not fit in rigorously in the Lündqvist quantum mechanical formulation, it explains satisfactorily the static and dynamical properties of ionic solids. The three body shell model is also applied to transition metal compounds [71], however, the anomalous behaviour of phonon spectrum is not explained. Perhaps Sinha's work may further be extended to see the justification of modified SM equations. Certainly DSM has one additional degree of freedom and explains the results better. The adddition of supershell long range Coulomb interactions may further improve the results.

In addition, the deformation SM and breathing SM are also used to explain the phonon spectrum of transition metal compounds [71]. An abinito derivation of these models will be given in Chapter 5.

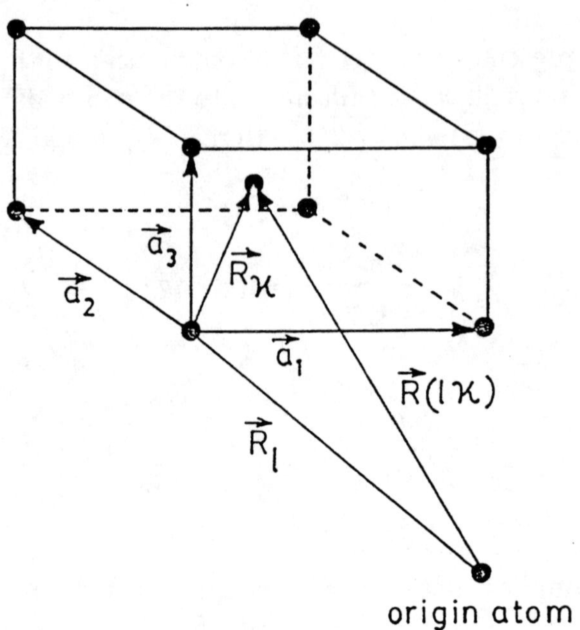

Figure 2.1: Unit cell of the crystal and equilibrium positions of atoms in the unit cell.

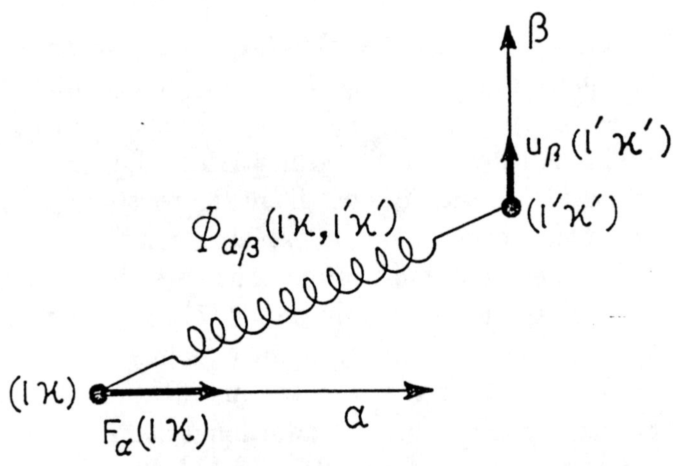

Figure 2.2: Spring constants $\Phi_{\alpha\beta}(l\kappa, l'\kappa')$ of the springs joining atoms $(l\kappa)$ and $(l'\kappa')$. $u_\beta(l'\kappa')$ is the displacement of $(l'\kappa')$ atom in the β direction and $F_\alpha(l\kappa)$ is the force on $(l\kappa)$ atom in the α direction.

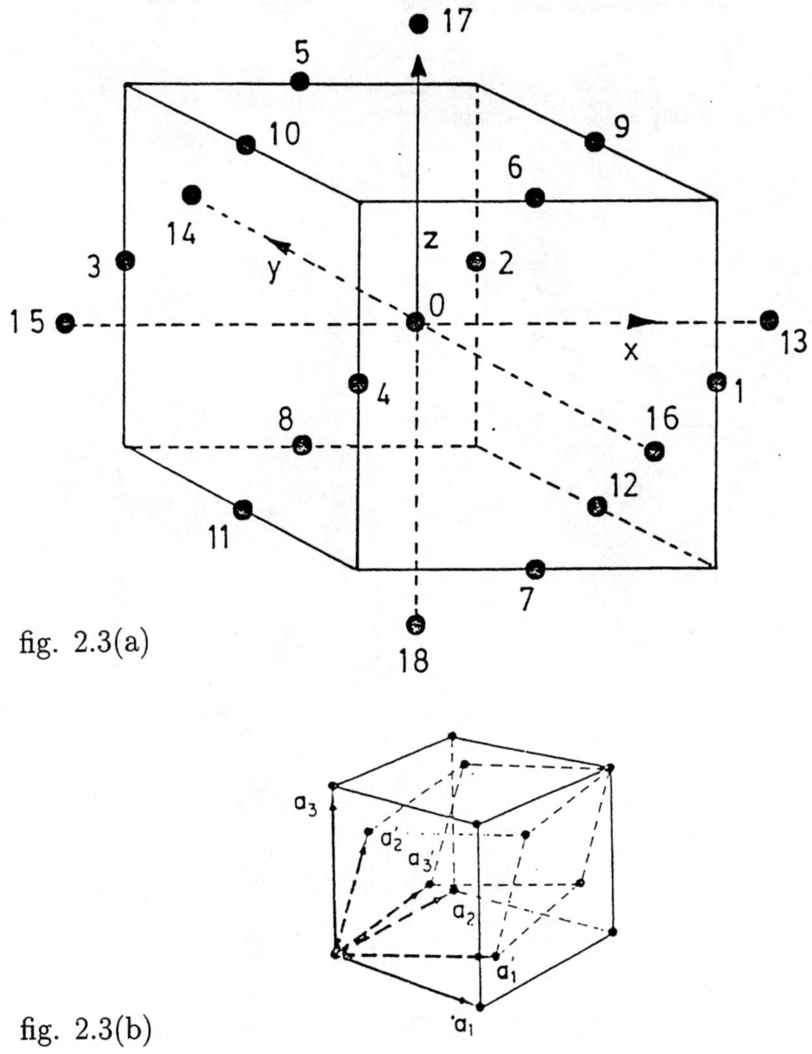

Figure 2.3: (a) Unit cell and the first and second nearest neighbour atomic positions in the fcc lattice. (1-12) are 1NNs with coordinates {110} and (13-18) are 2NNs with coordinates {200}. (b) Unit cell, primitive cell and primitive translation vectors of the fcc lattice. $\vec{a_1}, \vec{a_2}, \vec{a_3}$ are unit cell translation vectors and $\vec{a_1}', \vec{a_2}', \vec{a_3}'$ are primitive cell translation vectors. In Cartesian coordinates $\vec{a_1}' = (a/2)(\hat{i} + \hat{j}), \vec{a_2}' = (a/2)(\hat{j} + \hat{k}), \vec{a_3}' = (a/2)(\hat{k} + \hat{i})$.

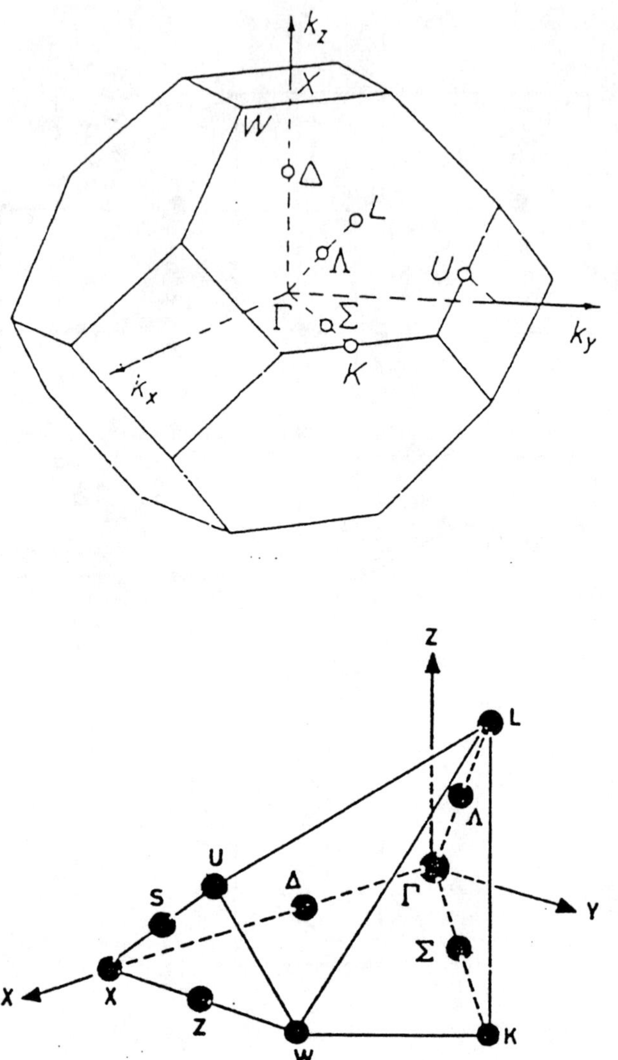

Figure 2.4: Brillouin zone and its 1/48th irreducible part of fcc lattice. $\Gamma\Delta X = [100], \Gamma\Sigma K = [110]$ and $\Gamma\Lambda L = [111]$.

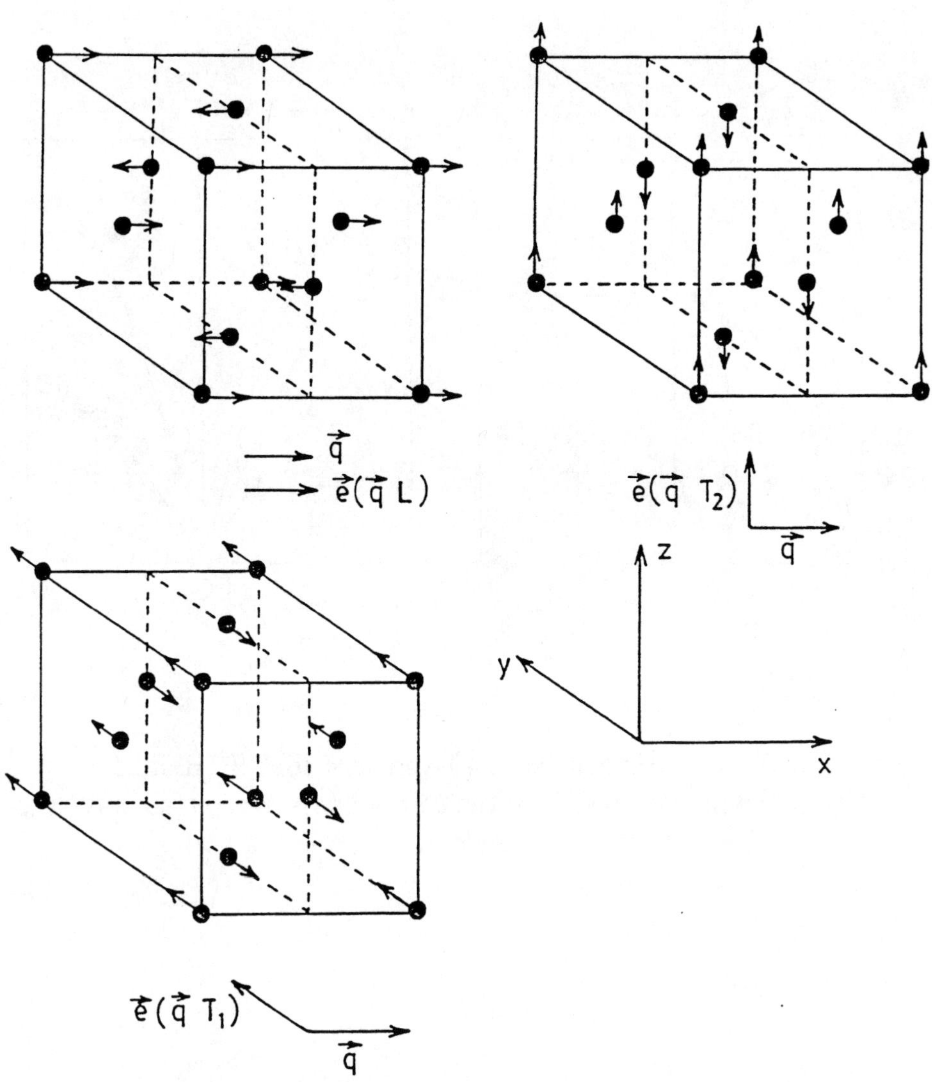

Figure 2.5: Pattern of atomic displacements in the fcc lattice for $\vec{q} = (2\pi/a)(100)$. The entire plane of atoms move in one phase and neighbouring plane in the opposite phase.

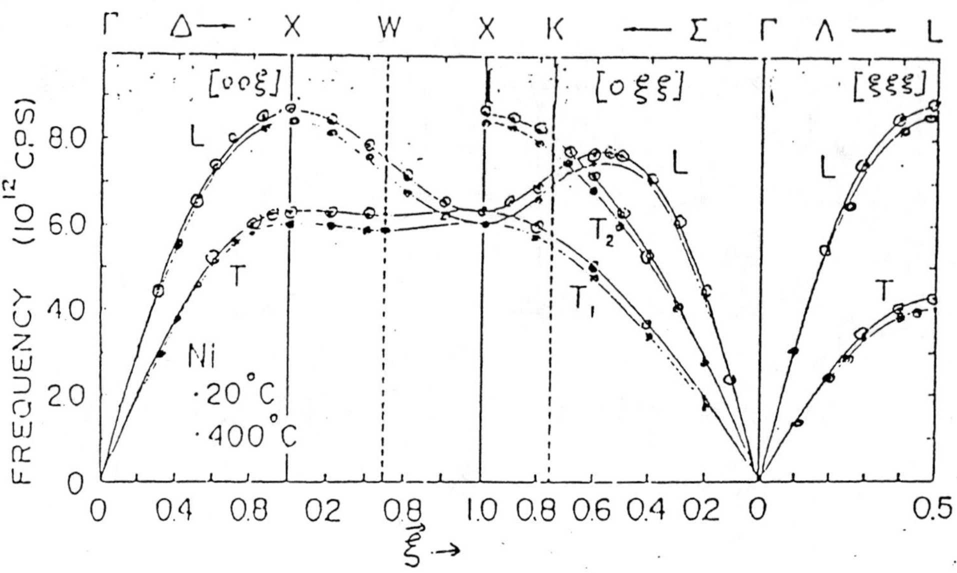

Figure 2.6: Phonon frequencies $\omega(\vec{q})$ versus $\vec{q}$ for Ni (Ref.[26]). $\vec{q} = (2\pi/a)(\xi_1\xi_2\xi_3)$ where ξ_1, ξ_2, ξ_3 are the components of reduced wave vector $\vec{\xi}$.

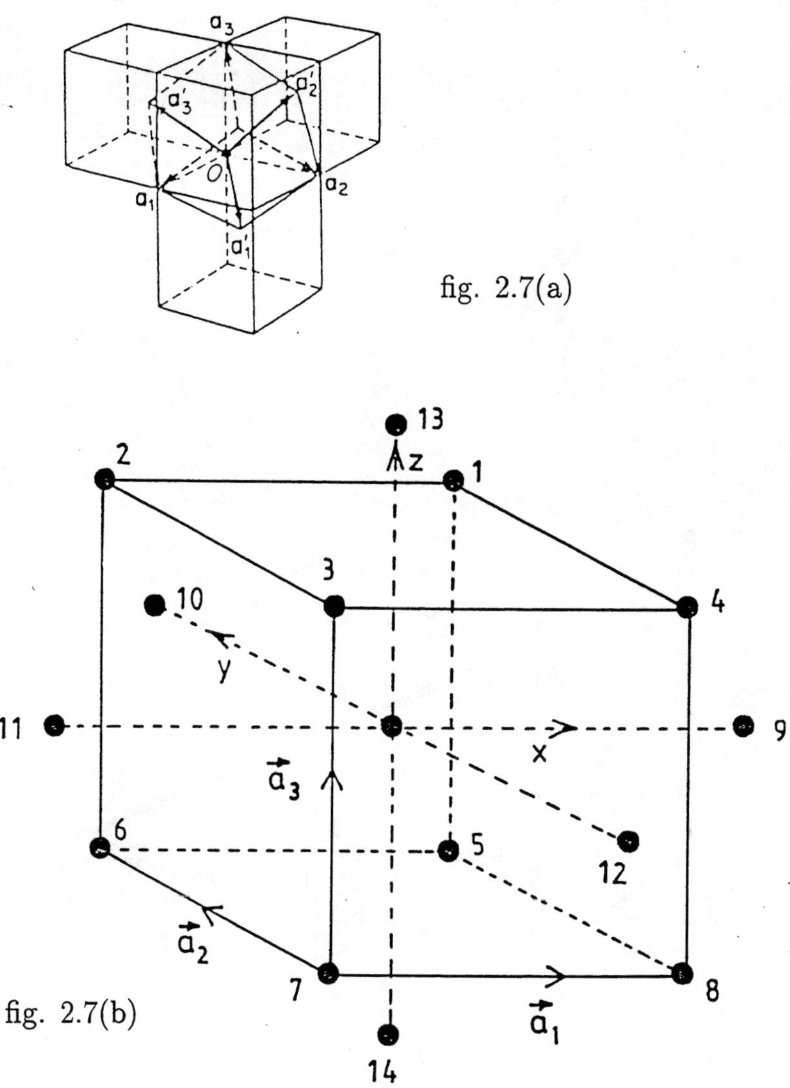

Figure 2.7: (a) Unit cell translation vectots $\vec{a_1}, \vec{a_2}, \vec{a_3}$ and primitive cell translation vectors $\vec{a_1}', \vec{a_2}', \vec{a_3}'$ for bcc lattice. In Cartesian coordinates $\vec{a_1}' = (a/2)(-\hat{i}+\hat{j}+\hat{k})$., $\vec{a_2}' = (a/2)(\hat{i}-\hat{j}+\hat{k})$, and $\vec{a_3}' = (a/2)(\hat{i}+\hat{j}-\hat{k})$. (b) First and second NNs atomic positions in bcc lattice. (1-8) are 1NNs $\{111\}$ and (9-14) are 2NNs $\{200\}$.

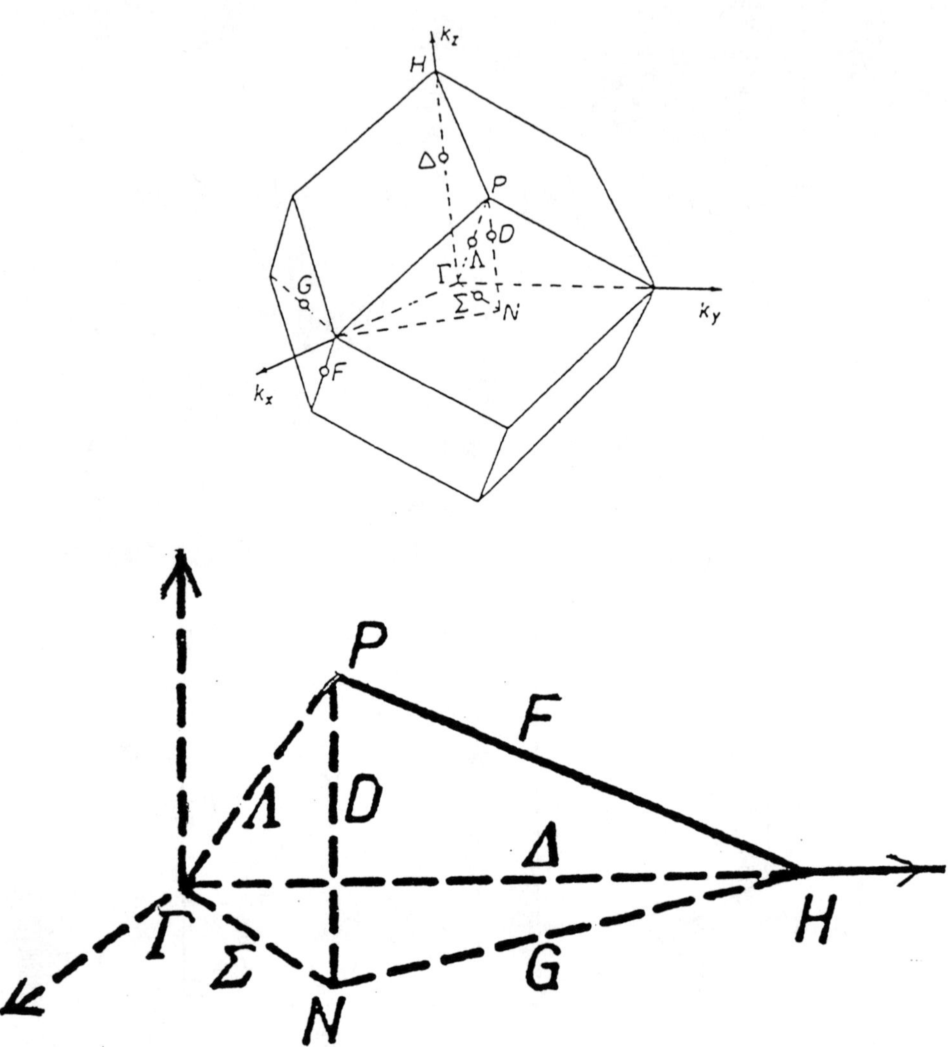

Figure 2.8: Brillouin zone and its 1/48th irreduciable part of bcc lattice. $\Gamma\Delta H$ = [100], $\Gamma\Sigma N$ =[110], $\Gamma\Lambda$ P=[111].

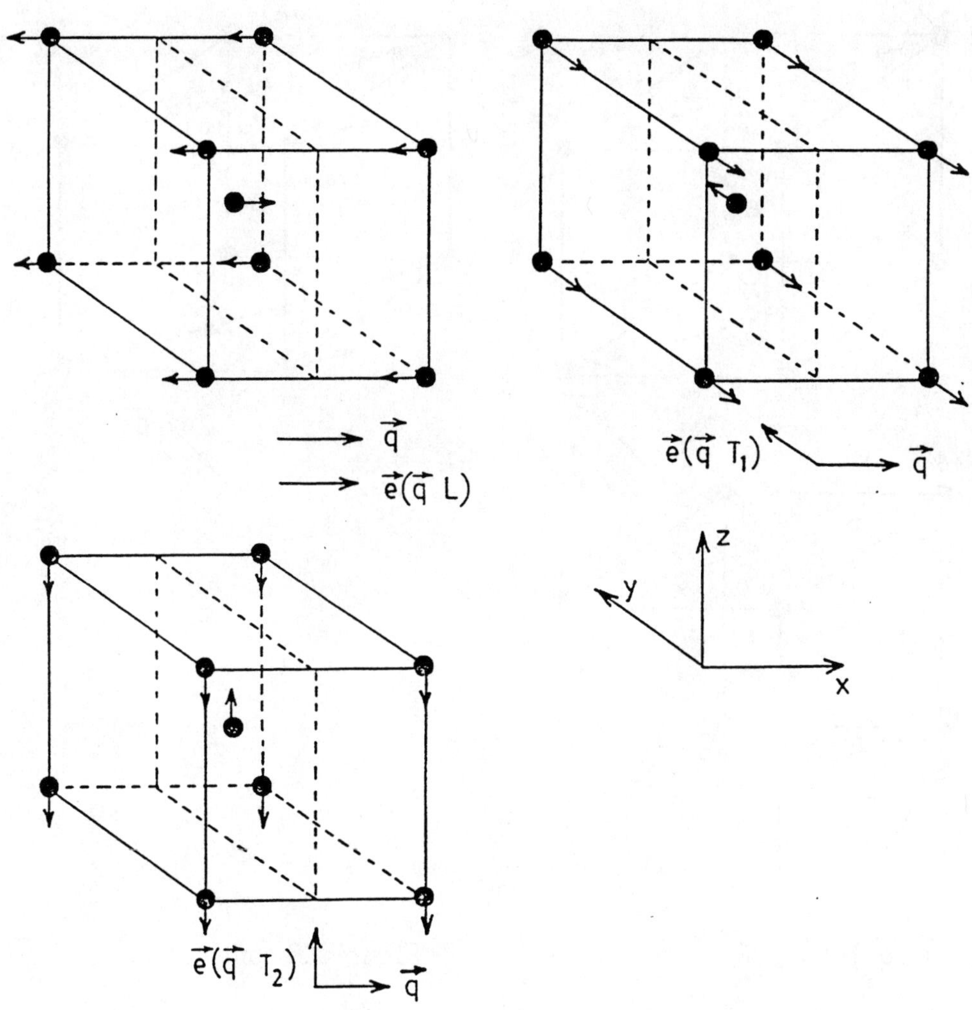

Figure 2.9: Pattern of atomic displacements in bcc lattice. (a) $\vec{q} = (2\pi/a)(100)$. The entire plane of atoms move in one phase and neighbouring plane in the opposite phase with wavelength a. (b) $\vec{q} = (2\pi/a)(110)$, the description is the same as for (a) except the wavelength = 1.141a.

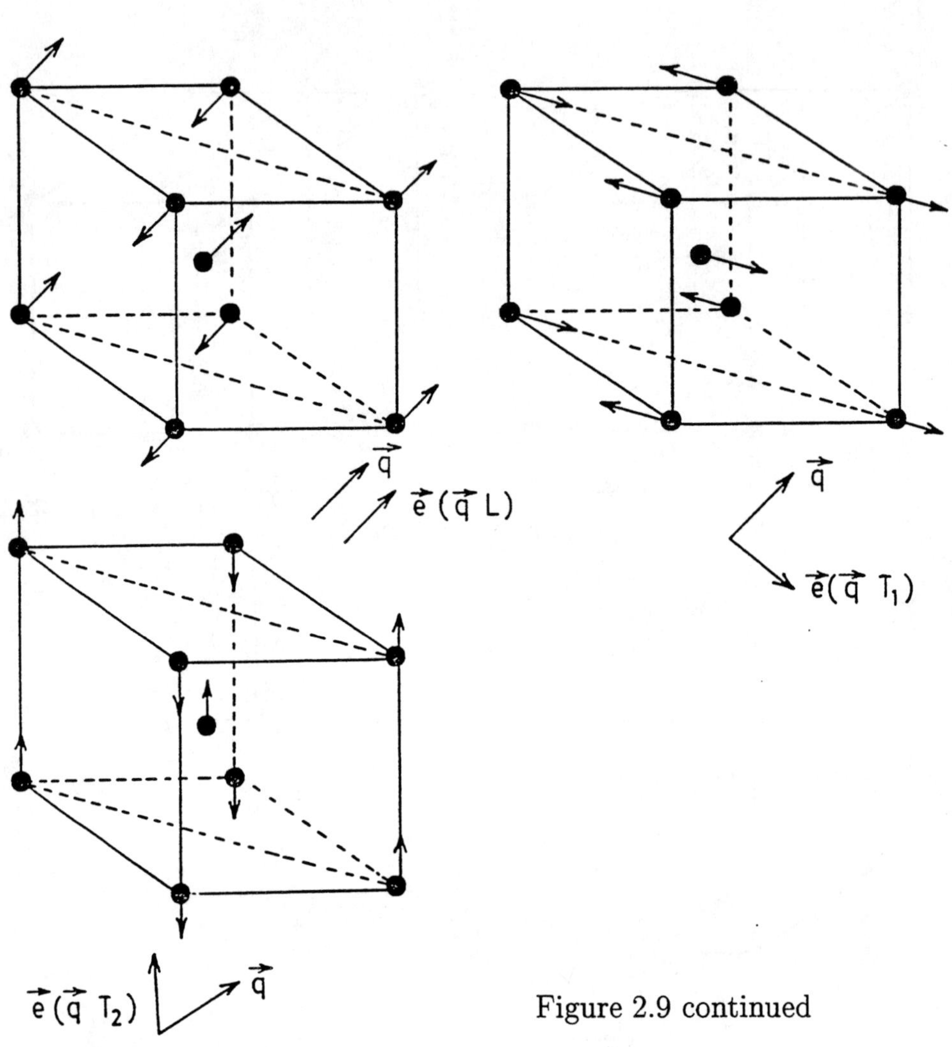

Figure 2.9 continued

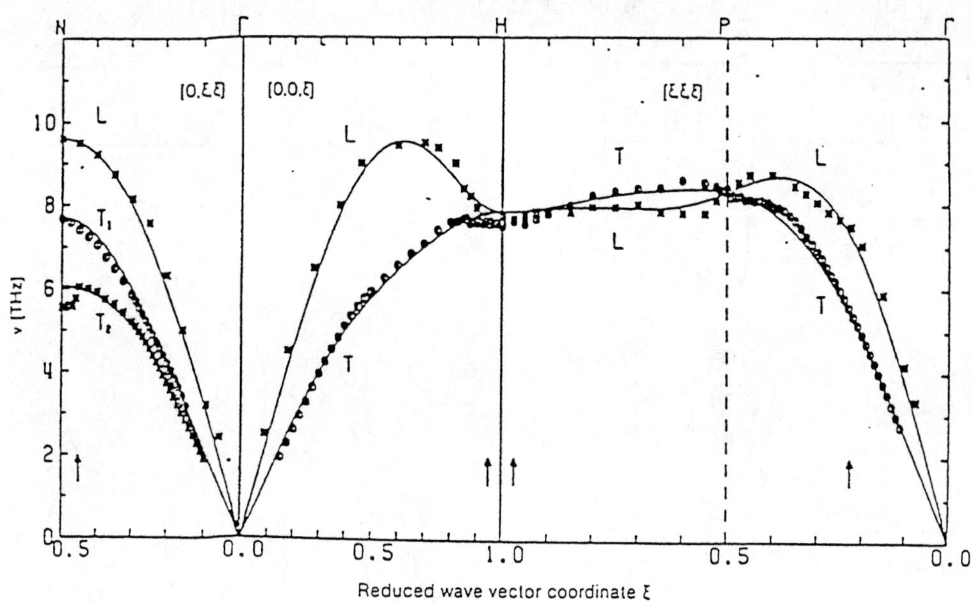

Figure 2.10: The measured and calculated phonon frequencies of anti-ferromagnetic Cr. The curves exhibit anomalies (shown by arrows) in the T_2 branch at the N point and in its vicinity, in both the longitudinal and transverse branches in the vicinity of H point and in the longitudinal branch at about 0.25[111]. Solid lines represent the calculations of 4NN BvK model.

Table 2.1: Force constant matrices $\Phi(0, l')$ corresponding to $1NNs$ (1-12) and $2NNs$ (13-18) of atom at the origin in the fcc lattice.

Atom pair	$\Phi(0, l')$	Atom pair	$\Phi(0, l')$
$\begin{aligned}0-1\\0-3\end{aligned}$	$-\begin{pmatrix} \alpha_1 & \gamma_1 & 0 \\ \gamma_1 & \alpha_1 & 0 \\ 0 & 0 & \beta \end{pmatrix}$	$\begin{aligned}0-6\\0-8\end{aligned}$	$-\begin{pmatrix} \beta_1 & 0 & 0 \\ 0 & \alpha_1 & -\gamma_1 \\ 0 & -\gamma_1 & \alpha_1 \end{pmatrix}$
$\begin{aligned}0-2\\0-4\end{aligned}$	$-\begin{pmatrix} \alpha_1 & -\gamma_1 & 0 \\ -\gamma_1 & \alpha_1 & 0 \\ 0 & 0 & \beta_1 \end{pmatrix}$	$\begin{aligned}0-9\\0-11\end{aligned}$	$-\begin{pmatrix} \alpha_1 & 0 & \gamma_1 \\ 0 & \beta_1 & 0 \\ \gamma_1 & 0 & \alpha_1 \end{pmatrix}$
$\begin{aligned}0-5\\0-7\end{aligned}$	$-\begin{pmatrix} \beta_1 & 0 & 0 \\ 0 & \alpha_1 & \gamma_1 \\ 0 & \gamma_1 & \alpha_1 \end{pmatrix}$	$\begin{aligned}0-10\\0-12\end{aligned}$	$-\begin{pmatrix} \alpha_1 & 0 & -\gamma_1 \\ 0 & \beta_1 & 0 \\ -\gamma_1 & 0 & \alpha_1 \end{pmatrix}$
$\begin{aligned}0-13\\0-15\end{aligned}$	$-\begin{pmatrix} \alpha_2 & 0 & 0 \\ 0 & \beta_2 & 0 \\ 0 & 0 & \beta_2 \end{pmatrix}$	$\begin{aligned}0-14\\0-16\end{aligned}$	$-\begin{pmatrix} \beta_2 & 0 & 0 \\ 0 & \alpha_2 & 0 \\ 0 & 0 & \beta_2 \end{pmatrix}$
$\begin{aligned}0-17\\0-18\end{aligned}$	$-\begin{pmatrix} \beta_2 & 0 & 0 \\ 0 & \beta_2 & 0 \\ 0 & 0 & \alpha_2 \end{pmatrix}$		

Table 2.2: Atomic force constants (in units of 10^4 dynes/cm) for fcc paramagnetic (400^0C) and ferromagnetic (22^0C) Ni determined by BvK fit upto $5NNs$ (Ref.[26]).

nNN	Force Constants	
	400^0C	22^0C
$1NN$ (110)	$\alpha_1 = 1.625$	1.7319
	$\beta_1 = -0.097$	-0.0436
	$\gamma_1 = 1.939$	1.910
$2NN$ (200)	$\alpha_2 = 0.1070$	0.1044
	$\beta_2 = 0.0056$	-0.0780
$3NN$ (211)	$\alpha_3 = 0.0963$	0.0842
	$\beta_3 = 0.0449$	0.0263
	$\gamma_3 = 0.0458$	0.0424
	$\delta_3 = -0.0391$	-0.0109
$4NN$ (220)	$\alpha_4 = 0.0115$	0.0402
	$\beta_4 = -0.0457$	-0.0185
	$\gamma_4 = 0.0222$	0.0660
$5NN$ (310)	$\alpha_5 = -0.0256$	-0.0085
	$\beta_5 = -0.0063$	0.007
	$\gamma_5 = -0.0040$	0.0018
	$\delta_5 = -0.0072$	-0.0035

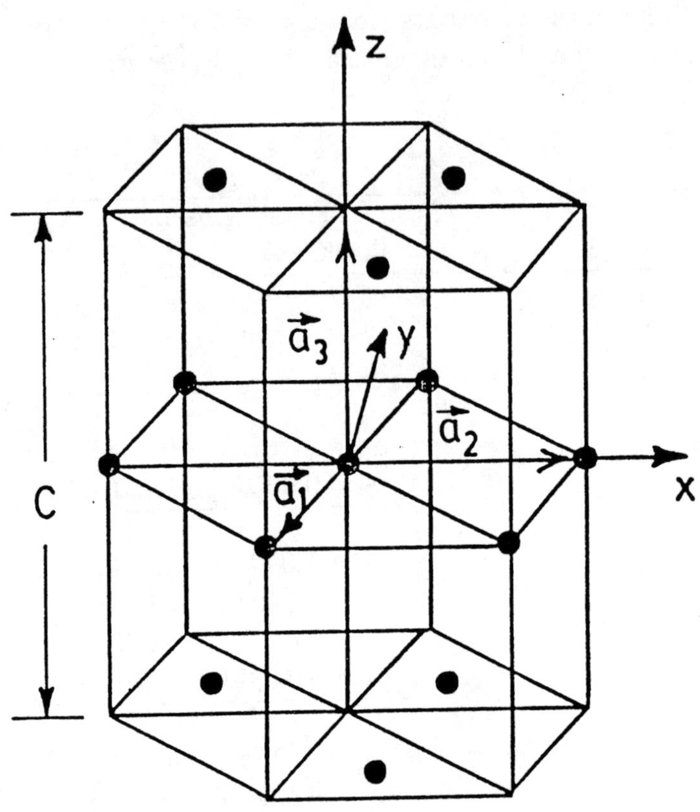

Fig.2.11 a

Figure 2.11: (a) Primitive translation vectors $\vec{a_1}, \vec{a_2}, \vec{a_3}$ of hcp lattice. $\vec{a_1}$ and $\vec{a_2}$ make an angle of $(2\pi/3)$ and $\vec{a_3}$ is perpendicular to the plane of $\vec{a_1}$ and $\vec{a_2}$. Here $|\vec{a_1}| = |\vec{a_2}| = a$, and $|\vec{a_3}| = c$. The Cartesian coordinates are shown in the figure; x-axis is along $\vec{a_2}$ and z-axis is along $\vec{a_3}$. (b) The first and second NNs of basis atoms at O and O'. The coordinates of the atoms in the crystal axis and Cartesian axes are given in Tables 2.6 and 2.8 respectively. (c) Brillouin zone and its reduced part of hcp lattice.

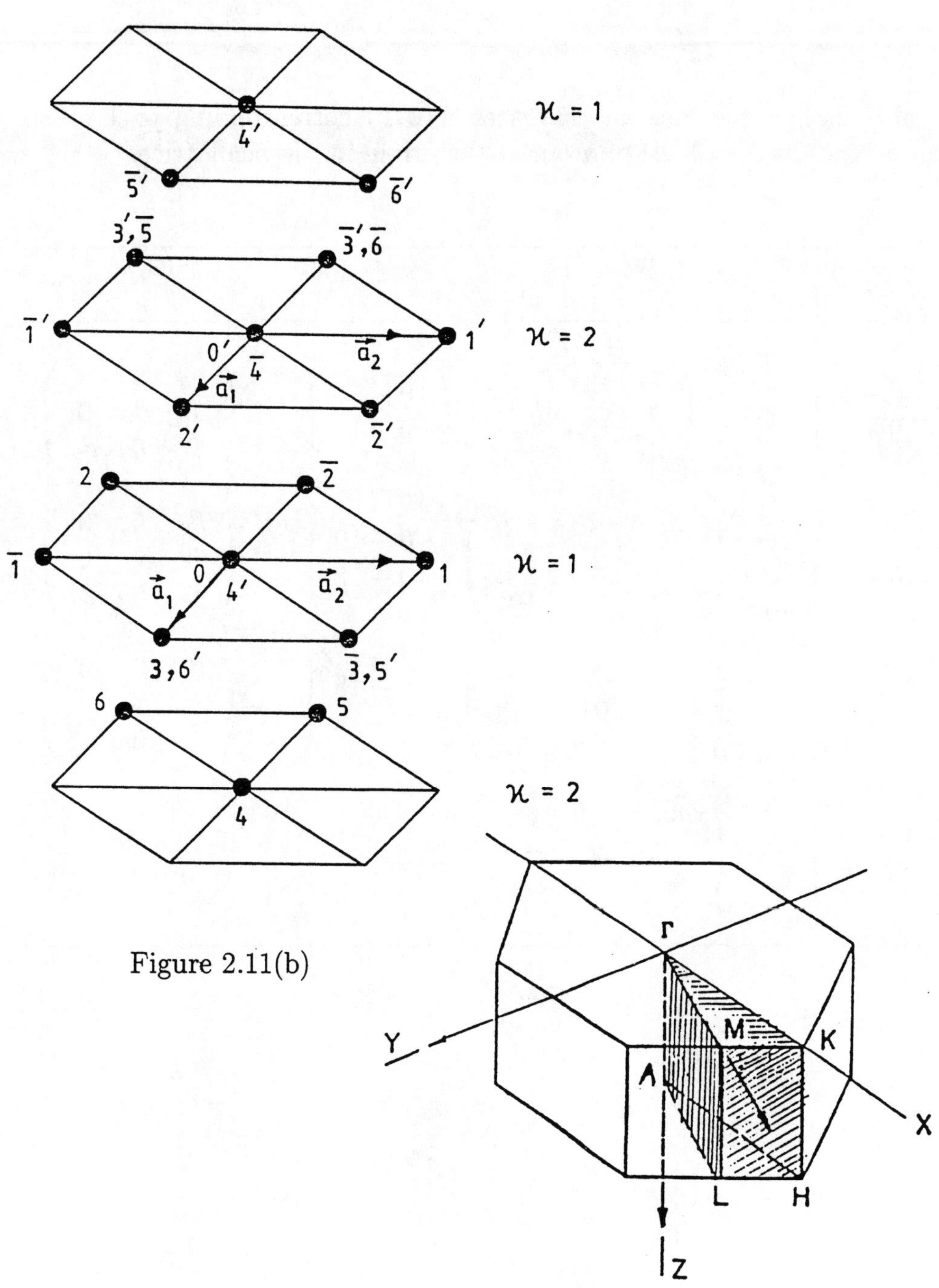

Figure 2.11(b)

Figure 2.11(c)

Table 2.3: Force constant matrices $\Phi(0,l')$ corresponding to $1NNs$ (1-8) and $2NNs$ (9-14) of atom at the origin in the bcc lattice.

Atom pair	$\Phi(0,l')$	Atom pair	$\Phi(0,l')$
$\begin{aligned}0-1\\0-7\end{aligned}\Big\}$	$-\begin{pmatrix} \alpha_1 & \beta_1 & \beta_1 \\ \beta_1 & \alpha_1 & \beta_1 \\ \beta_1 & \beta_1 & \alpha_1 \end{pmatrix}$	$\begin{aligned}0-9\\0-11\end{aligned}\Big\}$	$-\begin{pmatrix} \alpha_2 & 0 & 0 \\ 0 & \beta_2 & 0 \\ 0 & 0 & \beta_2 \end{pmatrix}$
$\begin{aligned}0-2\\0-8\end{aligned}\Big\}$	$-\begin{pmatrix} \alpha_1 & -\beta_1 & -\beta_1 \\ -\beta_1 & \alpha_1 & \beta_1 \\ -\beta_1 & \beta_1 & \alpha_1 \end{pmatrix}$	$\begin{aligned}0-10\\0-12\end{aligned}\Big\}$	$-\begin{pmatrix} \beta_2 & 0 & 0 \\ 0 & \alpha_2 & 0 \\ 0 & 0 & \beta_2 \end{pmatrix}$
$\begin{aligned}0-3\\0-5\end{aligned}\Big\}$	$-\begin{pmatrix} \alpha_1 & \beta_1 & -\beta_1 \\ \beta_1 & \alpha_1 & -\beta_1 \\ -\beta_1 & -\beta_1 & \alpha_1 \end{pmatrix}$	$\begin{aligned}0-13\\0-14\end{aligned}\Big\}$	$-\begin{pmatrix} \beta_2 & 0 & 0 \\ 0 & \beta_2 & 0 \\ 0 & 0 & \alpha_2 \end{pmatrix}$
$\begin{aligned}0-4\\0-6\end{aligned}\Big\}$	$-\begin{pmatrix} \alpha_1 & -\beta_1 & \beta_1 \\ -\beta_1 & \alpha_1 & -\beta_1 \\ \beta_1 & -\beta_1 & \alpha_1 \end{pmatrix}$		

Table 2.4: Phonon dispersion relations for monatomic bcc lattice along principal symmetry directions. The interatomic interactions are extended upto $2NNs$.

$\vec{q}$	$\vec{e}(\vec{q}j)$	$M\omega_j^2(\vec{q})$
	$\vec{e}(\vec{q}L)$	$8\alpha_1\left\{1 - \cos\left(\frac{qa}{2}\right)\right\} + 2\alpha_2\left\{1 - \cos(qa)\right\}$
[100]	$\vec{e}(\vec{q}T_1)$	$8\alpha_1\left\{1 - \cos\left(\frac{qa}{2}\right)\right\} + 2\beta_2\left\{1 - \cos(qa)\right\}$
	$\vec{e}(\vec{q}T_2)$	$8\alpha_1\left\{1 - \cos\left(\frac{qa}{2}\right)\right\} + 2\beta_2\left\{1 - \cos(qa)\right\}$
	$\vec{e}(\vec{q}L)$	$8(\alpha_1 + \beta_1)\left\{1 - \cos^2\left(\frac{qa}{2\sqrt{2}}\right)\right\} + 2(\alpha_2 + \beta_2)\left\{1 - \cos\left(\frac{qa}{\sqrt{2}}\right)\right\}$
[110]	$\vec{e}(\vec{q}T_1)$	$8(\alpha_1 - \beta_1)\left\{1 - \cos^2\left(\frac{qa}{2\sqrt{2}}\right)\right\} + 2(\alpha_2 + \beta_2)\left\{1 - \cos\left(\frac{qa}{\sqrt{2}}\right)\right\}$
	$\vec{e}(\vec{q}T_2)$	$8\alpha_1\left\{1 - \cos^2\left(\frac{qa}{2\sqrt{2}}\right)\right\} + 4\beta_2\left\{1 - \cos\left(\frac{qa}{\sqrt{2}}\right)\right\}$
	$\vec{e}(\vec{q}L)$	$8\alpha_1\left\{1 - \cos^3\left(\frac{qa}{2\sqrt{3}}\right)\right\} + 16\beta_1\sin^2\left(\frac{qa}{2}\right)\cos\left(\frac{qa}{\sqrt{3}}\right)$ $+ (2\alpha_2 + 4\beta_2)\left\{1 - \cos\left(\frac{qa}{\sqrt{3}}\right)\right\}$
[111]	$\vec{e}(\vec{q}T_1)$	$8\alpha_1\left\{1 - \cos^3\left(\frac{qa}{2\sqrt{3}}\right)\right\} - 8\beta_1\sin^2\left(\frac{qa}{2}\right)\cos\left(\frac{qa}{\sqrt{3}}\right)$ $+ (2\alpha_2 + 4\beta_2)\left\{1 - \cos\left(\frac{qa}{\sqrt{3}}\right)\right\}$
	$\vec{e}(\vec{q}T_2)$	same as that of $\vec{e}(\vec{q}T_1)$

Table 2.5: Atomic force constants (in units of dynes/cm) for bcc Cr determined by BvK fit upto $4NNs$ (Ref.[31]).

nNN	Force Constants
$1NN(111)$	$\alpha_1 = 13526$ $\beta_1 = 6487$
$2NN(200)$	$\alpha_2 = 35915$ $\beta_2 = -1564$
$3NN(220)$	$\alpha_3 = 2042$ $\beta_3 = -50$ $\gamma_3 = 2871$
$4NN(311)$	$\alpha_4 = -1257$ $\beta_4 = 432$ $\gamma_4 = 7$ $\delta_4 = 516$

$$
\left.\begin{array}{c}\Phi^{1}\\ \Phi^{1'}\end{array}\right\} = -\begin{pmatrix} \alpha_1+2\beta_1 & \delta_1 & 0 \\ -\delta_1 & \alpha_1-2\beta_1 & 0 \\ 0 & 0 & \gamma_1 \end{pmatrix},
\qquad
\left.\begin{array}{c}\Phi^{\bar{1}}\\ \Phi^{\bar{1}'}\end{array}\right\} = -\begin{pmatrix} \alpha_1+2\beta_1 & -\delta_1 & 0 \\ \delta_1 & \alpha_1-2\beta_1 & 0 \\ 0 & 0 & \gamma_1 \end{pmatrix},
$$

$$
\left.\begin{array}{c}\Phi^{2}\\ \Phi^{2'}\end{array}\right\} = -\begin{pmatrix} \alpha_1-\beta_1 & \delta_1-\sqrt{3}\beta_1 & 0 \\ -\delta_1-\sqrt{3}\beta_1 & \alpha_1+\beta_1 & 0 \\ 0 & 0 & \gamma_1 \end{pmatrix},
\qquad
\left.\begin{array}{c}\Phi^{\bar{2}}\\ \Phi^{\bar{2}'}\end{array}\right\} = -\begin{pmatrix} \alpha_1-\beta_1 & -\delta_1+\sqrt{3}\beta_1 & 0 \\ \delta_1+\sqrt{3}\beta_1 & \alpha_1+\beta_1 & 0 \\ 0 & 0 & \gamma_1 \end{pmatrix},
$$

$$
\left.\begin{array}{c}\Phi^{3}\\ \Phi^{3'}\end{array}\right\} = -\begin{pmatrix} \alpha_1-\beta_1 & \delta_1+\sqrt{3}\beta_1 & 0 \\ -\delta_1+\sqrt{3}\beta_1 & \alpha_1+\beta_1 & 0 \\ 0 & 0 & \gamma_1 \end{pmatrix},
\qquad
\left.\begin{array}{c}\Phi^{\bar{3}}\\ \Phi^{\bar{3}'}\end{array}\right\} = -\begin{pmatrix} \alpha_1-\beta_1 & -\delta_1-\sqrt{3}\beta_1 & 0 \\ \delta_1-\sqrt{3}\beta_1 & \alpha_1+\beta_1 & 0 \\ 0 & 0 & \gamma_1 \end{pmatrix},
$$

$$
\left.\begin{array}{c}\Phi^{4}\\ \Phi^{4'}\end{array}\right\} = -\begin{pmatrix} \lambda_1+2\mu_1 & 0 & 0 \\ 0 & \lambda_1-2\mu_1 & 2\sigma_1 \\ 0 & 2\sigma_1 & \nu_1 \end{pmatrix},
\qquad
\left.\begin{array}{c}\Phi^{\bar{4}}\\ \Phi^{\bar{4}'}\end{array}\right\} = -\begin{pmatrix} \lambda_1+2\mu_1 & 0 & 0 \\ 0 & \lambda_1-2\mu_1 & -2\sigma_1 \\ 0 & -2\sigma_1 & \nu_1 \end{pmatrix},
$$

$$
\left.\begin{array}{c}\Phi^{5}\\ \Phi^{5'}\end{array}\right\} = -\begin{pmatrix} \lambda_1-\mu_1 & -\sqrt{3}\mu_1 & -\sqrt{3}\sigma_1 \\ -\sqrt{3}\mu_1 & \lambda_1+\mu_1 & -\sigma_1 \\ -\sqrt{3}\sigma_1 & -\sigma_1 & \nu_1 \end{pmatrix},
\qquad
\left.\begin{array}{c}\Phi^{\bar{5}}\\ \Phi^{\bar{5}'}\end{array}\right\} = -\begin{pmatrix} \lambda_1-\mu_1 & \sqrt{3}\mu_1 & -\sqrt{3}\sigma_1 \\ \sqrt{3}\mu_1 & \lambda_1+\mu_1 & \sigma_1 \\ -\sqrt{3}\sigma_1 & \sigma_1 & \nu_1 \end{pmatrix},
$$

$$
\left.\begin{array}{c}\Phi^{6}\\ \Phi^{6'}\end{array}\right\} = -\begin{pmatrix} \lambda_1-\mu_1 & \sqrt{3}\mu_1 & \sqrt{3}\sigma_1 \\ \sqrt{3}\mu_1 & \lambda_1+\mu_1 & -\sigma_1 \\ \sqrt{3}\sigma_1 & -\sigma_1 & \nu_1 \end{pmatrix},
\qquad
\left.\begin{array}{c}\Phi^{\bar{6}}\\ \Phi^{\bar{6}'}\end{array}\right\} = -\begin{pmatrix} \lambda_1-\mu_1 & -\sqrt{3}\mu_1 & \sqrt{3}\sigma_1 \\ -\sqrt{3}\mu_1 & \lambda_1+\mu_1 & \sigma_1 \\ \sqrt{3}\sigma_1 & \sigma_1 & \nu_1 \end{pmatrix},
$$

Table 2.6: Relation between indices p ánd $(l, \kappa\kappa')$. These are the coordinates of NNs in the crystal axis system. The coordinates of atoms $(1, 2, 3, \bar{1}, \bar{2}, \bar{3})$ are with respect to atom at O. The coordinates of atoms $(1', 2', 3', \bar{1}', \bar{2}', \bar{3}')$ are with respect to atom at O'. The coordinates of atoms $(4, 5, 6, \bar{4}, \bar{5}, \bar{6})$ are with respect to atom at $\bar{4}(O')$ and of atoms $(4', 5', 6', \bar{4}', \bar{5}', \bar{6}')$ are with respect to atom at $4'(O)$. Here $\kappa = 1$ is for the planes containing atoms $(4', 5', 6')$ and $(\bar{4}', \bar{5}', \bar{6}')$ and $\kappa = 2$ is for the planes containing atoms $(4, 5, 6)$ and $(\bar{4}, \bar{5}, \bar{6})$.

p	1	2	3	4	5	6
l_1	0	-1	1	0	-1	-1
l_2	1	-1	0	0	0	-1
l_3	0	0	0	-1	-1	-1
$(\kappa\kappa')$	(11)	(11)	(11)	(21)	(21)	(21)

p	$\bar{1}$	$\bar{2}$	$\bar{3}$	$\bar{4}$	$\bar{5}$	$\bar{6}$
l_1	0	-1	1	0	-1	-1
l_2	-1	0	1	0	-1	0
l_3	0	0	0	0	0	0
$(\kappa\kappa')$	(11)	(11)	(11)	(21)	(21)	(21)

p	$1'$	$2'$	$3'$	$4'$	$5'$	$6'$
l_1	0	1	-1	0	1	1
l_2	1	0	-1	0	1	0
l_3	0	0	0	0	0	0
$(\kappa\kappa')$	(22)	(22)	(22)	(12)	(12)	(12)

p	$\bar{1}'$	$\bar{2}'$	$\bar{3}'$	$\bar{4}'$	$\bar{5}'$	$\bar{6}'$
l_1	0	1	-1	0	1	1
l_2	-1	1	0	0	0	1
l_3	0	0	0	1	1	1
$(\kappa\kappa')$	(22)	(22)	(22)	(12)	(12)	(12)

Table 2.7: Force constant matrices $\boldsymbol{\Phi}^{\mathbf{p}}$ corresponding to the $1NNs$ of atoms at O and O' in Fig. 2.11b for hcp lattice. Symbol p is explained in Table 2.6.

Table 2.8: Cartesian componets (x, y, z) of vector $\vec{R}^0_\rho = \vec{R}^0_l + \vec{R}^0_{\kappa\kappa'}$ with $\vec{R}^0_{\kappa\kappa'} = \vec{R}^0_\kappa - \vec{R}^0_{\kappa'} = (a/\sqrt{3})\hat{j} - (c/2)\hat{k}$ for hcp lattice.

p	1	2	3	$\bar{1}$	$\bar{2}$	$\bar{3}$	ρ	4	5	6	$\bar{4}$	$\bar{5}$	$\bar{6}$
$\frac{1}{a}x(l,11)$	1	$-\frac{1}{2}$	$-\frac{1}{2}$	-1	$\frac{1}{2}$	$\frac{1}{2}$	$\frac{1}{a}x(l,21)$	0	$\frac{1}{2}$	$-\frac{1}{2}$	0	$-\frac{1}{2}$	$\frac{1}{2}$
$\frac{1}{a}y(l,11)$	0	$\frac{\sqrt{3}}{2}$	$-\frac{\sqrt{3}}{2}$	0	$\frac{\sqrt{3}}{2}$	$-\frac{\sqrt{3}}{2}$	$\frac{1}{a}y(l,21)$	$-\frac{1}{\sqrt{3}}$	$\frac{1}{2\sqrt{3}}$	$\frac{1}{2\sqrt{3}}$	$-\frac{1}{\sqrt{3}}$	$\frac{1}{2\sqrt{3}}$	$\frac{1}{2\sqrt{3}}$
$\frac{1}{c}z(l,11)$	0	0	0	0	0	0	$\frac{1}{c}z(l,21)$	$-\frac{1}{2}$	$-\frac{1}{2}$	$-\frac{1}{2}$	$\frac{1}{2}$	$\frac{1}{2}$	$\frac{1}{2}$

Table 2.9 Atomic force constants with estimated error (in units of 10^4 dynes/cm) for $Ni_{0.3}Fe_{0.7}$ with fcc structure (Ref. [38]).

nNN	Force constants
$1NN(110)$	$\begin{cases} \alpha_1 = 1.567 \pm 0.014 \\ \beta_1 = -0.005 \pm 0.018 \\ \gamma_1 = 1.771 \pm 0.020 \end{cases}$
$2NN(200)$	$\begin{cases} \alpha_2 = -0.146 \pm 0.026 \\ \beta_2 = 0.091 \pm 0.015 \end{cases}$
$3NN(211)$	$\begin{cases} \alpha_3 = 0.012 \pm 0.011 \\ \beta_3 = 0.020 \pm 0.007 \\ \gamma_3 = -0.008 \pm 0.009 \\ \delta_3 = 0.041 \pm 0.007 \end{cases}$
$4NN(220)$	$\begin{cases} \alpha_4 = 0.048 \pm 0.008 \\ \beta_4 = -0.001 \pm 0.012 \\ \gamma_4 = 0.038 \pm 0.015 \end{cases}$
$5NN(310)$	$\begin{cases} \alpha_5 = 0.009 \pm 0.007 \\ \beta_5 = -0.006 \pm 0.003 \\ \gamma_5 = -0.008 \pm 0.003 \\ \delta_5 = 0.006 \pm 0.003 \end{cases}$

Table 2.10: Atomic force constants (in units of dynes/cm) for Y obtained in the MAS model, fitted upto $6NNs$. The symbols are the same as used in Ref. [34].

nNN	MAS notation
$1NN$	$\begin{cases} \delta_1 = 2047.2 \\ \epsilon_{1x} = -1628.0 \\ \epsilon_{1z} = -3640.7 \end{cases}$
$2NN$	$\begin{cases} \alpha_2 = 10124.0 \\ \beta_{2x} = 1455.7 \\ \beta_{2z} = 510.36 \end{cases}$
$3NN$	$\begin{cases} \delta_3 = -273.15 \\ \epsilon_{3x} = 1211.5 \\ \epsilon_{3z} = 1510.6 \end{cases}$
$4NN$	$\begin{cases} \beta_{4x} = -178.3 \\ \alpha_4 + \beta_{4z} = -83.0 \end{cases}$
$5NN$	$\begin{cases} \delta_5 = 39.31 \\ \epsilon_{5x} = 456.3 \\ \epsilon_{5z} = -582.0 \end{cases}$
$6NN$	$\begin{cases} \alpha_6 = 1855.9 \\ \beta_{6x} = -93.0 \\ \beta_{6z} = 592.7 \end{cases}$

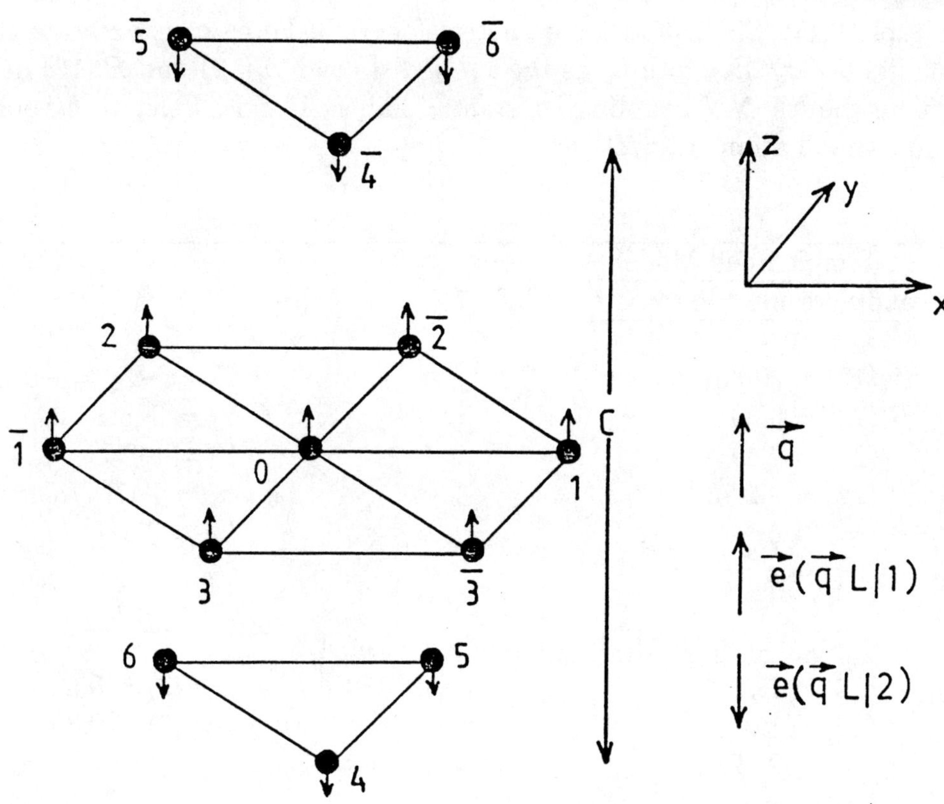

Figure 2.12: Pattern of atomic displacements in the hcp lattice for $\vec{q} = (2\pi/c)(001)$ in the longitudinal acoustic mode. The entire plane of atoms move in one phase and the neighbouring plane in the opposite phase with wavelength c.

Table 2.11: Shell model parameters for TaC. Force constants are in units of $(e^2/2d_0)$ with d_0 as the $1NN$ distance. $A_l(12)$ and $B_l(12)$ denote the lth NN coupling constants. Labels 1 and 2 refer to carbon and metal atoms respectively.

A. <u>Simple Shell Model</u>
<u>and Screening Parameters</u> (Ref. [73])

$A_1(12) = 20.38$ $Z_1 = -Z_2 = -0.75e$

$B_1(12) = 6.39$ $Y_1 = 0$

$A_2(22) = 9.49$ $Y_2 = -1.52e$

$B_2(22) = -2.73$ $k_s = 0.47(\pi/d_0)$

$K_1 = 228.5$ $k_F = 0.37(\pi/d_0)$

$K_2 = 184.4$

B. <u>Extended Shell Model Parameters</u> (Ref. [73])

$A_1(12) = 1.79$ $B_2(22) = 0.54$

$B_1(12) = 1.06$ $A_3(12) = 0.30$

$A_2(22) = 2.39$

C. <u>Double Shell Model Parameters</u> (Ref. [76])

$A_1^s(12) = -0.96$ $B_2^s(12) = -0.12$

$A_2^s(22) = -1.30$ $K^s = 8.0$

$B_1^s(12) = -0.64$

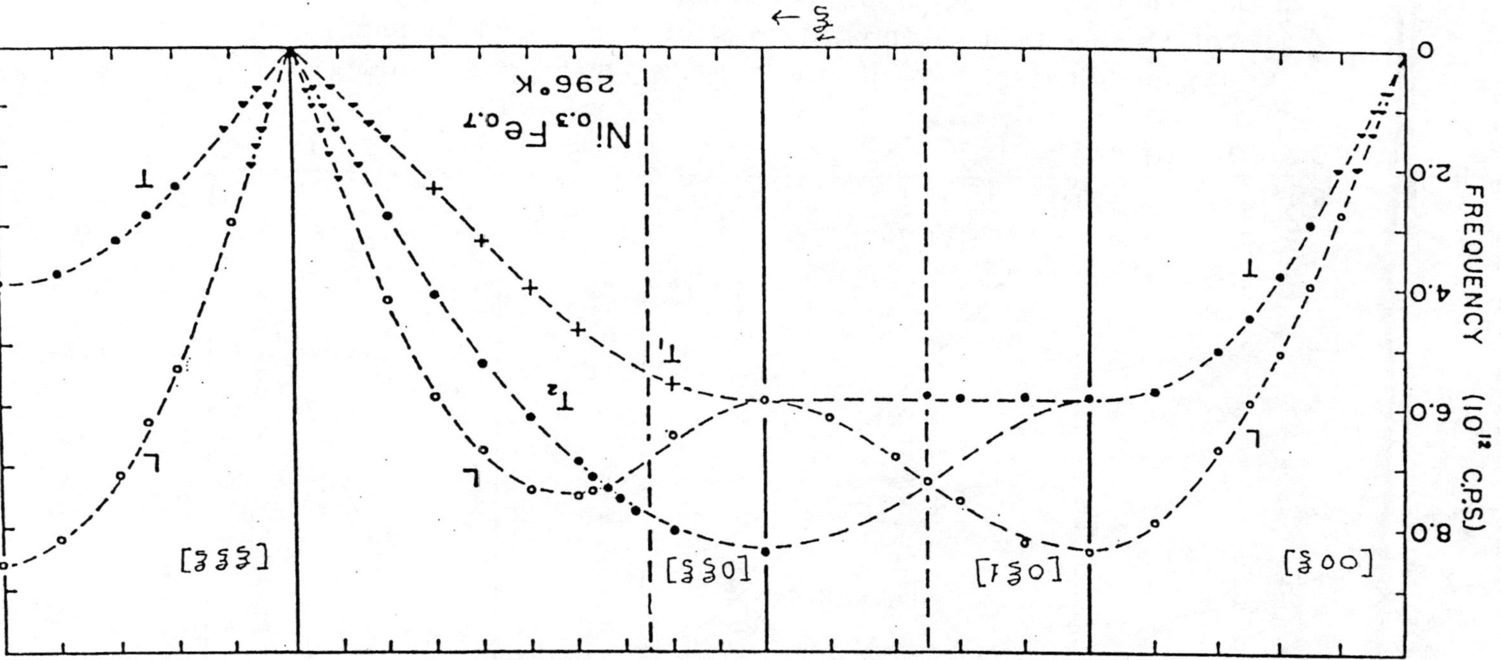

Figure 2.13: Phonon dispersion curves $\omega(\vec{q})$ versus $\vec{q}$ for $Ni_{0.3}Fe_{0.7}$ of average fcc structure along different symmetry directions. The open and solid circles are experimental points and dashed lines represent the 5NN BvK fitting (Ref.[38]).

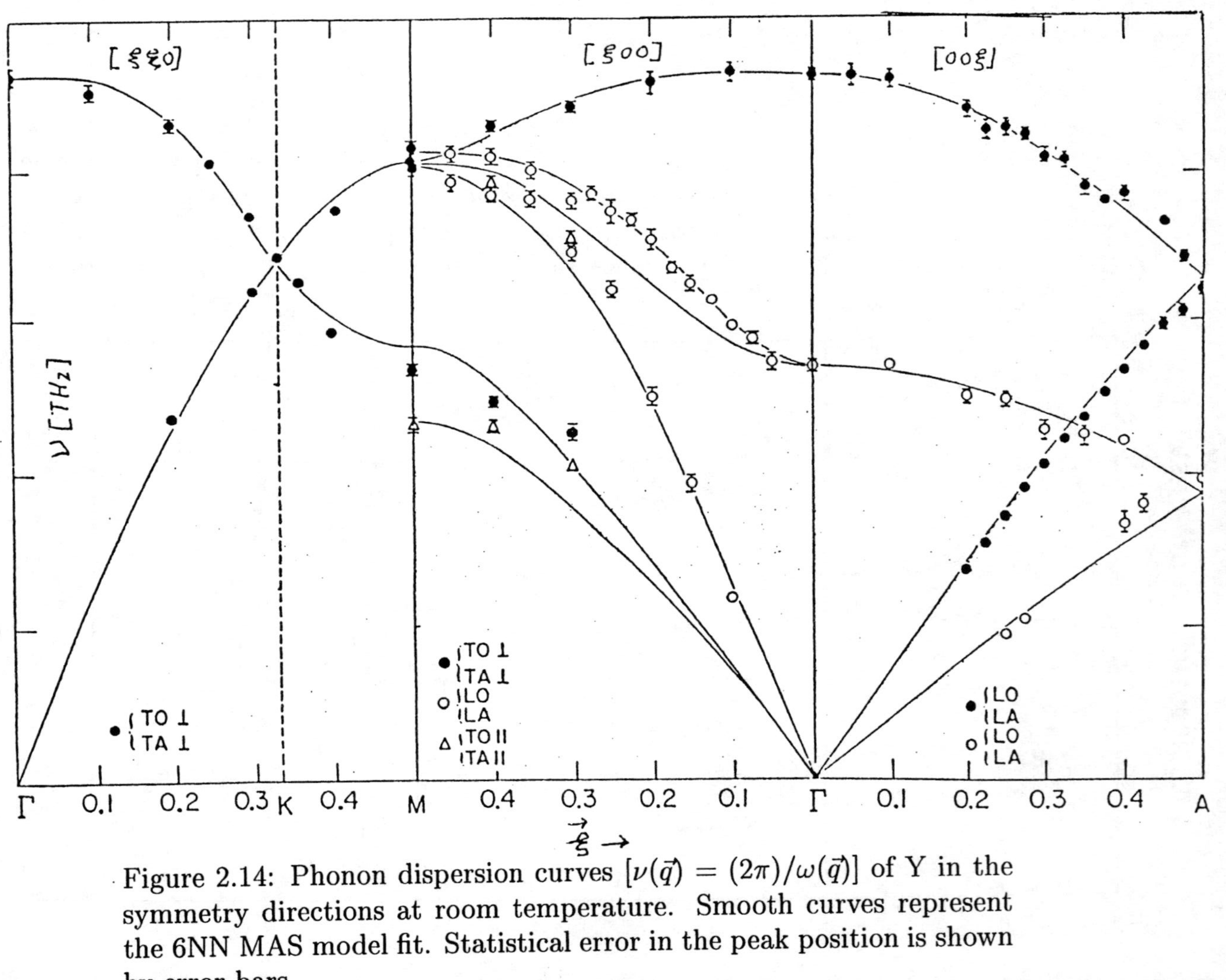

Figure 2.14: Phonon dispersion curves $[\nu(\vec{q}) = (2\pi)/\omega(\vec{q})]$ of Y in the symmetry directions at room temperature. Smooth curves represent the 6NN MAS model fit. Statistical error in the peak position is shown by error bars.

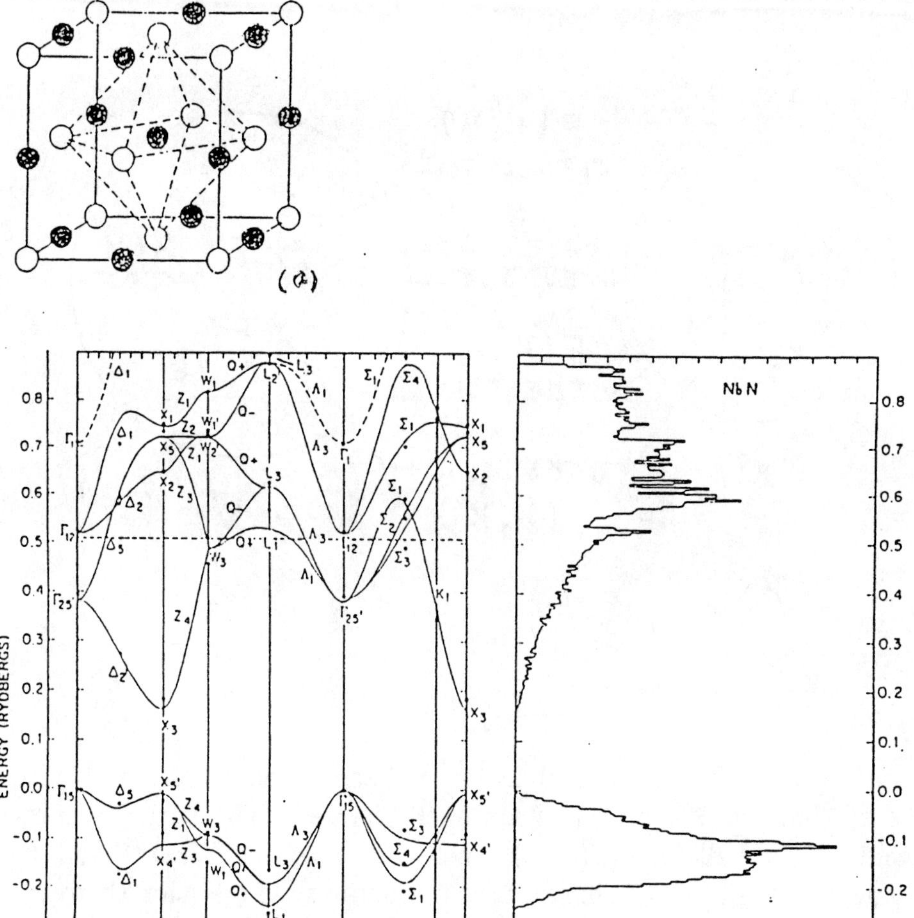

Figure 2.15: (a) Cubic unit cell of rocksalt structure NbN. The solid and open circles represent Nb and N atoms respectively. Dashed lines indicate the octahedral coordination of metal atom with its neighbours. (b) Electron energy bands and density of states(per-spin per- Rydberg -unit cell). Symmetry points are the same as in the B.Z. of fcc lattice.

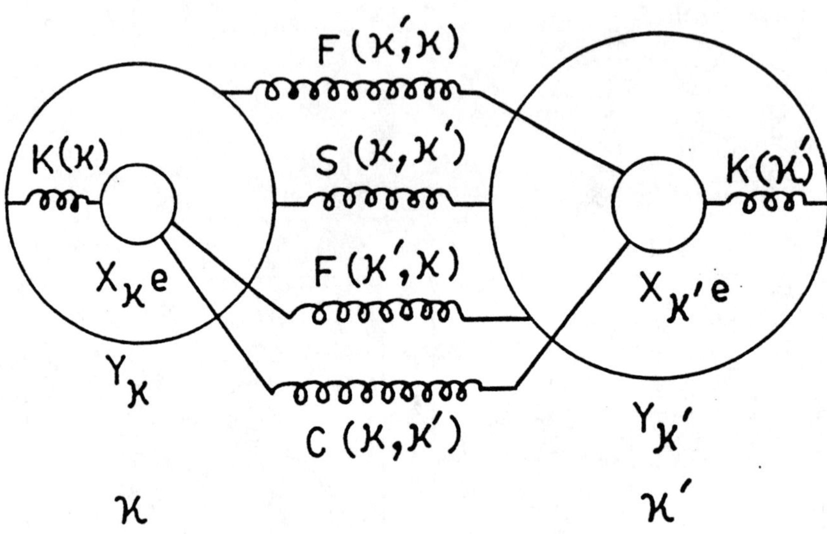

Figure 2.16: Cochran's generalized simple shell model. $X_\kappa e$ and $X_{\kappa'} e$ are core charges, $Y_\kappa e$ and $Y_{\kappa'} e$ are shell charges. The springs represent interionic shell-shell $[S(\kappa', \kappa)]$, shell-core $[F(\kappa', \kappa), F(\kappa, \kappa')]$ and intraionic core-shell $[K(\kappa)]$ short range interactions. In addition there are also interionic long range Coulomb interactions represented by $C(\kappa, \kappa')$.

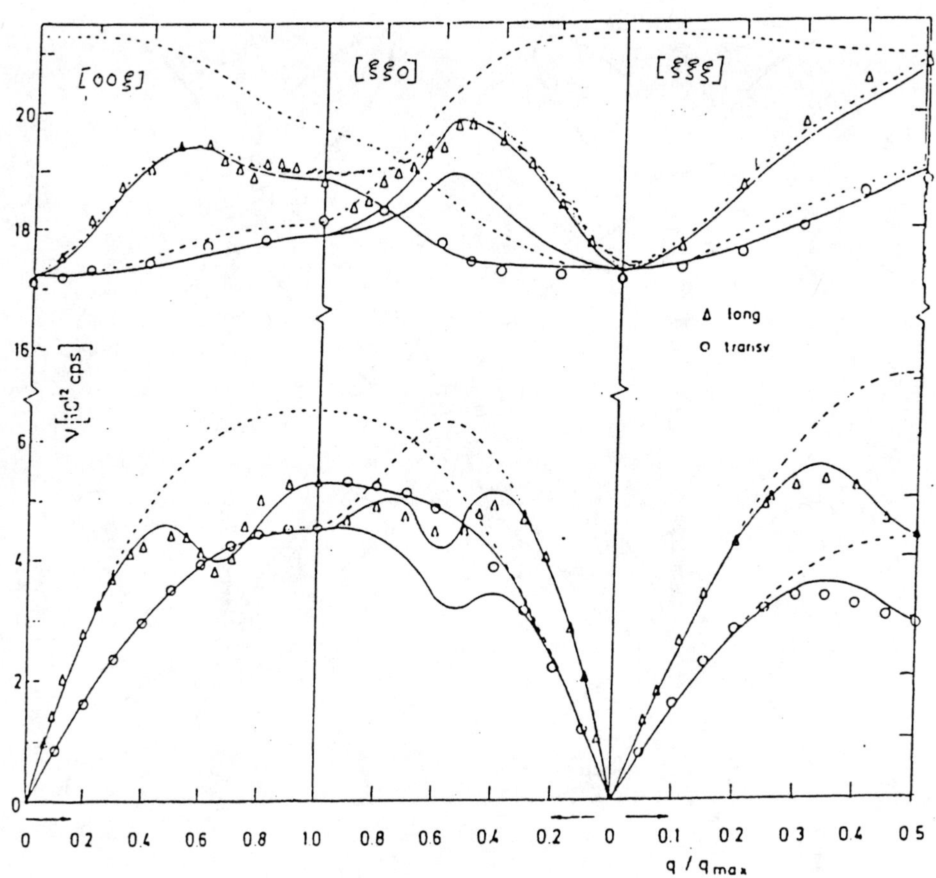

Figure 2.17: Phonon dispersion curves of TaC. Triangles and circles represent experimental values(Ref.[36]). Dashed, dotted and solid lines show the results obtained from simple shell model, screened shell model and extended shell model respectively. Dotted lines almost coincide with the solid lines (Ref.[37]).

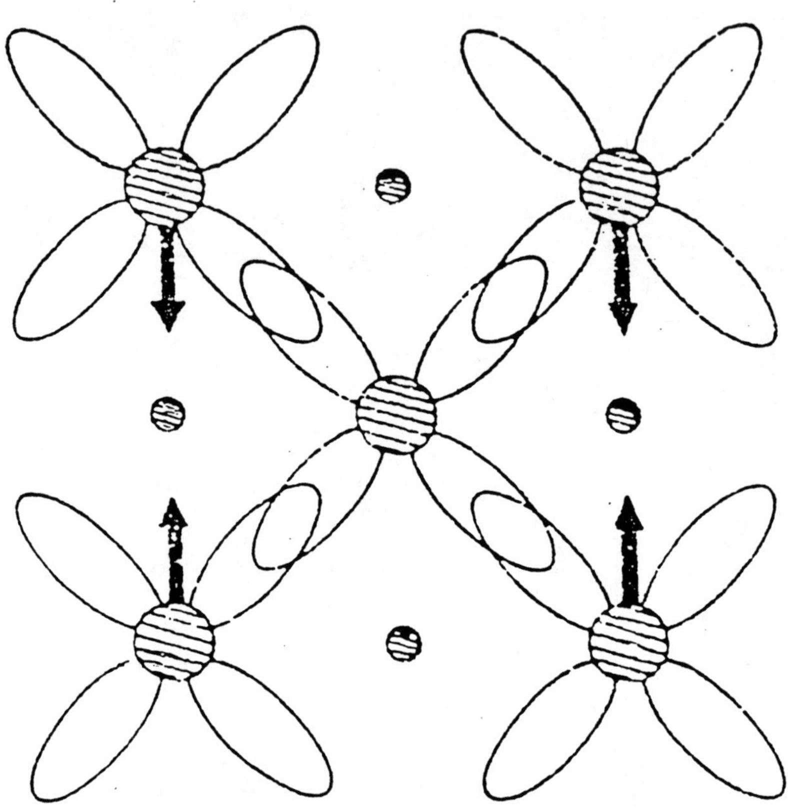

Figure 2.18: Pattern of motion of metallic ions for acoustic phonons with $\vec{q} \simeq (0.5, 0, 0)q_m$. The loops denote the distribution of charge density for metallic d-electrons with T_{2g} symmetry.

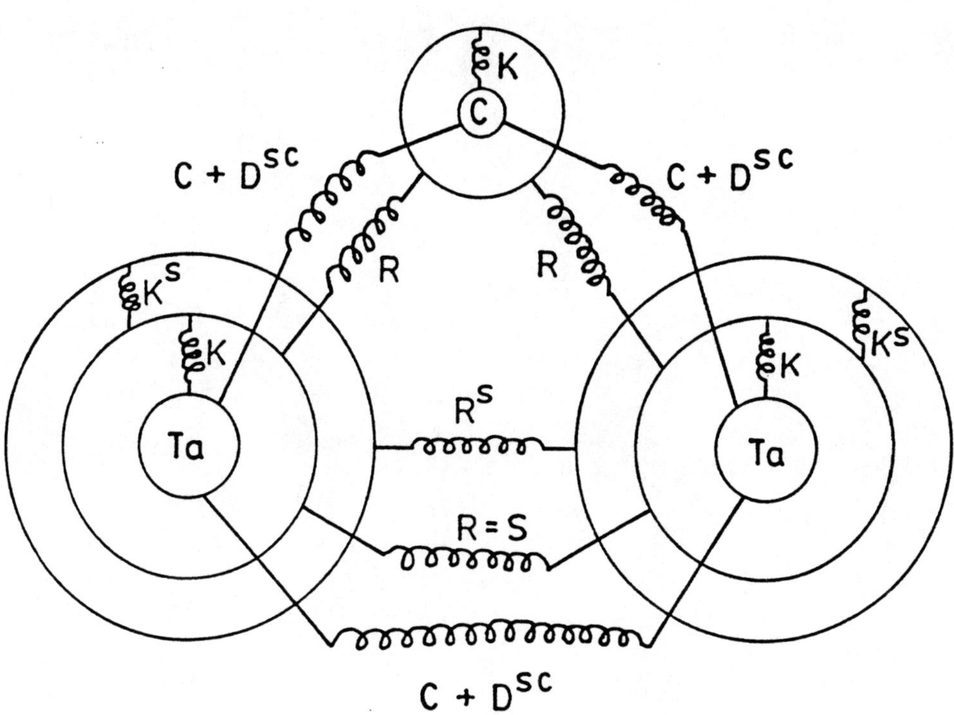

Figure 2.19: Weber's double shell model for TaC . The short range interactions are represented collectively by $\mathbf{R} = \mathbf{S}$. The parameter R^s represents supershell-supershell interaction and K^s represents the interaction of supershell with core and inner shell. Other parameters are the same as defined for screened shell model and extended shell model.

Bibliography

[1] M.Born and K.Haung, "Dynamical Theory of Crystal Lattice" (Oxford Univ., Press, Oxford, 1954).

[2] A.A. Maradudin, E.W.Montroll, G.H. Weiss and I.P.Ipatova, "Solid State Physics, Suppl.3," 2nd Ed. (Acad. Press, New York, 1971).

[3] G. Venkataraman,L.A.Feldkamp and V.C.Sahni, "Dynamics of Perfect Crystals" (MIT Press, London, 1975) p.20

[4] P.Brüch, "Phonons: Theory and Experiment I" (Springer Verlag, Heidelberg, 1982) p.1.

[5] L.J.Sham and J.M. Ziman, "Solid State Physics" Vol. 15 (Acad.Press, New York, 1965)p.221.

[6] C.Kittel, "Introduction to Solid State Physics" (J.Wiley and Sons, Inc. 1971).

[7] S.K.Joshi and A.K. Rajagopal, "Solid State Physics," Vol.22,(Acad. Press,New York,1969)p.159.

[8] G.V.Chester, Adv. Phys. 10, 357 (1961).

[9] J.M.Ziman, "Principles of Theory of Solids(Cambridge Univ.Press, Cambridge, 1972)p.200.

[10] E.G.Brownman, Yu M.Kagan, Sovt.Phys.JETP 25, 365 (1967).

[11] P.Brüsch, "Phonons: Theory and Experiment II" (Springer-Verlag Heidelberg, 1983).

[12] J.M.Ziman, "Electrons and Phonons" (Oxford Univ. Press,Oxford, 1972).

[13] H.Goldstein, "Classical Mechanics" (Addison-Wesley 1980)p.154.

[14] M.Lax, in "Lattice Dynamics" Ed. R.F.Wallis (Pergamon Press, Oxford, 1965) p.523.

[15] R.Courant and D.Hilbert, "Methods of Mathematical Physics" Vol.1 (Interscience, New York, 1951).

[16] M.Born and Th. von Kármán, Z.Physik 13, 297(1972).

[17] P.P.Ewald, Z.Krist. 56, 129(1921).

[18] C.Kittel, "Introduction to Solid State Physics" (J. Wiley & Sons, Inc., New York, 1971) Chap. 2.

[19] C.V.Raman, Proc. Ind. Acd. Sc. 13, 1(1941); 14A, 317 (1941).

[20] J.R.Hardy, Phil. Mag. 7, 315(1962).

[21] A.J.E.Foreman and W.H.Lomer, Proc. Phys. Soc.B70, 1148 (1957).

[22] G.H.Begbie and M.Born, Proc. Roy. Soc.(London) A188, 179 (1947); G.H.Begbie, Proc. Roy. Soc.(London) A188, 189(1947).

[23] A.H.A.Penny, Proc. Cambridge Phil.Soc. 44, 423(1948).

[24] A.Czachor, Phys. Stat. Sol. 53, K65(1972).

[25] B.N.Brockhouse, T.Arase, G.Caglioti, K.R.Rao and A.D.B. Woods, Phys. Rev. 128, 1099(1962).

[26] G.A.DeWit and B.N.Brockhouse, J.Appl. Phys. 39, 451(2968).

[27] G.Liebfried, in "Theory of Condensed Matter" (IAEA, Vienna, 1968) p.175.

[28] P.K.Iyenger, G.Venkataraman, P.R.Vijayaraghavan and A.P.Roy, in "Inelastic Scattering of Neutrons in Solids and Liquids" Vol.1 (IAEA, Vienna, 1965) p.153.

[29] A.D.B.Woods, B.N.Brockhouse, R.H.March, A.T.Stewart and R.Bowers, Phys. Rev. 128, 1112(1962).

[30] G.L. Squires, Arkiv. Für Physik, 25, 21(1965).

[31] W.M. Shaw and L.D. Mühlestein, Phy. Rev. B4, 969(1971).

[32] C.Czachor, in "Inelastic Scattering of Neutrons in Solids and Liquids" (IAEA, Vienna, 1965) Vol.1, p.181.

[33] M.F. Collins, Proc. Phys. Soc. 80, 362 (1962).

[34] S.K. Sinha, T.O.Brun, L.D. Mühlestein and J. Sakurai, Phys. Rev. B1, 2430 (1969).

[35] R. M. Nicklow, N. Wakabayashi and P.R. Vijayaraghavan, Phys. Rev. B3, 1229 (1971).

[36] H.G. Smith and W.Gläser, Phys. Rev. Lett. 25, 1611 (1970).

[37] H.G.Smith, Phys. Rev. Lett. 29, 353 (1972).

[38] E.D. Hallman and B.N. Brockhouse, Cand. J.Phys. 47, 1117 (1969).

[39] G.W. Lehman, T.Wolfram and R.E. deWames, Phys. Rev. 128, 1593 (1962).

[40] Y.P. Varshni and R.C. Shukla, J. Chem. Phys. 43, 3966 (1965).

[41] R.J. Birgeneau, J. Cordes, G.Dolling and A.D.B.Woods, Phys. Rev. 136, A1359 (1964).

[42] G.W.Lehman, T.Wolfram and R.E.deWames, in "Lattice Dynamics", Ed. R.F.Wallis (Pergamon Press, Oxford, 1965) p.101.

[43] R.E. de Wames, T.Wolfram and G.W.Lehman, Phys. Rev. 138, A717 (1965).

[44] A.B.Bhatia, Phys. Rev. 97, 363 (1955).

[45] J.deLaunay, "Solid State Physics", Vol. 2(Acad.Press,New York, 1956) p.276.

[46] P.K. Sharma and S.K.Joshi, J.Chem Phys. 39, 2633 (1963).

[47] K.Krebs, Phys. Rev. 138, A143 (1965).

[48] S.K.Sinha, Phys. Rev. 138, 2430 (1965).

[49] M. Born and K. Huang, " Dynamical Theory of Crystal Lattices " (Oxford Univ. Press, Oxford, 1954) p.217-233.

[50] J.deLaunay, Solid State Physics, Vol. 2(Acad. Press, New York, 1956) p.248-251.

[51] B.C.Clark, D.C.Gazis and R.F.Wallis, Phys.Rev. 134, A1486 (1964).

[52] J.C.Upadhayaya, Proc. Roy. Soc., A356, 345 (1977).

[53] L.J. Slutsky and C.W. Garland, J. Phys. Chem. Solids,26,787 (1957).

[54] E.A.Metzbower, Phys. Stat. Sol.(b)20, 681 (1967).

[55] S.Prakash, in "Current Trends in Lattice Dynamics," Ed. K.R. Rao (BARC,India, 1979).p.197.

[56] L.E. Toth, "Transition Metal Carbides and Nitrides" (Acad. Press, New York, 1973) ; W.M.Müller, J.P. Blackledge and G.G. Libowitz, "Metal Hydrides"(Acad. Press, New York, 1968).

[57] P.B. Allen, in " Dynamical Properties of Solids," Eds. G.K. Horton and A.A. Maradudin (North-Holland Publ. Co., 1980)p.97.

[58] G.Hägg, Z.Phys. Chem. B.12, 33 (1931).

[59] R.G. Lye and E.M. Logothetis, Phys. Rev. 147, 622 (1966).

[60] H. Bilz, Z.Phys, 153, 338 (1958).

[61] L.F. Mattheiss, Phys. Rev. B5, 315 (1972).

[62] R.Caudron, J.Castaing, and P. Costa, Solid State Commun. 8, 621 (1970).

[63] B.G.Dick and A.W.Overhauser, Phys. Rev. 112, 90 (1958); J.E. Hanlon and A.W.Lawson, Phys. Rev. 113, 472 (1959); A.D.B. Woods, W.Cochran and B.N.Brockhouse, Phys. Rev. 119, 980 (1960).

[64] E.W. Kellermann, Phil. Trans. Roy. Soc.(London) 238, 513 (1940).

[65] Ref.[3], p.234.

[66] E.E.Havinga, Phys. Rev. 119, 1193 (1960).

[67] W.Cochran, Adv.Phys. 10, 401 (1961).

[68] R.A.Cowley, Proc. Roy. Soc. (London) A268, 121 (1962).

[69] J.G. Traylor, H.G. Smith, R.M.Nicklow and M.K.Wilkinson, Phys. Rev. B3, 3457 (1971).

[70] S.K.Sinha, Phys. Rev. 177, 1256 (1969).

[71] R.K.Singh, Phys. Reports, 85, 259(1969).

[72] W.Cochran and R.A.Cowley, J.Phys. Chem. Solids 23, 447 (1962).

[73] W.Weber, H.Bilz and U.Schröder, Phys. Rev.Lett. 28, 600 (1972).

[74] W.Cochran, R.A. Cowley, G.Dolling and M.M. Elcombe, Proc. Roy Soc. A293, 433 (1966).

[75] L.E. Toth, Transition Metal Carbides and Nitrides (Acad. Press, New York, 1971).

[76] W.Weber, Phys. Rev. B8, 5082 (1973).

[77] W.Weber, Ph.D. Thesis (Technische Univ. Munchen, 1972) (Unpublished).

[78] W.Kress, P.Roedhammer, H.Bilz, W.D. Teuchert, and A.N. Christensen, Phys. Rev. B17, 111 (1978).

[79] A.N. Christensen, O.W.Dietrich, W.Krees and W.D.Teuchert, Phys. Rev. B19, 5699 (1979).

[80] A.N.Christensen, O.W.Dietrich, W.Kress, W.D.Teuchert and R.Currat, Solid State Commun. 31, 795 (1979).

[81] B.T.Mattheiss, H.Suhl and C.S.Ting, Phys. Rev. Lett. 27, 245 (1971).

[82] S.O.Lündqvist, Ark. Fysk. 12, 263 (1957).

[83] L.A.Feldkamp, J.Phys. Chem. Solids, 33, 711 (1972).

[84] R.K. Singh, in "Current Trends in Lattice Dynamics" Ed. K. R. Rao (BARC, 1979) p. 137 and references therein.

Chapter 3

EXPERIMENTAL SURVEY

Contents

List of Figures

Chapter 3

EXPERIMENTAL SURVEY

The interest in the experimental study of phonon dispersion relations in metals is basically for two reasons: (i) To know the nature and spacial variation of interatomic forces and (ii) to know the relation between the interatomic forces, phonon anomalies and superconducting transition temperature. The bulk phonons are measured by thermal neutron scattering[1] and diffuse X-ray scattering[2]. However, the surface phonons are measured by low energy electron diffraction[3] and high resolution elastic scattering of He-atoms [4]. We first describe the principles of these methods and then tabulate the literature and discuss the special features of phonon dispersion relations of these materials.

3.1 Thermal Neutron Scattering

The energy range of thermal neutrons is from 1-100 meV. A neutron energy of 10 meV has a speed of 1.38 Km/sec and a wavelength of $2.68 A^0$ which is of the order of interatomic spacing in a solid. In the simplest case of a monatomic metal, the atomic nuclei are arranged in the periodic lattice. The neutron waves which are scattered from one nucleus may interfere with neutron waves which are scattered from other nuclei. If the atomic nuclei have zero spins and no isotopes, the scattering is coherent. If the atomic nuclei have spins and exist in different isotopic states, these atomic nuclei scatter independently because of their random distribution and this scattering is incoherent.

To begin with we represent the neutrons by a plane wave and consider the interaction between neutron wave $\exp(ikz)$ which is propagating in the z-direction and an isolated bound atomic nucleus. As a result of neutron-nucleus interaction, a scattered wave of the form $(A/R)\exp(i\vec{k}.\vec{R})$ is produced where A is the scattering amplitude. If there are several nuclei, there will be several scattered waves and these may interfere. If the atoms are frozen, the interference pattern will be governed by Bragg's law. But at any finite temperature, the atoms are vibrating about their mean positions. These periodic motions about the mean positions give rise to their own characteristic reflections in addition to the Bragg reflections. Thus one finds satellites around the Bragg peaks. By measuring the Doppler shift associated with these satellites, the phonon wave vector $\vec{q}$ and the corresponding phonon frequency $\omega_j(\vec{q})$ are determined.

The differential cross section, for an assembly of N fixed nuclei having same atomic number, can be written as

$$
\begin{aligned}
(d\sigma/d\Omega) &= \mid \sum_{l=1}^{N} A_l \exp(-i\vec{Q}\cdot\vec{R}_l) \mid^2 \\
&= \sum_l A_l^2 + \sum_{l>l'} A_l A_{l'} \exp[-i\vec{Q}\cdot(\vec{R}_l - \vec{R}_{l'})]
\end{aligned}
\tag{3.1}
$$

where $\vec{R}_l$ is the position vector of the lth nucleus, $\vec{Q} = \vec{k} - \vec{k}'$ and $\vec{k}$ and $\vec{k}'$ are incident and scattered neutron wave vectors respectively. Here the scattering amplitude from the lth nucleus A_l is assumed to be real. Averaging the scattered amplitude for each nucleus over the spin states and over the distribution of different isotopes, Eq. (3.1) can be rearranged as

$$
\begin{aligned}
(d\sigma/d\Omega) &= N <A^2> + \sum_{l>l'} <A>^2 \exp[-i\vec{Q}\cdot(\vec{R}_l - \vec{R}_{l'})] \\
&= \mid \sum_l^{N} A_c \exp(-i\vec{Q}\cdot\vec{R}_l) \mid^2 + N A_{in}^2.
\end{aligned}
\tag{3.2}
$$

where

$$
A_{in} = (<A^2> - <A>^2)^{(1/2)}, \quad A_c = <A> .
\tag{3.3}
$$

Here the scattering amplitudes of different nuclei are assumed uncorrelated. It is evident from the first term of Eq.(3.2) that in the total coherent scattering, the scattered wave amplitudes of different nuclei are involved. This remains true even if the nuclei are vibrating about their mean positions. Therefore coherent scattering reveals the information about the cooperative motion of different nuclei. The second term of Eq.(3.2) does not involve any phase relationship between scattering nuclei. Therefore the total incoherent scattering gives the information about the motion of one nucleus only.

Thermal neutrons interact with the lattice both elastically and inelastically. Consequently both the coherent and incoherent scatterings consist of elastic and inelastic parts. Elastic coherent scattering provides the information about crystal structure including its magnetic state. The temperature variation of elasic incoherent scattering is used to test the validity of a physical model for frequency distribution function. Inelastic coherent scattering determine the individual phonons and thus provides the mapping of phonon dispersion relations. The inelastic incoherent scattering determine directly the frequency distribution function. However from the experimental point of view, it is not always possible to make the clear cut separation between the above scattering processes. We discuss here the inelastis coherent and incoherent scatterings separately which are relevent to the study of phonons in transition metals and their compounds and alloys.

3.1.1 Inelastic Coherent Scattering

The neutron scattering is governed by the energy transfer

$$\hbar\omega = (\hbar^2/2m_N)(\vec{k'}^2 - \vec{k}^2)$$
$$= E' - E \tag{3.4}$$

and the momentun transfer

$$\hbar\vec{Q} = \hbar(\vec{k'} - \vec{k}), \tag{3.5}$$

where E and E' are the energies of incident and scattered neutrons. The detailed theoretical analysis [5] shows that the scattering cross section of these neutrons from a single crystal for coherent scattering is governed by the equation

$$(d^2\sigma/d\Omega d\omega) = (d^2\sigma/d\Omega d\omega)^{(0)} + (d^2\sigma/d\Omega d\omega)^{(1)}$$
$$+ \dots \tag{3.6}$$

The first term represents zero phonon contribution, second term represents one phonon contribution and so on. Here it is assumed that the crystal is at a temperature where nuclear spin or isotopic degenarcy is absent. van-Hove[6] has shown that,

$$(d^2\sigma/d\Omega d\omega)^{(0)} = (1/\hbar)(\kappa'/\kappa)N^2 \mid f^0(\vec{Q}) \mid^2 \Delta(\vec{Q})\delta(\omega), \tag{3.7}$$

where $\Delta(\vec{Q}) = 1$ if $\vec{Q} = \vec{G}$ is a reciprocal lattice vector, otherwise zero.

$$f^0(\vec{Q}) = \sum_{\kappa=1}^{r} A_c^\kappa \exp[-W_\kappa(\vec{Q})] \exp[-i\vec{Q} \cdot \vec{R}_0(\kappa)] \tag{3.8}$$

and

$$W_\kappa(\vec{Q}) = (1/2) < [\vec{Q} \cdot \vec{u}(l\kappa)]^2 > . \tag{3.9}$$

Here A_c^κ is the coherent scattering amplitude of nuclear species on the κth site and $\exp(-W_\kappa(\vec{Q}))$ is Debye-Waller factor. Equation (3.7) is the Bragg equation and is the basis for determining chemical and magnetic structures by neutron scattering method.

The first order term of Eq.(3.6) is expressed as

$$(d^2\sigma)/d\Omega d\omega)^{(1)} = (1/\hbar)(k'/k)N \sum_{\vec{q}j} \mid f^{(1)}(\vec{Q},\vec{q}j) \mid^2 \Delta(\vec{Q} - \vec{q})$$
$$\times([1+ < n(\vec{q}j) >_T]\delta(\omega + \omega_j(\vec{q}))$$
$$+ < n(\vec{q}j) >_T \delta(\omega - \omega_j(\vec{q}))), \tag{3.10}$$

where

$$f^{(1)}(\vec{Q},\vec{q}j) = \sum_\kappa (\hbar/2M_\kappa\omega_j(\vec{q}))^{1/2} [\vec{Q} \cdot \vec{e}(\vec{q}j,\kappa)]A_c^\kappa$$
$$\times \exp[-W_\kappa(\vec{Q})] \exp[-i\vec{Q} \cdot \vec{R}_0(\kappa)]. \tag{3.11}$$

Equation (3.10) is similar to Eq.(3.7) except that the singular scattering is governed by the equations,

$$\vec{k} - \vec{k}' = \vec{Q},$$

$$\omega = (1/\hbar)(E - E') = \pm\omega_j(\vec{q}). \qquad (3.12)$$

These are the conservation laws for momentum and energy transfer. Since the dispersion relations are invariant with respect to translation by $\vec{G}$, therefoe, $\vec{Q}$ is often reduced to a phonon wave vector $\vec{q}$ in the first Brillouin zone (reduced wave vector). Experimentally observed phonon frequencies are plotted against the components of the reduced wave vector defined as $\vec{q} = (2\pi/a)\vec{\xi}$ along the principal symmetry directions. Here $\vec{\xi}$, for example, along [110] direction is $\vec{\xi} = [\xi, \xi, 0]$.

In the first order process given by the second term of Eq.(3.6) the intensity is governed by $\mid f^{(1)} \mid^2$ either by one phonon creation ($\omega = -\omega_j(\vec{q})$) or by one phonon annihilation ($\omega = \omega_j(\vec{q})$). The temperature dependent factors associated with these two processes are different. The polarization of phonons involved in the scattering process is included through the factor $[\vec{Q} \cdot \vec{e}(\vec{q}j, \kappa)]$ in $f^{(1)}$. The coherently scattered neutrons involving the absorption or emission of one phonon stand out as sharp peaks over the smooth background which arises from incoherent processes and multiphonon coherent processes. The study of energy distribution of coherently scattered neutrons gives direct information about the phonon dispersion relations.

Another important feature of the experimentally determined phonon dispersion relations is the presence of anomalies at particular values of $\vec{q}$. The phonon anomalies may occur at the extermal portions of the Fermi surface [7] (Kohn anomalies) or at the separation of the nesting portions of the Fermi surface [8]. The discontinuities in the slope of the dispersion curves are expected to occur at points where the condition

$$\vec{Q} = \vec{q} + \vec{G} = \vec{k}_1 - \vec{k}_2 \qquad (3.13)$$

is satisfied. Here $\vec{k}_1$ and $\vec{k}_2$ define two points on the Fermi surface with parallel tangent planes. The locus of the phonon wave vectors satisfying Eq.(3.13) defines the so-called Kohn surface. The relative size and other experimental aspects of Kohn anomalies have been discussed by Stedman et al [9] and Weymouth and Stedman [10]. The weak anomalies are studied by plotting the mean slope $(\Delta\omega/\Delta q)$ of the measured phonon dispersion curves as a function of $\vec{q}$. These graphs render the kinks clearly visible for steep curves than a direct plot of the measured phonon frequencies. Taylor suggested that [11] the ionic valency is the dominant factor in the determination of magnitude of the phonon anomalies. However, Vosko et al [12] made an extensive stydy of phonon dispersion relations of monovalent, trivalent and tetravalent metals (Na,Al,Pb) and concluded that the magnitudes of Kohn anomalies depend mainly on crystal structure and to a lesser degree on the ionic valency.

3.1.2   Experimental Methods

(a) Triple Axis Spectrometer

The geometry of neutron scattering process for a particular lattice is shown in Fig. 3.1 and the schematic diagram of the triple axis spectrometer is shown in Fig.3.2 [13]. The incident wave vector $\vec{k}$ is varied by changing the angle Θ_M. The desired angle between $\vec{k}$ and a certain (hkl) plane of the crystal under study is obtained by properly setting the sample table i.e. by varying angle Ψ. By rotating the second arm, the angle Φ between $\vec{k}$ and $\vec{k}'$ is varied and by rotating the third arm, the scattered neutron beam in the direction $\vec{k}'$ is analysed. In this process neutron beam is scattered by three crystals, therefore the final intensity measured by the counter is not very high. Therefore many precautions are to be taken to keep the background low for an accurate determination of peak position.

The direction of phonons is randomly distributed in the reciprocal space and their energies may not cover the range which is necessary to draw the dispersion curves. Therefore, several methods are adopted to observe one-phonon coherent scattering process. The one, most often used, is the constant $\vec{Q}$ method. In this method a particular phonon $(\vec{q}j)$ in the irreducible part of the Brillouin zone is selected. This choice of $(\vec{q}j)$ decides which one of the $\vec{Q}(=\vec{q}+\vec{G})$ point in the reciprocal space is optimum for its observation. The choice of this $\vec{Q}$ point is governed by $\vec{k}$ and $\vec{k}'$ which is fixed by the choice of Θ_M and Θ_A respectively and also by the factor $f^{(1)}$ which is estimated from some simple model for lattice dynamics. The experimental variables $\vec{k}$ and $\vec{k}'$ and the crystal orintations are adjusted in such a way that $\vec{Q}$ remains constant. The scattered intensity is measured at series of ω values. When the value of ω matches with $\omega_j(\vec{q})$, both the conservation conditions (3.12) are satisfied and the scattering occures with an intensity proportional to $\mid f^{(1)} \mid^2$ which results in a peak in the scattered intensity as shown in Fig.3.3. The width of the peak is related to the resolution of the instrumrnt and in some cases by other effects such as anharmonicity. By repeating the experiment for the various positions of the crystal and measuring the scattered neutron intensity for various scattering angles one determines $\omega_j(\vec{q})$ for various values of $\vec{q}$. These measurements give the phonon dispersion relations.

The conservation conditions (3.12) specify only the wave vector and phonon frequency. The information about the phonon polarization is contained in the structure factor $f^{(1)}$ and therefore, it is fixed by measuring the intensity in different directions. However, as discussed in Chapter 2, while assigning the correct polarization vector to any particular branch of dispersion relations one often uses available theoretical results or makes use of some general results such as for small $\vec{q}$, the longitudinal acoustic waves have higher frequency than the transverse acoustic waves.

(b) Time of Flight (TOF) Method

The neutron wavelength $\lambda[= (h/m_N v) = (h/m_N l)t]$ is proportional to time t needed by the neutron to travel a distance l. This feature is used to measure the wavelength distribution of the scattered neutrons in this method. The schematic diagram for this method is shown in Fig.3.4. The pulsed monoenergetic neutron beam, scattered by the specimen, is further reflected by a single crystal to increase the resolution. These scattered neutrons are received by a neutron counter which is connected to a multichannel time analyser which records the intensity distribution of the scattered neutrons. The full spectral distribution of scattered neutrons is measured by continuously changing the wave vector of incident neutron beam.

Both the methods have merits and demerits [13]. However, their choice is decided by the problem of interest. In general, the triple axis spectrometer is more favourable for phonon dispersion relation measurements while the time of flight method is more useful for frequency distribution measurements.

3.1.3   Inelastic Incoherent Scattering

In nature hydrogen and vanadium are incoherent scatterers. Therefore incoherent scattering has been utilised to study the narrow bands in the frequency spectra or the local modes of the materials which contain these elements [14]. To understand the principle of this method, we consider the following expansion of the scattering cross section which is similar to Eq.(3.6) [15] i.e.

$$\left(\frac{d^2\sigma}{d\Omega d\omega}\right)_i = \left(\frac{d^2\sigma}{d\Omega d\omega}\right)_i^{(0)} + \left(\frac{d^2\sigma}{d\Omega d\omega}\right)_i^{(1)} + \ldots \qquad (3.14)$$

where

$$\left(\frac{d^2\sigma}{d\Omega d\omega}\right)_i^{(0)} = \frac{1}{\hbar}\frac{k'}{k}N\sum_\kappa \mid A_i^\kappa \mid^2 \exp[-2W_\kappa(\vec{Q})]\delta(\omega), \qquad (3.15)$$

$$\begin{aligned}
\left(\frac{d^2\sigma}{d\Omega d\omega}\right)^{(1)} &= \frac{1}{\hbar}\frac{k'}{k}N\sum_\kappa \mid A_i^\kappa \mid^2 \exp[-2W_\kappa(\vec{Q})]\left(\frac{3r\hbar}{2M_\kappa \mid \omega \mid}\right)\\
&\times \left([\vec{Q}\cdot\vec{G}_+^\kappa(\omega)\cdot\vec{Q}]\frac{1}{\exp(x)-1}\right.\\
&\quad + \left.[\vec{Q}\cdot\vec{G}_-^\kappa(\omega)\cdot\vec{Q}]\frac{\exp(x)}{\exp(x)-1}\right),
\end{aligned}$$

$$(3.16)$$

and $x = (\hbar \mid \omega \mid /k_B T)$. The elements of tensor $G_\pm^\kappa(\omega)$ are defined as,

$$\left(G_\pm^\kappa\right)_{\alpha\beta} = \frac{1}{3rN}\sum_{\vec{q}j}e_\alpha^*(\vec{q}j,\kappa)e_\beta(\vec{q}j,\kappa)\delta(\omega \mp \omega_j(\vec{q})), \qquad (3.17)$$

which possess the following properties

$$\left(G_+^{\kappa}\right)_{\alpha\beta} = \left(G_-^{\kappa}(-\omega)\right)_{\alpha\beta}, \tag{3.18}$$

and

$$\sum_{\kappa\alpha}\left(G_+^{\kappa}\right)_{\alpha\alpha} = \frac{1}{3rN}\sum_{\vec{q}j}\delta(\omega - \omega_j(\vec{q}))$$

$$= g(\omega). \tag{3.19}$$

The zero phonon term (3.15) represents elastic scattering and does not provide any information about the lattice dynamics except Debye-Waller factor which is very useful in certain instances [16]. The one phonon incoherent scattering is proportional to $G^{\kappa}(\omega)$ which is related to the phonon density of states $g(\omega)$ as evident from Eq.(3.19). Thus the characteristics of $g(\omega)$ may be studied by incoherent scattering. Figure 3.5 shows an example of incoherent scattering spectra. Figure 3.5a shows the $g(\omega)$ versus $\hbar\omega$ for vanadium [17] and Fig.3.5b for $\beta-NbH_{0.95}$ [18].

3.2   Thermal Diffuse X-Ray Scattering

The principles of X-ray scattering and neutron scattering are very similar [2]. Let $\vec{k}$ is the wave vector associated with the incident X-rays of energy E and wave vector $\vec{k}'$ is associated with the scattered X-rays of energy E'. In the inelastic scattering process, the momentum conservation condition is the same as given in Eqs.(3.5) and (3.12). However, the energy conservation condition becomes,

$$E' - E = \hbar c(\vec{k}' - \vec{k}) = \pm\hbar\omega, \tag{3.20}$$

where c is the velocity of electromagnetic radiation. In the interaction of X-rays with the isolated atoms, the X-rays are mainly scattered by the core electrons. Thus the X-ray scattering by an atom is characterized as the scattering from the rigid charge distribution for the unpolarized incident radiation. The amplitude of scattered radiations $A^{\kappa}(\vec{Q})$ is given as [19]

$$A^{\kappa}(\vec{Q}) = \frac{e^2}{mc^2}\left(\frac{1 + \cos^2\theta}{2}\right)f^{\kappa}(\vec{Q}), \tag{3.21}$$

where

$$f^\kappa(\vec{Q}) = \int \rho^\kappa(\vec{r}) \exp(-i\vec{Q}\cdot\vec{r})d\vec{r} \qquad (3.22)$$

is the Fourier transform of electron density $\rho^\kappa(\vec{r})$ and θ is the scattering angle. The X-ray scattering cross section is the same as given in Eq.(3.6) except that $A_c^\kappa(\vec{Q})$ is replaced by $A^\kappa(\vec{Q})$. Thus the one phonon part is the same as given in Eq.(3.10). For X-rays $\hbar\omega \approx 10^{-6}E$, therefore from Eq.(3.4), the scattered wave vector $\vec{k}'$ has very closely the same magnitude as $\vec{k}$ because $(E'-E)$ is very small. The scattering surface is then the Ewald's sphere (Fig.3.1), and the intensities corresponding to phonon creation and annihilation processes appear at the same point in the reciprocal space. Moreover the phonons of all the modes subject to the conservation laws (3.12) contribute towards the intensity of diffuse Bragg spots. For thsese reasons, the analysis of the intensity of diffuse Bragg spots to obtain the phonon dispersion relations becomes quite difficult

. Using Eq.(3.21) in Eqs.(3.10) and (3.11) the expression for X-ray intensity due to one phonon scattering process (per unit cell in electron volt units, i.e. $(e^2/mc^2)^{1/2}(1+\cos^2\theta)$), is written as [20]

$$
I(eV/cell) \;=\; \sum_{j=1}^{3r} \frac{E_{\vec{q}j}}{\omega_j^2(\vec{q})} \; | \sum_\kappa \frac{\vec{Q}\cdot\vec{e}(\vec{q}j,\kappa)}{M_\kappa^{(1/2)}} f^\kappa(\vec{Q}) \exp(-W_\kappa(\vec{Q}))
$$
$$
\times \;\; \exp(-i\vec{Q}\cdot\vec{R}_0(\kappa)) \;|^2 \qquad (3.23)
$$

where

$$E_{\vec{q}j} = \hbar\omega_j(\vec{q})[(1/2) + <n(\vec{q}j)>_T]. \qquad (3.24)$$

The values of $E_{\vec{q}j}$ are obtained by iterative procedure. Assuming $E_{\vec{q}j} = k_BT$, Eq.(3.23) is solved for $\omega_j(\vec{q})$ which is further substituted in Eq.(3.24) for new value of $E_{\vec{q}j}$. This $E_{\vec{q}j}$ is again inserted in Eq.(3.23) and it is solved for $\omega_j(\vec{q})$. This procedure is repeated till two successive values of $\omega_j(\vec{q})$ agree within the defined error limits. In addition to the above given first order term, the total intensity of diffuse Bragg spots also consists of Compton scattering [21] and higher order phonon contributions [22]. These contributions are calculated and subtracted from the total intensity before using Eq.(3.23) to compute phonon dispersion curves.

Equation (3.23) is greately simplified for a given crystal structure and for a particular mode of vibration. For example for a monatomic cubic crystal and for longitudinal wave along [100] direction, Eq.(3.23) simplify as

$$I_1 = \frac{1}{M} \mid f \mid^2 \exp(-2W) \mid \vec{Q} \mid^2 \left(E_{\vec{q}l}/\omega_l^2(\vec{q})\right) \tag{3.25}$$

and thereby gives the phonon dispersion curves for longitudinal waves. The transverse modes are obtained by measuring intensities along an appropriate direction in the reciprocal space at right angle to the symmetry axis. In this case $\vec{Q}$ makes an angle ϕ relative to the original longitudinal direction and Eq.(3.23) becomes

$$I_1 = \frac{1}{M} \mid f \mid^2 \exp(-2W) \mid \vec{Q} \mid^2 \left(E_{\vec{q}l} \sin^2 \phi/\omega_l^2(\vec{q}) + E_{\vec{q}t} \cos^2 \phi/\omega_t^2(\vec{q})\right). \tag{3.26}$$

With the help of previously determined dispersion curves for the longitudinal modes, one may obtain the corresponding dispersion curve for the transverse modes.

The typical measurements for [100]L mode for V metal are shown in Fig.3.6 [23].

3.3   Low Energy Electron Thermal Diffuse Scattering

Thermal diffuse scattering of low energy electrons is used to study the phonon dispersion relations from the surface of crystalline solids. In the thermal diffuse scattering process, it is assumesd that the atoms only in the outer most layer (i.e.on the surface) of the sample contribute towards the scattering [24], [25]. The appropriate lattice is then a family of rods (hk) normal to the crystal surface. The geometry of the surface scattering is shown in Fig.3.7. Let the energy of the incident electron be E and the wave vector component parallel to the surface be $\vec{k}_{||}$. Let the electron strikes the crystal and scatter by a process that involves emission of a phonon of frequency $\omega_j(\vec{q}_{||})$ with phonon wave vector component $\vec{q}_{||}$ parallel to the surface. The momentum and enegy conservation laws demand

$$\vec{k}'_{||} = \vec{k}_{||} - \vec{q}_{||} + \vec{G}_{||} \tag{3.27}$$

and

$$E' = E - \hbar\omega_j(\vec{q}_{||}). \tag{3.28}$$

E' and $\vec{k}'_{||}$ are the energy and wave vector of the scattered electron. $\vec{G}$ is the appropriate two dimensional reciprocal lattice vector which keeps $\vec{k}'_{||}$ in the first Brillouin zone. The state $|\vec{k}'_{||} >$ will also consist of linear combination of all possible plane waves with energy E' and wave vector $(\vec{k}'_{||} + \vec{G}_{||})$ parallel to the surface. A plane wave charaterized by $\vec{G}_{||}$ such that

$$E' - \frac{1}{2m}\left(\vec{k}'_{||} + \vec{G}\right)^2 > 0 \tag{3.29}$$

may also occur in the propagating part of the outgoing wave provided it is allowed by the boundary conditions to get mixed with the wave function. For a given phonon wave vector $\vec{q}_{||}$ and phonon frequency $\omega_j(\vec{q}_{||})$ involved in the scattering geometry, the direction of the outgoing wave is determined by satisfying the momentum and energy conservation conditions (3.27) and (3.28) along with the condition

$$\vec{k}'_z(\vec{G}'_{||}) = [2mE' - |\vec{k}'_{||} + \vec{G}'_{||}|^2]^{1/2}, \tag{3.30}$$

and the direction of incident beam is determined with the condition

$$k_z = -[2mE - |\vec{k}_{||}|^2]^{1/2}. \tag{3.31}$$

The subscript z denotes the z-component of the wave vector. With these considerations, which are similar to those proposed by Mills et al [26], Roundy and Mills [27] gave a theoretical discussion for the inelastic scattering of low energy electrons by crystal surface involving one-phonon processes.

Let the incident and scattered wave functions be defined as

$$\psi^{(0)}_{\vec{k}_{||},E}(\vec{r}_{||}, z) = \exp(i\vec{k}_{||} \cdot \vec{r}_{||})u_{\vec{k}_{||},E}(\vec{r}_{||}, z) \tag{3.32}$$

and

$$\psi^s_{\vec{k}_{||},E_s}(\vec{r}_{||},z) = \exp(i\vec{k}'_{||}\cdot\vec{r}_{||})u_{\vec{k}'_{||},E_s}(\vec{r}_{||},z) \tag{3.33}$$

respectively where $\vec{r}_{||}$ is the position vector of the electron in the metal surface and $u_{\vec{k}_{||},E}$ are the periodic parts of the wave function. Taking into account the umklapp processes, the scattered wave function outside the crystal is written as,

$$\psi^s_{\vec{k}'_{||}}(\vec{r}_{||},z) = \frac{1}{\sqrt{\Omega_s}}\sum_{\vec{G}_{||}}\exp\left(i(\vec{k}'_{||}+\vec{G}_{||})\cdot\vec{r}_{||}\right)\exp(ik'_z z)A_{\vec{k}'_{||}}(\vec{G}_{||}),$$

$$\tag{3.34}$$

where $A_{\vec{k}'_{||}}$ is the amplitude of $\psi^s_{\vec{k}'_{||}}(\vec{r}_{||},z)$ and Ω_s is the volume of the two dimensional crystal in question. Use of Eqs.(3.32) and (3.33) gives the scattering matrix element as

$$M_{lz}(\vec{k}'_{||},E_s;\vec{k}_{||},E_0) = \frac{i\Omega_s}{N_sA_s}\sum_{q_z}V_{lz}(\vec{K})\exp(-iq_z l_z)\int_{unit\ cell}\frac{d^2\vec{r}_{||}}{A_s}$$

$$\times\ \int_{-\infty}^0 dz u_{\vec{k}'_{||}}(\vec{r}_{||},z)\exp(iq_z z)\exp(i\vec{G}_{||}\cdot\vec{r}_{||})u_{\vec{k}_{||}}(\vec{r}_{||},z)$$

$$\tag{3.35}$$

where $V_{lz}(\vec{K})$ is the Fourier transform of electron-ion potential of metal surface, q_z is the z-component of the phonon wave vectors, A_s is the area of unit cell and N_s is the number of atoms in the surface layer. The components of the vector $\vec{K}$ are

$$\vec{K}_{||} = \vec{k}_{||}-\vec{k}'_{||}+\vec{G}'_{||}; K_z = q_z. \tag{3.36}$$

Use of Eqs.(3.34) and (3.35) gives the expression for the scattering cross section as

$$\left(\frac{d^2\sigma}{d\Omega dE'}\right) = \frac{mN_sk'}{8\pi^2 M}\mid A_{\vec{k}'_{||}}(\vec{G}_{||})\mid^2\sum_{l_z,l'_z}M_{lz}(\vec{k}'_{||},E';\vec{k}_{||},E)$$

$$\times\ M^*_{l'z}(\vec{k}'_{||},E';\vec{k}_{||},E)[1+<n(\vec{q}j)>_T]\frac{1}{\omega_j(\vec{q}_{||})}$$

$$\times\ [\vec{Q}\cdot\vec{e}(\vec{q}_{||}j,l_z)][\vec{Q}\cdot\vec{e}(\vec{q}_{||}j,l'_z)]\rho(\vec{q}_{||},\omega_j(\vec{q}_{||})), \tag{3.37}$$

where M is the ionic mass, $\vec{Q}(=\vec{q}_{||}+\hat{z}(k_z-k'_z))$ is the scattering vector and $\rho(\vec{q}_{||},\omega_j(\vec{q}_{||}))$ is the surface phonon density of states with frequency $\omega_j(\vec{q}_{||})$. The expression (3.37) is similar to Eq.(3.10).

Equation(3.37) is further simplified to analyse the shape of energy loss spectrum and its dependence on lattice dynamical properties of the surface region for a model crystal. It is suggested that the energy loss spectrum for a general value of momentum transfer consists of one or more lines which arise due to scattering from the surface modes and a band which has its origin in the scattering produced by the bulk modes. The shape of the band and the position of the peaks were found highly sensitive to atomic force constants in the surface layer. However, the resolution of the electron spectrometer remains a problem for observing these phonons.

The first high resolution (0.7 meV) loss spectra from Ni[100] surface with high impact energies (180 and 320 eV) was reported by Lehwald et al [28]. The experimental details of double-pass electron spectrometer are discussed in Ref.[3]. The sample spectra with $|\vec{q}_{||}| = |\vec{k}_{||}^{(0)} - \vec{k}' - \vec{G}_{||}^{(01)}| = 0.4A^{0-1}$ and $q_{||} = 1.26A^{0-1}$ ($\bar{X}$- point of the two dimensional Brillouin zone) are shown in Fig.3.8. The two peaks correspond to the phonon emission and absorption processes. The peak positions as a function of $q_{||}$ are shown in Fig.3.9. The data coincided with the predictions of Allen et al [29]. Similar measurements were also carried out for Cu[111] surface [30].

3.4   He Atom Inelastic Scattering

The measurements of surface phonons by He atom scattering is very similar to bulk phonon maping with neutron scattering [4]. The thermal beam of He atoms is scattered from the exponentially decaying outer fringes of the surface charge density profile. The conventional diffraction occures because the in-plane charge distribution reflects on the surface reciprocal net. If the energy of thermal He atoms is kept as low as 0.25 meV and inelastic scattering events are studied keeping the magnitude of parallel momentum transfer less than that of a surface reciprocal vector, the surface phonon peaks are observed between the

Bragg peaks. The momentum transfer is set by fixing incoming and outgoing beam directions with respect to the crystal normal. The kinetic energy of scattered atoms is measured by the time of flight analysis.

A typical time of flight spectra for Ag[111] [31] taken on both the sides of the specular angle peak at $\theta_i = 45^0$ along the $< 11\bar{2} >$ azimuth is shown in Fig.3.10. The arrows above the spectra indicate the significant inelastic structures and the arrows at the bottom indicate the incoherent elastic peaks which are attributed to the crystal imperfections. The first sharp inelastic peak is attributed to the surface Rayleigh mode and the other maxima lie in the bulk region. As the angle moves away from the specular angle ($\theta_i = 45^0$), the regions of the phonon spectrum of larger $\mid \vec{Q} \mid$ and $\mid \hbar\omega \mid$ are probed. The complete measurements for surface phonons along with the calculations of surface phonon spectrum of Ag(111) are shown in Fig.3.11. The phonon measurements for Ni(111) surface are carried out by Feuerbacker and Wills [32].

3.5   Tabulation of Experimental Data

The compilation of experimental data of phonon dispersion curves is presented by Kress [33]. Here we tabulate the list of references and the prominent features of the phonon dispersion curves of transition metals and their compounds and alloys. The description of the tabulation is as follows:

1. Before the first slash: Name of the material with references in the brackets.

2. Between the first and second slashes: Symmetry directions along which the measurements are available.

3. Between the second and third slashes: Symmetry directions and the phonon modes which exhibit anomalies.

4. Between the third and fourth slashes: Lattice dynamical models which are used to fit the experimental data.

5. After the fourth slash: additional comments.

Most of the bulk phonon measurements are carried out by thermal neutron scattering technique except a few one. For example, the phonons in Cu and V are also measured by diffue X-ray scattering. The surface phonons are measured by low energy electron diffuse scattering (LEEDS) and He atom scattering (HEAS). The different lattice dynamical models which are used to fit the experimental phonon dispersion curves are abbreviated as follows:

GTF: Generalized tensor force model (BvK model).

AS: Axially symmetric model.

MAS: Modified axially symmetric model.

FA: Fourier analysis (planer force constant model).

CFM: Central force model.

CF: Charge fluctuation model.

SM: Shell model.

SSM: Screened shell model.

DSM: Double shell model.

BSM: Breathing shell model.

RIM: Rigid ion model.

3.5.1 Elemental Metals

(i) fcc Structure

Ag [34]-[36]/ [001], [110], [111], [012] /-/GTF.

Au [37]/ [100], [110], [111], [201]/ [100]T_1 / GTF, AS, combination of GTF and AS.

$\gamma - Ce$[38]/ [001], [110], [111], [012]/ [110]L, [110]T_1, T_2, [111]T / GTF/ fcc to hcp phase transition at T$\approx$ 260^{0}K.

Cu[39] -[47]/ [001], [110], [111]/ [001]L, [110]T/ GTF, AS, GTF combined with AS/ About 20 weak anomalies are observed.

La[48]/ [001], [110], [111], [012]/ [110]L, [110]T/ GTF/ fcc to hcp phase transition at 583^{0}K.

Ni[49]- [51]/ [001], [011], [111], [012]/ [011]T_1/ GTF, AS/ Exists in both the paramagnetic and ferromagnetic phases.

Pd[46], [47]/ [100], [110],[111]/ [110]T_1/ GTF/ paramagnetic phase.

Pt[52], [53]/ [100], [110], [111]/ [110]T_1/ GTF/ paramagnetic phase.

Th[54]/ [001], [011], [111], [012]/ [011]L, [011]T_1, T_2/ GTF,FA/ .

Yb[55]/ [011], [110], [111], [012]/ [110]T_1/ GTF/ .

(ii) bcc Structure

Cr[56]- [58]/ [001], [011], [111]/ [011]T, [011]T_2, [111]L, [111]T/ GTF/ Exists in both the para and antiferromagnetic phases with transition at 311^0K.

Fe[59]- [61]/ [100], [110], [111], [221], [0.50.51]/ -/ GTF, CFM/ .

Mb[62]- [64]/ [001], [110], [111]/ [001]L, [001]T, [110]T, [111]L, [111]T/ GTF, AS/ Phonon anomalies due to nesting of the Fermi surface.

Nb[63], [65]- [68]/ [100], [110], [111], [112], [0.50.51]/ [100]L, [110]T_2, [111]L/ GTF/ .

Ta[69]/ [001],[110], [111]/ [001]L, [001]T.[110]T_2/ GTF, AS/ .

Tl[70]/ [001], [110], [111]/ [110]L, [110]T_1, [111]L/ -/ bcc to hcp phase transition at T $\approx 503^0$K.

V[71]- [77]/ [100], [110], [111]/ [100]L, [110]T_2, [111]L, [111]T/ GTF/ .

W[78]/ [001], [110], [111]/ [110]T_2/ GTF.

$\beta - Zr$[79]/ [001], [110], [111]/ [111]L/ -/ bcc to hcp phase transition at T=1340^0K.

(iii) hcp Structure

Cd[80]/ [100], [110], [001]/ [110], [100]/ Model pseodpotentials are used to explain the experimental data.

Co[81]/ [100], [110], [001]/ -/ GTF, CFM/ .

Dy[82]/ [001], [110]/ -/ -/ Ferromagnetic phase, phonon and magnon modes intersect.

Hf [81], [83]/ [001], [110], [100]/ [001]LO/ MAS.

Ho[84], [85]/ [100], [110], [001], [0.501], [110.5], [100.5], $[\frac{1}{3}, \frac{1}{3}\, 1]$ / GTF/
.

Re[81], [86]/ [001]/ [001]LO / CFM/ .

Ru[86]/ [001]/ [001]LO/ CFM/ .

Sc[87]/ [001], [100], [110]/ [001]LA/ MAS/ Acoustic isotropy observed.

Tb[88]/ [001], [100], [110], [110.25]/ [001]LA/ GTF/ .

Tc[81], [86], [89], [90]/ [001], [110], [100]/ [001]LO, [001]TO/ GTF,
CFM/ .

Ti[81], [91]/ [001], [100], [110]/ [001]LO/ MAS, CFM/ .

Tl[88], [92], [93]/ [100], [001], [110]/ -/ GTF, MAS/ .

Y[94]/ [001], [100], [110]/ [001]LO/ MAS/ Acoustic isotropy observed.

Zn[92], [95]/ [001], [100], [110]/ -/ MAS/ .

Zr[96], [97]/ [001], [100], [110]/ [001]LO, [110]T_1/ MAS/ .

(iv) Orthorhombic Structure

$\alpha - Ce$[98]/ [001], [110], [111], [012]/ [111]T/ GTF, CFM/ Exhibits fcc
to dhcp phase transition at 260^0K.

$\alpha - U$[99], [100]/ [100], [010], [001]/ [100]LO, [100]T/ SM/ .

3.5.2   Metallic Compounds

(i) NaCl Structure

LaS[101]/ [001], [110], [111]/ In all the longitudinal modes / DSM/ Optical modes have very small dispersion. LO and TO modes are nearly
degenerate.

NbO[102]/ [100], [110]/ -/ GTF/ .

$NdSb$[103]/ [001], [110], [111]/ [001]L/ SM/ .

SmS[104] / [001], [011], [111]/ [111]L/ -/ .

$Sm_{0.75}Y_{0.25}S$[105], [106]/ [001], [110], [111]/ Softening in all the longidu-
dinal modes/ BSM/ .

ThP[107]/ Observed few phonon peaks/ -/ -/ .

ThS[107]/ Observed few phonon peaks/ -/ -/ .

$TmSe$[108], [109]/ [001], [110], [111]/ -/ GTF/ .

UAs[110]/ [001], [110], [111]/ -/ AS/ . The frequencies of transverse
optical modes are higher than the frequencies of longitudinal modes.

UP[107]/ Observed few phonon peaks/ -/ -/ .

US[111]/ [001], [110], [111]/ -/ -/ .

USb[110]/ [001], [110], [111]/ -/ AS/ Phonon frequencies do not change
appreciably with temperature.

USe[112]/ Observed few phonon modes/ -/ -/ .

UTe[112]/ [111]/ -/ -/ .

YS[113]/ [001], [110], [111]/ [001]L, [110]L, [111]L/ DSM/ .

(ii) Cu$_3$Au Structure (Interpenetrating fcc)

$CePd_3$[114]/ [100], [110], [111]/ -/ GTF/ .

$CeSn_3$[115]/ [001], [110], [111]/ -/ BSM/ .

Cu_3Au[116], [117]/ [001], [011], [111], [012]/ -/ GTF/ .

$Cu_{0.7}Fe_{0.3}$[118]/ [001], [110], [111]/ -/ GTF/ No evidence of strong
magnetic flux.

Cu_3Zn[119]/ [001], [011], [111], [012]/ -/ GTF/ .

$FePt_3$[120], [121]/ [001], [110], [111]/ [111]L, [111]T/ GTF/ Paramagnetic phase.

$FePt_3$[120], [121]/ [001], [110], [111]/ [111]L, [111]T/ GTF/ Ferromagnetic phase.

$LaSn_3$[122]/ [001], [110], [111]/ -/ BSM/ .

Ni_3Al[123]/ [100], [110], [111]/ - / GTF/.

$Ni_3Fe, Ni_{0.3}Fe_{0.7}, Ni_{0.5}Fe_{0.5}$[119]/ [001], [011], [111], [012]/ -/ GTF/ .

Pd_3Fe [124]/ [001], [110]/ [110]L/ GTF/ .

Pt_3Mn[125]/ [100], [110], [111]/ [111]L, [111]T/ GTF/ .

(iii) Other Compounds

$CeAl_2$[126]/ [100], [110], [111]/ [111]L, [111]T/ AS/ .

$Cu_2Mo_6S_8$[127]/ [100], [110], [111], [1$\bar{1}$ 0]/ [111]L, [111]T/ -/ .

$Cu_{0.53}Zn_{0.47}, \beta - Brass$ [128]/ [001], [110], [111], [0.50.51]/ -/ GTF/ .

Fe_3Al[129]/ [100], [110], [111]/ [111]L, [111]T/ GTF/ .

$Fe_{0.40}Zr_{0.60}$ [130]/ Few phonon peaks are measured/ -/ -/ .

$Fe_{0.5}Ti_{0.5}$ [131]/ [100], [110], [0.510], [0.50.51], [111]/ -/ GTF/ .

$LaAg$[132]/ [001], [110], [111]/ [110]L, [110]T/ GTF/ .

$LaAg_{0.8}In_{0.2}$ [132]/ [110]/ -/ -/ Cubic to tetragonal phase transition observed near T=142^{0}K.

$LaAl_2$[133], [134]/ [100], [011], [111]/ -/ GTF.

LaB_6, YB_6[135]/ Few phonons observed/ -/ -/ .

$Mn_{0.85}Ni_{0.09}Co_{0.06}$ [136]/ [100], [110], [111]/ -/ -/ .

Mo_6Se_8 [127]/ [001], [110], [1 $\bar{1}$0], [111]/ -/ -/ .

$NiAl$ [137]/ Few phonon peaks observed/ -/ -/ .

$NiTi$ [138], [139]/ [100], [110], [111]/ [100]T, [110]T/ GTF/ .

$Ni_{46.8}Ti_{50.0}Fe_{3.2}$ [140]/ [110]/ [110]T/ -/ .

Ti_2O_3 [141]/ [100], [001], [110]/ [110]TA/ -/ Metal to semimetal phase transition between $T = 600^0K$ to 300^{0}K.

$Tl_{0.33}WO_3$ [133], [134]/ [001], [110]/ [001]/ -/ .

UAl_2 [142]/ [100], [110], [111]/ -/ -/ .

V_2O_3 [143]/ [100], [110], [112]/ [110]T/ -/ Metal to insulator phase transition at $T = 160^0K$.

YAl_2 [133], [134]/ [100], [011], [111]/ -/ GTF/ .

YZn [144]/ [001], [110], [0.50.51], [111]/ -/ GTF/ .

(iv) Layered and Linear (A-15) Conductor Compounds

Cr_3Si [145], [146]/ [100], [011], [111]/ -/ AS/ .

Nb_3Al [147]/ Few phonon peaks observed/ -/ -/ .

Nb_3Sb [148]/ [110], [001], [111]/ [110] optical modes/ GTF/ .

$Nb_{3.1}Ga_{0.9}$ [149]/ [100], [110], [111]/ -/ -/ .

Nb_3Ge [150]/ Few phonon peaks observed/ -/ -/ .

$NbSe_2$ [151]/ [100], [001], / [100]/ -/ .

Nb_3Sn[152], [153]/ [100], [110], [111]/ [111]/ GTF/ The martensitic transition temprature $T = 46^0K$ for cubic to slightly distorted tetragonal phase transition.

TaS_2 [154]/ [001], [100]/ [100]L/ -/ .

$TaSe_2$ [151]/ [100], [001], [100.5]/ [100]/ -/ .

$TiSe_2$ [155]/ [001], [100], [110]/ -/ GTF/ .

V_3Ga [156]/ Few phonon peaks observed/ -/ -/ .

V_3Ge [157]/ [100], [111]/ [111]L/ -/ .

$V_3Ge_{0.75}Al_{0.25}$/ Few phonon peaks observed/ -/ -/ .

V_3Si [145], [146]/ [100], [110]/ -/ AS/ .

3.5.3　Metallic Carbides, Nitrides and Hydrides

(i) Carbides

HfC [159]/ [001], [110]/ [001]LA, [110]LA, [110]TA, [110]TO/ GTF, SSM, Pseudopotential/ .

NbC [160]/ [001], [110], [111]/ [001]LA, [110]LA, [111]LA/ DSM/ .

TaC [159], [166]/ [001], [110]/ [001]LA, [110]LA, [110]TA, [110]TO/ GTF, SSM, pseudopotential/ .

$ThC_x, x = 0.063$ [161]/ [001], [011], [111], [012]/ -/ GTF/ .

$TiC_x, x = 0.95, 0.89$ [162]/ [100], [110], [111]/ [001]LA, [110]LA, [111]LA/ SSM/ .

UC [163], [164]/ [001], [110], [111]/ [110]/ GTF/.

VC [165]/ [100], [011], [111]/ [011]TA, [011]TO, [100]T/ SSM/ .

ZrC [166]/ [001], [110], [111]/ [001]LA, [110]LA, [111]LA/ GTF/.

(ii) Nitrides

HfN [167]/ [001], [110], [111]/ [001]LA, [110]LA, [110]TA, [111]LA / DSM/ The frequencies of TO mode in the [111] direction are higher than the frequencies of LO modes.

$\delta - NbN_x, x = 0.93$ [168]/ [001], [110], [111]/ [001]LA, [001]TA, [110]LA, [110]TA/ SSM, DSM/.

ThN [169]/ Few phonon peaks are measured/ .

$TiN_x, x = 0.98$ [170]/ [001], [110], [111]/ [001]LA, [110]LA, [111]LA, [111]TA/ SSM, DSM/.

UN [171]/ [001], [110], [111]/ [110]LA/ .

$VN_x, x = 0.86$ [172]/ [100], [110], [111]/ [100]LA, [100]TA, [110]LA, [110]TA/ -/.

$ZrN_x, x = 0.93$ [173]/ [001], [110], [111]/ [001]LA, [110]LA, [111]LA/ DSM/.

(iii) Hydrides

$CeD_x, x = 2.12, 2.72$ [174]/ [001], [110], [111]/ -/ GTF/ .

$NbH_x, x = 0.78, 0.82$ [175], [176]/ [100], [110], [111]/ [100], [110]/ GTF/

.

$NbD_x, x = 0.15, 0.40, 0.60, 0.85$ [177]/ [100], [110], [111]/ [100]L, [100]T, [110]L, [110]T/ GTF/ .

$NbD_x, x = 0.2, 0.71, 0.75$ [178], [179]/ Description is the same as in ref.[177].

$PdH_x, x = 0.03$ [180]/ [110]/ [110]T_1/ -/ .

$PdD_x, x = 0.63, 0.74, 0.78$ [181], [182]/ [100], [110], [111]/ [100]TO,

[110]TO/ GTF/ .

$TaH_x, x = 0.18$ [183], [184]/ [100], [110], [111]/ [100]L, [100]T, [110]L/.

$TaD_x, x = 0.18, 0.22$ [175], [183], [184]/ [100], [110], [111]/ [100]L, [100]T, [110]L, [110]T/ GTF/ .

$VD_x, x = 0.70$ [185]/ [001], [110], [111]/ -/ -/.

(iv) Others

In the following hydrides only local phonon modes are observed.

$FeTiH_x, x = 0.06, 0.94, 1.4$ [186].

$LaD_x, x = 2.9$ [187].

$LaNiH_x(D_x)$ [188].

$NdH_x, x = 2, 06, 2.62$ [189].

$Pd_{1-x}Ag_xH_y, x = 0.06, y = 0.4$ [190].

$Pd_{1-x}Ce_xH_y, x = 0.06, y = 0.4$ [190].

$PrH_x, x = 1.98$ [189].

$PrD_x, x = 1.7, 2.3$ [190].

ThH_2, Th_4H_{15} [191], [192].

$TiH_x, x = 0.05, 0.07, 0.14$ [193], [194].

$TiD_x, x = 0.09$ [193], [194].

$TiCuH_x$ [195].

$YH_x, x = 0.2, 2, 3$ [196], [197].

$YbH_x, x = 1.97$ [198].

$YD_x, x = 2, 3$ [197].

3.5.4   Disordered Alloys

(i) fcc Host Lattice (Solid Solution Phase)

(Directions are assigned in the virtual crystal approximation)
$Au_{1-x}Ni_x, x = 0.4$ [199]/ [100], [110], [111]/ [100]T/ GTF/ .
$Co_{1-x}Fe_x, x = 0.08$ [200]/ [001], [110], [111]/ -/ GTF, AS/ .
$Cu_{1-x}Au_x, x = 0.03$ [201]/ Few phonon peaks are observed/ .
$Cu_{1-x}Ni_x, x = 0.1$ [202]/ [001], [110], [111]/ -/ -/ .
$Cu_{1-x}Pt_x, x = 0.05, 0.25$ [203]/ [001]/ -/ -/ Double peak structure is
observed in the T-modes.
$Ni_{1-x}Co_x, x = 0.10, 0.25$ [204]/ [001], [110], [111]/ [001]L, [111]L/ FA/
.

$Ni_{1-x}Cr_x, x = 0.12$ [205]/ [001], [110]/ [110]L/ -/ .
$Ni_{1-x}Pd_x, x = 0.45$ [206]/ [001], [110], [111]/ [001]T, [110]T_1, [111]T/
GTF/.
$Ni_{1-x}Pt_x, x = 0.05, 0.3, 0.5, 0.75, 0.95$ [203], [207]/ [001], [110], [111]/
-/ -/ Resonance modes dominate in the [001] and [110] directions.
$Pd_{1-x}Ag_x, x = 0.04$ [208]/ [001], [110], [111]/ [110]T_2/ GTF/ .
$Pd_{1-x}Fe_x, x = 0.04, 0.10$ [209]/ [001], [110], [111]/ [110]T_1 / GTF/ .
$Pd_{1-x}Pt_x, x = 0.16$ [210]/ [001], [110]/ -/ GTF/ .
$Pd_{1-x}Rh_x, x = 0.05, 0.10$ [211]/ [001], [110], [111]/ [110]T_1/ GTF/ .

(ii) bcc Host Lattice (Solid Solution Phase)

(Directions are assigned in the virtual crystal approximation)
$Cr_{1-x}Fe_x, x = 0.01$ [212]/ [011]/ [011]T_2/ -/ .
$Fe_{1-x}Mn_x, x = 0.30$ [213]/ [001], [110], [111]/ -/ GTF/ Measurements
exist above as well as below Néel temperature.
$Fe_{1-x}Ni_x, x = 0.35$ [214]/ [001], [110], [111]/ -/ GTF, BSM/ .
$Fe_{1-x}Pd_x, x = 0.28$ [215]/ [100], [110], [111]/ -/ -/ Measurements exist

in both the paramagnetic and ferromagnetic phases.

$Fe_{1-x}Pt_x, x = 0.278$ [216]/ [001], [110]/ -/ -/ .

$Mo_{1-x}Rh_x, x = 0.15$ [217]/ [001], [110], [111]/ [001]T, [111]L, [111]T, [110]L, [110]T/ -/ .

$Mo_{1-x}Ru_x, x = 0.044$ [218]/ [001], [110], [111]/ [110]T, [111]L, [111]T/ GTF/ .

$Nb_{1-x}Mo_x, x = 0.15, 0.35.0.41.0.56, 0.75, 0.91, 1.0$ [63]/ [001], [110], [111]/ [001]L, [111]L/ -/ The magnitude and position of phonon anomalies change with x.

$Nb_{1-x}Zr_x, x = 0.12, 0.5$ [219]- [221]/ [001], [110], [111], [112]/ [112]T/ CF/ .

$Nb_{0.88}Zr_{0.03}Mo_{0.09}$ [221]/ [001], [110], [111]/ [001]L, [110]T/ CF/ .

(iii) Others

(In the following alloys few local modes are measured).

$Cu_{1-x}Au_x$ [222].

$Cu_{1-x}Ba_x$ [223].

$Ti_{1-x}Zr_x, x = 0.38$ [224].

$V_{1-x}Be_x$ [225].

$V_{1-x}Cr_x, x = 0.25$ [226].

$V_{1-x}Ni_x, x = 0.05$ [227].

$V_{1-x}O_x$ [228].

$V_{1-x}Pt_x, x = 0.05$ [229].

$V_{1-x}Ta_x, x = 0.057$ [230].

$V_{1-x}Ti_x, x = 0.25$ [226].

$V_{1-x}U_x, x = 0.03$ [231].

$V_{1-x}W_x, x = 0.055$ [230].

$Y_{1-x}Tb_x, x = 0.1$ [232].

3.5.5   Surface Phonons

(i) Electron Scattering Measurements

$Ag(111)$ [30], [233]$/ < 11\bar{2} >, < 110 >/ -/ -/$.
$Cu(100)$ [234], [235]$/ < 100 >/ -/$ CFM$/$.
$Cu(110)$ [236], [237], [30]$/ < 11\bar{2} >/ -/$ CFM$/$.
$HfC_x(100)$ [238]$/ < 11\bar{2} >, < 100 >/ -/$ SSM$/$.
$NbC_x(100)$ [239]$/ < 11\bar{2} >/ -/$ DSM$/$.
$Ni(100)$ [240], [241]$/ < 100 >, < 11\bar{2} >/ -/$ CFM$/$.
$TaC_x(100)$ [239]$/ < 11\bar{2} >/ -/$ DSM$/$.
$TiC_x(100)$ [238], [239], [242]$/ < 11\bar{2} >/ -/ -/$.
$W(001)$ [243]$/$ Few phonon peaks are observed.

(ii) He-Atom Scattering Measurements

$Ag(111)$ [236], [244]$/ < 11\bar{2} >, < 110 >/ -/ -/$.
$Au(111)$ [236], [244]$/ < 11\bar{2} >, < 110 >/ -/ -/$.
$Cu(111)$ [236], [244]$/ < 11\bar{2} >, < 110 >/ -/ -/$.
$Ni(111)$ [245] .
$Pt(111)$ [246] .

3.6   Discussion

We describe here the important observations made in different groups of d and f-band materials.

3.6.1   Elemental Metals

Cu, Ag, Au

Figure 3.12 shows the phonon dipersion curves for Cu along with the phonon anomalies. The notable features of phonon dispersion curves of Cu, Ag, Au are as follows:

The phonon frequencies of Cu and Ag satisfy homology criterian

$$\frac{\omega(Ag)}{\omega(Cu)} = \frac{(Ma^2)_{Cu}^{1/2}}{(Ma^2)_{Ag}^{1/2}} \approx 0.679 \qquad (3.38)$$

which is not satisfied by the phonon frequencies of Au. The temperature variation of phonon frequencies is of the order of 3-4 % which is consistent with the estimated isothermal elastic constants [247]. About 20 anomalies are found in the phonon dispersion curves of Cu[45]. The anomalies in the $[\xi\xi 0]L$ and $[\xi\xi 0]T_1$ modes are most pronounced and these arise due to Fermi surface nesting[248].

The force constant analysis shows that the first nearest neighbour (1NN) interaction dominates. However, the weak but long range interactions also exist. The force constants of Cu and Au are not symmetric due to dominance of anisotropic nature of d-electrons. However, the force constants of Ag are nearly axially symmetric. The similar pattern is found in the temperature and pressure dependence of elastic constants and Debye temperature of these metals[37], [249].

Ni, Pd, Pt

The phonon dispersion curves of Ni, Pd, and Pt resemble with each other and also with those of Cu, Ag and Au. The dispersion curves for Ni are shown in Fig.2.6. The noteworthy features are as follows:

The phonon frequencies of Ni, Pd and Pt are in the inverse ratio of square root of their masses. The phonon frequencies in the para and ferromagnetic phases of Ni are nearly the same. An anomalous wiggle about the line representing the velocity of sound is found in the $[0, \xi, \xi]T_1$ mode in Ni, Pd and Pt. The possible positions of the anomalies in Ni, Pd and Pt are shown in Fig.3.13. The position and the strength of anomalies depend upon the topology of Fermi surface arising from the 5th electronic energy band of Ni, Pd and Pt [250], [251]. With the increase in temperature, the strngth of the anomalies decreases and these get broadened due to thermal smearing of the 5th band Fermi surface. The phonon anomalies predict the presence of long range forces (upto 8NNs) in these metals.

Nb, Ta, V (group V B)

This group comprises of superconducting metals. The phonon dispersion relations for Nb,Ta and V are similar and these are shown in Figs.
4.5 and 5.4. The main features are as follows:

The L and T modes in the [001] direction and T_1 and T_2 modes in
the [110] direction intersect each other. There are extra maxima and
minima in the [001]L mode. The presence of phonon anomalies in the
[001]L, [111]L and [111]T modes indicate the presence of long range
forces in Nb, Ta and V [252], [253]. These forces are asymmetric in
nature, particularly in V, which indicates the pronounced asphericity
of conduction electron charge distribution [253] and which is related to
Friedel oscillations [254]. With the increase in temperature, the phonon
anomalies and hence the interatomic forces get reduced [63], [68].

Cr, Mo, W (Group V I B)

The phonon dispersion curves of Cr, Mo and W resemble with each
other. However, these are different from those of Nb, Ta and V. The
phonon dispersion curves of Cr are shown in Fig.2.10. The phonon
anomalies in these metals are found near N and H symmetry points.
The phonon anomalies in the antiferromagnetic Cr are more pronounced
as compared to those in the paramagnetic Cr. The differences in the
anomalous behaviour of the phonon spectra predict the different electronic properties of (V B) and (VI B) group elements. This fact is
supported by the differences in the superconducting properties of the
metals in these two groups [255].

The force constant analysis predicts that the partially localized d-
electrons are involved in the quasi-covalent bonding and the interatomic
forces are long ranged. An intercomparison of the force constants in the
fcc and bcc metals shows that in the bcc metals the forces at the 1NN
sites are attractive while in the fcc metals these forces are repulsive.

Fe

The phonon dispersion curves of Fe do not show any significant anomalous behaviour and the interatomic forces are nearly axially symmetric.

Sc, Y

The phonon dispersion relations in Sc and Y are similar to each other which is due to their similar electronic structure [256]. The phonon dispersion curves for Y are given in Fig.2.14. The phonon frequencies of Sc and Y are in the ratio of their ion-plasma frequencies. The TA modes show considerable isotropy which is the characteristic of all the hcp metals with c/a ratio close to its ideal value (=1.633). The force constant analysis shows that the forces acting normal to the basal plane are short ranged while parallel to the basal plane are long ranged [87], [94], [257].

Ti, Zr, Hf, Tc, Re, Ru

The phonon dispersion curves of superconducting hcp metals Ti, Zr, Hf, Tc, Re and Ru are similar to each other and also to those of Sc and Y. In this group of metals the phonon frequencies increase with the increase of temperature in all the modes except in the $[00\xi]LO$ mode. With the decrease of temperature, $[00\xi]LO$ mode softens significantly and finally exhibits a dip at the zone center. In Ru such an anomaly is not observed [86].

For comparison, the phonon frequencies of Ti, Tc and Zr along $[00\xi]$ direction are shown in Fig. 3.14. Liu et al [258] showed that the anomalous behaviour of $[00\xi]LO$ phonons is caused by the splitting of doubly degenerate electronic bands near the Fermi surface. The splitting is caused by the lattice distortion which corresponds to the zone centre $[00\xi]$ LO mode. Wakabayashi et al [81] pointed out that the softening of $[00\xi]LO$ mode near Γ symmetry point is related to an incipient insatbility in the electronic charge density which leades to the formation of a dipolar charge density wave.

Tl, Co

The phonon anomalies are not observed in the phonon dispersion curves of these metals. The measurements on Tl [88], [92], [93] indicate that the phonons in the low symmetry directions are needed to obtain the force constants. The temperature dependence of the line width and phonon frequencies suggest that the hcp to bcc phase transition is of the first order.

Tb, Ho, Dy

The phonon dispersion curves for Dy are characteristically different from those of Tb and Ho which is due to its ferromagnetic character and which leades to magnon-phonon interaction. The typical phonon dispersion curves for Dy are shown in Fig.3.15. The magnon acoustic (MA) modes intersacts with the TA phonon mode in the $[00\xi]$ direction. The MA modes also tends to intersact with both the LA and T_1A phonon modes in the $[\xi\xi 0]$ direction. The force constants of Tb and Ho satisfy rotational invariance properties.

$La, \gamma - Ce, Yb, Th$

The phonon frequencies of La,γ-Ce and Yb are similar to each other. γ-Ce shows anomalous phonon behaviour in the $[\xi\xi\xi]T$ mode which is related to its fcc to dhcp phase transition while Yb shows the anomalous behaviour in the $[\xi\xi 0]T_1$ mode. The phonon anomalies are the consequence of the electronic structure of these metals [54], [258].

$\alpha - U$

The neutron scattering experiments show the presence of charge density waves [99], [100] which correspond to the condensation of LO phonons near $\vec{q}=[0.500]$. The calculated polarizability function does not show any structure which may be attributed to the phonon peaks. The detailed investigations of phonon anomalies in α-U are needed.

3.6.2   Metallic Compounds

YS, LaS, SmS, ThS

The metallic sulphides show the anomalous phonon structure near zone
boundary. In LaS, the LO and TO modes are nearly degenerate and are
almost dispersionless. Fig.3.16 shows the dispersion curves of SmS both
in the semiconducting and metallic mixed valence phases. The anomalous behaviour is almost absent in the semiconducting phase while the
phonon softening occures in the $[\xi\xi\xi]$LA mode in the metallic mixed
valence phase. In *US* a band of weak scattering of unknown origin is
also observed [111].

(NbO), (ThP, UP), (NdSb, USb), (TmSe, USe), (UTe)

NbO has NaCl structure with 25 percent vacancies in both the sublattices. The dispersion curves of the acoustic modes of NbO are in close
similarity with those of VN and NbN. In the phosphate group, the
acoustic peaks of ThP are sharper than those of UP because magnetic
excitations also contribute in UP. The phonon frequencies of NdSb and
USb are similar and exhibit little variation with temperature.

In TmSe, in the mixed valence phase, the phonon frequencies are
non-degenerate at the zone boundary in the $[00\xi]$ direction. In TmSe,
the LA and TA modes cross over in the $[\xi\xi\xi]$ and $[\xi\xi 0]$ directions which
predicts the strong effects of conduction electrons. The features of
phonon dispersion curves of UAs and UTe are similar to those of USb
and TmSe, respectively. Figure 3.17 shows the phonon frequencies of
UAs.

Jackman et al [259] fitted the phonon dispersion relations of UX
(X = C, N, As, Sb, S, Se, Te) series using rigid-ion and shell models. The U-X force constants are found nearly constant for the series
whereas U-U force constants vary systematically from being large and
positive for compounds with smallest lattice parameter to negative for
the chalcogenide series. The negative U-U force constant is identified
with destablizing f-d interaction.

$(Cu_3Au), (LaSn_3, CeSn_3), (Ni_3Al, Ni_3Fe, Ni_{0.3}Fe_{0.7})$
$(FePt_3, Fe_3Pt), (Pd_3Fe, Pt_3Mn)$

Phonon dospersion relations of Cu_3Au do not exhibit phonon anomalies. The calculated elastic constants with CFM agree with the ultrasonic measurements and hence the d-band effects are negligible.

$CeSn_3$ and $LaSn_3$ are isostructural compounds and their phonon spectra are similar and structureless. One should note that $CeSn_3$ is in the mixed valence phase while $LaSn_3$ is not. The phonon spectrum of Ni_3Al predicts its fcc structure while the phonon spectra of $Ni_3Fe, Ni_{0.3}Fe_{0.7}$ predict their averaged fcc structure.

The measurements of $FePt_3(T_N = 170^0K)$ are in the paramagnetic phase while those of Fe_3Pt are in the ferromagnetic phase. The cross-over of LA and LO modes shows the strong electronic effect while the softening of the phonon modes near zone boundary is related to the structural phase transition. The results of Pd_3Fe and Pt_3Mn are similar to those of Fe_3Pt.

$YAl_2, LaAl_2, CeAl_2, UAl_2$

The phonon dispersion curves of these materials are similar and have band gap between two sets of optical modes. The phonon anomalies in the transverse modes of $CeAl_2$ predict the existence of long range forces. In other materials, the observed phonon softening in the transverse acoustic modes is associated with the cubic to tetragonal phase transition.

$Cu_{0.53}Zn_{0.47}(\beta - brass), NiTi, Mn_{0.85}Ni_{0.09}Co_{0.06}, Ti_2O_3, V_2O_3$

All these materials undergo the phase transition which is reflected in their phonon dispersion relations. The phonon dispersion relations of $Cu_{0.53}Zn_{0.47}$ are those of an averaged CsCl structure and their force constant analysis predicts long range forces. The temperature variation of phonon frequencies shows that at about 727^0K, the CsCl structure transforms into disordered phase. The temperature variation of phonon frequencies of NiTi shows a martensitic phase transition at

$T_M = 308^0 K$. The phonon anomalies in the $[\xi\xi 0]$ direction are found to be strongly temperature dependent [139]. The condensation of $[\xi\xi 0]T_2$ acoustic phonons at $(2/3)q_{max}$ into a premartensitic phase shows a second order displacive phase transformation as shown in Fig. 3.18. The softening of acoustic modes in the $Mn_{0.85}Ni_{0.09}Co_{0.06}$ correspond to cubic to tetragonal phase transition at $T = 174^0 K$.

The measured phonons of Ti_2O_3 exhibit a gradual metal to semimetal phase transition as the temperature is lowered from $600^0 K$ to about $300^0 K$. In V_2O_3 the low lying transverse optical mode in the $[\xi\xi 0]$ direction disappears abruptly at the metal to insulator phase transition at $T = 160^0 K$.

$TiSe_2, TaS_2, TaSe_2, NbSe_2$

In these materials the phonon peaks are observed and these are associated with the charge density waves [260], [261].

$(V_3Ga, V_3Si, V_3Ge), (Nb_3Al, Nb_{3.1}Ga_{0.9},$
$Nb_3Ge, Nb_3Sn, Nb_3Sb), (Cr_3Si)$

These are low temperature superconducting materials and the exhaustive measurements of temperature variation of the phonons in these materials are carried out. For $V_3Ge, T_c = 6^0 K$ while for $V_3Ge_{0.75}Al_{0.25}, T_c = 12^0 K$ which shows the enhancement of T_c with the introduction of Al in V_3Ge and this also softens whole of the phonon spectrum. For $Nb_{3.1}Ga_{0.9}, T_c = 12^0 K$ which enhances to $21^0 K$ for stoichiometric Nb_3Ga. Further the phonon softening is observed when the temperature is lowered from $285^0 K$ to $80^0 K$. The typical measurements of $Nb_3Sn(T_c = 18^0 K)$ are shown in Fig.3.19. The pronounced anomalies in the $[\xi\xi\xi]LA$ mode are associated with the strong electron-phonon (e-p) coupling [262]. However, such anomalous features are not found in the phonon dispersion curves of Nb_3Sb and there is no significant softening in the phonon spectrum of Cr_3Si.

3.6.3   Carbides, Nitrides and Hydrides

$(TiC, ZrC, HfC, TaC, NbC, VC, UC, ThC)$
$(TiN, ZrN, \delta - NbN, HfN, VN_x, UN)$

The TM carbides and nitrudes are superconducting materials and can be devided into three groups. The 8 valence electron (VE) compounds ZrC, TiC, HfC, TaC, NbC, VC, UC, and ThC have low $T_c (\leq 0.05^0 K)$. The other two groups are 9VE (TiN, ZrN and HfN) and 10VE ($\delta - NbN, VN_x$ and UN) compounds which have higher T_c. As an illustration, the phonon dispersion curves of TaC, TiN and VN are shown in Fig.3.20.

The phonon spectra of 9VE and 10VE compounds are similar to each other but different from those of 8VE compounds. For example, the phonon anomalies in TiN are confined in a narrow region in $\vec{q}$-space than in the isoelectronic NbC. The phonon structure is more pronounced in the acoustic modes than in the optiocal modes. The phonon spectrum remains unchanged during the superconducting transition. The gap between the acoustic and optical modes of TM carbides indicates large screening effects. In the $\vec{q} \to 0$ limit, the elastic constants [263], [264] are reproduced from the measured phonon frequencies.

The Fourier analysis of the data shows that the metal-metal interactions are dominent while C-C interactions are small but not negligible in the carbides [264]. The metal-metal interactions decrease in going from HfC to ZrC on to TiC but the metal-nonmetal interactions remain almost the same. The DSM analysis shows that the shift in the soft mode region in the nitrides is due to the strengthening of the 1NN coupling in the metal sublattice. The e-p interactions in 9VE and 10VE systems are significantly different from each other. The superconducting behaviour of 9VE and 10VE TMCs and the non-superconducting behaviour of 8VE TMCs does not depend upon the change in the number of valence electrons which is due to the change in metal and non-metal components or due to stoichiometry [162], [173].

$NbD_x, TaH_x, TaD_x, VD_x$

These are nonstoichiometric solid solutions of hydrogen(H) and deutrium(D) [265], [266] which occupy the tetrahedral interstitial sites in the bcc lattice [267]- [269]. The phonon frequencies of $TaD_{0.22}(H_{0.18})$ and Ta are shown in Fig.3.21. There is a tendency for an overall increase in the phonon frequencies as D is added and the phonon anomalies are also become weak. Shapiro et al [175] observed new peaks in $NbD_{0.85}, NbH_{0.82}$ and $TaD_{0.78}$ having energy between 15-19 meV and Magerl et al [180] argued that these peaks are attributed to the presence of flat optical modes in the β-phase of NbD. The existence of the well defined acoustic modes in $NbD_{0.45}$ is probably the reflection of the weak coupling between the light interstitials and heavy metal ions. The lattice structure and the position of interstitials are found to change with temperature.

PdH_x, PdD_x

In these nonstoichiometric solid solutions, H(D) occupy octahedral position [270]. The 3 percent addition of H does not change significantly the phonon frequencies and the strength of phonon anomalies [181]. The phonon frequencies of PdD_x are measured for different values of x [181], [182] but the phonon anomaly is not observed in the $[\xi\xi0]T_1$ mode. Further the acoustic and optical modes in PdD_x are found to be nearly insensitive to the ordering transition occuring at nearly 80^0K [182]. The analysis of the phonon spectra shows that the second-like nearest neighbour D-D interactions are comparable to the first-like nearest neighbour Pd-Pd interactions. But these interactions are much weaker than the first-like nearest neighbour Pd-D interactions.

$CeD_x, YH_x(D_x), LaH_x(D_x), PrH_x(D_x), NdH_x(D_x), YbH_x$

The phonon dispersion relations of CeD_x are similar to those of PdH_x. However, the optical modes are well dispersed. In the $YH_x(D_x)$, the analysis of local mode frequencies shows that H(D) occupy partially octahedral and partially tetrahedral sites. Similar predictions are made for $LaH_x(D_x), PrH_x(D_x), NdH_x(D_x)$ and YbH_x from the neutron scattering measurements.

ThH_x, TiH_x, ZrH_x

The value of T_c for Th_4H_{15} is 7.6^0K while ThH_2 is non-superconducting above 1^0K. Figure 3.22 shows the phonon density of states $g(\omega)$ for Th_4H_{15} and ThH_2 for their optical hydrogen modes. The density of states for these modes broaden considerably in going from ThH_2 to Th_4H_{15}. The broadening is due to the enhancement of electron-phonon interaction which increases T_c. The similar behaviour of $g(\omega)$ is also found for TiH_x and ZrH_x solid solutions. In the hcp α-phase of ZrH_x, H is found on the regular tetrahedral sites and in other phases H is distributed on other interstitial sites also.

$Pd_{1-x}Ag_xH_y, TiCuH_x, LaNi_5H_x, FeTiH_x$

In these high T_c and hydrogen storage materials, the measurements of local modes of H are reported. An understanding of these materials is still awaited.

3.6.4   Disordered Alloys

$Ni_{1-x}Pd_x, Ni_{1-x}Pt_x, Ni_{1-x}Co_x$

The phonon frequencies of Cu, Ni and Pd based alloys are similar to those of the fcc metals. The phonon dispersion curves of $Ni_{0.55}Pd_{0.45}$ are similar to those of Ni. In $Ni_{1-x}Pd_x$, the phonon modes show the significant broadening while in $Ni_{1-x}Pt_x$ these are separated into two parts. The general shape and concentration dependence in Ni alloys is explained by Taylor's coherent potential approximation (CPA) theory [271]- [273]. In $Ni_{1-x}Co_x$, the phonon frequencies increase with the increase of x particularly at large $\vec{q}$-values.

$Pd_{1-x}Fe_x, Pd_{1-x}Rh_x$

The phonon spectra of nonsuperconducting alloys $Pd_{1-x}Fe_x$, $Pd_{1-x}Rh_x$ show an increase in the phonon frequencies with the increase in x particularly at small $\vec{q}$-values. In $Pd_{1-x}Rh_x$, the phonon anomaly in the $[\xi\xi0]T_1$ branch is more pronounced as compared to that in Pd. In $Pd_{1-x}Fe_x$, with the increase of x, the slopes of $[\xi\xi0]T_1$ mode at $\xi = 0.27$, 0.41, 0.52 and 0.77 show the strong electronic effect.

$Cu_{0.875}Ga_{0.125}, Cu_{0.917}Ge_{0.083}, Cu_{0.75}Zn_{0.25}$

These alloys are the isoelectronic solid solutions and their dispersion curves are nearly identical because the elements Cu, Zn, Ge, and Ga have similar masses and coupling constants. However, the dispersion curves differ from those of pure Cu.

$Fe_{0.70}Mn_{0.30}, Fe_{0.65}Ni_{0.35}, Fe_{0.72}Pd_{0.28}, Co_{0.92}Fe_{0.08}$

These are invar alloys with base element Fe (bcc). The phonon dispersion curves of these alloys exhibit fcc like symmetry and are analyzed by GTF model for the fcc lattice. An exception is $Co_{0.92}Fe_{0.08}$ alloy which also exhibits fcc like phonon dispersion relations.

$Nb_{1-x}Mo_x, Nb_{1-x}Zr_x$

The phonon dispersion curves of V, Nb, Mo, Ta and Cr based alloys show the bcc symmetry. The phonon spectra are analyzed using GTF model to predict the force constants between different constituents. In $Nb_{1-x}Mo_x$, for small values of x, the anomalies corresponding to Nb spectrum persist although weaker than in pure Nb. At about $x = 0.75$ the anomalies corresponding to Nb disappear and with the further increase of x, the phonon dispersion curves tend to become similar to those of Mo. This behaviour is related to the change in the electron-phonon interaction which involves change in the scattering mechanism of conduction electrons near the Fermi surface. The increase in the phonon frequencies with x is partially due to the decrease in lattice parameter and partially due to the change in electronic structure [274].

Similar trend of phonon frequencies is also found in NbD_x inspite of increasing the lattice parameter. In $NbD_{0.45}$, the anomalies in the $[\xi\xi\xi]L$ mode disappear but not in $Nb_{1-x}Mo_x$ of the same concentration. The rigid-ion model does not explain these results.

At higher temperatures, $Nb_{1-x}Zr_x$, for $0.8 < x < 1.0$ makes a random bcc phase. The phonon spectrum softens with increase of x. The phonon dispersion relations of $Nb_{0.08}Zr_{0.92}$ exhibit a soft-mode zone boundary mode [219] which implies the existence of incipient lattice instability concerning the ω-phase transformation. The bcc-ω phase transformation has also been observed in $Ti_{1-x}Mo_x$, for $0.07 \leq x \leq 0.2$ [220]. The softening of modes originates from the strong electron-phonon coupling [275].

3.6.5   Surface Phonons

The surface phonon spectrum consists of surface modes (S_i), also known as Rayleigh modes. Toennies and co-workers [236], [243] observed new longitudinal resonance (R_i) modes above Rayleigh modes on (111) surface of noble metals in the He scattering experiments. These results are interpretted by Bortolani et al [276] by reducing the lateral force constants in the upper most surface layer of Cu(111), Ag(111) and Au(111) by 50 to 70 percent. In these experiments, the surface modes which are predicted to lie in the gaps of the bulk phonon bands, are not probed. Later Mohamed et al [30] observed these gap modes near $\bar{M}$ symmetry point in the surface Brillouin zone for Cu(111) surface as shown in Fig.3.23. The dispersion curves are analyzed in the CFM by assuming 15 percent softening of the force constants as compared to that in the bulk material. On the other hand the measurements on Cu(111) and Ni(111) surfaces do not show any softening of the intralayer force constants, rather interlayer force constants are stiffened by about 20 percent [234], [235], [240], [241]. This shows that the interatomic forces on the surface depend strongly on its structure.

Reider and Drexel [275] were pioneer in measuring the phonons in TiN micro-crystals using the neutron scattering technique. Later the measurements were carried out by the high resolution electron energy loss spectroscopy for the acoustic and optical surface dispersion curves of $NbC_x(100), TaC_x(100), TiC(100)$ and $HfC_x(100)$. The typical phonon dispersion curves of $NbC_x(100)$ and $TaC(100)$ are shown in Fig.3.24 which have the follwing features: (i) In contrast to bulk phonons, the longitudinal modes do not exhibit anomalies which may be due to the difference in the electronic structure of the surface and bulk. (ii) The vibrational energy ratio of NbC_x and TaC_x agrees with the square root of their mass ratios (i.e. reduced mass) which shows that the force constants of TaC_x and NbC_x surfaces are very close to each other. It is also suggested that the nearest neighbour C-Nb and C-Ta force constants are enhanced as compared to bulk force constants.

In $HfC_x(100)$, the surface mode $S_{2'}$ is explained by decreasing the force constants between the first carbon layer and second metal layer while the S_2 mode is explained by increasing this force constant. Similar tendency of force constants is also found on $TiC(100)$ surface.

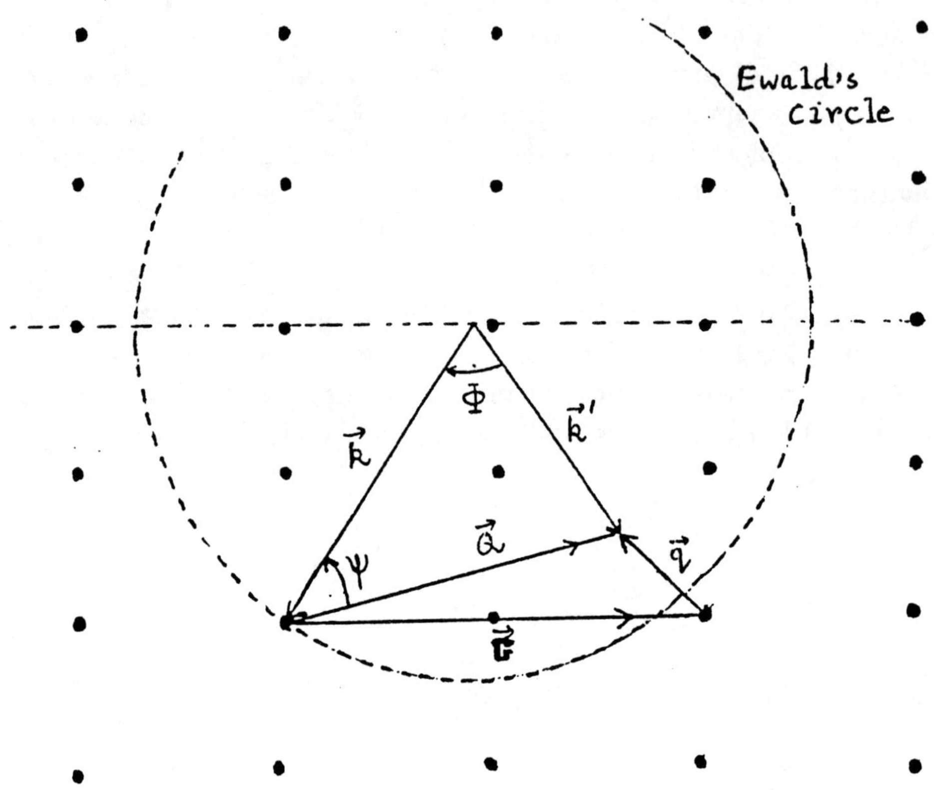

Figure 3.1: Ewald's construction for neutron inelastic scattering. Dots are the reciprocal lattice points. $\vec{k}$ and $\vec{k}'$ are the wave vectors of incident and scattered neutrons respectively. $\vec{Q} = \vec{q} + \vec{G}$ where $\vec{q}$ is phonon wave vector which takes part in the scattering process and $\vec{G}$ is reciprocal lattice vector.

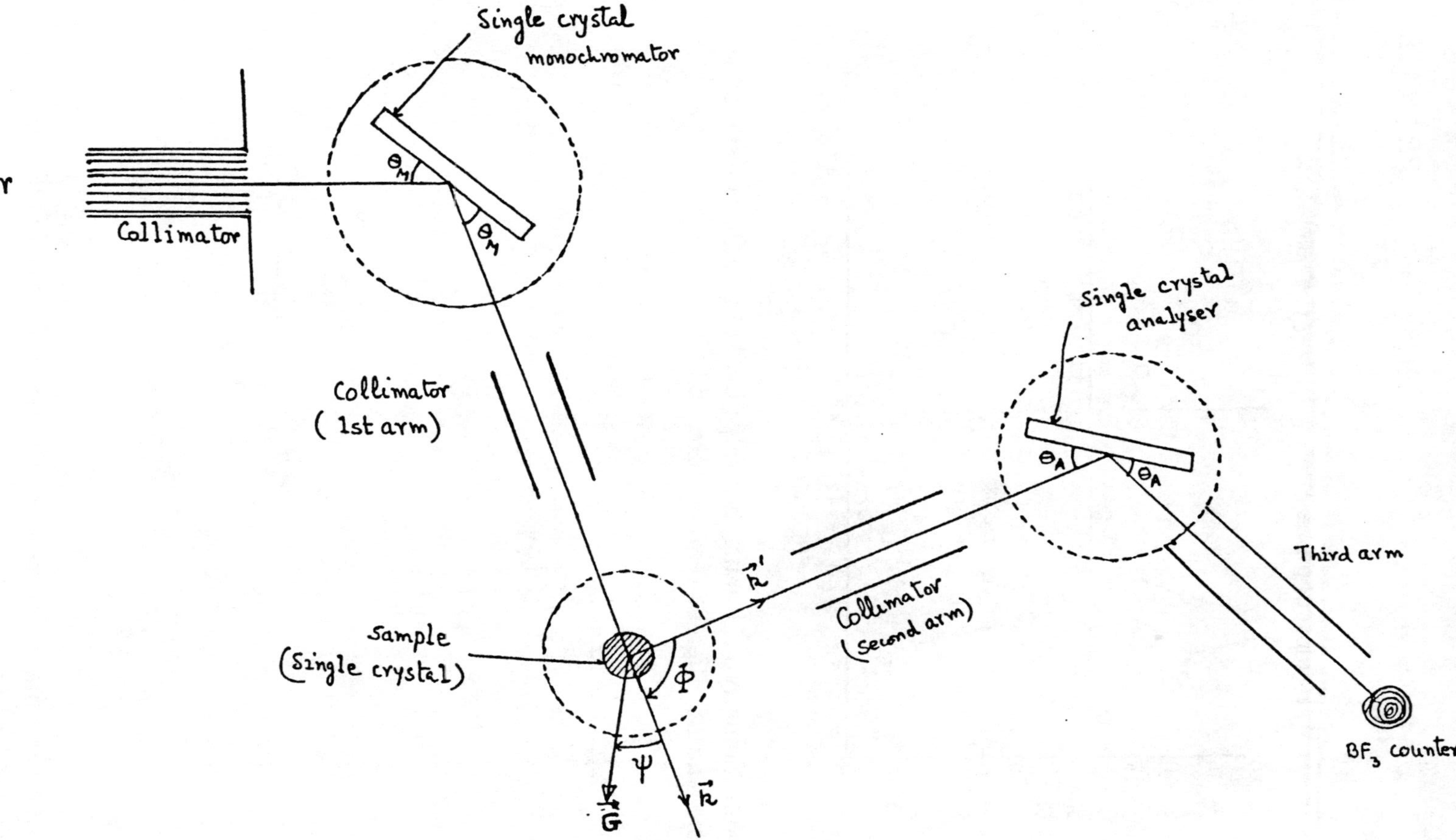

Figure 3.2: Schematic diagram of triple axis spectrometer.

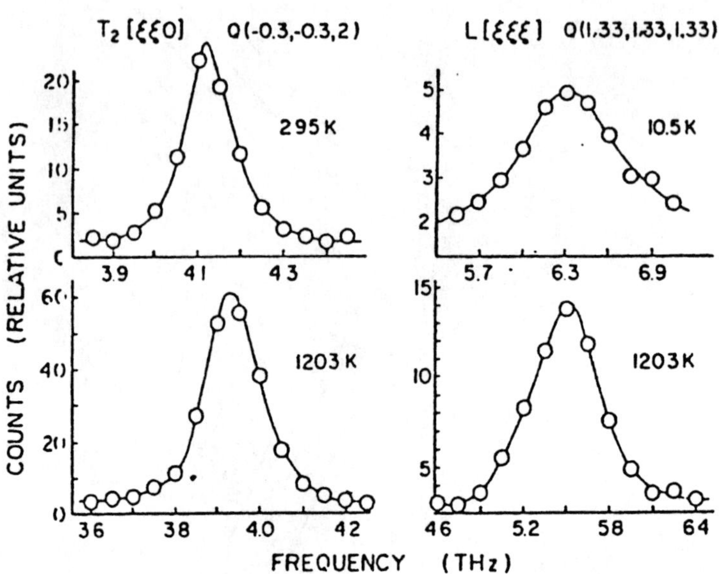

Figure 3.3: Some typical neutron groups for Mo in different symmetry directions and at different temperatures [62].

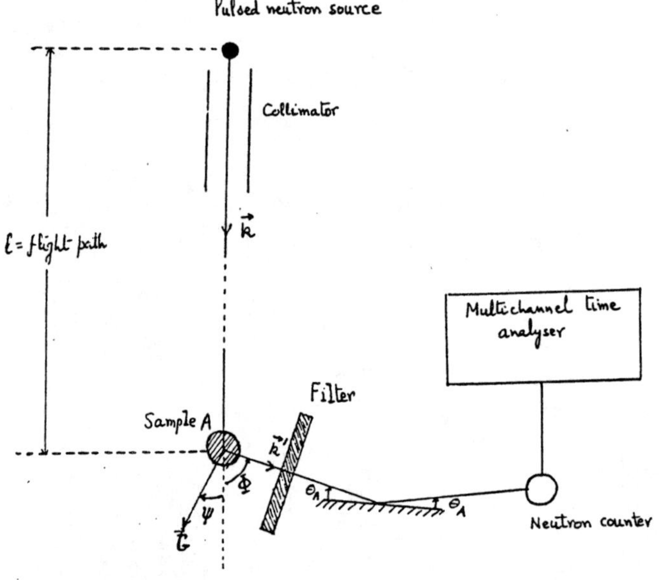

Figure 3.4: Schematic diagram for the time of flight (TOF) set up for neutron inelastic scattering.

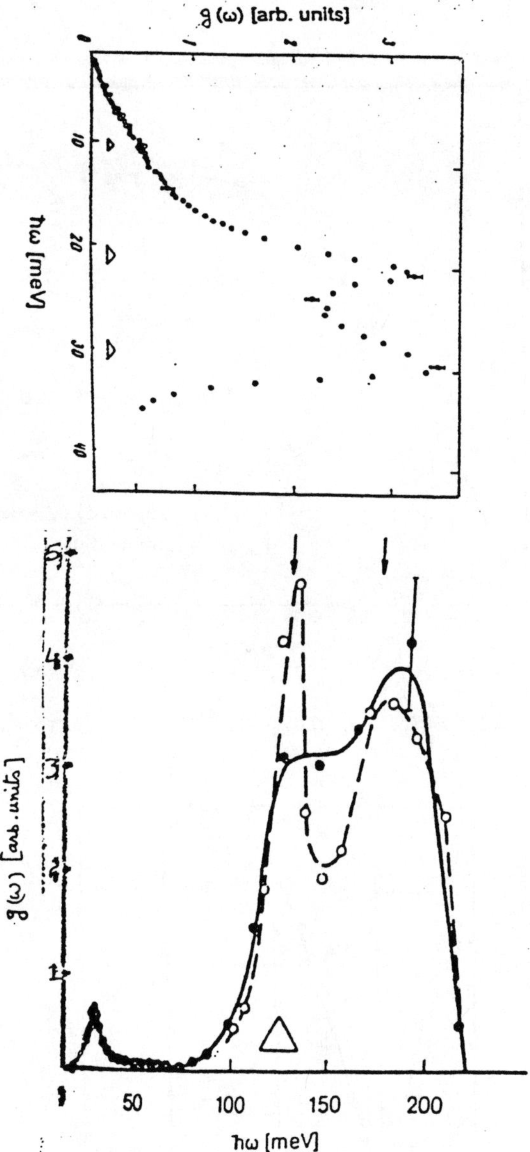

Figure 3.5: (a) One phonon density of states $g(\omega)$ versus $\hbar\omega$ for V metal [17] obtained by TOF method. The bars indicate the experimental errors. (b) $g(\omega)$ versus $\hbar\omega$ for $\beta - NbH_{0.95}$ [18] with TOF method. Full circles and solid lines are for $30\mu s/m$ resolution while open circles and dashed lines are for $15\mu s/m$ resolution. The centers of inelastic peaks which are due to optical phonon modes of H atoms, are at $\hbar\omega = 120meV$ and 170 meV.

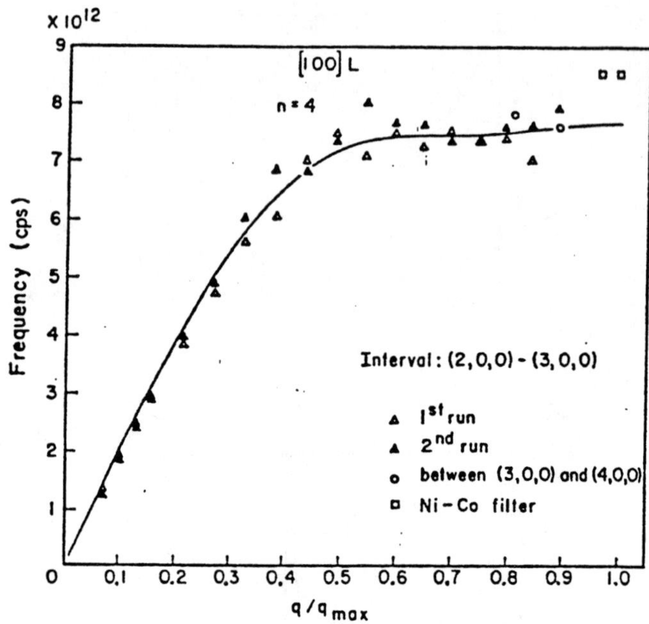

Figure 3.6: Frequency versus q/q_{max} in the $[100]L$ mode for V metal [23].

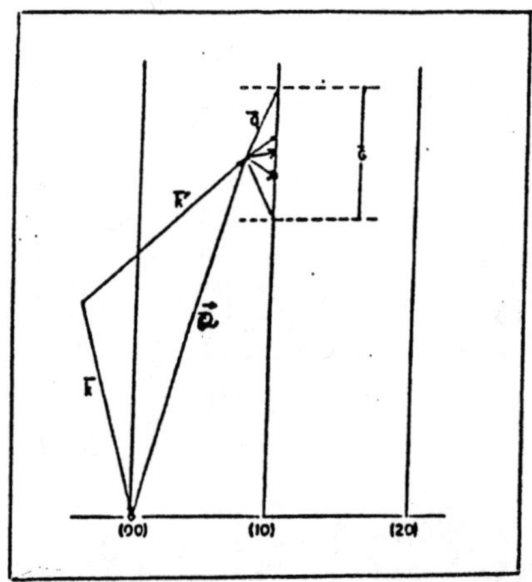

Figure 3.7: Thermal diffuse scattering from a surface. $\vec{k}$ and $\vec{k}'$ are wave vectors of incident and scattered electrons respectively. $\vec{q} = \vec{k}' - \vec{k}$. $\vec{G}$ is basis vector of reciprocal lattice of the bulk structure and $\vec{q}$ is phonon wave vector. All the phonons for which $\vec{q}$ reaches the tip of $\vec{Q}$ to a reciprocal rod, contribute to thermal diffuse scattering at $\vec{Q}$ [25].

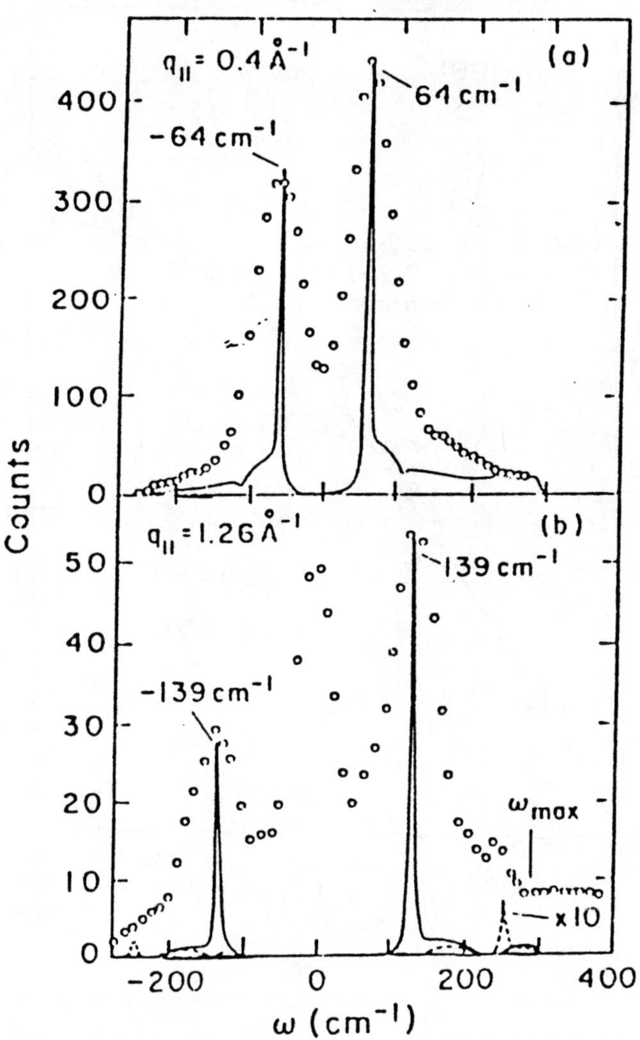

Figure 3.8: Sample spectra for $q_\parallel = 0.04A^{0-1}$ and $1.26A^{0-1}(\bar{X}$ point) obtained with a beam energy of 180 eV. The solid lines in (a) and (b) are the spectral densities for vertical motion of atoms in the first layer. The dashed line in (b) represents the spectral density of parallel motion weighted with the ratio of $(\vec{k}_\parallel - \vec{k}_\parallel')^2/(\vec{k}_\perp - \vec{k}_\perp')^2$ [28].

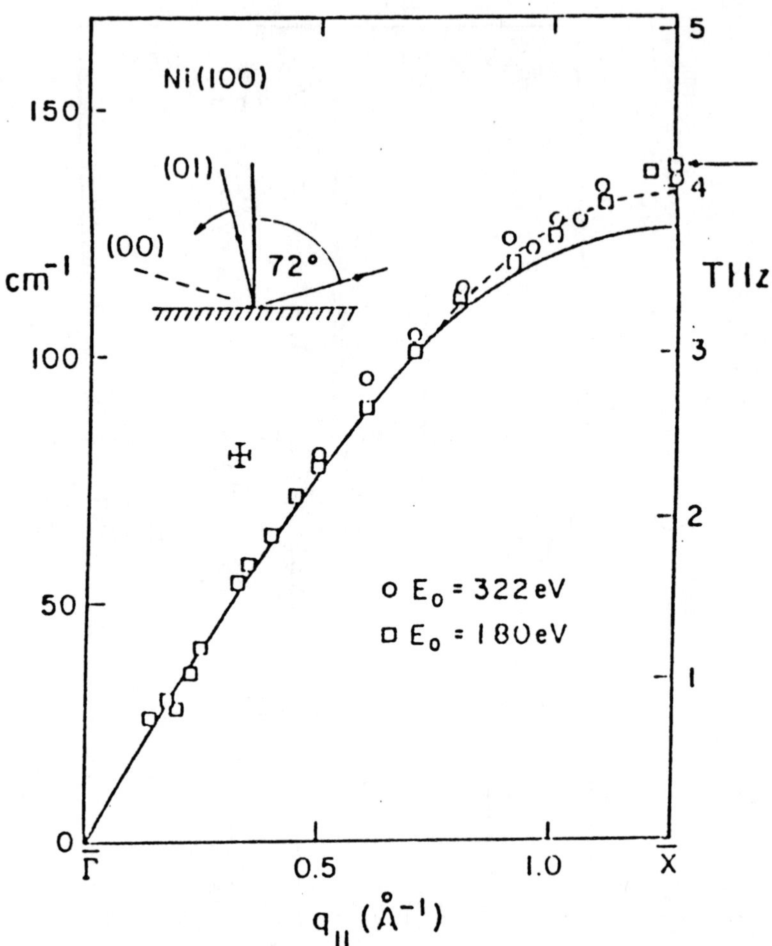

Figure 3.9: Peak positions of the loss spectra versus $q_\parallel$ for two impact energies of 180 and 332 eV. The solid line represents the dispersion curves calculated with the help of nearest neighbour CFM where the force constants are chosen which fit the bulk phonon spectrum. The dashed line is with a 20 percent increase in the force constants between the first and second layer [28].

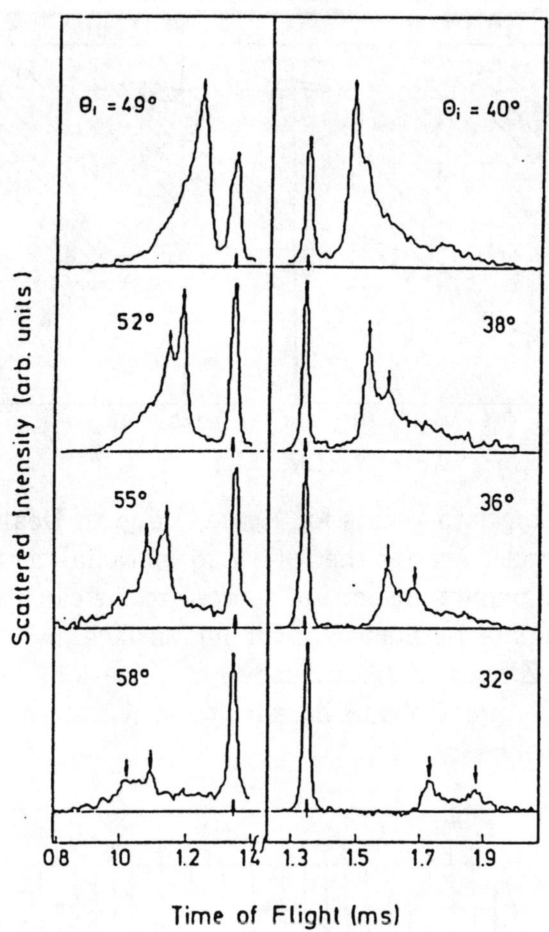

Figure 3.10: TOF spectra for $< 11\bar{2} >$ azimuth for $Ag(111)$ surface with a target at $140^0 K$ [31]. The actual signal count per channel is plotted against the measured time of flight.

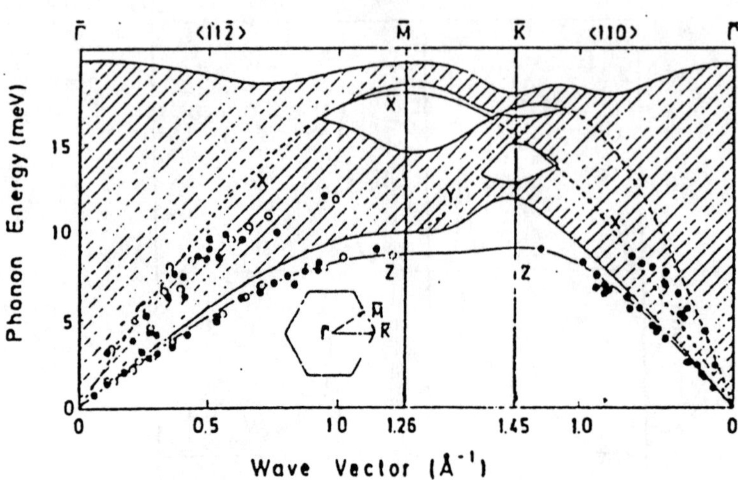

Figure 3.11: Phonon data points for Ag(111) due to Doak et.al. [31]. Filled and open circles are for the solid and epitaxial crystals respectively. The surface phonon dispersion curves are the calculated results. The dashed curves are resonances involving surface and bulk modes. The symbols X,Y,Z refer to surface mode polarization: OZ is normal to surface and OX and OY are parallel to $< 11\bar{2} >$ and $< 110 >$ azimuths respectively.

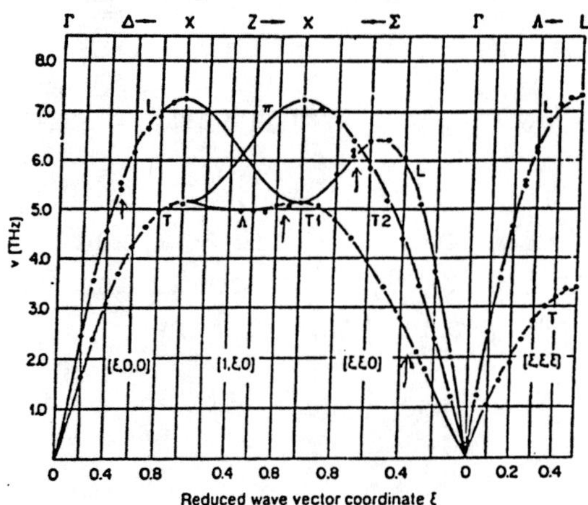

Figure 3.12: Phonon dispersion curves for Cu at 49^0K [44]. The solid lines show 6NN-AS model fit of the measured values. Arrows indicate the prominent anomalies.

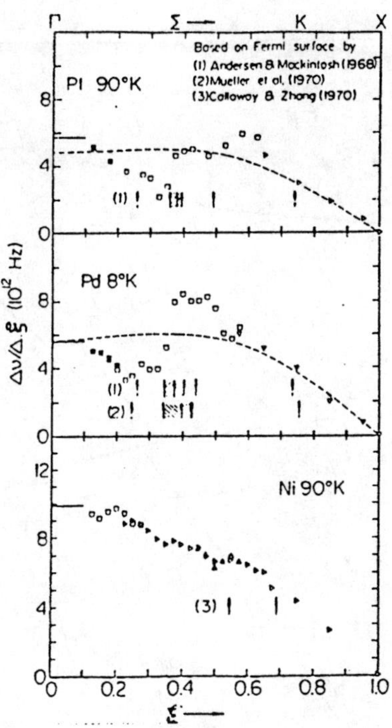

Figure 3.13: $(\Delta\nu)/(\Delta\xi)$ of the $[0\xi\xi]T_1$ modes in Pt, Pd, and Ni. The arrows indicate the position of Kohn anomalies. The dashed curves give the slopes of phonon dispersion of the unperturbed frequencies of these modes.

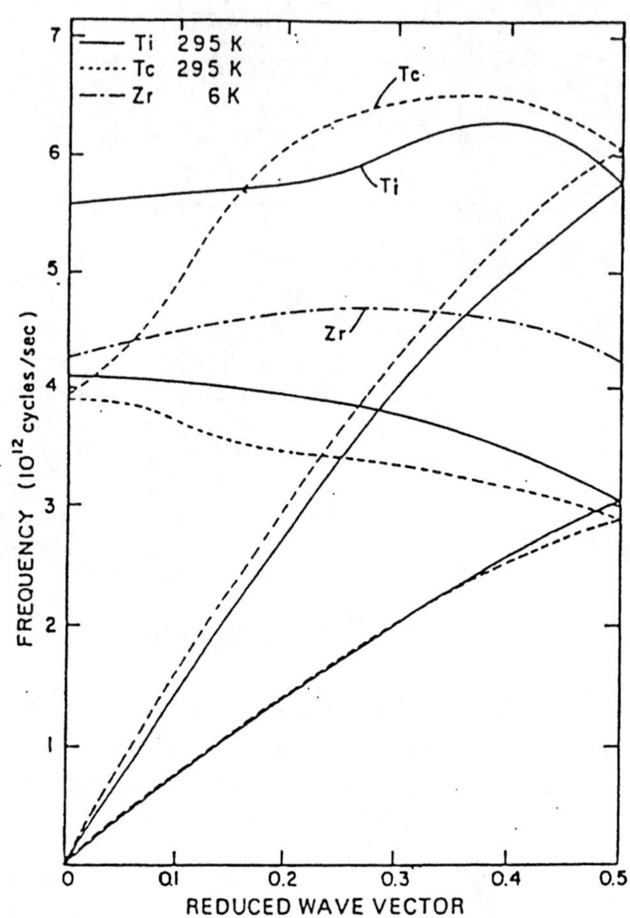

Figure 3.14: Phonon dispersion curves of Ti, Tc and Zr along the $[00\xi]$ direction.

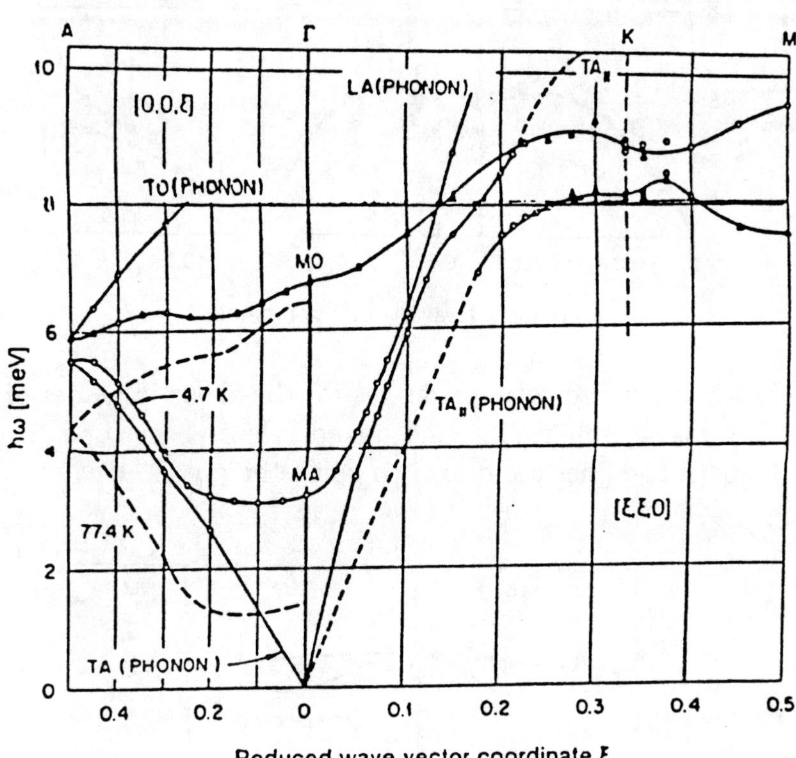

Figure 3.15: Phonon and magnon dispersion curves for ferromagnetic Dy [82]. Magnon optical and acoustic modes are represented by (MO) and (MA) respectively. The lines are drawn to guide the eye. The dashed lines in the [00ξ] direction represent the magnon dispersion at $77.4^0 K$.

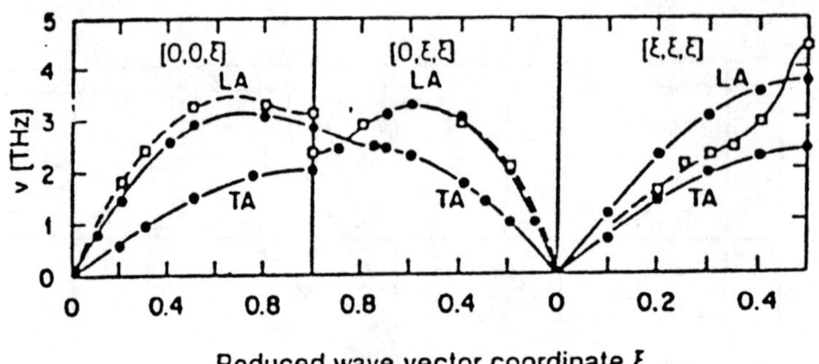

Figure 3.16: Phonon dispersion curves for the acoustic modes of semi-conducting phase (full circles) and mixed valent phase (open squares) of SmS [104]. The lines are drawn to guide the eye.

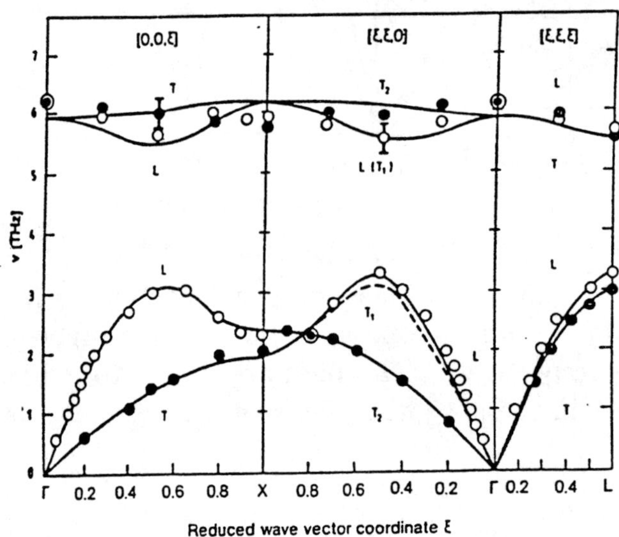

Figure 3.17: Phonon frequencies of UAs [110]. Open and solid circles are for longitudinal and transverse modes respectively. The solid lines show the 3NN-AS model fit of the measured results. The dashed lines represent the calculations for T_1 branch which is not measured experimentally.

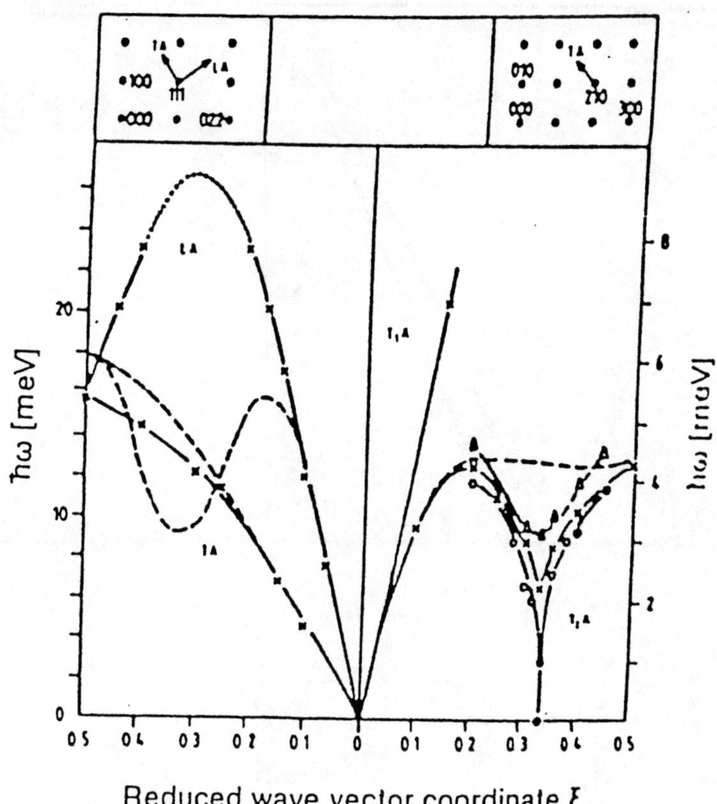

Figure 3.18: Phonon frequencies of the acoustic modes in the $[\xi\xi\xi]$ and $[\xi\xi 0]$ directions of NiTi [139]. The open triangles, crosses and open circles are the results at $T = 523^0 K, 423^0 K$, and $338^0 K$ respectively. Dashed lines show 2NN-GTF fitting retaining only the short range part. The solid and dotted lines are drawn through the experimental data.

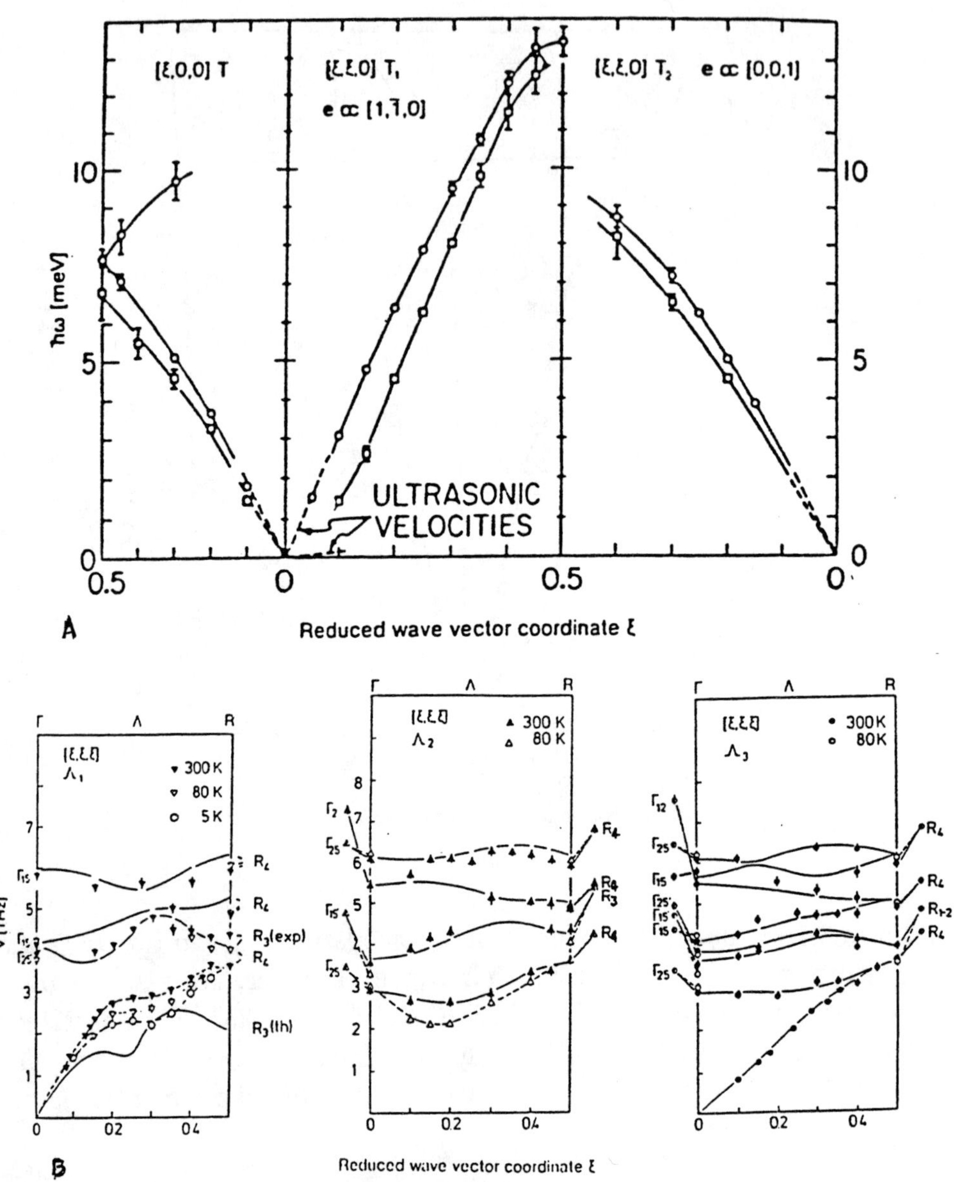

Figure 3.19: (a) Phonon frequencies of Nb_3Sn [152] for transverse modes in the [100] and [110] directions. Open circles and squares represent the data at $T = 295^0 K$ and $46^0 K$ respectively. Dashed lines show the elastic slopes derived from ultra-sound measurements. (b) Phonon frequencies of Nb_3Sn in the [111] direction at different temperatures. Solid lines are the calculated results in the modified tight-binding sheme. Dashed and dotted lines join the data points.

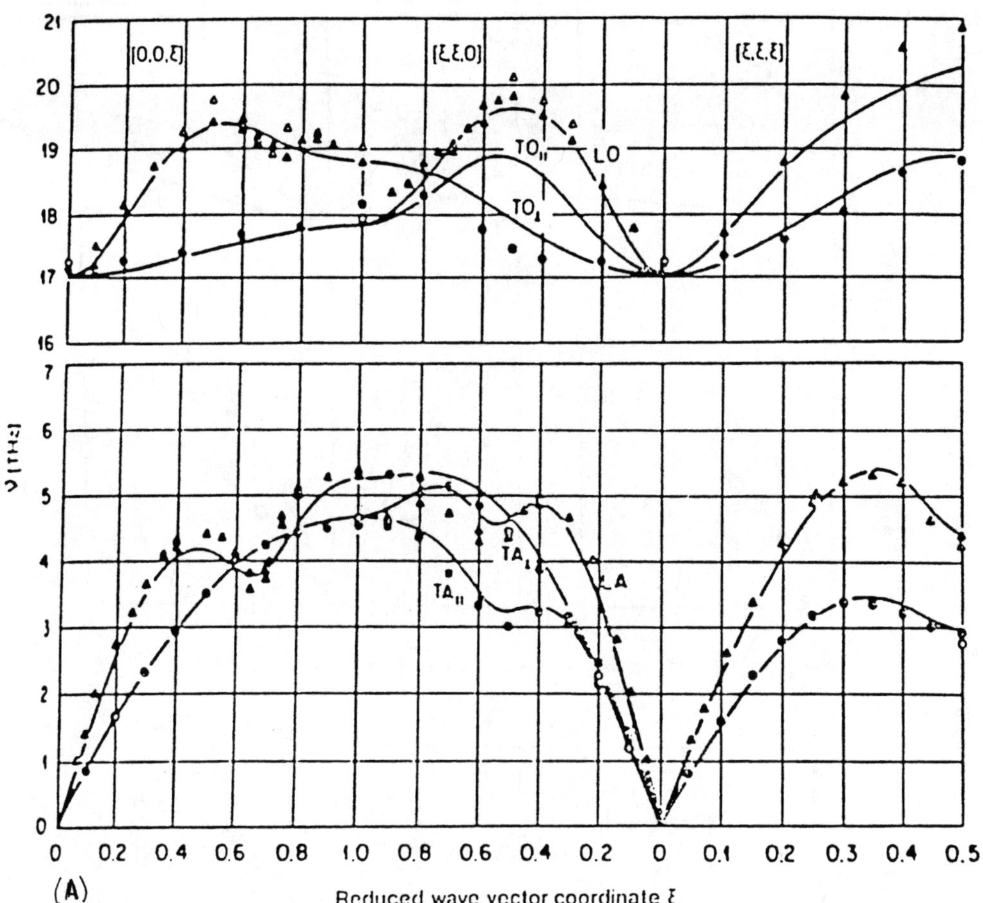

Figure 3.20: (a) Phonon dispersion curves of TaC [159], [166] at $T = 4.2^0 K$ (open symbols) and at $T = 298^0 K$ (filled symbols). Triangles denote the longitudinal modes and circles denote the transverse modes. Solid lines represent the calculations of DSM. (b) Phonon dispersion curves of $TiN_{0.98}$ [170]. Full lines represent the calculated results of the DSM and the dashed lines are obtained from the SM calculations. (c) Phonon dispersion curves for $VN_{0.86}$ [172]. Solid lines represent the theoretical results obtained from Varma-Wever formalism (Chapter 5). Dotted lines represent the phonon frequencies for TiN which are shown for comparison. The dashed lines are drawn through the experimental data.

Figure 3.20 continued

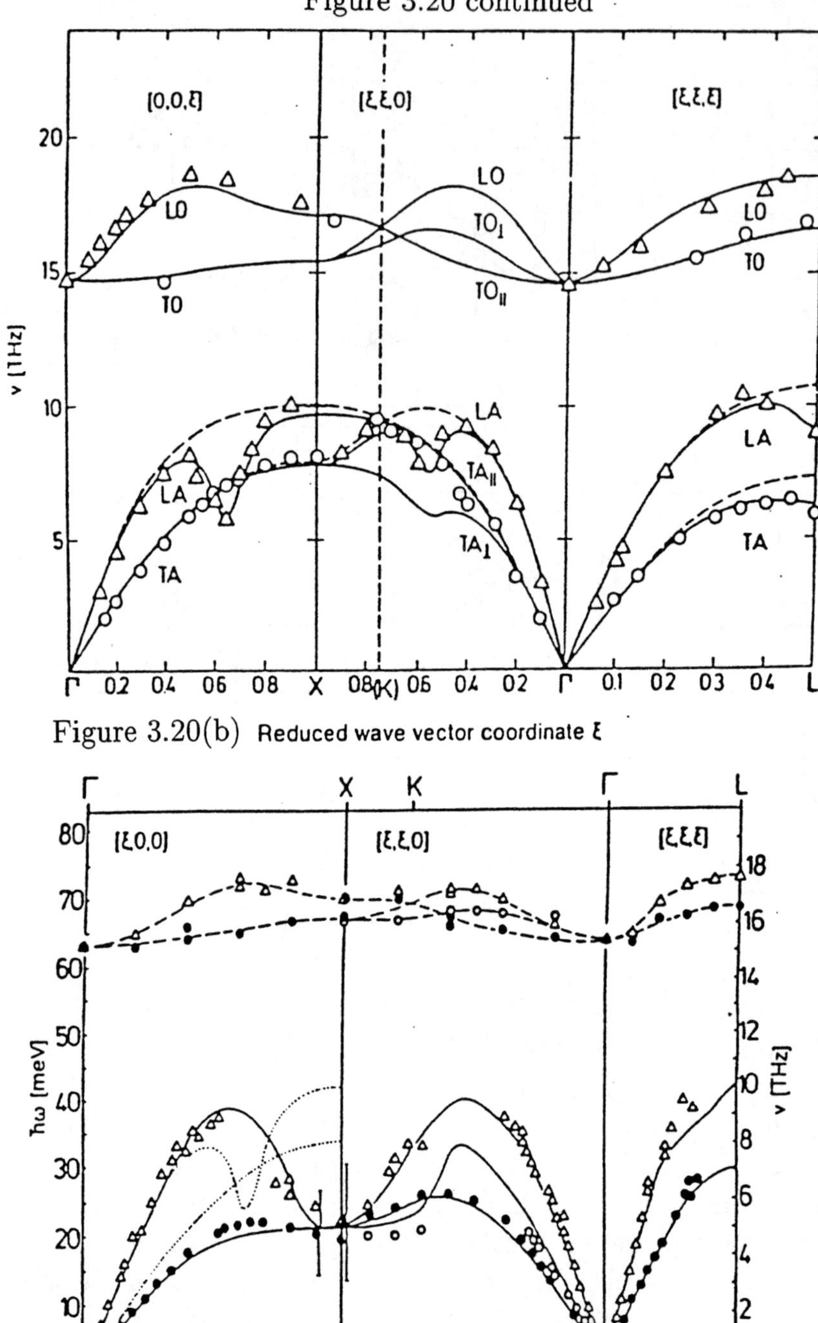

Figure 3.20(b) Reduced wave vector coordinate ξ

Figure 3.20(c) Reduced wave vector coordinate ξ

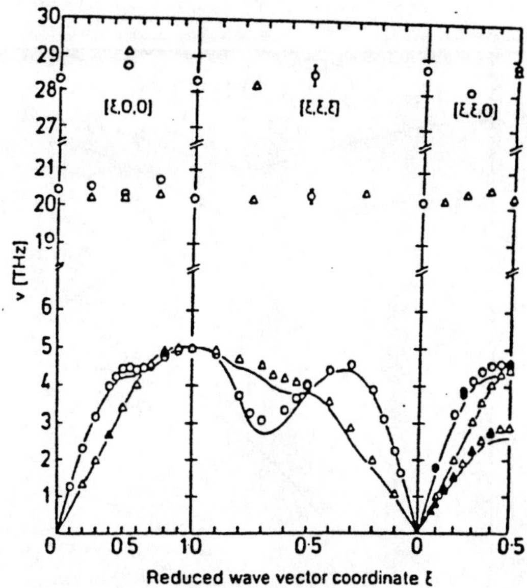

Figure 3.21: Phonon dispersion curves for Ta (full curves), $TaD_{0.22}$ (open symbols) and $TaH_{0.18}$ (full symbols) [183], [184]. Longitudinal and transverse modes are represented by circles and triangles respectively.

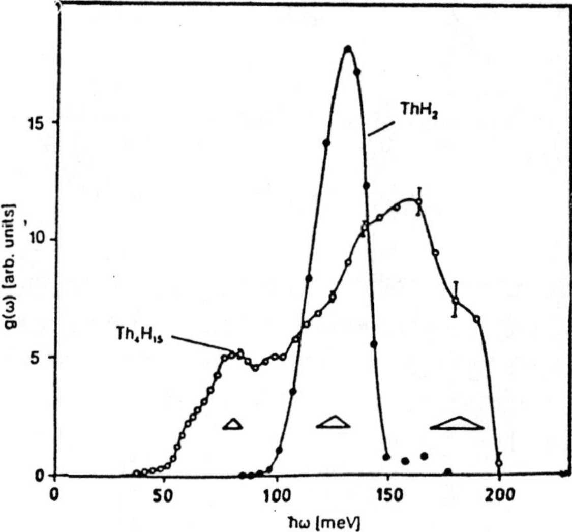

Figure 3.22: Density of states $g(\omega)$ versus (ω) in the range of the optical hydrogen modes for ThH_2 (filled circles) and Th_4H_{15} (open circles) [191]. The bars show the experimental error. Solid lines through the experimental data are drawn for the guidance.

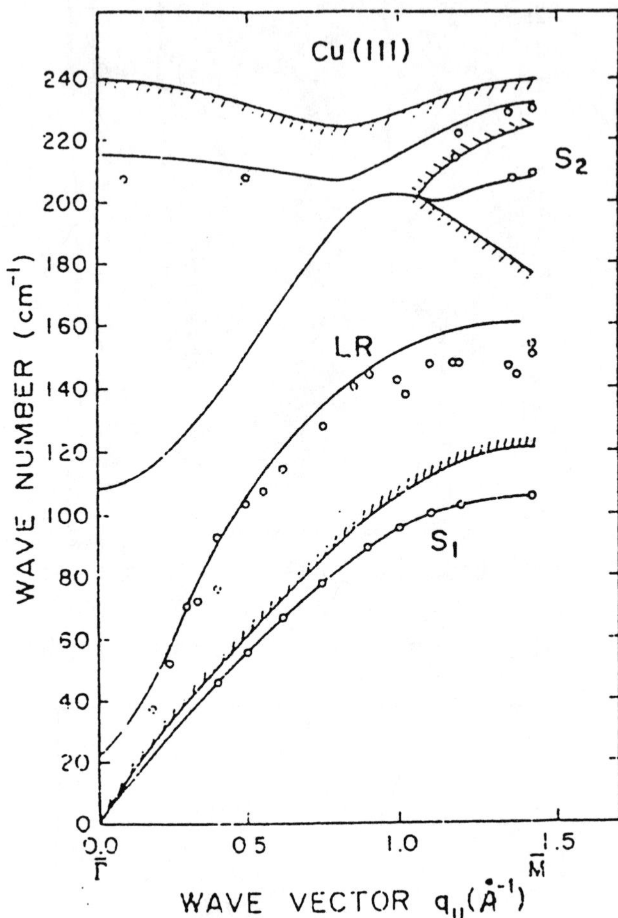

Figure 3.23: Experimental and theoretical (15 percent softening model) dispersion curves for surface modes (S_1 and S_2) and resonance modes (LR) on Cu(111) surface [30]. Experimental data are represented by open circles and calculations by solid lines. Bulk mode boundaries are shown as cross-hatched curves.

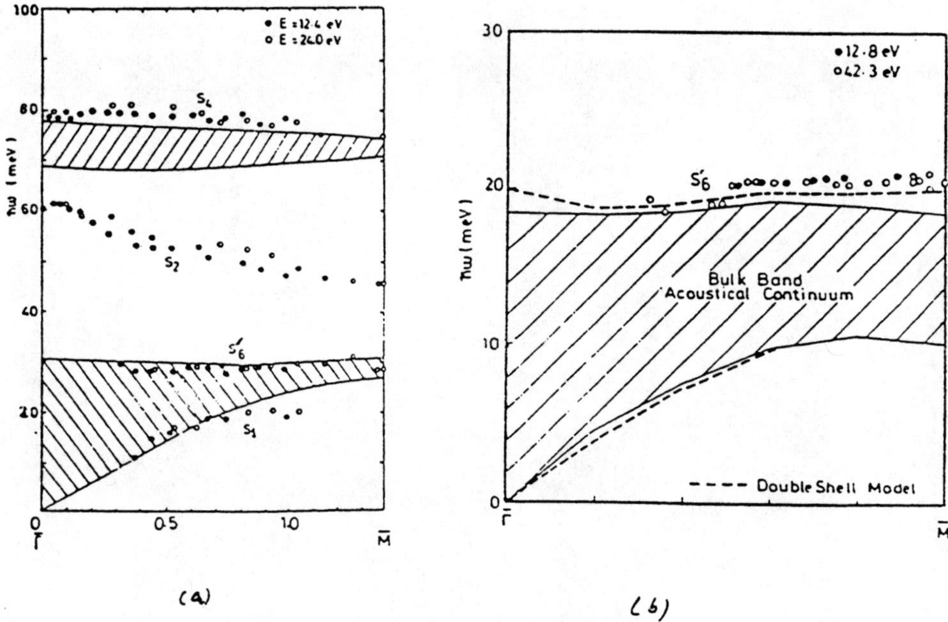

Figure 3.24: (a) Experimental dispersion curves of the surface phonons in the $< 11\bar{2} >$ direction for NbC(100) surface [239] for $S_1, S_{6'}, S_2$ and S_4 modes. The hatched regions are the bulk phonon bands. (b) Experimental data points (open and filled circles) for $S_{6'}$ mode and theoretical dispersion curves (dashed lines) along $< 11\bar{2} >$ direction for TaC(100) surface [239]. The hatched region is the bulk phonon band.

Bibliography

[1] G. Venkataraman, L.A.Feldkamp and V.C.Sahni, " Dynamics of Perfect Crystals" (MIT Press, Cambridge, 1975) Chap.5.

[2] W.Cochran, Rep. Prog. Phys. 26, 1(1963).

[3] "Electron Energy Losses and Surface Vibrations" Eds.H. Ibach and D.L.Mills (Academic Press, San Francisco, 1982) Chap.2.

[4] G.Brusdeylines, R.B.Doaks and J.P.Toennies, Phys. Rev.Lett. 46, 437(1981); Phys. Rev. B27, 3662(1983).

[5] W.Marshall and S.Lovesey, " Theory of Thermal Neutron Scattering" (Oxford Uni. Press, London, 1971); I.I.Gurevich and L.V.Tarasov, "Low Energy Neutron Physics" (North-Holland, Amsterdam, 1968).

[6] L.van Hove, Phys. Rev. 89, 1189(1953); Phys. Rev. 95,249 (1954).

[7] W.Kohn, Phys. Rev. Lett. 2,393(1959); E.J.Woll and W.Kohn, Phys. Rev. 126, 169(1962).

[8] L.M.Roth, H.J.Zeiger and T.A.Kaplan, Phys. Rev. 149, 519(1966).

[9] R.Stedman, L.Almqvist, G.Nilsson and G.Raunio, Phys. Rev. 163, 567(1967).

[10] J.Weymouth and R.Stedman, Phys. Rev. B2, 4743(1970).

[11] P.L.Taylor, Phys. Rev. 131, 1995(1963).

[12] S.H.Vosko, R.Taylor and G.H.Keech, Can. J. Phys. 43, 1187 (1965).

[13] B.Buras, in "International Course on the Theory of Condensed Matter" (ICTP, Treste, 1967).

[14] H.Boütin and S.Yip " Molecular Spectroscopy with Neutrons" (MIT Press, cambridge, 1969).

[15] G.C.Summerfield, J.M.Carpenter and N.A.Lurie, "Lecture Notes on Thermal Neutron Scattering" (University of Michigan, 1968).

[16] G.Venkataraman, K.Usha Deniz, P.K.Iyengar, A.P.Roy and P.R.Vijayaraghavan, J. Phys. Chem. Solids, 27, 1103(1966).

[17] G.F.Syrykh, A.P.Zhernov, M.G.Zemlyanov, S.P.Mironov, N.A. Chernoplekov, Yu. L.Shifikov, Soviet Phys. JEPT, Solid State Physics. 43, 183(1976).

[18] N.A.Chernoplekov, M.G.Zemlyanov, V.A.Somenkov and A.A. Chertkov, Soviet Phys. JEPT, Solid State Physics, 11, 2343 (1970).

[19] R.W.James, " Optical Principles of the Diffraction of X-Rays" (Bell, London 1950) ps. 31,109.

[20] M.Born, Proc. Roy. Soc.(London) A180, 397(1942).

[21] A.J.Freeman, Acta Cryst. 15, 682(1962).

[22] C.B.Walker, Phys. Rev. 103, 547(1956).

[23] R.Colella and B.W.Batterman, Phys. Rev. B1, 3913 (1970).

[24] E.R.Jones, J.T.McKinney and M.B.Webb, Phys. Rev. 151, 476(1964).

[25] J.T.McKinney, E.R.Jones, and M.B.Webb, Phys. Rev. 160, 523(1967) and references therein.

[26] D.L.Mills, A.A.Maradudin and E.Burstein, Ann. Phys. (N.Y.) 56, 504(1970).

[27] V.Roundy and D.L.Mills, Phys. Rev. B5, 1347(1972).

[28] S.Lehwald, L.M.Szeflel, H.Ibach, T.J.Rahman and D.L. Mills, Phys. Rev. Lett. 50, 518(1983).

[29] R.E.Allan, G.P.Alldredge and F.W.deWette, Phys. Rev. B4, 1661(1971).

[30] M.H.Mohmad, L.L.Kesmodel, B.M.Hall and D.L.Mills, Phys. Rev. B37, 2763(1988).

[31] R.B.Doak, V.Harten and J.P.Toennies, Phys. Rev. Lett. 51, 578(1983).

[32] B.Feuerbacker and R.F.Wills, Phys. Rev. Lett. 47, 526 (1981).

[33] Physics Data No. 26, 1(1987) (Ed. W.Kress).

[34] W.Drexel, W.Gläser and F.Gompf, Phys. Lett. 28A, 537 (1969); F.Gompf, W.Reichardt, W.Gläser and K.H.Beckurts, in "Neutron Inelastic Scattering" (IAEA,Vienna, 1978) Vol. 2, p.417.

[35] W.Drexel, Z.Physik 255, 281(1972).

[36] W.A.Kamitakahara and B.N.Brockhouse, Phys. Lett. 29A, 639(1969).

[37] J.W.Lynn, H.G.Smith and R.M.Nicklow, Phys. Rev. B8, 3493(1973).

[38] C.Stassis, T.Gould, O.D.McMasters, K.A.Gschneider and R.M.Nicklow, Phys. Rev. B19, 5746(1979).

[39] E.H.Jacobsen, Phys. Rev. 97, 654(1955).

[40] E.Z.Vintaikin, V.V.Gorbachev and P.L.Gruzin, Soviet Physics-Solid State Physics 7, 296(1965).

[41] T.Schneider and E.Stoll, Solid State Commun. 4, 79 (1966).

[42] S.K.Sinha, Phys. Rev. 143, 422(1966); S.K.Sinha and G.L. Squires, in "Lattice Dynamics" Ed. R.F.Wallis (Pergamon Press, 1965) p. 53.

[43] E.C.Sevenson, B.N.Brockhouse and J.M.Rowe, Phys. Rev. 155, 619(1967).

[44] R.M.Nicklow, G.Gilat, H.G.Smith, L.J.Raubinheimer and M.K. Wilkinson, Phys. Rev. 164, 922(1967).

[45] G.Nilsson and S.Rolandson, Phys. Rev. B7, 2393(1973); Phys. Rev. B9, 3278(1974).

[46] A.P.Miller and B.N.Brockhouse, Phys. Rev. Lett. 20, 798 (1968); Cand. J. Phys. 49, 704(1970).

[47] A.P.Miller, Cand. J. Phys. 53, 2491(1975).

[48] C.Stassis, C.K.Loong and J.Zarestky, Phys. Rev. B26,5426 (1982); W.B.Holzapfel, Phys. Rev. B30, 2232(1984).

[49] S.Hautecler and W.vanDingenen, J. Phys.(Paris) 25, 653 (1964).

[50] R.J.Birgeneau, J.Cordes, G.Dolling and A.D.B.Woods, Phys. Rev. 136, A1359(1964).

[51] G.A.deWit and B.N.Brockhouse, J. Appl. Phys. 39, 451(1968).

[52] R.Orlich and W.Drexel, in "Neutron Inelastic Scattering" (IAEA, Vienna, 1968) Vol. 1, p.203.

[53] D.H.Dutton, B.N.Brockhouse and A.P.Miller, Cand. J.Phys. 50, 2915(1972).

[54] R.A.Reese, S.K.Sinha and D.T.Peterson, Phys. Rev. B8, 1332 (1973).

[55] C.Stassis, C.K.Loong, C.Theisen and R.M.Nicklow, Phys. Rev. B26, 4106(1982).

[56] H.B.Möller and A.R.Mackintosh, in "Inelastic Scattering of Neutrons" (IAEA, Vienna, 1965) Vol. 1, p. 95.

[57] W.M.Shaw and L.D.Mühlestein, Phys.Rev. B4. 969(1971).

[58] L.D.Mühlestein, E.Gürmen and R.M.Cunningham, in " Neutron Inelastic Scattering (IAEA, Vienna, 1972) p. 53.

[59] G.C.E.Low, Proc. Phys. Soc. 79, 480(1962).

[60] G.Shirane, V.J.Minkiewicz, R.Nathan, H.A.Alperin and S.J. Pickart, Bull. Am. Phys. Soc. 11, 759(1966).

[61] B.N.Brockhouse, H.E.Abou-Helal and E.D.Hallman, Solid State Commun. 5, 211(1967); J.Zarestky and C.Stassis, Phys. Rev. B35, 4500(1987).

[62] J.Zarestky, C.Stassis, B.N.Harmon, K.M.Ho and C.L.Fu, Phys. Rev. B28, 697(1983).

[63] B.M.Powell, P.Martel and A.D.B.Woods, Phys. Rev. 171, 727(1968).

[64] C.B.Walker and P.A.Egelstaff, Phys. Rev. 177, 1111 (1969).

[65] Y.Nabagawa and A.D.B.Woods, Phys. Rev. Lett. 11, 271(1963).

[66] A.D.B.Woods and B.M.Powell, Phys. Rev. Lett. 15, 778(1965).

[67] R.I.Sharp, J.Phys. C2, 421,432(1969).

[68] B.M.Powell, A.D.B.Woods and P.Martel, in " Neutron Inelastic Scattering" (IAEA, Vienna, 1972) p.43.

[69] A.D.B.Woods, Phys. Rev. 136, A781(1964).

[70] M.Iizumi, J. Phys. Soc.Japan 52, 549(1983).

[71] J.Peretti, I.Pelah and W.Kley, Phys. Lett. 3, 105(1962).

[72] I.Pelah, R.Hass, W.Kley, K.H.Kreb, J.Peretti and R.Rubin, in " Inelastic Scattering of Neutrons in Solids and Liquids" (IAEA, Vienna, 1963) Vol.2, p.155.

[73] G.Dolling and A.D.B.Woods, in "Thermal Neutron Scattering" Ed. P.A.Egelstaff (Academic Press, New York, 1965) p.210.

[74] D.J.Page, Proc. Phys. Soc.(London) 91, 76(1967).

[75] J.Pons-Corbeau and J.Jouffroy, Bull. Soc. Fran. Mineral Crist. 90, 498(1967).

[76] B.M.S.Kashyap and B.W.Batterman, Bull. Am. Phys. Soc. 12, 282(1967).

[77] R.Collela and B.W.Batterman, Phys. Rev. B1, 3913(1970).

[78] S.H.Chen and B.N.Brockhouse, Solid State Commun. 2, 73 (1964).

[79] C.Stassis, J.Zarestky and N.Wakabayashi, Phys. Rev. Lett. 41, 1726(1978); C.Stassis and J.Zarestky, Solid State commun. 52, 9(1984).

[80] B.Dorner, A.A.Chernyskov, V.V.Pushkarev, A. Yu. Rumyantsev and R.Pynn, J.Phys. F11, 365(1981).

[81] N.Wakabayashi, R.H.Scherm and H.G.Smith, Phys. Rev. B25, 5122(1982).

[82] R.M.Nicklow, N.Wakabayashi, in "Neutron Inelastic Scattering" (IAEA, Vienna, 1972) p. 611.

[83] C.Stassis, D.Arch, O.D.McMasters and B.N.Harmon, Phys. Rev. B24, 730(1981).

[84] J.A.Leake, V.J.Minkiewicz and G.Shirane, Solid State Commun. 7, 535(1969).

[85] R.M.Nicklow, N.Wakabayashi and P.R.Vijayaraghavan, Phys. Rev. B3, 1229(1971).

[86] H.G.Smith and N.Wakabayashi, Solid State Commun. 39,371 (1981).

[87] N.Wakabayashi, S.K.Sinha and F.H.Spedding, Phys. Rev. B4, 2398(1971).

[88] J.C.G.Houmann and R.M.Nicklow, Phys. Rev. B1, 3943(1970).

[89] H.G.Smith, N.Wakabayashi, R.M.Nicklow and S.Mihailovich, in "Proc. Low Temperature Physics" LT-13 Ed. Timmerhaus, W.J.O. Sullivan and E.Hammel (Plenum Press, New York, 1974) Vol.3.

[90] H.G.Smith, N.Wakabayashi and M.Möstöller, in "Super- conductivity in d and f-band Metals" Ed. D.H.Doughlass (Plenum Press, New York, 1976) p.223.

[91] C.Stassis, D.Arch, B.N.Harmon and N.Wakabayashi, Phys. Rev. B19, 181(1979).

[92] D.L.McDonald, M.M.Elcombe and A.W.Pryor, J.Phys. C2, 1857(1969).

[93] T.G.Worlton and R.E.Schmunk, Phys. Rev. B3, 4115(1971).

[94] S.K.Sinha, T.O.Brun, L.D.Mühlestein and J.Sakurai, Phys. Rev. B1, 2430(1969).

[95] L.Almqvist and R.Stedman, J. Phys. F1, 785(1971).

[96] H.E.Bezdek, R.E.Schmunk and L.Finegold, Phys. Stat. Sol. 42, 275(1971).

[97] C.Stassis, J.Zarestky, D.Arch, O.D.McMasters and B.N.Harmon, Phys. Rev. B18, 2632(1978).

[98] C.Stassis, T.Gould, O.D.McMasters, K.A.Schneider Jr. and R.M.Nicklow, Phys. Rev. B25, 64(1982).

[99] W.P.Crummett, H.G.Smith, R.M.Nicklow and N.Wakabayashi, Phys. Rev. B19, 6028(1979).

[100] H.G.Smith, N.Wakabayashi, W.P.Crummett, R.M.Nicklow, G.H.Lander and E.S.Fisher, Phys. Rev. Lett. 44, 1612(1980).

[101] W.Reichardt, P.Roedhammer and F.Holtzberg, Prog. Rept. Teilinstitute Nukleare Festkörperphysik FK-2670 Ed. K.Käfer (1978) p.1.

[102] W.Reichardt, P.Roedhammer and Hufnagel, in "Physics of Transition Metals" Ed. P.Rhodes (Institute of Physics, London, 1981) p. 519.

[103] N.Wakabayashi and A.Furrer, Phys. Rev. B13, 4343(1976).

[104] H.A.Mook, D.R.McWhan and F.Holtzberg, Phys. Rev. B25, 4321(1982).

[105] H.A.Mook, R.M.Nicklow, T.Penney, F.Holtzberg and M.W.Shafer, Phys. Rev. B18, 2925(1978).

[106] S.Ichinose and Y.Kuroda, Phys. Rev. B25, 2550(1980).

[107] F.A.Wedgewood, J. Phys. C7, 3203(1974).

[108] W.Bührer, A.Furrer, P.Wachter, in " Valence Instabilities" Eds.P.Wachter et al (North-Holland, Amsterdam, 1982) p.103.

[109] S.Ichinose and I.Tamura, Phys. Stat. Sol.(b) 120, 703(1983).

[110] W.G.Stirling, G.H.Lander and O.Vogt, J. Phys. C16, 4093(1983).

[111] P.de V.DuPlessis, T.M.Holden, W.J.L.Buyers, J.A.Jackson, A.F. Murray and C.F.vanDoorn, J. Phys. C18, 2809(1985).

[112] W.J.L.Buyers, A.F.Murray, J.A.Jackson,T.M.Holden, P.de V.Du Plessis and O.Vogt, J.Appl. Phys. 52, 2222(1981). Phys. Rev. B33, 7114(1986).

[113] P.Roedhammer, W.Reichardt and F.Holtzberg, Phys. Rev. Lett. 40, 465(1978); in "Lattice Dynamics" Ed. M.Balkanski (Flammarion, Paris, 1978) p.84.

[114] A.Severing, W.Reichardt, E.Holland-Moritz, D.Wohlleben and W.Assmus, Phys. Rev. B38, 1773(1988).

[115] L.Pintschovius et al, Solid State Commun. 47, 663(1983).

[116] E.D.Hallman, Cand. J. Phys. 52, 2235(1974).

[117] S.Katano, M.Iizumi and Y.Noda, J. Phys. F18, 2195(1988).

[118] R.G.Lloyd, L.Cussen and P.W.Mitchell, Solid State Commun. 66, 109(1988).

[119] E.D.Hallman and B.N.Brockhouse, Cand. J. Phys.47,1117 (1969).

[120] Y.Noda, Y.Endoh, S.Katano and M.Iizumi, Physica B+C 120, 317(1983).

[121] Y.Noda and Y.Endoh, J. Phys. Soc. Japan 57, 4225(1988).

[122] C.Stassis, C.K.Loong, J.Zarestky, O.D.McMasters and R.M.Nicklow, Solid State Commun. 36, 677(1980).

[123] C.Stassis, F.X.Kayser, C.K.Loong and D.Arch, Phys. Rev. B24, 3048(1981).

[124] W.G.Stirling, R.A.Cowley and N.W.Stringfellow, J.Phys. F2, 421(1972).

[125] D.M.Paul, R.A.Cowley and B.W.Lucas, J. Phys. F9, 39(1979).

[126] D.Ritcher, J.J.Rush and J.M.Rowe, Phys. Rev. B27, 6227 (1983).

[127] B.P.Schweiss, B.Renker and R.Flükinger, in "Ternary Superconductors" Eds G.K.Shenoy, B.D.Dunlop and F.Y.Fardin (North-Holland, New York, 1980) p.29.

[128] G.Dolling and G.Gilat, in "Inelastic Scattering of Neutrons"
(IAEA, Vienna, 1965) p. 343.

[129] C.vanDizk, Phys.Lett. 32A, 255(1970).

[130] F.Gompf, H.J.Schmid and B.Renker, J. Phys.(Paris) 42,
49(1981).

[131] J.M.Wetter, G.Arnold and R.Wanger, Phys. Rev. B27, 955
(1983).

[132] K.Knorr, B.Renker, W.Assums, B.Lüthi, R.Takke and
H.J.Lauter, Z. Phys. B29, 151(1980).

[133] W.A.Kamitakahara, K.Scharnberg and H.R.Schanks, Phys. Rev.
Lett. 43, 1607(1979).

[134] C.T.Yeh, W.Reichardt, N.Nücker and M.Loewenhaupt, J.
Phys.(Paris) 42, 371(1981).

[135] G.Schell, H.Winter, H.Rietschel and F.Gompf, Phys. Rev. B25,
1589(1982).

[136] R.D.Lowde, R.J.Harlly, G.A.Saunders, M.Sato, R.Scherm and
C.Underhill, Proc. Roy. Soc.(London) A374, 87(1981).

[137] S.M.Shapiro, J.L.Larese, Y.Noda, S.C.Moss and L.E. Tanner,
Phys. Rev. Lett. 57, 3199(1986).

[138] R.Brüinsma, Phys. Rev. B25, 2951(1982).

[139] H.Tietze, M.Müller and B.Renker, J. Phys. C17,L529 (1984).

[140] S.K.Satiza, S.M.Shapiro, M.B.Salemon and C.M.Wayman, Phys.
Rev. B29, 6031(1984).

[141] M.Yetheraj, S.A.Werner and W.B.Yelon, in " Symposium on
Neutron Scattering" (AIP Conf.Proc. New York) 89(1982) p. 343.

[142] B.Renker, K.Käfer, J.Burkin "Prog. Rept. Teilinstitut Nukleare Festkörperphysik, KFK-2881" Eds. K.P.Bohner and M.Kobbelt (1979) p. 28.

[143] A.Okamoto, Y.Fujita and C.Tasuyama, J. Phys. Soc. Japan 52, 312(1983).

[144] T.S.Prevender, S.K.Sinha and J.F.Smith, Phys. Rev. B6, 4438(1972).

[145] W.Kobbelt, N.Nücker, W.Reichardt and B.Schearer, in " Superconductivity in d and f-band Metals", Eds. W.Bückel and W.Weber (Kernforschungszentrum, Karlsruhe, 1982) p. 119.

[146] J.E.Jorgensen, J.D.Axe, L.M.Corliss and J.M.Hastings, Phys. Rev. B25, 5856(1982).

[147] B.P.Schweiss, B.Renker, E.Schneider and W.Reichardt, [90] p. 189.

[148] L.Pintschovius, H.G.Smith, W.Weber, W.Reichardt, N.Wakabayashi, G.Webb, Z.Fisk, Y.K.Chang, E.Aker and C.Polifis, [145] p. 9 ; Phys. Rev. B28, 5866(1983).

[149] L.Pintschovius, W.Reichardt, E.Aker, C.Polifis and H.G.Smith, "Proc. 17th Int. Conf. on Low Temperature Physics" Eds. U.Eckern, A.Schmid, W.Weber,H.Wühl (North-Holland, Amsterdam, 1984) p. 587.

[150] P.Müller, U.Buchenau, N.Nücker, B.Renker and A. Miller, [149], p. 599.

[151] D.E.Moncton, J.D.Axe and F.J.DiSalvo, Phys. Rev. B16, 801(1977).

[152] J.D.Axe and G.Shirane, Phys. Rev. B8, 1965(1973); Phys. Rev. B18, 3742(1978).

[153] L.Pintschovius, H.Takei and N.Toyota, Phys. Rev. Lett. 54, 1260(1985).

[154] K.R.A.Ziebeck, B.Dorner, W.G.Stirling and R.Schöllhorn, J. Phys. F7, 1139(1977).

[155] N.Wakabayashi, H.G.Smith, K.C.Woo and F.C.Brown, Solid State Commun. 28, 923(1978).

[156] E.Schneider, P.Schweiss and W.Reichardt, "Proc. of Conf. on Neutron Scattering" Ed. R.M.Moon (Gatlinburg, 1976) p.223.

[157] W.Reichardt, H.G.Smith and Y.K.Chang, [149], p. 587.

[158] G.F.Syrykh, M.G.Zemlyanov and N.A.Chernoplekov, Soviet Physik: Solid State, 22, 789(1980).

[159] H.G.Smith and W.Gläser, Phys. Rev. Lett. 25,1611 (1970); Phys. Rev. Lett. 29, 353(1972).

[160] H.G.Smith and W.Gläser, in "Proc. Int. Conf. on Phonons" Rennes, France, Ed. M.A.Nusivovici (Flammarion, Paris, 1971) p. 145.

[161] L.D.Mühlestein, E.Gürmen and R.M.Cunningham, [68] p. 53.

[162] L.P.Pintschovius, W.Reichardt and B.Scheerer, J. Phys. C11, 1557(1978).

[163] H.G.Smith, [90] p. 321.

[164] F.A.Wedgwood, J. Phys. C7, 3203(1974).

[165] L.Pintschovivus, W.Reichardt and B.Scheerer, [101], KFK-2538, Eds. J.Greek and G.Linker (1977) p. 4.

[166] H.G.Smith, N.Wakabayashi and M.Möstoller, [90] p. 223.

[167] A.N.Christensen, W.Kress, M.Miura and N.Lehner, Phys. Rev. B22, 977(1983).

[168] A.N.Christensen, O.W.Dietrich, W.Kress, W.D.Teuchert and R.Currat, Solid State Commun. 31, 795(1979).

[169] F.A.Wedgwood, J. Phys. C7, 320(1974).

[170] W.Kress, P.Roedhammer, H.Bilz, W.D.Teuchert and A.N. Christensen, Phys. Rev. B17, 111(1978).

[171] G.Dolling, T.M.Holden, E.C.Sevensson, W.J.L.Buyers and G.H.Lander, [113] p. 81; Physica B102, 291 (1980).

[172] W.Weber, P.Roedhammer, L.Pintschovius, W.Reichardt, F. Gompf and A.N.Christensen, Phys. Rev. Lett. 43, 868(1979).

[173] A.N.Christensen, O.W.Dietrich, W.Kress and W.D.Teuchert, Phys. Rev.B19, 5699(1979).

[174] C.J.Glinka, J.M.Rowe, J.J.Rush, C.G.Libowitz and A.J. Maeland, Solid State Commun. 22, 541(1977); P.Vorderwisch, S.Hautecler and B.Dorner, Solid State Commun. 34, 853(1980).

[175] S.M.Shapiro, D.Ritcher, Y.Noda and H.Birnbaum, Phys. Rev. B23, 1594(1981).

[176] J.Eckert, J.A.Goldstone, D.Tonks and D.Richter, Phys. Rev. B27, 1980(1983).

[177] J.M.Rowe, N.Vagelatos, J.J.Rush and H.E.Flotow, Phys. Rev. B12, 2959(1975).

[178] E.Burkel, H.Behr, H.Metzger, J.Peisel and G.Eckhold, Z. Phys. B53, 27(1983); D.Richter, J. Less Comm. Met. 89, 292(1983).

[179] J.J.Rush, A.Magrel, J.M.Rowe, J.M.Harris and J.L.Provo, Phys. Rev. B24, 4903(1981).

[180] A.Magrel, J.M.Rowe and D.Richter, Phys. Rev.B23, 1605(1981).

[181] J.M.Rowe, J.J.Rush and H.G.Smith, Phys. Rev. B8, 6013(1975).

[182] O.Blaschko, R.Klamencic, P.Weinzierl and L.Pintschovius, Phys. Rev. B24, 1552(1981).

[183] A.Magrel, N.Stump, W.D.Teuchert, G.Alefeld and V.Wagner, J. Phys. C10, 2783(1977).

[184] A.Magrel, W.D.Teuchert and R.Scherm, J. Phys. C11, 2175 (1978).

[185] J.J.Rush, J.M.Rowe, C.J.Glinka, N.Vagelatos and H.E.Flotow, Phys. Rev. B21, 5613(1980).

[186] S.M.Shapiro, F.Reindinger and J.F.Lynch, J. Phys. F12, 1869(1982).

[187] P.P.Parshin, M.G.Zemlyanov, M.E.Kost, A.Yu Rumyantsev and N.A.Chernoplekov, Inorg. Mater. 14, 1288(1978).

[188] W.Bührer, A.Fürrer, W.Hälg and L.Schlapbach, J. Phys. F9, L141(1978).

[189] P.P.Parshin, M.G.Zemlyanov, M.E.Kost, A.Yu Rumyantsev and N.A.Chernoplekov, Soviet Physik- Solid State 22, 275(1980).

[190] D.G.Hunt and D.K.Ross, J.Less Comm. Met. 49, 169(1976).

[191] M.Deitrich W.Reichardt and W.Rietschel, Solid State Commun. 21, 603(1971).

[192] J.F.Miller, C.B.Satterthwaite, T.O.Brun, J.D.Jorgensen, and K.Sköld, "Proc. Conf. Neutron Scattering" Ed. R.M.Moon (Gatlinberg, 1976) Vol. 1, p. 529.

[193] R.Hempelmann, D.Richter and B.Stritzker, J. Phys. F12, 79(1982).

[194] N.Kunitomi, H.Shiraishi, Y.Tsunoda and K.Osamura, Solid State Commun. 44, 767(1982).

[195] J.J.Rush, J.M.Rowe and A.J.Maeland, J. Phys. F10, L283(1980).

[196] J.E.Bonnet, S.K.P.Wilson and D.K.Ross, in "Electronic Structure and Properties of Hydrogen in Metals" Eds. P.Jena and C.B.Satterthwaite (Plenum Press, New York, 1983) p. 183.

[197] J.J.Rush, H.E.Flotow, D.W.Conner and C.L.Thaker, J. Chem. Phys. 45, 3817(1966).

[198] A.J.Maeland, J. Chem. Phys. 52, 3952(1970).

[199] T.B.Wu and W.Yelon, Acta Metall. 30, 2065(1982).

[200] E.C.Sevenson, B.M.Powell, A.D.B.Woods and W.D. Teuchert, Cand. J. Phys. 57, 253(1979).

[201] E.C.Sevenson and B.N.Brockhouse, Phys. Rev. Lett. 18, 858(1967).

[202] M.Sakamoto and Y.Hamaguchi, in "Neutron Inelastic Scattering" (IAEA, Vienna, 1968) p.181.

[203] N.Kunikomi, Y.Tsunoda and H.Shiraishi, Solid State Commun. 34, 519(1980).

[204] F.Menzinger, F.Sacchetti and M.C.Spinelli, Phys. Rev. B12, 2253(1975).

[205] P.Bosi, F.Menzinger and F.Sacchetti, J. Phys. F10, 1673(1980).

[206] W.A.Kamitakahara and B.N.Brockhouse, Phys. Rev. B10, 1200(1974).

[207] Y.Tsunoda, N.Kunitomi, N.Wakabayashi, R.M.Nicklow and H.G.Smith, Phys. Rev. B19, 2876(1979).

[208] E.Maliszewski, S.Bednarski, A.Czachor and J.Sosnowski, J. Phys. F9, 2335(1979).

[209] E.Maliszewski, J.Sosnowski, S.Bednarski, A.Czachor and A.Holas, J. Phys. F5, 1455(1975).

[210] P.Bosi, F.Dupre, F.Menzinger, F.Sacchetti and M.C. Spinelli, Nouvo Cimento B46, 337(1978).

[211] E.Maliszewski, S.Bednarski, A.Czachor and J.Sosnowski, Phys. Rev. B16, 770(1977); Z. Phys. B36, 127(1979).

[212] K.Mikke and J.J.Jankowska, Phys. Stat. Sol.(b) 55, K1(1973).

[213] Y.Endoh, Y.Noda and M.Iizumi, J. Phys. Soc. Japan 50, 469(1981).

[214] Y.Endoh and Y.Noda, J. Phys. Soc. Japan 46, 806 (1979); ibid 57, 4225(1988).

[215] M.Sato, B.H.Grier, S.M.Shapiro and M.Miyajima, J. Phys. F12, 2117(1982).

[216] K.Tajma, Y.Endoh, Y.Ishikawa and W.G.Stirling, Phys. Rev. Lett. 37, 519(1976).

[217] H.G.Smith, N.Wakabayashi and M.Möstoller, [90] p.223.

[218] R.J.Arsenault, D.M.Esterling, R.M.Nicklow and M. Möstoller,J. Phys. F14, 1619(1984).

[219] J.D.Axe, D.T.Keating and S.C.Moss, Phys. Rev. Lett. 35,530(1975).

[220] H.Terauchi, K.Sakaue and H.Hida, J. Phys. Soc. Japan 50. 3932(1981).

[221] R.L.Cappelletti, N.Wakabayashi, W.Kamitakahara, J.C. Traylor and A.J.Bevolo, Phys. Rev. B25, 6090(1982).

[222] E.C.Sevenson and W.A.Kamitakahara, Cand. J. Phys. 49, 2291(1971).

[223] N.A.Chernoplekov, Soviet Phys. JETP 53, 369(1981).

[224] N.A.Chernoplekov, M.G.Zemlyanov, E.G.Brovman and A.G.Chicherin, in "Inelastic Scattering of Neutrons in Solids and Liquids"(IAEA, Vienna, 1963) Vol.2 p.173.

[225] Yu. L.Shitikov, N.A.Chernoplekov, M.G.Zemlyanov and A.P. Zhernov, J. Phys. C16. 2471(1983).

[226] G.F.Syrykh, M.G.Zemlyanov, N.A.Chernoplekov and B.I.Savelev, Soviet Phys. JETP 54, 165(1981).

[227] B.Mozer, in "Neutron Inelastic Scattering"(IAEA, Vienna, 1968) Vol. 1 p.53.

[228] S.A.Danilkin, V.V.Zakurkin, S.I.Morozov and V.V. Sumin, Soviet Phys. Solid State 22, 1948(1980).

[229] G.F.Syrykh, M.G.Zemlyanov, N.A.Chernoplekov and V.M.Koltygin, Soviet Phys.JETP 46, 162(1977).

[230] G.F.Syrykh, A.P.Zhernov, M.G.Zemlyanov, S.P. Mironov, N.A.Chernoplekov and Yu.L.Shitikov, Soviet Phys. JETP 43, 183(1976).

[231] G.F.Syrykh, A.P.Zhernov, M.G.Zemlyanov, N.A. Chernoplekov and G.I.Solovev, Phys. Stat. Sol.(b)79, 105 (1977).

[232] M.Möstoller, T.Kaplan, N.Wakabayashi and R.M. Nicklow, Phys. Rev. B10, 3144(1974).

[233] J.T.McKinney, E.R.Jones and M.B.Webb, Phys. Rev. 160, 523(1967).

[234] B.F.Mason and B.R.Williams, Phys. Rev. Lett. 46, 1138(1981); J. Chem. Phys. 75, 2199(1981).

[235] L.L.Kesmodel, M.L.Xu and S.Y.Tong, Phys. Rev. B34 2010(1986).

[236] U.Harten, J.P.Toennies and Ch.Wolt, Faraday Discuss. Chem. Soc. 80, 137(1985).

[237] M.Wuttig, C.Oshima, T.Aizawa, R.Souda, O.S.Otani and Y.Ishizawa, Surf. Sci.192. 573(1987).

[238] C.Oshima, R.Souda, M.Aono, S.Otani and Y.Ishizawa, Phys. Rev Lett. 56, 240(1986).

[239] S.Lehwald, J.M.Szeftel, H.Ibach, T.S.Rehman and D.L. Mills, Phys. Rev. Lett. 50, 581(1983).

[240] M.Rocca, S.Lehwald and H.Ibach, Surf. Sci. 138, L123(1984).

[241] C.Oshima, R.Souda,M. Aono, S. Otani and Y. Ishizawa, (private communication).

[242] A.Adnot, and J.D. Carette, Phys. Rev. Lett. 38, 1084 (1977).

[243] R.B.Doak, V.Harten and J.P. Toennies, Phys. Rev. Lett. 51, 575(1983).

[244] B.Feuerbacher and R.F.Wallis, Phys. Rev. Lett.47, 526(1981).

[245] K.Kern, R.David, R.L.Palmer, G.Cosma and T.S.Rehman, Phys. Rev. B33,4334(1986).

[246] W.C.Overtone and J.Gaffney, Phys. Rev. 98, 969(1955).

[247] B.Segall, Phys. Rev. 125, 109(1962).

[248] J.R.Neighbours and G.A.Alers, Phys. Rev. 111, 707 (1958); Y.A.Chang, J. Appl. Phys. 37, 3567(1966); W.B.Daniels and C.S.Smith, Phys. Rev. 111, 713(1958).

[249] O.K.Anderson and A.R.Mackintosh, Solid State Commun. 6, 285(1968); J.B.Ketterson, F.M.Müller and L.R.Windmiller, Phys. Rev. 186, 656(1969).

[250] J.Callaway and H.M.Zhang, Phys. Rev.B1,305(1970).

[251] A.D.B.Woods, B.N.Brockhouse, N.H.March and A.T. Stewart, Phys. Rev.128,1112(1962).

[252] R.J.Weiss and J.J.DeMarco, Phys. Rev. 140, A1223 (1965).

[253] S.H.Koening, Phys. Rev.135, A1693(1964).

[254] B.T.Matthias, T.H.Geballe and V.B.Compton, Rev. Mod. Phys.,35, 1(1963).

[255] G.F.Fleming and T.L.Loucks, Phys. Rev. 173. 685(1968).

[256] T.L.Loucks, Phys. Rev. 144, 504(1966).

[257] P.K.Iyenger, G.Venkataraman, Y.H.Gameel and K.R.Rao, in "Neutron Inelastic Scattering"(IAEA, Vienna, 1968) Vol.1, p.195.

[258] S.H.Liu, R.P.Gupta and S.K.Sinha, Phys. Rev. B4, 1100(1971).

[259] J.A.Jackman, T.M.Holden, W.J.L.Buyers, P. de V.DuPlessis, O.Vöigt and J.Genossor, Phys.Rev.B33, 7144(1986).

[260] D.E.Monclon, F.J. DiSalvo and J.D.Axe, [113] p.561.

[261] W.Weber, [145] p.15.

[262] H.L.Brown, P.E.Armstrong and C.P.Kempter, J. Chem. Phys. 45, 547(1966).

[263] R.W.Bartlett and C.W.Smith, J.Appl.Phys. 38, 5428(1967).

[264] R.J.Walter and W.T.Chandler, Trans, AIME, 233, 762(1965).

[265] C.Wert, D.O.Thompson and O.Buck, J. Phys. Chem. Solids 31, 1973(1970).

[266] H.D.Carstenjen and R.Sizmann, Ber.Bunsenges Phys. Chem. 76, 1223(1872).

[267] V.A.Somenkov, A.V.Gurskaya, M.G.Zemlyanov, M.E.Kost, N.A.Chernoplekov and A.A.Chertkov, Sovt. Phys.-Solid State 10, 1076(1968); Sovt. Phys.-Solid State 10, 2133(1969).

[268] V.A.Somenkov, I.R.Entin, A.Yu.Chervyakov, S.Sh.Shilshstein and A.A.Chertkov, Sovt.Phys.-Solid State 13, 2187(1972).

[269] J.M.Rowe, J.J.Rush, M.Möstoller, H.G.Smith and H.E.Flotow, Phys. Rev. Lett. 33, 1297(1974).

[270] J.M.Rowe, J.J.Rush, L.A. deGraaf and G.A.Ferguson, Phys. Rev. Lett. 29, 1250(1972).

[271] D.W.Taylor, Phys.Rev. 156, 1017(1967).

[272] W.A.Kamitakahara and D.W.Taylor, Phys. Rev. B10, 1190(1974).

[273] E.Colavita, A.Franciosi, R.Rosei, F.Sacchetti, E.S.Giuliano, R.Ruggeri and D.W.Lynch, Phys. Rev. B20, 4866(1979).

[274] F.Gompf, W.Richter, B.Scheerer and W.Weber, Physica B108, 1337 (1981).

[275] K.H.Reider and W.Drexel, Phys. Rev. Lett 34, 148(1975).

[276] V.Bortolani, F.Nizzoli and G.Santoro in "Lattice Dynamics" Ed. M.Balkanski, (Flammarion Science, 1977) p.302.

Chapter 4

MICROSCOPIC THEORIES OF LATTICE DYNAMICS

Contents

List of Figures

List of Tables

Chapter 4

MICROSCOPIC THEORIES OF LATTICE DYNAMICS

In the BvK theory of lattice dynamics, the force constants are just the parameters. Therefore BvK theory does not provide the information about the fundamental nature of forces and their origin in the solid. For the detailed understanding of these forces, the lattice dynamical studies must start from the consideration of electronic wave functions and their readjustment to the nuclear motions. Born and Oppenheimer [1] were the first to formulate the microscopic theory of lattice dynamics of solids (Chapter 2). Later on many authors have developed the microscopic theories of lattice dynamics of solids and of metals in particular [2]-[13].

The lattice dynamics of simple (s-band) metals is more or less understood using the following approximations: (i) The conduction electrons are uniformly distributed in the metal, (ii) the ion-cores are small in size and are rigid and move as a whole and (iii) the conduction electrons readjust to the ionic motions adiabatically. In the transition metals (TM) and rare earth metals (REM), approximations (i) and (ii) are not valid. The d and f-electrons in these metals are quasilocalized [14] and the d and f-shells suffer distortion in the presence of thermal field and these shells move along with the rigid ion-cores.

In the microscopic theories of lattice dynamics, screening of ion-ion interaction through conduction electrons, is a central phenomenon. In this chapter we present the general formulation of dielectric screening and dynamical matrix starting from the first principles. The application to TMs, TMCs, REMs and their compound and alloys will also be discussed in detail.

4.1   Longitudinal Dielectric Tensor

4.1.1   Linear Response Theory

Consider a system of large number of electrons subjected to a static external scalar potential $\Delta V_{ext}(\vec{r})$ due to a set of external charges. The electons will redistribute themselves in accordance with $\Delta V_{ext}(\vec{r})$ and therefore there will be a change in the electron density $\Delta\rho(\vec{r})$ which in turn will produce an additional screening potential $\Delta V_{sc}(\vec{r})$. Thus the total change in the potential $\Delta V_t(\vec{r})$ inside the system is

$$\Delta V_t(\vec{r}) = \Delta V_{ext}(\vec{r}) + \Delta V_{sc}(\vec{r}) \tag{4.1}$$

where

$$\begin{aligned}\Delta V_t(\vec{r}) &= e^2 \int \frac{\Delta\rho(\vec{r}\,')}{|\,\vec{r} - \vec{r}\,'\,|} d\vec{r}\,' \\ &= \int \Delta\rho(\vec{r}\,')v(\vec{r} - \vec{r}\,')d\vec{r}\,' \end{aligned} \tag{4.2}$$

and $v(\vec{r}-\vec{r}\,') = e^2/\,|\,\vec{r}-\vec{r}\,'\,|$ is the electron-electron Coulomb interaction.

If it is assumed that $\Delta V_{sc}(\vec{r})$ is small, it can be expressed as a linear function of $\Delta V_{ext}(\vec{r})$ i.e.

$$\Delta V_{sc}(\vec{r}) = \int f(\vec{r}, \vec{r}\,')\Delta V_{ext}(\vec{r}\,')d\vec{r}\,' \tag{4.3}$$

where $f(\vec{r}, \vec{r}\,')$ is a nonlocal response function and represents the induced potential at $\vec{r}$ due to an unit external potential at $\vec{r}\,'$. We write Eq.(4.1) as

$$\Delta V_t(\vec{r}) = \int \epsilon^{-1}(\vec{r}, \vec{r}\,')\Delta V_{ext}(\vec{r}\,')d\vec{r}\,' \tag{4.4}$$

where $\epsilon^{-1}(\vec{r}, \vec{r}\,')$ is known as the response function which is equal to $\Delta V_t(\vec{r})$ for an unit external potential at $\vec{r}\,'$. Combining Eqs.(4.1), (4.3) and (4.4) one gets

$$\epsilon^{-1}(\vec{r}, \vec{r}\,') = \delta(\vec{r} - \vec{r}\,') + f(\vec{r}, \vec{r}\,'). \tag{4.5}$$

The dielectric function $\epsilon(\vec{r}, \vec{r}')$ is the inverse of response function $\epsilon^{-1}(\vec{r}, \vec{r}')$ and satisfies the orthogonality relation

$$\int \epsilon(\vec{r}, \vec{r}'')\epsilon^{-1}(\vec{r}'', \vec{r}')d\vec{r}'' = \int \epsilon^{-1}(\vec{r}\vec{r}'')\epsilon(\vec{r}''\vec{r}')d\vec{r}''$$
$$= \delta(\vec{r} - \vec{r}'). \tag{4.6}$$

The above expressions are quite general as nothing has been said about the source and probe. If the external charge (source) and the test charge (probe) are represented as e and t, it leads to four different response functions [15]-[17] namely $\epsilon_{tt}^{-1}, \epsilon_{et}^{-1}, \epsilon_{te}^{-1}$ and ϵ_{ee}^{-1}. In the lattice dynamical studies, both the source and probe are the atomic nuclei of solid, therefore, we study ϵ_{tt}^{-1} which is written as ϵ^{-1} throughout.

In the Fourier space $\epsilon^{-1}(\vec{r}, \vec{r}')$ can be expanded as

$$\epsilon^{-1}(\vec{r}, \vec{r}') = \frac{1}{\Omega_0} \sum_{\vec{Q},\vec{Q}'} \epsilon^{-1}(\vec{Q}, \vec{Q}') \exp(i\vec{Q} \cdot \vec{r}) \exp(-i\vec{Q}' \cdot \vec{r}') \tag{4.7}$$

where

$$\epsilon^{-1}(\vec{Q}, \vec{Q}') = \frac{1}{\Omega_0} \int d\vec{r} \int d\vec{r}' \epsilon^{-1}(\vec{r}, \vec{r}') \exp(-i\vec{Q} \cdot \vec{r}) \exp(i\vec{Q}' \cdot \vec{r}'). \tag{4.8}$$

Here $\vec{Q}$ is a vector in the Fourier space, Ω_0 is atomic volume and all the physical quantities are normalized to atomic volume. Use of Eqs.(4.7) and (4.8) in Eq.(4.6) gives

$$\sum_{\vec{Q}''} \epsilon(\vec{Q}, \vec{Q}'')\epsilon^{-1}(\vec{Q}'', \vec{Q}') = \sum_{\vec{Q}''} \epsilon^{-1}(\vec{Q}, \vec{Q}'')\epsilon(\vec{Q}'', \vec{Q}')$$
$$= \delta_{\vec{Q},\vec{Q}'}. \tag{4.9}$$

The Fourier transform of inverse dielectric function given in Eq.(4.5) is

$$\epsilon^{-1}(\vec{Q}, \vec{Q}') = \delta_{\vec{Q},\vec{Q}'} + f(\vec{Q}, \vec{Q}') \tag{4.10}$$

where $f(\vec{Q}, \vec{Q}')$ is the Fourier transform of $f(\vec{r}, \vec{r}')$. Similarly, the Fourier transform of $\Delta V_t(\vec{r})$ with the help of Eqs.(4.1) to (4.4) is defined as

$$\Delta V_t(\vec{Q}) = \Delta V_{ext}(\vec{Q}) + v(\vec{Q})\Delta\rho(\vec{Q}), \tag{4.11}$$
$$= \sum_{\vec{Q}'} \epsilon^{-1}(\vec{Q}, \vec{Q}')\Delta V_{ext}(\vec{Q}'). \tag{4.12}$$

The linear part of the electron density perturbation $\Delta\rho(\vec{r})$ is defined as

$$\Delta\rho(\vec{r}) = \int \chi(\vec{r}, \vec{r}')\Delta V_{ext}(\vec{r}')d\vec{r}' \tag{4.13}$$

where $\chi(\vec{r}, \vec{r}')$ is the response to $\Delta V_{ext}(\vec{r})$. The Fourier transform of $\Delta\rho(\vec{r})$ is

$$\Delta\rho(\vec{Q}) = \sum_{\vec{Q}'} \chi(\vec{Q}, \vec{Q}')\Delta V_{ext}(\vec{Q}'). \tag{4.14}$$

Using Eq.(4.14) in Eq.(4.11) and comparing it with Eq.(4.12), one gets

$$\epsilon^{-1}(\vec{Q}, \vec{Q}') = \delta_{\vec{Q},\vec{Q}'} + v(\vec{Q})\chi(\vec{Q}, \vec{Q}'). \tag{4.15}$$

From Eqs.(4.10) and (4.15), the function $f(\vec{Q}, \vec{Q}')$ is related to $\chi(\vec{Q}, \vec{Q}')$ as

$$f(\vec{Q}, \vec{Q}') = v(\vec{Q})\chi(\vec{Q}, \vec{Q}'). \tag{4.16}$$

In the study of electronic properties of solids, one often encounters with the dielectric function $\epsilon(\vec{r}, \vec{r}')$. To understand the different physical properties of $\epsilon(\vec{r}, \vec{r}')$ and $\epsilon^{-1}(\vec{r}, \vec{r}')$, we write $\epsilon(\vec{r}, \vec{r}')$ in a more general form as

$$\epsilon(\vec{r}, \vec{r}') = \delta(\vec{r}, \vec{r}') - g(\vec{r}, \vec{r}') \tag{4.17}$$

where $g(\vec{r}, \vec{r}')$ is another nonlocal function. The inverse of Eq.(4.4), with the help of Eq.(4.6), can be written as

$$\Delta V_{ext}(\vec{r}) = \int \epsilon(\vec{r}, \vec{r}')\Delta V_t(\vec{r}')d\vec{r}'. \tag{4.18}$$

Using Eq.(4.17) in (4.18), one obtains Eq.(4.1) where the screening potential is defined as

$$\Delta V_{sc}(\vec{r}) = \int g(\vec{r}, \vec{r}')\Delta V_t(\vec{r}')d\vec{r}'. \tag{4.19}$$

Therefore the function $g(\vec{r}, \vec{r}')$ is the screening potential at $\vec{r}$ due to a unit potential inside the crystal at $\vec{r}'$. One can also write $\Delta\rho(\vec{r})$ as

$$\Delta\rho(\vec{r}) = \int \chi^0(\vec{r}, \vec{r}')\Delta V_t(\vec{r}')d\vec{r}' \qquad (4.20)$$

where $\chi^0(\vec{r}, \vec{r}')$ is known as the polarizability function and is different from $\chi(\vec{r}, \vec{r}')$ (see Eq.(4.13)). Using Eqs. (4.1) and (4.2) in Eq.(4.20), we get

$$\Delta\rho(\vec{r}) = \int \chi^0(\vec{r}, \vec{r}') \left[\Delta V_{ext}(\vec{r}') + \int \Delta\rho(\vec{r}'')v(\vec{r}' - \vec{r}'')d\vec{r}''\right] d\vec{r}'. \qquad (4.21)$$

The above equations give the inter-relations between the different response functions. From Eqs.(4.3), (4.4), (4.18) and (4.19), it is evident that $f(\vec{r}, \vec{r}')$ and $\epsilon^{-1}(\vec{r}, \vec{r}')$ are the response functions to an external scalar potential while $g(\vec{r}, \vec{r}')$ and $\epsilon(\vec{r}, \vec{r}')$ are the response functions to the total scalar potential acting in the medium. Therefore $\epsilon^{-1}(\vec{r}, \vec{r}')$ can be experimentally measured as a response to an external potential while $\epsilon(\vec{r}, \vec{r}')$ is a purely theoretical physical quantity as it is too difficult to measure the total potential inside the solid. In a theoretical formulation, it is more convenient to deal with $\epsilon(\vec{r}, \vec{r}')$ rather than its inverse. Once $\epsilon(\vec{r}, \vec{r}')$ is known, the inverse can be obtained with the help of Eq.(4.6).

The Fourier transfrom of $\epsilon(\vec{r}, \vec{r}')$ is

$$\epsilon(\vec{r}, \vec{r}') = \frac{1}{\Omega_0} \sum_{\vec{Q}, \vec{Q}'} \epsilon(\vec{Q}, \vec{Q}') \exp(i\vec{Q} \cdot \vec{r}) \exp(-i\vec{Q}' \cdot \vec{r}') \qquad (4.22)$$

where

$$\epsilon(\vec{Q}, \vec{Q}') = \frac{1}{\Omega_0} \int d\vec{r} \int d\vec{r}' \epsilon(\vec{r}, \vec{r}') \exp(-i\vec{Q} \cdot \vec{r}) \exp(i\vec{Q}' \cdot \vec{r}'). \qquad (4.23)$$

With these definitions, Eqs.(4.17) to (4.19) and (4.21) can be written in the Fourier space as

$$\Delta V_{ext}(\vec{Q}) = \sum_{\vec{Q}'} \epsilon(\vec{Q}, \vec{Q}')\Delta V_t(\vec{Q}'), \qquad (4.24)$$

$$\Delta V_{sc}(\vec{Q}) = \sum_{\vec{Q}'} g(\vec{Q}, \vec{Q}')\Delta V_t(\vec{Q}'), \qquad (4.25)$$

$$\epsilon(\vec{Q}, \vec{Q}') = \delta_{\vec{Q}, \vec{Q}'} - g(\vec{Q}, \vec{Q}'), \qquad (4.26)$$

and

$$\Delta\rho(\vec{Q}) \;=\; \int \chi^0(\vec{Q},\vec{Q}') \left[\Delta V_{ext}(\vec{Q}') + \Delta\rho(\vec{Q}')v(\vec{Q}')\right] d\vec{Q}'.$$

$$(4.27)$$

Equation (4.27) has to be solved self-consistently as $\Delta\rho(\vec{Q})$ appears on both the sides of equation. $\epsilon^{-1}(\vec{Q},\vec{Q}')$ and $\epsilon(\vec{Q},\vec{Q}')$ given in Eqs.(4.10) and (4.26) are infinite dimensional matrices in $\vec{Q}$ and $\vec{Q}'$. In the crystalline solids $\vec{Q}$ and $\vec{Q}'$ differ only by a reciprocal lattice vector due to translational symmetry. Therefore one can write $\vec{Q} = \vec{q} \pm \vec{G}$ and $\vec{Q}' = \vec{q}' \pm \vec{G}'$ where $\vec{G}$ and $\vec{G}'$ are reciprocal lattice vectors and $\vec{q}$ is reduced wave vector.

4.1.2   Hartree Dielectric Function

The quantum mechanical description of dielectric tensor is given by many authors [15], [18], [19]. We shall describe here the general formalism for the dielectric matrix $\epsilon(\vec{Q},\vec{Q}')$ where electron-electron interaction is included in the Hartree approximation and the linear response to the external field is accounted for [18].

In the Hartree approximation the electronic wavefunction for a particular configuration of the lattice ions denoted by $\vec{R} = \{\vec{R}\}$, satisfies the self-consistent Schrödinger equation

$$\left[-\frac{\hbar^2}{2m}\nabla_r^2 + V_b(\vec{r},\vec{R}) \;+\; e^2\sum_{\vec{k}'\mu'}\int d\vec{r}' \frac{|\,\psi_{\vec{k}'\mu'}(\vec{r}',\vec{R})\,|^2}{|\,\vec{r}-\vec{r}'\,|^2}\right] \psi_{\vec{k}\mu}(\vec{r},\vec{R})$$

$$= E_{\vec{k}\mu}\psi_{\vec{k}\mu}(\vec{r},\vec{R}) \qquad\qquad (4.28)$$

where $V_b(\vec{r},\vec{R})$ is the sum of bare rigid ion potentials $v_b(\vec{r}-\vec{R}_l)$. Here $\vec{R}_l$ are the displaced positions of the ions and $\mu(= lm\sigma)$ is the band index (l, m and σ are the orbital, magnetic and spin quantum numbers respectively). The term with $\vec{k} = \vec{k}'$ is retained in the Hartree potential to avoid difficulties concerning the orthogonalization and the sum is over all the occupied states $|\,\vec{k}'\mu' >$.

Thermal field makes the ions to vibrate about their mean positions which causes a change in the bare-ion potential. For small displacements, $V_b(\vec{r}, \vec{R})$ is expanded in the powers of $\vec{u}_l$ and the terms linear in $\vec{u}_l$ are retained. Thus

$$
\begin{aligned}
V_b(\vec{r}, \vec{R}) &= \sum_l v_b(\vec{r} - \vec{R}_l) \\
&= \sum_l v_b(\vec{r} - \vec{R}_l^0) + \sum_l \vec{u}_l \cdot \frac{\partial v_b}{\partial \vec{R}_l} \Big|_{\vec{R}_l^0} + \dots \\
&= V_b^0(\vec{r}) + \Delta V_b(\vec{r}).
\end{aligned}
\tag{4.29}
$$

$V_b^0(\vec{r})$ is the unperturbed electron-ion potential and the change in potential $\Delta V_b(\vec{r})$ will produce a change in the electron wave function which is written for a configuration $\{\vec{R}\}$ as

$$
\psi_{\vec{k}\mu}(\vec{r}, \vec{R}) = \psi_{\vec{k}\mu}(\vec{r}) + \Delta \psi_{\vec{k}\mu}(\vec{r}).
\tag{4.30}
$$

In the first order perturbation theory

$$
\Delta \psi_{\vec{k}\mu}(\vec{r}, \vec{R}) = \sum_{\vec{k}'\mu'} \frac{< \psi_{\vec{k}'\mu'} \mid \Delta V_t(\vec{r}) \mid \psi_{\vec{k}\mu} >}{E_{\vec{k}\mu} - E_{\vec{k}'\mu'}} \psi_{\vec{k}'\mu'}(\vec{r}).
\tag{4.31}
$$

The change in electron wave function causes the change in the electron denisty $\Delta\rho(\vec{r})$ which in turn yields an additional screening potential $\Delta V_{sc}(\vec{r})$ which is given in Eq.(4.3). Therefore

$$
\Delta V_t(\vec{r}) = \Delta V_b(\vec{r}) + \Delta V_{sc}(\vec{r}).
\tag{4.32}
$$

Equation (4.32) is similar to Eq.(4.1) except that here the external potential is $\Delta V_b(\vec{r})$ which arises due to thermal motion of ions. The total electron density in the perturbed crystal is given as

$$
\begin{aligned}
\rho(\vec{r}, \vec{R}) &= e^2 \sum_{\vec{k}\mu} f_{\vec{k}\mu} \psi^*_{\vec{k}\mu}(\vec{r}, \vec{R}) \psi_{\vec{k}\mu}(\vec{r}, \vec{R}) \\
&= e^2 \sum_{\vec{k}\mu} f_{\vec{k}\mu} \psi^*_{\vec{k}\mu}(\vec{r}) \psi_{\vec{k}\mu}(\vec{r}) + e^2 \sum_{\vec{k}\mu} f_{\vec{k}\mu} \Big[\psi^*_{\vec{k}\mu}(\vec{r}) \Delta\psi_{\vec{k}\mu}(\vec{r}) \\
&\quad + \Delta\psi^*_{\vec{k}\mu}(\vec{r}) \psi_{\vec{k}\mu}(\vec{r}) \Big] + \dots \\
&= \rho(\vec{r}) + \Delta\rho(\vec{r})
\end{aligned}
\tag{4.33}
$$

244   *S. PRAKASH*

where $f_{\vec{k}\mu}$ is Fermi-Dirac distribution function. Equation (4.30) is used to obtain Eq.(4.33). The potential $\Delta V_{sc}(\vec{r})$ is related to $\Delta\rho(\vec{r})$ through the Poisson equation

$$
\begin{aligned}
\nabla^2\left[\Delta V_{sc}(\vec{r})\right] &= -4\pi\Delta\rho(\vec{r}) \\
&= -4\pi e^2 \sum_{\vec{k}\mu} f_{\vec{k}\mu}\left[\psi^*_{\vec{k}\mu}(\vec{r})\Delta\psi_{\vec{k}\mu}(\vec{r}) + \Delta\psi^*_{\vec{k}\mu}(\vec{r})\psi_{\vec{k}\mu}(\vec{r})\right].
\end{aligned}
$$

$$(4.34)$$

Substituting Eq.(4.31) in Eq.(4.34) and rearranging the terms one gets

$$
\begin{aligned}
\nabla^2\left[\Delta V_{sc}(\vec{r})\right] &= -4\pi e^2 \sum_{\vec{k}\mu}\sum_{\vec{k}'\mu'} \frac{f_{\vec{k}\mu}-f_{\vec{k}'\mu'}}{E_{\vec{k}\mu}-E_{\vec{k}'\mu'}} < \psi_{\vec{k}'\mu'}\mid\Delta V_t(\vec{r})\mid\psi_{\vec{k}\mu}> \\
&\quad \times \psi^*_{\vec{k}\mu}(\vec{r})\psi_{\vec{k}'\mu'}(\vec{r}).
\end{aligned}
$$

$$(4.35)$$

If $\Delta V_b(\vec{r}),\Delta V_{sc}(\vec{r})$ and hence $\Delta V_t(\vec{r})$ are slowly varying functions of $\vec{r}$, these can be expanded in the Fourier space as

$$
\begin{aligned}
\Delta V_t(\vec{r}) &= \sum_{\vec{Q}}\Delta V_t(\vec{Q})\exp(i\vec{Q}\cdot\vec{r}), \\
\Delta V_{sc}(\vec{r}) &= \sum_{\vec{Q}}\Delta V_{sc}(\vec{Q})\exp(i\vec{Q}\cdot\vec{r}), \\
\Delta V_b(\vec{r}) &= \sum_{\vec{Q}}\Delta V_b(\vec{Q})\exp(i\vec{Q}\cdot\vec{r}).
\end{aligned}
$$

$$(4.36)$$

Using Eqs.(4.36) in (4.35) one gets

$$
\begin{aligned}
\sum_{\vec{Q}'}\Delta V_{sc}(\vec{Q}')\mid\vec{Q}'\mid^2\exp(i\vec{Q}'\cdot\vec{r}) &= -4\pi e^2 \sum_{\vec{Q}'}\sum_{\vec{k}\mu}\sum_{\vec{k}'\mu'}\frac{f_{\vec{k}\mu}-f_{\vec{k}'\mu'}}{E_{\vec{k}\mu}-E_{\vec{k}'\mu'}} \\
&\quad \times <\psi_{\vec{k}'\mu'}\mid\exp(i\vec{Q}'\cdot\vec{r})\mid\psi_{\vec{k}\mu}> \\
&\quad \times \psi^*_{\vec{k}\mu}(\vec{r})\psi_{\vec{k}'\mu'}(\vec{r})\Delta V_t(\vec{Q}').
\end{aligned}
$$

$$(4.37)$$

Post multiplying Eq.(4.37) by $\exp(-i\vec{Q}\cdot\vec{r})$ and integrating over the atomic volume, one gets

$$
\Delta V_{sc}(\vec{Q}) = v(\vec{Q})\sum_{\vec{Q}'}\sum_{\vec{k}\mu}\sum_{\vec{k}'\mu'}\frac{f_{\vec{k}\mu}-f_{\vec{k}'\mu'}}{E_{\vec{k}\mu}-E_{\vec{k}'\mu'}}
$$

$$\times \;\; < \psi_{\vec{k}'\mu'} \mid \exp(i\vec{Q}' \cdot \vec{r}) \mid \psi_{\vec{k}\mu} >$$

$$\times \;\; < \psi_{\vec{k}\mu} \mid \exp(-i\vec{Q} \cdot \vec{r}) \mid \psi_{\vec{k}'\mu'} > \Delta V_t(\vec{Q}') \qquad (4.38)$$

where $v(\vec{Q}) = 4\pi e^2/|\vec{Q}|^2$ and

$$\frac{1}{\Omega_0} \int \exp(i(\vec{Q}' - \vec{Q}) \cdot \vec{r}) d\vec{r} \;=\; \delta_{\vec{Q}',\vec{Q}}. \qquad (4.39)$$

In the present calculations, Eq.(4.24) is

$$\Delta V_b(\vec{Q}) \;=\; \sum_{\vec{Q}'} \epsilon(\vec{Q}, \vec{Q}') \Delta V_t(\vec{Q}'). \qquad (4.40)$$

Equations (4.36) and (4.40) are used in Eq.(4.32) to get

$$\Delta V_{sc}(\vec{Q}) \;=\; \sum_{\vec{Q}'} \left[\delta_{\vec{Q},\vec{Q}'} - \epsilon(\vec{Q}, \vec{Q}') \right] \Delta V_t(\vec{Q}'). \qquad (4.41)$$

The comparison of Eqs.(4.38) and (4.41) gives the dielectric matrix

$$\epsilon(\vec{Q}, \vec{Q}') = \delta_{\vec{Q},\vec{Q}'} - v(\vec{Q})\chi^0(\vec{Q}, \vec{Q}') \qquad (4.42)$$

where the polarizability tensor $\chi^0(\vec{Q}, \vec{Q}')$ is defined as

$$\chi^0(\vec{Q}, \vec{Q}') \;=\; \sum_{\vec{k}\mu} \sum_{\vec{k}'\mu'} \frac{f_{\vec{k}\mu} - f_{\vec{k}'\mu'}}{E_{\vec{k}\mu} - E_{\vec{k}'\mu'}} < \psi_{\vec{k}'\mu'} \mid \exp(i\vec{Q}' \cdot \vec{r}) \mid \psi_{\vec{k}\mu} >$$

$$\times < \psi_{\vec{k}\mu}(\vec{r}) \mid \exp(-i\vec{Q} \cdot \vec{r}) \mid \psi_{\vec{k}'\mu'} > . \qquad (4.43)$$

$\chi^0(\vec{Q}, \vec{Q}')$ is also known as the susceptibility tensor. Comparing Eqs.(4.26) and (4.42) one can write the relation between the function $g(\vec{Q}, \vec{Q}')$ and the polarizability function as

$$g(\vec{Q}, \vec{Q}') = v(\vec{Q})\chi^0(\vec{Q}, \vec{Q}'). \qquad (4.44)$$

For a crystalline solid Eqs.(4.42) and (4.43) are written in terms of wave vector $\vec{q}$, restricted in the first BZ, as

$$\epsilon(\vec{q}+\vec{G}, \vec{q}+\vec{G}') \;=\; \delta_{\vec{G},\vec{G}'} - v(\vec{q}+\vec{G})\chi^0(\vec{q}+\vec{G}, \vec{q}+\vec{G}') \qquad (4.45)$$

where

$$\chi^0(\vec{q}+\vec{G},\vec{q}+\vec{G}') \;=\; \sum_{\vec{k}\mu}\sum_{\vec{k}'\mu'}\left[\frac{f_{\vec{k}\mu}-f_{\vec{k}'\mu'}}{E_{\vec{k}\mu}-E_{\vec{k}'\mu'}}\right]$$

$$\times <\psi_{\vec{k}'\mu'}\mid e^{(i(\vec{q}+\vec{G}')\cdot\vec{r})}\mid\psi_{\vec{k}\mu}>$$

$$\times <\psi_{\vec{k}\mu}\mid e^{(-i(\vec{q}+\vec{G})\cdot\vec{r})}\mid\psi_{\vec{k}'\mu'}> . \qquad (4.46)$$

Here

$$\vec{k}' \;=\; \vec{k}+\vec{q}\pm\vec{H}, \qquad (4.47)$$

and $\vec{H}=\vec{G}\pm\vec{G}'$ is also a reciprocal lattice vector. On the right side of Eqs.(4.43) and (4.46), the term in the square brackets, is called energy band structure part of the polarizability tensor and the remaining factor is called overlap part.

The full quantum mechnical treatment of frequency and wave vector dependent dielectric matrix is given by Nozieres and Pines [20] and by Ehrenreich and Cohen [21]. These authors obtained the explicit expressions for longitudinal component of dielectric matrix and evaluated it in the free electron approximation. Adler [22] and Wiser [23] extended the formalism for full dielectric matrix and applied it to insulating cubic solids.

The dielectric matrix given in Eq.(4.45) is applicable to all kinds of solids. For insulators

$$\lim_{\vec{q}\to 0}\epsilon^{-1}(\vec{q},\vec{q})=\frac{1}{\epsilon_0}, \qquad (4.48)$$

with $\vec{G}=\vec{G}'=0$. ϵ_0 is the low frequency optical dielectric constant. The off-diagonal elements $(\vec{G}\neq\vec{G}')$ of $\epsilon(\vec{q}+\vec{G},\vec{q}+\vec{G}')$ are nonzero for the insulators and these are necessary to satisfy the acoustic sum rule [10], [12] which is the charge neutrality condition and ensures that all the acoustic frequencies approach to zero as $\mid\vec{q}\mid\to 0$. The dielectric function for the metals diverges as $\mid\vec{q}\mid\to 0$ and then the acoustic sum rule has no restriction on the dielectric function. The nearly free electron approximation gives a satisfactory description of simple metals as the off-diagonal elements of dielectric matrix are small as compared to diagonal elements. Neglecting the off-diagonal elements and restricting the wave vector $\vec{q}$ in the first BZ, Eq.(4.45) simplifies as

$$\epsilon(\vec{q}, \vec{q}) = 1 - v(\vec{q}) \sum_{\vec{k}\mu} \sum_{\vec{k}'\mu'} \left[\frac{f_{\vec{k}\mu} - f_{\vec{k}'\mu'}}{E_{\vec{k}\mu} - E_{\vec{k}'\mu'}} \right]$$
$$\times \ |< \psi_{\vec{k}\mu} | \exp(-i\vec{q}\cdot\vec{r}) | \psi_{\vec{k}'\mu'} >|^2 . \qquad (4.49)$$

In the free electron approximation, the Bloch wave function

$$\psi_{\vec{k}\mu}(\vec{r}) = \frac{1}{\sqrt{\Omega_0}} \exp(i\vec{k}\cdot\vec{r}) \qquad (4.50)$$

and energy eigenvalues

$$E_{\vec{k}\mu} = \frac{\hbar^2 k_\mu^2}{2m_\mu} \qquad (4.51)$$

where $\vec{k}_\mu$ and m_μ are the electron wave vector and electron mass in the band μ respectively. Thus Eq.(4.49) simplifies as

$$\epsilon(\vec{q}) = 1 - v(\vec{q}) \sum_{\vec{k}\mu} \frac{f_{\vec{k}\mu} - f_{\vec{k}+\vec{q}\mu}}{E_{\vec{k}\mu} - E_{\vec{k}+\vec{q}\mu}} \qquad (4.52)$$

because the overlap matrix elements reduce to unity. If it is further assumed that the conduction electrons occupy only one partially filled s-band and adjust to the nuclear motion through the intraband transitions, Eq.(4.52) further simplifies as (Appendix A)

$$\epsilon(\vec{q}) = 1 - v(\vec{q}) \sum_{\vec{k}} \frac{f_{\vec{k}} - f_{\vec{k}+\vec{q}}}{E_{\vec{k}} - E_{\vec{k}+\vec{q}}}$$
$$= 1 + \frac{me^2}{2\pi k_F \hbar^2 \eta^2} \left[1 + \frac{1-\eta^2}{2\eta} ln \left| \frac{1+\eta}{1-\eta} \right| \right] \qquad (4.53)$$

with $\eta = q/2k_F$. The Lindhard function $\epsilon(q)$ is a well behaved function except at $q = 2k_F$. The first derivative of $\epsilon(q)$ diverges to minus infinity logarithmically at $q = 2k_F$. In the limit $q \to 2k_F$, Eq.(4.53) simplifies as

$$\epsilon(2k_F) = 1 + \frac{me^2}{2\pi k_F \hbar^2}. \qquad (4.54)$$

One finds a very small inflection in $\epsilon(q)$ verses q curve in the neighbourhood of $q = 2k_F$. After a huge magnification of the curve, the infinite slope becomes apparent. As $q \to 0$,

$$\epsilon(q) = \frac{4me^2 k_F}{\pi \hbar^2 q^2} \tag{4.55}$$

which diverges as $(1/q^2)$. This is in agreement with the fact that the screening is complete at the large distances in a metal. As $q \to \infty$, $\epsilon(q)$ approaches unity as

$$\epsilon(q) \cong 1 + \frac{16me^2 k_F^3}{3\pi \hbar^2 q^2}. \tag{4.56}$$

4.1.3   Exchange and Correlation Interactions

The calculation of dielectric function in the above section is in the Hartree approximation. The electrons are assumed non-interacting and single particle wave functions are used. In reality, the electrons are interacting and many body interactions which are known as exchange and correlation interactions among the conduction electrons must be accountd for in the calculation of $\epsilon(q)$ [20]. This is a complete subject in itself, therefore we introduce here only a qualitative description of the modifications of dielectric function due to exchange and correlation interactions. These interactions reduce the Coulomb interaction between the conduction electrons. Therefore the effective electron-electron interaction in the Fourier spacee is written as

$$v^*(q) = v(q)\left[1 - f_{xc}(q)\right]. \tag{4.57}$$

The function $f_{xc}(q)$ accounts for the exchange and correlation interactions. If we rewrite Eq.(4.52) as

$$\epsilon(q) \;=\; 1 - v(q)\chi^0(q), \tag{4.58}$$

the dielectric function $\epsilon_{te}(q)$ which includes the exchange and correlation interactions, is modified as [15]

$$\begin{aligned}
\epsilon_{te}(q) \;&=\; 1 - v(q)\left[1 - f_{xc}(q)\right]\chi^0(q) \\
&=\; 1 - v^*(q)\chi^0(q) \\
&=\; \epsilon^*(q).
\end{aligned} \tag{4.59}$$

However, in the ion-ion interaction the diagonal part of the particle response function in Eq.(4.15) is modified by including the exchange and correlation interactions as

$$
\begin{aligned}
\chi^*(q) &= \frac{1}{v^*(q)}\left[\frac{1}{\epsilon^*(q)} - 1\right] \\
&= \frac{1}{v^*(q)}\frac{1 - \epsilon^*(q)}{\epsilon^*(q)}.
\end{aligned}
\tag{4.60}
$$

Using $\epsilon^*(q)$ from Eq.(4.59) in Eq.(4.60), one gets

$$
\chi^*(q) = \frac{\chi^0(q)}{1 - v^*(q)\chi^0(q)}.
\tag{4.61}
$$

Using Eq.(4.61) for $\chi(q)$ in Eq.(4.15) one gets the modified inverse dielectric function as

$$
\begin{aligned}
\epsilon^{*-1}(q) &= 1 + v(q)\chi^*(q) \\
&= \frac{1 + v(q)f_{xc}(q)\chi^0(q)}{1 + v(q)[1 - f_{xc}(q)]\chi^0(q)} \\
&= \frac{1 + v(q)f_{xc}(q)\chi^0(q)}{\epsilon_{te}(q)}.
\end{aligned}
\tag{4.62}
$$

Therefore the dielectric function $\epsilon^*(q)$ $[= 1/\epsilon^{*-1}(q)]$ becomes

$$
\epsilon^*(q) = 1 - \frac{v(q)\chi^0(q)}{1 + v(q)[1 - f_{xc}(q)]\chi^0(q)}.
\tag{4.63}
$$

The relation between $\epsilon^*(q)$ and $\epsilon_{te}(q)$ becomes more transparent if we multiply and devide the right side of Eq.(4.63) by $(1 - f_{xc}(q))$ and rearrange the terms as

$$
\epsilon^*(q) = \frac{[1 - f_{xc}(q)]\epsilon_{te}(q)}{1 - f_{xc}(q)\epsilon_{te}(q)}.
\tag{4.64}
$$

In all the cases $f_{xc}(q) \to 0$ as $q \to 0$, i.e. the exchange and correlation effects become unimportant at small values of q and this makes $\epsilon^*(q) = \epsilon_{te}(q)$ as $q \to 0$.

The inbuilt philosophy in the calculations of the effects of exchange and correlation interactions has been to reduce the Coulomb interaction between the conduction electrons. This modified dielectric function has been used to calculate the phonon frequencies of simple metals. Toya [24] used Slater exchange potential

$$v_{ex}(r) = -3e^2 \left[\frac{3}{8\pi} \rho(r) \right]^{1/3}. \tag{4.65}$$

In the local density approximation, use of Eq.(4.65) gives

$$f_{xc}(q) = \tfrac{5}{16} \frac{q^2}{k_F^2} \qquad \text{if} \quad 5q^2 > 16k_F^2$$
$$= 0 \quad \text{otherwise.} \tag{4.66}$$

The exchange and correlation interactions are investigated extensively for the free electron gas using many body perturbation theory [25]-[36]. Hubbard [35] gave the expression for $f_{xc}(q)$ as

$$f_{xc}(q) = \frac{q^2}{2(q^2 + Ik_F^2)} \tag{4.67}$$

where I is an adjustable parameter. Singwi et al [31] calculated $f_{xc}(q)$ using self-consistent screening theory by satisfying the compressibility sum rule and achieving a positive pair-correlation function. These authors suggested $f_{xc}(q)$ in the analytic form as

$$f_{xc}(q) = A \left[1 - \exp(-Bq^2/k_F^2) \right]. \tag{4.68}$$

The prameters A and B depend upon the interelectronic distance. Although the nonlinear screening calculations have become available, these are yet not applied to the calculations of phonons frequencies.

In a TM, the conduction electrons consist of both the s and d-characters. The exchange and correlation corrections should be included in the interactions among the core and conduction electrons as well as in the interactions among the conduction electrons. The latter will consist of exchange and correlation interactions among the s-s, s-d and d-d electron interactions. The most important exchange interaction among d-electrons is the exclusion of two electrons with the same spin from occupying the same orbital at the same site. The exchange and correlation corrections given in Eqs.(4.67) and (4.68) cannot be adopted as such for quasilocalized d-electrons. Lindgren and Schwarz [33] derived the exchange corrections for d-electrons using Slater exchange potential. The correlations among d-electrons are often neglected in the calculations of phonon frequencies as these are not yet known explicitly.

4.1.4 Few Observations

Equation (4.24) defines the dielectric matrix as

$$\Delta V_{ext}(\vec{Q}) \;=\; \sum_{\vec{Q}'} \epsilon(\vec{Q},\vec{Q}')\Delta V_t(\vec{Q}'). \tag{4.69}$$

This equation is equivalent to the equation

$$\vec{D}(\vec{Q}) \;=\; \sum_{\vec{Q}'} \epsilon(\vec{Q},\vec{Q}')\vec{E}(\vec{Q}'), \tag{4.70}$$

if $\vec{D}(\vec{Q})$ and $\vec{E}(\vec{Q})$ are longitudinal fields. Equation (4.70) in the matrix form is

$$\vec{D} \;=\; \epsilon\vec{E}, \tag{4.71}$$

which is generalized form of one of the Maxwell's equations for dielectrics and it is valid for all type of solids. If the solid is homogeneous and isotropic

$$\epsilon(\vec{Q},\vec{Q}') \;=\; \delta_{\vec{Q},\vec{Q}'}\epsilon(\vec{Q},\vec{Q}'). \tag{4.72}$$

Thus $\epsilon(\vec{Q},\vec{Q}')$ becomes scalar and one obtains the Maxwell equation [37]. It is because of this similarity of Eqs.(4.69) and (4.70) that $\epsilon(\vec{Q},\vec{Q}')$ is usually called dielectric matrix. The local field (LF) in a dielectric is the net field acting at the ion site inside the solid and is different from the macroscopic field [38]. The LF includes the additional contribution arising from the dipoles induced on the ions which makes the electron density anisotropic and nonuniform. The nonuniformity of the electron density in a solid makes the off-diagonal terms of $\epsilon(\vec{Q}, \vec{Q}')$ finite. Therefore the off-diagonal elements of $\epsilon(\vec{Q}, \vec{Q}')$ are manifestations of the LFs. $\epsilon(\vec{Q}, \vec{Q}')$ is connected to the dielectric constant $\epsilon(\infty)$ as [15]

$$\lim_{Q \to 0} \frac{1}{\epsilon^{-1}(\vec{Q}, \vec{Q})} = \sum_{\alpha\beta} \hat{Q}_\alpha \epsilon_{\alpha\beta}(\infty) \hat{Q}_\beta \qquad (4.73)$$

where $\hat{Q}_\alpha$ and $\hat{Q}_\beta$ are unit vectors.

4.2 Dynamical Matrix

Toya [24] was the first to put forward a microscopic theory of lattice dynamics for monovalent simple metals. Toya separated the interactions in a metal into three parts: (i) Direct ion-ion Coulomb interaction, (ii) ion-ion short range repulsive interaction arising from the overlap of ionic charge distribution (a consequence of Pauli's exclusion principle) and (iii)ion-ion interaction via conduction electrons. Later on several authors [2] - [13] developed the microscopic theories of lattice dynamics for different class of materials. The most concise and general theories are due to Pick et al [11], [12] and Sham [9], [10], [13] which we describe here.

A metal consists of an ordered assembly of ions through which the conduction electrons permeate. The Hamiltonian for such a system is written as in Eqs.(2.1)- (2.3). In the adiabatic approximation, the equation of motion of ions involves the electronic energy $E_n(\vec{R})$ as a part of potential energy of the lattice. If the potential energy terms in Eqs.(2.1)-(2.3) are grouped together, one can write

$$\Phi(\vec{R}) = \sum_{l'\kappa'>l\kappa} v_I(\vec{R}(l\kappa) - \vec{R}(l'\kappa')) + \sum_{l\kappa i} v_b(\vec{r}_i - \vec{R}(l\kappa))$$

$$+ \sum_{i>j} \frac{e^2}{|\vec{r}_i - \vec{r}_j|}. \qquad (4.74)$$

The first term in Eq. (4.74) is the total repulsive ion-ion interaction which is the sum of Coulomb interaction $\Phi^c(\vec{R}(l\kappa) - \vec{R}(l'\kappa'))$ and the short range repulsive overlap interaction $\Phi^r(\vec{R}(l\kappa) - \vec{R}(l'\kappa'))$. The second term is the attractive electron-ion interaction and the last term is the repulsive electron-electron Coulomb interaction. The last two terms can be combined together to get screened electron-ion interaction $\Phi^e(\vec{R})$. Here $\Phi^e(\vec{R})(= E_n(\vec{R})$ in the adiabatic approximation) also includes the exchange and correlation interactions among the core and conduction electrons and among the conduction electrons. Therefore we write Eq. (4.74) as

$$\Phi(\vec{R}) \;=\; \Phi^c(\vec{R}) + \Phi^r(\vec{R}) + \Phi^e(\vec{R}). \qquad (4.75)$$

Thus the force constants $\Phi_{\alpha\beta}(l\kappa, l'\kappa')$ are also the sum of three contributions i.e.

$$\Phi_{\alpha\beta}(l\kappa, l'\kappa') \;=\; \Phi^c_{\alpha\beta}(l\kappa, l'\kappa') + \Phi^r_{\alpha\beta}(l\kappa, l'\kappa') + \Phi^e_{\alpha\beta}(l\kappa, l'\kappa')$$

$$(4.76)$$

where

$$\Phi^c_{\alpha\beta}(l\kappa, l'\kappa') \;=\; \nabla_{l\kappa\alpha}\nabla_{l'\kappa'\beta}\Phi^c(\vec{R}), \qquad (4.77)$$

$$\Phi^r_{\alpha\beta}(l\kappa, l'\kappa') \;=\; \nabla_{l\kappa\alpha}\nabla_{l'\kappa'\beta}\Phi^r(\vec{R}), \qquad (4.78)$$

$$\Phi^e_{\alpha\beta}(l\kappa, l'\kappa') \;=\; \nabla_{l\kappa\alpha}\nabla_{l'\kappa'\beta}\Phi^e(\vec{R}),$$

$$\;=\; \nabla_{l\kappa\alpha}\nabla_{l'\kappa'\beta}E^e(\vec{R}). \qquad (4.79)$$

The subscripts c,r,e denote the Coulomb, repulsive overlap and the electronic contributions in Eqs.(4.75) - (4.79).

The force constants due to ion-ion Coulomb interaction and ion-electron-ion interaction can be obtained with the help of bare-ion potential and the dielectric screening matrix. The potential seen by an ion at $\vec{R}(l\kappa)$ due to another ion at $\vec{R}(l'\kappa')$, in analogy with Eq.(4.4), is

$$\Phi(l\kappa, l'\kappa') \;=\; \int \epsilon^{-1}(\vec{R}_{l\kappa}, \vec{r})v_b(\vec{r} - \vec{R}(l'\kappa'))d\vec{r} \qquad (4.80)$$

where $v_b(\vec{r} - \vec{R}(l'\kappa'))$ is the bare pair potential experienced by an unit charge at $\vec{r}$ (here a test charge) due to an ion at $\vec{R}(l'\kappa')$. $\Phi(l\kappa, l'\kappa')$ is the total ion-ion interaction potential. Here the response of conduction electrons is included through dielectric function $\epsilon(\vec{R}(l\kappa), \vec{r})$. We write the Fourier transforms of $v_b(\vec{r} - \vec{R}(l'\kappa'))$ and $\epsilon^{-1}(\vec{R}(l\kappa), \vec{r})$ as

$$v_b(\vec{r} - \vec{R}(l'\kappa')) = \sum_{\vec{q}\vec{G}''} v_b(\vec{q} + \vec{G}'') \exp\left[(i(\vec{q} + \vec{G}'') \cdot (\vec{r} - \vec{R}(l'\kappa'))\right] \qquad (4.81)$$

and

$$\epsilon^{-1}(\vec{R}(l\kappa),\vec{r}) = \sum_{\vec{q}\vec{G}\vec{G}'} \epsilon^{-1}(\vec{q}+\vec{G},\vec{q}+\vec{G}')\exp(i(\vec{q}+\vec{G})\cdot\vec{R}(l\kappa))$$
$$\times \exp(-i(\vec{q}+\vec{G}')\cdot\vec{r}) \qquad (4.82)$$

Use of Eqs.(4.81) and (4.82) in Eq.(4.80) gives

$$\Phi(l\kappa,l'\kappa') = \sum_{\vec{q}\vec{G}\vec{G}'\vec{G}''} \epsilon^{-1}(\vec{q}+\vec{G},\vec{q}+\vec{G}')v_b(\vec{q}+\vec{G}'')$$
$$\times \exp(i(\vec{q}+\vec{G})\cdot\vec{R}(l\kappa))\exp(-i(\vec{q}+\vec{G}'')\cdot\vec{R}(l'\kappa'))$$
$$\times \int \exp(i(\vec{G}''-\vec{G}')\cdot\vec{r})d\vec{r}. \qquad (4.83)$$

The integral in Eq.(4.83) is Dirac delta function $\delta_{\vec{G}'',\vec{G}'}$. Therefore

$$\Phi(l\kappa,l'\kappa') = \sum_{\vec{q}\vec{G}\vec{G}'} \epsilon^{-1}(\vec{q}+\vec{G},\vec{q}+\vec{G}')v_b(\vec{q}+\vec{G}')$$
$$\times \exp(i(\vec{q}+\vec{G})\cdot\vec{R}(l\kappa))\exp(-i(\vec{q}+\vec{G}')\cdot\vec{R}(l'\kappa')).$$
$$(4.84)$$

Differentiating Eq.(4.84) with respect to $u_\alpha(l\kappa)$ and $u_\beta(l'\kappa')[\vec{R}(l\kappa) = \vec{R}^0(l\kappa)+\vec{u}(l\kappa)]$ and evaluting the derivatives at the equilibrium positions, we get the force constants as

$$\Phi_{\alpha\beta}(l\kappa,l'\kappa') = \sum_{\vec{q}\vec{G}\vec{G}'} (\vec{q}+\vec{G})_\alpha(\vec{q}+\vec{G}')_\beta\epsilon^{-1}(\vec{q}+\vec{G},\vec{q}+\vec{G}')v_b(\vec{q}+\vec{G}')$$
$$\times \exp(i(\vec{q}+\vec{G})\cdot\vec{R}^0(l\kappa))\exp(-i(\vec{q}+\vec{G}')\cdot\vec{R}^0(l'\kappa')).$$
$$(4.85)$$

Using $\Phi_{\alpha\beta}(0\kappa,l\kappa')$ from Eq.(4.85) in Eq.(2.63), the dynamical matrix becomes (normalized to one unit cell)

$$\bar{D}_{\alpha\beta}(\vec{q},\kappa\kappa') = \frac{1}{(M_\kappa M_{\kappa'})^{1/2}} \sum_{\vec{G}\vec{G}'} (\vec{q}+\vec{G})_\alpha(\vec{q}+\vec{G}')_\beta$$
$$\times\epsilon^{-1}(\vec{q}+\vec{G},\vec{q}+\vec{G}')v_b(\vec{q}+\vec{G}')\exp(i(\vec{G}\cdot\vec{R}^0_\kappa - \vec{G}'\cdot\vec{R}^0_{\kappa'})).$$
$$(4.86)$$

If the resultant ion-electron-ion interaction is Coulombic in nature i.e.

$$Z_\kappa v_b(\vec{q}+\vec{G}') \;=\; v_I(\vec{q}+\vec{G}')$$

$$=\; \frac{4\pi Z_\kappa Z_{\kappa'} e^2}{\Omega_0 \mid \vec{q}+ +\vec{G}' \mid^2}, \tag{4.87}$$

use of Eq.(4.87) in Eq.(4.86) gives

$$\bar{D}_{\alpha\beta}(\vec{q},\kappa\kappa') \;=\; \frac{4\pi Z_\kappa Z_{\kappa'} e^2}{\Omega_0 (M_\kappa M_{\kappa'})^{1/2}} \sum_{\vec{G}\vec{G}'} \frac{(\vec{q}+\vec{G})_\alpha (\vec{q}+\vec{G}')_\beta}{\mid \vec{q}+\vec{G}' \mid^2}$$

$$\times \epsilon^{-1}(\vec{q}+\vec{G},\vec{q}+\vec{G}') \exp(i(\vec{G}\cdot\vec{R}^0_\kappa - \vec{G}'\cdot\vec{R}^0_{\kappa'})). \tag{4.88}$$

$Z_\kappa e$ and $Z_{\kappa'} e$ are the charges on the ions at $\vec{R}^0_\kappa$ and $\vec{R}^0_{\kappa'}$ respectively. Equation (4.88) is the primary result [7] - [12] and it can be rearranged to write

$$\bar{D}_{\alpha\beta}(\vec{q},\kappa\kappa') \;=\; \frac{\Omega_0}{4\pi e^2} \frac{1}{(M_\kappa M_{\kappa'})^{1/2}} \sum_{\vec{G}\vec{G}'} (\vec{q}+\vec{G})_\alpha v_b^\kappa(\vec{q}+\vec{G}) \mid \vec{q}+\vec{G} \mid^2$$

$$\times \epsilon^{-1}(\vec{q}+\vec{G},\vec{q}+\vec{G}') v_b^{\kappa'}(\vec{q}+\vec{G}')(\vec{q}+\vec{G}')_\beta$$

$$\times \exp(i(\vec{G}\cdot\vec{R}^0_\kappa - \vec{G}'\cdot\vec{R}^0_{\kappa'})) \tag{4.89}$$

where $v_b^\kappa(\vec{q}+\vec{G})$ is the Fourier transform of bare electron-ion potential for the κth type of ion and is defined as

$$v_b^\kappa(\vec{q}+\vec{G}) \;=\; \frac{-4\pi Z_\kappa e^2}{\Omega_0 \mid \vec{q}+\vec{G} \mid^2}. \tag{4.90}$$

Using $\epsilon^{-1}(\vec{q}+\vec{G},\vec{q}+\vec{G}')$ from Eq.(4.15) in Eq.(4.89) we get the general expression for dynamical matrix as

$$\bar{D}_{\alpha\beta}(\vec{q},\kappa\kappa') \;=\; \bar{D}^c_{\alpha\beta}(\vec{q},\kappa\kappa') + \bar{D}^E_{\alpha\beta}(\vec{q},\kappa\kappa') \tag{4.91}$$

where

$$\bar{D}^c_{\alpha\beta}(\vec{q},\kappa\kappa') \;=\; \frac{4\pi Z_\kappa Z_{\kappa'} e^2}{\Omega_0 (M_\kappa M_{\kappa'})^{1/2}} \sum_{\vec{G}} \frac{(\vec{q}+\vec{G})_\alpha (\vec{q}+\vec{G})_\beta}{\mid \vec{q}+\vec{G} \mid^2}$$

$$\times \exp(i\vec{G}\cdot(\vec{R}^0_\kappa - \vec{R}^0_{\kappa'})) \tag{4.92}$$

and

$$\bar{D}_{\alpha\beta}^{E}(\vec{q},\kappa\kappa') \;=\; \frac{1}{(M_{\kappa}M_{\kappa'})^{1/2}} \sum_{\vec{G}\vec{G}'} (\vec{q}+\vec{G})_{\alpha} v_{b}^{\kappa}(\vec{q}+\vec{G}) \chi(\vec{q}+\vec{G},\vec{q}+\vec{G}')$$
$$\times v_{b}^{\kappa'}(\vec{q}+\vec{G}')(\vec{q}+\vec{G}')_{\beta} \exp(i(\vec{G}\cdot\vec{R}_{\kappa}^{0} - \vec{G}'\cdot\vec{R}_{\kappa'}^{0})). \tag{4.93}$$

The first term in Eq.(4.91) $\bar{D}_{\alpha\beta}^{c}(\vec{q},\kappa\kappa')$ is a consequence of bare ion-ion Coulomb interaction which is evaluated with the help of Ewald's Θ-function transformation (see Appendix B). The second term $\bar{D}_{\alpha\beta}^{E}(\vec{q},\kappa\kappa')$ is a consequence of ion-electron-ion interaction which is an indirect interaction between the ions via conduction electrons. This interaction is illustrated in Fig.4.1. The movement of ion A($l\kappa$) induces a local change in electron charge density at a point C. Consequently an additional potential is created and theorefore a force is exerted on the moving ion B($l'\kappa'$). Thus the ionic motions are correlated through the conduction electrons. An alternate method for obtaining $\bar{D}_{\alpha\beta}^{E}(\vec{q},\kappa\kappa')$ is given by Sham [10] using the self-consistent first order perturbation theory (see Appendix C).

Equation (4.88) is obtained for the Coulomb interaction between the ions. The calculation of short range repulsive overlap interaction $\Phi^{r}(\vec{R})$ from the first principles is nontrivial. In general, one uses Born-Mayer type pair potential i.e.

$$\Phi^{r}(R) = b' \exp[-(R-R_{0})/\rho'] \tag{4.94}$$

where b' and ρ' are the constants which are determined with the help of lattice parameter and the compressibility data. Here R_{0} is the equilibrium separation between the pair of ions. The force constant

$$\Phi_{\alpha\beta}^{r}(0\kappa,l\kappa') \;=\; \frac{\partial^{2}}{\partial R'_{\alpha}\partial R'_{\beta}} \Phi_{\alpha\beta}^{r}(\vec{r}-\vec{R}_{l}^{0})\,|_{\vec{r}=\vec{R}_{\kappa\kappa'}^{0}} \tag{4.95}$$

where $\vec{R}' = \vec{r} - \vec{R}_{l}^{0}$. Using Eq.(4.94) in Eq.(4.95) we get

$$\Phi_{\alpha\beta}^{r}(0\kappa,l\kappa') \;=\; \frac{b'}{\rho'}\left[\left(\frac{1}{|\vec{R}_{\kappa\kappa'}^{0}-\vec{R}_{l}^{0}|}+\frac{1}{\rho'}\right)\frac{\left(\vec{R}_{\kappa\kappa'}^{0}-\vec{R}_{l}^{0}\right)_{\alpha}\left(\vec{R}_{\kappa\kappa'}^{0}-\vec{R}_{l}^{0}\right)_{\beta}}{|\vec{R}_{\kappa\kappa'}^{0}-\vec{R}_{l}^{0}|^{2}}\right.$$
$$\left.-\frac{\delta_{\alpha\beta}}{|\vec{R}_{\kappa\kappa'}^{0}-\vec{R}_{l}^{0}|^{2}}\right]\exp\left[-\left(\vec{R}_{\kappa\kappa'}^{0}-\vec{R}_{l}^{0}-R_{0}\right)/\rho'\right]. \tag{4.96}$$

Hence the dynamical matrix from Eq.(2.63) can be written as

$$
\begin{aligned}
\bar{D}^{R}_{\alpha\beta}(\vec{q},\kappa\kappa') \;=\; & \frac{1}{(M_{\kappa}M'_{\kappa})^{1/2}} \sum_{l} \frac{b'}{\rho'} \left[\left(\frac{1}{\mid \vec{R}^{0}_{\kappa\kappa'} - \vec{R}^{0}_{l} \mid} + \frac{1}{\rho'} \right) \right. \\
& \times \frac{\left(\vec{R}^{0}_{\kappa\kappa'} - \vec{R}^{0}_{l} \right)_{\alpha} \left(\vec{R}^{0}_{\kappa\kappa'} - \vec{R}^{0}_{l} \right)_{\beta}}{\mid \vec{R}^{0}_{\kappa\kappa'} - \vec{R}^{0}_{l} \mid^{2}} - \frac{\delta_{\alpha\beta}}{\mid \vec{R}^{0}_{\kappa\kappa'} - \vec{R}^{0}_{l} \mid^{2}} \left. \right] \\
& \times \exp\left[- \left(\vec{R}^{0}_{\kappa\kappa'} - \vec{R}^{0}_{l} - R_{0} \right)/\rho' \right] \exp\left[i\vec{q} \cdot \left(\vec{R}^{0}_{\kappa\kappa'} - \vec{R}^{0}_{l} \right) \right].
\end{aligned}
$$

$$(4.97)$$

Combining Eqs.(4.91) and (4.97), the microscopic dielectric method permits the complete solution of lattice dynamical problem within the harmonic and adiabatic approximations provided the electron-ion matrix elements and the properly defined dielectric matrix are known.

Thus a unified theoretical description of lattice dynamics of crystalline solids must start with an unified description of dielectric screening and the electron-ion matrix elements in the solid. There have been various attempts to calculate the electron-ion matrix elements for TMs [38] - [40]. However, in the calculations of phonon frequencies, the local model potentials are frequently used. The structure of Eq.(4.89) is more or less preserved if the electron-ion potential is replaced by a local pseudopotential for the bare ions [6]. The electrons which are not included in the formation of ion-cores, take part in the dynamical screening process. Therefore the dielectric matrix and hence its inverse will depend upon the electronic structure of the metal.

Now we evaluate the electronic contribution to the dynamical matrix $\bar{D}^{FE}_{\alpha\beta}(\vec{q},\kappa\kappa')$ for a free electron metal. In these metals the LF effects are negligible. Therefore the dielectric matrix reduces to a diagonal (scalar) dielectric function $\epsilon(\vec{q}+\vec{G})$ (see Eq.4.60). Using Eq.(4.60) for $\chi(\vec{q}+\vec{G})$ in Eq.(4.93) one gets

$$
\begin{aligned}
\bar{D}^{FE}_{\alpha\beta}(\vec{q},\kappa\kappa') \;=\; & \frac{1}{(M_{\kappa}M_{\kappa'})^{1/2}} \sum_{\vec{G}} \frac{(\vec{q}+\vec{G})_{\alpha}(\vec{q}+\vec{G})_{\beta}}{v(\vec{q}+\vec{G})} v^{\kappa}_{b}(\vec{q}+\vec{G}) v^{\kappa'}_{b}(\vec{q}+\vec{G}) \\
& \times \left[1 - \frac{1}{\epsilon(\vec{q}+\vec{G})} \right] \exp\left[i\vec{G} \cdot (\vec{R}^{0}_{\kappa} - \vec{R}^{0}_{\kappa'}) \right].
\end{aligned}
$$

$$(4.98)$$

For a monatomic metal ($\kappa = \kappa'$) Eq.(4.98) further simplifies as

$$\bar{D}_{\alpha\beta}^{FE}(\vec{q},\kappa\kappa') = -\omega_p^2 \sum_{\vec{G}} \frac{(\vec{q}+\vec{G})_\alpha (\vec{q}+\vec{G})_\beta}{|\vec{q}+\vec{G}|^2} F_N(\vec{q}+\vec{G}) \quad (4.99)$$

where

$$F_N(\vec{q}+\vec{G}) = \left(\frac{4\pi Z e^2}{\Omega_0 |\vec{q}+\vec{G}|^2}\right)^{-2} \left[1 - \frac{1}{\epsilon(\vec{q}+\vec{G})}\right] |v_b^\kappa(\vec{q}+\vec{G})|^2$$

$$(4.100)$$

and $\omega_p = (4\pi Z^2 e^2/M\Omega_0)^{1/2}$ is the plasma frequency. Equation(4.100) is the expression for energy wave number characteristic function for a free electron metal and hence the conventional dynamical theory for simple metals is retrieved.

4.3   Microscopic Multipole Theories

The conduction electron density in d(f)-band materials is nonuniform due to quasilocalized character of d(f)- electrons. The density response function method described earlier is well suited to study the response of nonuniform electron density to the ionic motions. However, the two questions are to be resolved: How to account for the electronic band structure in the calculation of $\epsilon(\vec{q}+\vec{G}, \vec{q}+\vec{G}')$ and then how to invert this infinite dimensional dielectric matrix ?

These two questions will be addressed in this section for the model band structures of transition metals. Other detailed calculations of dielectric matrix will be discussed in chapter 6.

4.3.1   Wannier Representation

The Wannier functions are appropriate to represent the quasilocalized character of conduction electrons. Ferreira et al [41], Hanke and Bilz [42] and Hanke [43] used Wannier representation to formulate the lattice dynamics of metals, particularly for TMs. The Wannier functions are orthogonal localized functions which are defined as

$$| a_\mu(\vec{r} - \vec{R_t}) > \; = \; N^{-1/2} \sum_{\vec{k}} \exp(-i\vec{k} \cdot \vec{R_t}) \, | \psi_{\vec{k}\mu}(\vec{r}) > \quad (4.101)$$

where $| \psi_{\vec{k}\mu}(\vec{r}) >$ is the Bloch function for the kth state in the μth band. $\vec{R_t}$ is the index of tth cell in which the Wannier function $a_\mu(\vec{r} - \vec{R_t})$ for the μth band is localized. The Wannier functions are defined for each band and each unit cell and satisfy the orthogonality condition

$$< a_\mu(\vec{r} - \vec{R_t}) \, | \, a_{\mu'}(\vec{r} - \vec{R_{t'}}) > \; = \; \delta_{\mu\mu'} \delta_{tt'}. \quad (4.102)$$

Here the discussion is restricted for one atom per unit cell. Equation (4.102) also implies at least some localization. It is noted from Eq.(4.101) that the Wannier functions are the combination of Bloch functions of different wave vectors and therefore of different energies. Thus the Wannier functions are not the energy eigenfunctions [44].

Multiplying Eq.(4.101) by $\exp(i\vec{k'} \cdot \vec{R_t})$ and summing over $\vec{R_t}$ one finds the Bloch function

$$| \psi_{\vec{k}\mu}(\vec{r}) > \; = \; N^{-1/2} \sum_{t} \exp(i\vec{k} \cdot \vec{R_t}) \, | a_\mu(\vec{r} - \vec{R_t}) > . \quad (4.103)$$

The Bloch function in Eq.(4.103) is defined for a particular isolated μth band. In the d(f)-band metals, the s and d(f)-bands get splitted due to s-d (s-f) interaction as shown in Fig.4.2. The d(f)-bands have finite width and intersect the Fermi energy forming a band complex in which the bands are connected to each other. Such a band complex can be accounted for by adopting Slater-Koster interpolation criterion [45]. Thus we rewrite the Bloch function as

$$| \psi_{\vec{k}\mu}(\vec{r}) > \; = \; N^{-1/2} \sum_{\nu} e^{\nu}_{\vec{k}\mu} \sum_{t} \exp(i\vec{k} \cdot \vec{R_t}) \, | a_\mu(\vec{r} - \vec{R_t}) > .$$

$$(4.104)$$

Here the sum over ν corresponds to sum over all the Wannier functions of a band complex. $e^{\nu}_{\vec{k}\mu}$ are the vectors of the band complex which satisfy the following relations

$$\sum_{\nu} e^{\nu*}_{\vec{k}\mu} e^{\nu}_{\vec{k}\mu'} \; = \; \delta_{\mu\mu'} \quad (4.105)$$

$$\sum_{\mu} e^{\nu*}_{\vec{k}\mu} e^{\nu'}_{\vec{k}\mu} \; = \; \delta_{\nu\nu'}. \quad (4.106)$$

Thus $e^\nu_{\vec{k}\mu}$ are the elements of an unitary matrix in the band indices μ and ν [46], [47]. This generalized Wannier representation for Bloch functions allows the exact inversion of dielectric matix.

The Bloch functions given in Eq.(4.104) satisfy the Schrödinger equation

$$H \mid \psi_{\vec{k}\mu}(\vec{r}) > \ = \ E_{\vec{k}\mu} \mid \psi_{\vec{k}\mu}(\vec{r}) > \tag{4.107}$$

where H is Hamiltonian of the system and $E_{\vec{k}\mu}$ are the energy eigenvalues of a band complex which is disconnected from other band complexes. Using Eq.(4.104), the overlap matrix elements in the dielectric matrix can be written as

$$< \psi_{\vec{k}\mu} \mid \exp\left(-i(\vec{q}+\vec{G})\cdot\vec{r}\right) \mid \psi_{\vec{k}'\mu'} >$$
$$= \delta_{\vec{k}',\vec{k}+\vec{q}+\vec{G}} \sum_s \exp\left(-i(\vec{k}+\vec{q})\cdot\vec{R}_t\right) A_s(\vec{q}+\vec{G}) e^{\nu*}_{\vec{k}\mu} e^{\eta}_{\vec{k}'\mu'},$$
$$\tag{4.108}$$

where

$$A_s(\vec{q}+\vec{G}) \ = \ \exp(-i\vec{G}\cdot\vec{R}_t)\int a^*_\nu(\vec{r})\exp\left(-i(\vec{q}+\vec{G})\cdot\vec{r}\right) a_\eta(\vec{r}+\vec{R}_t)d\vec{r}.$$
$$\tag{4.109}$$

The index s stands for lattice vector index t and band indices ν and η i.e. $s = (t,\nu,\eta)$. Using Eq.(4.108) in Eq.(4.46), the polarizability tensor is obtained in the separable form as

$$\chi^0(\vec{q}+\vec{G},\vec{q}+\vec{G}') \ = \ \sum_{ss'} A_s(\vec{q}+\vec{G})N_{ss'}(\vec{q})A_{s'}(\vec{q}+\vec{G}'), \tag{4.110}$$

where

$$N_{ss'}(\vec{q}) \ = \ \sum_{\vec{k}\mu\mu'} \frac{f_{\vec{k}\mu} - f_{\vec{k}+\vec{q}\mu'}}{E_{\vec{k}\mu} - E_{\vec{k}+\vec{q}\mu'}} \exp(-i(\vec{k}+\vec{q})\cdot(\vec{R}_t - \vec{R}_{t'}))$$
$$\times e^{\nu'*}_{\vec{k}\mu} e^{\eta'*}_{\vec{k}+\vec{q}\mu'} e^{\eta}_{\vec{k}+\vec{q}\mu'} e^{\nu}_{\vec{k}\mu}. \tag{4.111}$$

These expressions are for monatomic lattice. The generalization for multiatomic lattice is straight forward. Using Eq.(4.110) in Eq.(4.45) the dielectric matrix becomes

$$\epsilon(\vec{q}+\vec{G},\vec{q}+\vec{G}') = \delta_{\vec{G}\vec{G}'} - v(\vec{q}+\vec{G})\sum_{ss'} A_s(\vec{q}+\vec{G})N_{ss'}(\vec{q})A_{s'}(\vec{q}+\vec{G}').$$

$$(4.112)$$

Thus the band structure part $N_{ss'}(\vec{q})$ and the overlap matrix element part $A_s(\vec{q}+\vec{G})A_{s'}(\vec{q}+\vec{G}')$ in Eq.(4.112) are factorized which is the main advantage of Wannier representation. This factorization makes possible to compute $\epsilon^{-1}(\vec{q}+\vec{G},\vec{q}+\vec{G}')$.

The dielectric matrix given in Eq.(4.112) can be inverted by matrix expansion as

$$\begin{aligned}
\epsilon^{-1} &= (\mathbf{I} - \mathbf{vANA}^\dagger)^{-1} \\
&= \mathbf{I} + \mathbf{vANA}^\dagger + \mathbf{vANA}^\dagger\mathbf{vANA}^\dagger + \dots \\
&= \mathbf{I} + \mathbf{vAN}\{\mathbf{I} + \mathbf{A}^\dagger\mathbf{vAN} + \mathbf{A}^\dagger\mathbf{vANA}^\dagger\mathbf{vAN} + \dots\}\mathbf{A}^\dagger \\
&= \mathbf{I} + \mathbf{vA}\{\mathbf{N}^{-1} - \mathbf{A}^\dagger\mathbf{vA}\}^{-1}\mathbf{A}^\dagger.
\end{aligned}$$
$$(4.113)$$

Equation (4.113) can be written with the subscripts as

$$\epsilon^{-1}(\vec{q}+\vec{G},\vec{q}+\vec{G}') = \delta_{\vec{G}\vec{G}'} + v(\vec{q}+\vec{G})\sum_{ss'} A_s(\vec{q}+\vec{G})S_{ss'}^{-1}(\vec{q})A_{s'}^*(\vec{q}+\vec{G}')$$

$$(4.114)$$

where

$$S_{ss'}^{-1}(\vec{q}) = N_{ss'}^{-1}(\vec{q}) - \sum_{\vec{G}''} A_s(\vec{q}+\vec{G}'')v(\vec{q}+\vec{G}'')A_{s'}(\vec{q}+\vec{G}'').$$

$$(4.115)$$

Thus in the localized description of conduction electrons, the inversion of infinite dimensional dielectric matrix reduces to the inversion of matrix $\mathbf{S}$. The dimension of $\mathbf{S}$ is greatly reduced for the well localized $\mid a_\nu(\vec{r} - \vec{R}_t) >$. Hence, in the localized description of d and f-band metals, the inversion of the dielectric matrix becomes possible.

The following two schemes are adopted to calculate $\epsilon(\vec{q}+\vec{G},\vec{q}+\vec{G}')$. The first is the linear combination of atomic orbital (LCAO) approximation and the second is mixed band structure scheme.

(a) LCAO Approximation

The Bloch functions $\mid \psi_{\vec{k}\mu}(\vec{r}) >$ in the LCAO approximation is

$$\mid \psi_{\vec{k}\mu}(\vec{r}) > \ = \ N^{-1/2} \sum_t \exp(-i\vec{k} \cdot \vec{R}_t) \mid \phi_\mu(\vec{r} - \vec{R}_t) >, \quad (4.116)$$

where $\mid \phi_\mu(\vec{r} - \vec{R}_t) >$ is the atomic wave function centered at the site $\vec{R}_t$ and the sum is over all the N lattice sites in the crystal. Equations (4.103) and (4.116) are similar except that the Wannier function is replaced by an atomic wave function. The use of atomic wave functions is more practical as these are known with reasonable precision while the knowledge of Wannier functions is not so sound for TMs.

Use of Eq.(4.116) will lead to the same result for dielectric matrix as given in Eq.(4.112) except that there will be no unitary matrices $e^\nu_{\vec{k}\mu}$ etc. and $\mid a_\mu(\vec{r} - \vec{R}_t) >$ will be replaced by $\mid \phi_\mu(\vec{r} - \vec{R}_t) >$. In this context, the second term of Eq.(4.114) will be the contribution due to multipole character of charge polarization at the lattice sites. The intraband ($\mu = \mu'$) transitions in various subbands (d-band has five subbands) produce monopoles. The interband ($\mu \neq \mu'$) transitions poduce dipoles at the atomic sites with form factor $A_s(\vec{q} + \vec{G})$. Therefore the second term of Eq.(4.114) represents the contributions from monopole-monopole, monopole-dipole and dipole-dipole interactions mediated by the resonant function $S^{-1}_{ss'}(\vec{q})$. Thus, Eq.(4.114) describes the screening similar to that of a dipolar model for insulators [48].

(b) Mixed Band Scheme

For d and f-band metals, it is useful to combine the LCAO description for d and f-band states with nonlocal orthogonalised plane wave (OPW) scheme for s and p-band states. This combined scheme [49], [50] is based on the fact that except for hybridization effects, d and f-bands closely resemble to those obtained in the tight-binding (TB) approximation, whereas s and p-bands are similar to those resulting from the nearly free electron approximation. In this combined band structure scheme, the dielectric function can be splitted up into two parts: (i) A purely diagonal part $\epsilon_0(\vec{q} + \vec{G})$ which arises from the smooth part of conduction electron density and (ii) a contribution which contains both the diagonal and nondiagonal parts and can be factorized in the form given in Eq.(4.110). Therefore the dielectric matrix is written as

$$
\begin{aligned}
\epsilon(\vec{q}+\vec{G},\vec{q}+\vec{G}') \;=\; & \epsilon_0(\vec{q}+\vec{G})\delta_{\vec{G}\vec{G}'} - v(\vec{q}+\vec{G}) \\
& \times \sum_{ss'} A_s(\vec{q}+\vec{G})N_{ss'}(\vec{q})A_{s'}(\vec{q}+\vec{G}').
\end{aligned}
$$

$$(4.117)$$

Here $\epsilon_0(\vec{q}+\vec{G})$ is analogous to the free electron like contribution to the dielectric matrix.

It is convenient to separate $\epsilon_0(\vec{q}+\vec{G})$ from the factorized contribution because in doing so the number of lattice vectors $\vec{R}_t$ and $\vec{R}_{t'}$ needed in the summation in Eq.(4.110) will greatly reduce. Thus the dimensionality of dielectric matrix will further reduce which will make the inversion of the dielectric matrix easier. Equation (4.117) is written in the matrix form as

$$
\epsilon \;=\; \epsilon_0 \mathbf{I} - \mathbf{vANA}^\dagger.
$$

$$(4.118)$$

The inverse of dielectric matrix can be evaluated in the same way as done in Eq.(4.113) and is given as

$$
\begin{aligned}
\epsilon^{-1} \;&=\; (\epsilon_0\mathbf{I} - \mathbf{vANA}^\dagger)^{-1} \\
&=\; (\mathbf{I} - \mathbf{v}\epsilon_0^{-1}\mathbf{ANA}^\dagger)^{-1}\epsilon_0^{-1} \\
&=\; \left[\mathbf{I} + \mathbf{v}\epsilon_0^{-1}\mathbf{ANA}^\dagger + \mathbf{v}\epsilon_0^{-1}\mathbf{ANA}^\dagger\mathbf{v}\epsilon_0^{-1}\mathbf{ANA}^\dagger + \ldots\right]\epsilon_0^{-1} \\
&=\; \epsilon_0^{-1} + \mathbf{v}\epsilon_0^{-1}\mathbf{AS}'^{-1}\mathbf{A}^\dagger\epsilon_0^{-1}
\end{aligned}
$$

$$(4.119)$$

where

$$
\mathbf{S}'^{-1} \;=\; \mathbf{N}^{-1} - \mathbf{A}^\dagger\mathbf{v}\epsilon_0^{-1}\mathbf{A}.
$$

$$(4.120)$$

The explicit expression for $\epsilon^{-1}(\vec{q}+\vec{G},\vec{q}+\vec{G}')$ becomes

$$
\begin{aligned}
\epsilon^{-1}(\vec{q}+\vec{G},\vec{q}+\vec{G}') \;=\; & \frac{1}{\epsilon_0(\vec{q}+\vec{G}')}\left[\delta_{\vec{G}\vec{G}'} + \frac{v(\vec{q}+\vec{G})}{\epsilon_0(\vec{q}+\vec{G})}\sum_{ss'} A_s(\vec{q}+\vec{G})\right. \\
& \times \left. S'^{-1}_{ss'}(\vec{q})A^*_{s'}(\vec{q}+\vec{G}')\right]
\end{aligned}
$$

$$(4.121)$$

where

$$S'^{-1}_{ss'}(\vec{q}) \;=\; N^{-1}_{ss'}(\vec{q}) - \sum_{\vec{G}''} A_s(\vec{q}+\vec{G}'')\frac{v(\vec{q}+\vec{G}'')}{\epsilon_0(\vec{q}+\vec{G}'')}A^*_{s'}(\vec{q}+\vec{G}'').$$

$$(4.122)$$

The use of Wannier functions in Eq.(4.121) further reduces the dimensions of matrix $\mathbf{S}'$.

In order to have some physical insight of the terms appearing in Eq.(4.121) we consider the response of elctron system to a weak external field. In RPA, the density response to a weak external perturbation is given by Eq.(4.14). Combining Eqs.(4.14) and (4.15) one finds

$$\delta\rho(\vec{q}+\vec{G}) \;=\; -\frac{1}{v(\vec{q}+\vec{G})}\sum_{\vec{G}'}\left[\delta_{\vec{G}\vec{G}'} - \epsilon^{-1}(\vec{q}+\vec{G},\vec{q}+\vec{G}')\right]\delta v(\vec{q}+\vec{G}').$$

$$(4.123)$$

Using Eq.(4.121) and defining

$$A_s(\vec{q}+\vec{G}) \;=\; \sum_{\alpha}(\vec{q}+\vec{G})_\alpha f^\alpha_s(\vec{q}+\vec{G}), \qquad (4.124)$$

Eq.(4.123) simplifies as

$$\delta\rho(\vec{q}+\vec{G}) \;=\; -\frac{1}{v(\vec{q}+\vec{G})}\left[1 - \frac{1}{\epsilon_0(\vec{q}+\vec{G})}\right]\delta v(\vec{q}+\vec{G})$$
$$+i\sum_{\alpha}(\vec{q}+\vec{G})_\alpha p_\alpha(\vec{q}+\vec{G}) \qquad (4.125)$$

where

$$p_\alpha(\vec{q}+\vec{G}) \;=\; -i\sum_{ss'}\frac{f^\alpha_s(\vec{q}+\vec{G})}{\epsilon_0(\vec{q}+\vec{G})}S'^{-1}_{ss'}(\vec{q})\sum_\beta\sum_{\vec{G}'}(\vec{q}+\vec{G}')_\beta$$
$$\times\frac{f^{\beta*}_{s'}(\vec{q}+\vec{G}')}{\epsilon_0(\vec{q}+\vec{G}')}\delta v(\vec{q}+\vec{G}'). \qquad (4.126)$$

The first term in Eq.(4.125) gives the free electron like charge distribution due to scalar dielectric function. The second term represents the charge distribution associated with the dipoles distribution. Therefore a nonuniform dipole charge ditribution is superimposed over a uniform charge distribution. The net nonuniform electron charge density is responsible for local field corrections [12], [51].

The dynamical matrix is obtained by substituting $\epsilon^{-1}(\vec{q}+\vec{G},\vec{q}+\vec{G}')$ from Eq.(4.121) in Eq.(4.88). It gives

$$
\begin{aligned}
\bar{D}_{\alpha\beta}(\vec{q},\kappa\kappa') &= \bar{D}^c_{\alpha\beta}(\vec{q},\kappa\kappa') + \bar{D}^{FE}_{\alpha\beta}(\vec{q},\kappa\kappa') + \frac{1}{(M_\kappa M_{\kappa'})^{1/2}} \\
&\quad \times \sum_{ss'} W^\kappa_{s\alpha}(\vec{q}) S'^{-1}_{ss'}(\vec{q}) W^{\kappa'*}_{s'\beta}(\vec{q})
\end{aligned}
\tag{4.127}
$$

where

$$
W^\kappa_{s\alpha}(\vec{q}) = \sum_{\vec{G}} (\vec{q}+\vec{G})_\alpha \frac{A_s(\vec{q}+\vec{G})}{\epsilon_0(\vec{q}+\vec{G})} v^\kappa_b(\vec{q}+\vec{G}) \exp(i\vec{G}\cdot\vec{R}_\kappa).
$$

$$
\tag{4.128}
$$

$W^\kappa_{s\alpha}(\vec{q})$ is the effective force experienced by the κth ion in the α-direction due to the coupling of density waves. Thus the electronic contribution consists of two parts: One part is free electron like $\bar{D}^{FE}_{\alpha\beta}(\vec{q},\kappa\kappa')$ and other part (last term of Eq.(4.127) is due to distribution of dipoles which are not necessarily centered on the ion sites. If the last term of Eq.(4.127) is neglected, $\bar{D}_{\alpha\beta}(\vec{q},\kappa\kappa')$ reduces to be the same as for simple metals. However, if $\epsilon_0(\vec{q}+\vec{G}) = 1$ and $\bar{D}^{FE}_{\alpha\beta}(\vec{q},\kappa\kappa') = 0$, Eq.(4.127) will represent the dynamical matrix for an insulator. For an ideal insulator, with dipoles centered at the ion sites, the contribution from the dipole distribution is finite which leads to the dipolar model for an insulator. But in the covalent solids and d(f)-band metals both the diagonal and nondiagonal contributions are finite in $\epsilon(\vec{q}+\vec{G},\vec{q}+\vec{G}')$, and hence one gets the screened dipole model [52] - [54].

4.3.2 Non-Interacting Band Representation

Although the theories of inverse dielectric matrix and dynamical matrix are exact in the Wannier representation, the actual calculations of phonon frequencies of transition metals and rare earth metals are non-tractable. As discussed in the earlier section, the mixed band scheme is more appropriate for these metals from the calculational point of view.

One of the simplest approximation is to assume that the energy bands are non-interacting (neglecting s-d (s-f) hybridization). In this model band scheme, as shown in Fig.4.2, Z_s electrons in the s-band and Z_d electrons in the d-band can be consvered separately such that $Z(= Z_s + Z_d)$, the number of available conduction electrons per atom remains constant. In this band scheme, the conduction electrons in the presence of perturbing field, readjust among themselves through the following intraband and interband transitions [55]: (i) From the filled and partially filled s-bands to partially filled s-bands and partially filled d-subbands (degeneracy of d-band is accounted for). (ii) From the filled and partially filled d-subbands to partially filled s-bands and partially filled d-subbands. Therefore the dielectric matrix given in Eq.(4.42) becomes

$$
\epsilon(\vec{q}+\vec{G}, \vec{q}+\vec{G}') = \delta_{\vec{G}\vec{G}'} - v(\vec{q}+\vec{G}) \sum_{\sigma} \Big[\epsilon_{ss}^{\sigma}(\vec{q}+\vec{G}, \vec{q}+\vec{G}')
$$
$$
+ \epsilon_{dd}^{\sigma}(\vec{q}+\vec{G}, \vec{q}+\vec{G}') + \epsilon_{ds}^{\sigma}(\vec{q}+\vec{G}, \vec{q}+\vec{G}')
$$
$$
+ \epsilon_{sd}^{\sigma}(\vec{q}+\vec{G}, \vec{q}+\vec{G}') \Big] \tag{4.129}
$$

where

$$
\epsilon_{ss}^{\sigma}(\vec{q}+\vec{G}, \vec{q}+\vec{G}') = \sum_{\vec{k}\vec{k}'} \frac{f_{\vec{k}s\sigma} - f_{\vec{k}'s\sigma}}{E_{\vec{k}s\sigma} - E_{\vec{k}'s\sigma}} < \psi_{\vec{k}'s\sigma} \mid e^{i(\vec{q}+\vec{G}'\cdot\vec{r})} \mid \psi_{\vec{k}s\sigma} >
$$
$$
\times < \psi_{\vec{k}s\sigma} \mid e^{-i(\vec{q}+\vec{G})\cdot\vec{r}} \mid \psi_{\vec{k}'s\sigma} >, \tag{4.130}
$$

$$
\epsilon_{dd}^{\sigma}(\vec{q}+\vec{G}, \vec{q}+\vec{G}') = \sum_{mm'} \sum_{\vec{k}\vec{k}'} \frac{f_{\vec{k}dm\sigma} - f_{\vec{k}'dm'\sigma}}{E_{\vec{k}dm\sigma} - E_{\vec{k}'dm'\sigma}}
$$
$$
\times < \psi_{\vec{k}'dm'\sigma} \mid e^{i(\vec{q}+\vec{G}')\cdot\vec{r}} \mid \psi_{\vec{k}dm\sigma} >
$$
$$
\times < \psi_{\vec{k}dm\sigma} \mid e^{-i(\vec{q}+\vec{G})\cdot\vec{r}} \mid \psi_{\vec{k}'dm'\sigma} >,
$$
$$
\tag{4.131}
$$

and

$$
\epsilon_{ds}^{\sigma}(\vec{q}+\vec{G}, \vec{q}+\vec{G}') = \sum_{m} \sum_{\vec{k}\vec{k}'} \frac{f_{\vec{k}dm\sigma} - f_{\vec{k}'s\sigma}}{E_{\vec{k}dm\sigma} - E_{\vec{k}'s\sigma}}
$$

$$\times <\psi_{\vec{k}'s\sigma} \mid e^{i(\vec{q}+\vec{G}')\cdot\vec{r}} \mid \psi_{\vec{k}dm\sigma}>$$

$$\times <\psi_{\vec{k}dm\sigma} \mid e^{-i(\vec{q}+\vec{G})\cdot\vec{r}} \mid \psi_{\vec{k}'s\sigma}> . \quad (4.132)$$

Here σ is spin index and magnetic quantum number $m = -2$ to 2. The expression for $\epsilon^{\sigma}_{sd}(\vec{q}+\vec{G}, \vec{q}+\vec{G}')$ is obtained by interchanging the subscripts s and dm in Eq.(4.132). Here $\epsilon^{\sigma}_{ss}(\vec{q}+\vec{G}, \vec{q}+\vec{G}')$, $\epsilon^{\sigma}_{dd}(\vec{q}+\vec{G}, \vec{q}+\vec{G}')$, $\epsilon^{\sigma}_{ds}(\vec{q}+\vec{G}, \vec{q}+\vec{G}')$ and $\epsilon^{\sigma}_{ds}(\vec{q}+\vec{G}, \vec{q}+\vec{G}')$ are the contributions to polarizability tensor which arise due to transitions from s-bands to s-bands, d-subbands to d-subbands, d-subbands to s-bands and s-bands to d-subbands respectively. Because of the orthogonality of spin wave functions, the transitions take place only in the bands of same spin quantum number. In Eq.(4.131) if $m = m'$, $\epsilon^{\sigma}_{dd}(\vec{q}+\vec{G}, \vec{q}+\vec{G}')$ is due to intraband transitions and if $m \neq m'$ it is due to interband transitions.

Equations (4.130) to (4.132) are further simplified using the free electron approximation for s-electrons and simple tight-binding approximation for d-electrons i.e.

$$\psi_{\vec{k}s\sigma}(\vec{r}) = (N\Omega_0)^{-1/2}\exp(i\vec{k}\cdot\vec{r}), \quad (4.133)$$

and

$$\psi_{\vec{k}dm\sigma}(\vec{r}) = N^{-1/2}\sum_t \exp(i\vec{k}\cdot\vec{R}_t)\phi_{dm\sigma}(\vec{r}-\vec{R}_t). \quad (4.134)$$

The energies of s-band and d-subbands are represented in the parabolic band approximation as

$$E_{\vec{k}s\sigma} = E^0_{s\sigma} + \hbar^2 k^2/2m_{s\sigma}, \quad (4.135)$$

and

$$E_{\vec{k}dm\sigma} = E^0_{dm\sigma} + \hbar^2 k^2/2m_{dm\sigma}. \quad (4.136)$$

$E^0_{s\sigma}$ and $E^0_{dm\sigma}$ are the energies in the s-band and d-subbands at $\vec{k} = 0$. $m_{s\sigma}$ and $m_{dm\sigma}$ are the effective masses of the electrons in the s-band and d-subbands respectively with spin σ. Equation (4.136) overestimates d-band energies near the zone boundary. If $k_{Fs\sigma}$ and $k_{Fdm\sigma}$ are the effective Fermi radii for s-band and d-subbands, the effective masses and Fermi radii for the s-band and d-subbands can be determined using isotropic non-interacting band model [55].

Using Eqs.(4.130) to (4.132) in Eq.(4.129) and combining the intra and interband transition contributions separately, the dielectric matrix is written as

$$\epsilon(\vec{q}+\vec{G},\vec{q}+\vec{G}') = \epsilon_0(\vec{q}+\vec{G})\delta_{\vec{G}\vec{G}'} - \epsilon_{intra}(\vec{q}+\vec{G},\vec{q}+\vec{G}')$$
$$-\epsilon_{inter}(\vec{q}+\vec{G},\vec{q}+\vec{G}'), \qquad (4.137)$$

where

$$\epsilon_0(\vec{q}+\vec{G}) = 1 - v(\vec{q}+\vec{G})\sum_\sigma \epsilon^\sigma_{ss}(\vec{q}+\vec{G}) \qquad (4.138)$$

is the Lindhard function given in Eq.(4.51) except that m and k_F are replaced by m_σ and $k_{Fs\sigma}$. $\epsilon_{intra}(\vec{q}+\vec{G},\vec{q}+\vec{G}')$ consists of d-d intraband transition contributions while $\epsilon_{inter}(\vec{q}+\vec{G},\vec{q}+\vec{G}')$ consists of d-d, d-s and s-d interband transition contributions. ϵ_{intra} and ϵ_{inter} contain both the diagonal and off-diagonal parts due to quasilocalized character of d-electrons.

Using Eq.(4.134) in Eq.(4.131) and neglecting the overlap of atomic orbitals at the different lattice sites, $\epsilon_{intra}(\vec{q}+\vec{G},\vec{q}+\vec{G}')$ is written in the separable form as

$$\epsilon_{intra}(\vec{q}+\vec{G},\vec{q}+\vec{G}') = v(\vec{q}+\vec{G})\sum_m\sum_{tt'} A^t_{dm}(\vec{q}+\vec{G})f_{tt'}(\vec{q})A^{t'*}_{dm}(\vec{q}+\vec{G}'),$$

$$(4.139)$$

where

$$A^t_{dm}(\vec{q}+\vec{G}) = \exp(-i\vec{G}\cdot\vec{R}_t)\int \phi^*_{dm}(\vec{r})\exp\left(-i(\vec{q}+\vec{G})\cdot\vec{r}\right)\phi_{dm}(\vec{r})d\vec{r}$$

$$(4.140)$$

and

$$f_{tt'}(\vec{q}) = \sum_\sigma\sum_{\vec{k}}\frac{f_{\vec{k}dm\sigma} - f_{\vec{k}+\vec{q}dm\sigma}}{E_{\vec{k}dm\sigma} - E_{\vec{k}+\vec{q}dm\sigma}}\exp(-i(\vec{k}+\vec{q})\cdot(\vec{R}_t - \vec{R}_{t'})).$$

$$(4.141)$$

In order to account for hybridization between the m-components of d-wave function and to satisfy the symmetry requirements, the matrix elements are evaluated for each m and these are averaged with equal weightage [56]. It gives

$$\begin{aligned} A_{dm}^{t}(\vec{q}+\vec{G}) &= A_t(\vec{q}+\vec{G}) \\ &= \exp(-i\vec{G}\cdot\vec{R}_t)\int j_0\left(\mid\vec{q}+\vec{G}\mid r\right)R_{nd}^2(r)dr, \end{aligned}$$

$$(4.142)$$

where $R_{nd}(r)$ is the radial part of nd-wave function and $j_0\left(\mid\vec{q}+\vec{G}\mid r\right)$ is spherical Bessel function of zeroth order.

In the interband part of dielectric matrix, the contribution $\epsilon_{dd}(\vec{q}+\vec{G},\vec{q}+\vec{G}')$ for $m\neq m'$ is in the separable form due to orthogonalization of atomic wave functions. However, the other two contributions ϵ_{ds} and ϵ_{sd} are not in the separable form because of the nonorthogonality of s and d-wave functions. The major interband contribution arises from $\epsilon_{dd}(\vec{q}+\vec{G},\vec{q}+\vec{G}')$ for $m\neq m'$. ϵ_{ds} and ϵ_{sd} are usually small. Therefore, for the purpose of inverting the dielectric matrix, the total interband contribution is approximated in the separable form as [57]

$$\epsilon_{inter}(\vec{q}+\vec{G},\vec{q}+\vec{G}') = v(\vec{q}+\vec{G})\sum_{ll'}B_l(\vec{q}+\vec{G})F_{ll'}(\vec{q})B_{l'}^*(\vec{q}+\vec{G}'),$$

$$(4.143)$$

where

$$\begin{aligned} B_l(\vec{q}+\vec{G}) &= B(\vec{q}+\vec{G})\exp(-i\vec{G}\cdot\vec{R}_l), & (4.144) \\ F_{ll'}(\vec{q}) &= F(\vec{q})\exp(-i\vec{q}\cdot(\vec{R}_l-\vec{R}_{l'})) & (4.145) \end{aligned}$$

and $\vec{R}_l$ denotes the suitable positions of the unit cells. Here $F(\vec{q})$ and $B(\vec{q}+\vec{G})$ are the band structure and overlap parts of the interband contribution respectively. The detailed numerical calculations of $F(\vec{q})$ and $B(\vec{q}+\vec{G})$ can be fitted in the appropriate analytical expressions. The function $F(\vec{q})$ must be finite at $\vec{q}=0$ but $B(\vec{q}+\vec{G})$ must approach zero as $\mid\vec{q}+\vec{G}\mid\to 0$ because the interband contribution vanishes at $\mid\vec{q}+\vec{G}\mid=0$ [57].

Combining Eqs.(4.137), (4.139), (4.142) and (4.143), we get

$$
\begin{aligned}
\epsilon(\vec{q}+\vec{G},\vec{q}+\vec{G}') \;=\;& \epsilon_0(\vec{q}+\vec{G})\delta_{\vec{G}\vec{G}'} \\
& -v(\vec{q}+\vec{G})\sum_{tt'} A_t(\vec{q}+\vec{G})f_{tt'}(\vec{q})A_{t'}^*(\vec{q}+\vec{G}') \\
& -v(\vec{q}+\vec{G})\sum_{ll'} B_l(\vec{q}+\vec{G})F_{ll'}(\vec{q})B_{l'}^*(\vec{q}+\vec{G}').
\end{aligned}
$$

$$(4.146)$$

Using the relation

$$
\sum_{\vec{G}''}\epsilon(\vec{q}+\vec{G},\vec{q}+\vec{G}'')\epsilon^{-1}(\vec{q}+\vec{G}'',\vec{q}+\vec{G}') \;=\; \delta_{\vec{G}\vec{G}'},
$$

$$(4.147)$$

and following the procedure described earlier (Eqs.(4.113) and (4.119)) we get

$$
\begin{aligned}
\epsilon^{-1}(\vec{q}+\vec{G},\vec{q}+\vec{G}') \;=\;& \frac{1}{\epsilon_0(\vec{q}+\vec{G})}\Bigg[\delta_{\vec{G}\vec{G}'} - \frac{v(\vec{q}+\vec{G})}{\epsilon_0(\vec{q}+\vec{G})} \\
& \times \Bigg\{\sum_{tl} B_l(\vec{q}+\vec{G})L_{lt}(\vec{q})A_t^*(\vec{q}+\vec{G}') \\
& + \sum_{tl} A_t(\vec{q}+\vec{G})L_{lt}^*(\vec{q})B_l^*(\vec{q}+\vec{G}') \\
& - \sum_{tt'} A_t(\vec{q}+\vec{G})T_{tt'}(\vec{q})A_{t'}^*(\vec{q}+\vec{G}') \\
& - \sum_{ll'} B_l(\vec{q}+\vec{G})S_{ll'}''(\vec{q})B_{l'}^*(\vec{q}+\vec{G}')\Bigg\}\Bigg]
\end{aligned}
$$

$$(4.148)$$

where

$$
\begin{aligned}
L_{lt}(\vec{q}) \;=\;& L_{tl}^*(\vec{q}) \\
\;=\;& -\sum_{t'l'} E_{ll'}(\vec{q})X_{l't'}(\vec{q})T_{t't}(\vec{q}) \\
\;=\;& -\sum_{l't'} S_{ll'}''(\vec{q})X_{l't'}(\vec{q})C_{t't}(\vec{q}),
\end{aligned}
$$

$$(4.149)$$

$$
T_{tt'}(\vec{q}) = \left[f_{tt'}^{-1}(\vec{q}) - U_{tt'}(\vec{q}) - \sum_{ll'} X_{tl}^*(\vec{q}) E_{ll'}(\vec{q}) X_{l't'}(\vec{q}) \right]^{-1},
$$

$$
= \sum_{ll'} \sum_{t''t'''} C_{tt''}(\vec{q}) X_{t''l}^*(\vec{q}) S_{ll'}''(\vec{q}) X_{l't'''}(\vec{q})
$$
$$
\times C_{t'''t'}(\vec{q}) + C_{tt'}(\vec{q}), \qquad (4.150)
$$

$$
S_{ll'}''(\vec{q}) = \left[F_{ll'}^{-1}(\vec{q}) - V_{ll'}'(\vec{q}) - \sum_{tt'} X_{lt}(\vec{q}) C_{tt'}(\vec{q}) X_{t'l'}^*(\vec{q}) \right]^{-1}
$$

$$
= \sum_{tt'} \sum_{l''l'''} E_{ll''}(\vec{q}) X_{l''t}(\vec{q}) T_{tt'}(\vec{q})
$$
$$
\times X_{t'l'''}(\vec{q}) E_{l'''l'}(\vec{q}) + E_{ll'}(\vec{q}), \qquad (4.151)
$$

$$
E_{ll'}(\vec{q}) = \left[F_{ll'}^{-1}(\vec{q}) - V_{ll'}'(\vec{q}) \right]^{-1}, \qquad (4.152)
$$

$$
C_{tt'}(\vec{q}) = \left[f_{tt'}^{-1}(\vec{q}) - u_{tt'}(\vec{q}) \right]^{-1}, \qquad (4.153)
$$

$$
V_{ll'}'(\vec{q}) = \sum_{\vec{G}} B_l^*(\vec{q}+\vec{G}) \frac{v(\vec{q}+\vec{G})}{\epsilon_0(\vec{q}+\vec{G})} B_{l'}(\vec{q}+\vec{G}), \qquad (4.154)
$$

$$
U_{tt'}(\vec{q}) = \sum_{\vec{G}} A_t^*(\vec{q}+\vec{G}) \frac{v(\vec{q}+\vec{G})}{\epsilon_0(\vec{q}+\vec{G})} A_{t'}(\vec{q}+\vec{G}), \qquad (4.155)
$$

$$
X_{lt}(\vec{q}) = X_{tl}^*(\vec{q})
$$

$$
= \sum_{\vec{G}} B_l^*(\vec{q}+\vec{G}) \frac{v(\vec{q}+\vec{G})}{\epsilon_0(\vec{q}+\vec{G})} A_t(\vec{q}+\vec{G}). \qquad (4.156)
$$

To have a physical insight into these expressions, we write the static density response function, defined by Eq.(4.15), as

$$
\chi(\vec{q}+\vec{G}, \vec{q}+\vec{G}') = -\frac{1}{v(\vec{q}+\vec{G})} \left[1 - \frac{1}{\epsilon_0(\vec{q}+\vec{G})} \right] \delta_{\vec{G}\vec{G}'}
$$
$$
+ \left\{ \sum_{tt'} \frac{A_t(\vec{q}+\vec{G})}{\epsilon_0(\vec{q}+\vec{G})} T_{tt'}(\vec{q}) \frac{A_{t'}^*(\vec{q}+\vec{G}')}{\epsilon_0(\vec{q}+\vec{G}')} \right.
$$
$$
+ \sum_{ll'} \frac{B_l(\vec{q}+\vec{G})}{\epsilon_0(\vec{q}+\vec{G})} S_{ll'}''(\vec{q}) \frac{B_{l'}^*(\vec{q}+\vec{G}')}{\epsilon_0(\vec{q}+\vec{G}')}
$$

$$- \sum_{lt} \frac{B_l(\vec{q}+\vec{G})}{\epsilon_0(\vec{q}+\vec{G})} L_{lt}(\vec{q}) \frac{A_t^*(\vec{q}+\vec{G'})}{\epsilon_0(\vec{q}+\vec{G'})}$$

$$- \sum_{tl} \frac{A_t(\vec{q}+\vec{G})}{\epsilon_0(\vec{q}+\vec{G})} L_{lt}^*(\vec{q}) \frac{B_l^*(\vec{q}+\vec{G'})}{\epsilon_0(\vec{q}+\vec{G'})} \Bigg\}.$$

$$(4.157)$$

Thus the electron density response function splits up into two parts: The first term of Eq.(4.157) is a purely diagonal part which is analogous to that for the free electron metal. The second part (in the curly brackets) corresponds to a set of monopole and dipole distributions produced by the polarization of d-electron charge distribution. The first term in the curly brackets represents a set of monopole distributions centered at the sites $\vec{R}_t$ with the effective form factor $A_t(\vec{q}+\vec{G})/\epsilon_0(\vec{q}+\vec{G})$. $U_{tt'}(\vec{q})$ given in Eq.(4.155) is the coupling coefficient between the monopole distributions centered at the sites $\vec{R}_t$ and $\vec{R}_{t'}$ which interact via screened electron-electron interaction $v(\vec{q}+\vec{G})/\epsilon(\vec{q}+\vec{G})$. The second term in the curly brackets represents a set of dipole distributions centered at the sites $\vec{R}_l$ with effective form factor $B_l(\vec{q}+\vec{G})/\epsilon_0(\vec{q}+\vec{G})$. $V_{ll'}'(\vec{q})$ given in Eq.(4.154) is the coupling coefficient between the dipole distributions centered at the sites $\vec{R}_l$ and $\vec{R}_{l'}$. These dipoles interact via screened electron-electron interaction potential. The remaining two terms in the curly brackets give the contributions due to monopole-dipole and dipole-monopole interactions and $X_{lt}(\vec{q})$ given in Eq.(4.156) is the coupling coefficient between the monopole and dipole distributions.

The physical picture which emerges out of the above discussion is that the monopole and dipole distributions, produced by the polarization of d-electrons, are screened by the uniform background of s-electrons. The whole lattice can be viewed as it consists of three sublattices. The first sublattice consists of ions, second sublattice consists of monopole distributions and third sublattice consists of dipole distributions and these sublattices interact with each other in the uniform back ground of s electrons. The sites $\vec{R}_l$ and $\vec{R}_t$ may be anywhere, not neccessarily the positions of the ions. Further the monopoles and dipoles are not necessarily the point monopoles and dipoles.

The linear perturbation of electron charge density can be calculated as given in Eq.(4.123) and it will have the contributions from the monopoles and dipoles. The intraband transitions correspond to the charge transfer process [53]. In the ionic crystals $\epsilon_0(\vec{q} + \vec{G}) = 1$ and $\vec{R}_t$ and $\vec{R}_l$ are the ionic sites. If for an ionic crystal, the interband transitions vanish, the remaining intraband transitions introduce charge transfer between the subcells of suitable volume around each ionic site in a unit cell.

The dynamical matrix is obtained by substituting Eq.(4.148) in Eq.(4.89) which is given as

$$
\begin{aligned}
\bar{D}_{\alpha\beta}(\vec{q}, \kappa\kappa') \;=\; & \bar{D}^c_{\alpha\beta}(\vec{q}, \kappa\kappa') + \bar{D}^{FE}_{\alpha\beta}(\vec{q}, \kappa\kappa') \\
& + \frac{1}{(M_\kappa M_{\kappa'})^{1/2}} \Bigg[\sum_{tt'} W^\kappa_{t\alpha}(\vec{q}) T_{tt'}(\vec{q}) W^{\kappa'*}_{t'\beta}(\vec{q}) \\
& + \sum_{ll'} Y^\kappa_{l\alpha}(\vec{q}) S''_{ll'}(\vec{q}) Y^{\kappa'*}_{l'\beta}(\vec{q}) \\
& - \sum_{tl} W^\kappa_{t\alpha}(\vec{q}) L^*_{lt}(\vec{q}) Y^{\kappa'*}_{l\beta}(\vec{q}) \\
& - \sum_{tl} Y^\kappa_{l\alpha}(\vec{q}) L_{lt}(\vec{q}) W^{\kappa'*}_{t\beta}(\vec{q}) \Bigg] ,
\end{aligned}
\tag{4.158}
$$

where

$$
W^\kappa_{t\alpha}(\vec{q}) \;=\; \sum_{\vec{G}} (\vec{q} + \vec{G})_\alpha A_t(\vec{q} + \vec{G}) \frac{v^\kappa_b(\vec{q} + \vec{G})}{\epsilon_0(\vec{q} + \vec{G})} \exp(i\vec{G} \cdot \vec{R}_\kappa),
\tag{4.159}
$$

$$
Y^\kappa_{l\alpha}(\vec{q}) \;=\; \sum_{\vec{G}} (\vec{q} + \vec{G})_\alpha B_l(\vec{q} + \vec{G}) \frac{v^\kappa_b(\vec{q} + \vec{G})}{\epsilon_0(\vec{q} + \vec{G})} \exp(i\vec{G} \cdot \vec{R}_\kappa).
\tag{4.160}
$$

Here $W^\kappa_t(\vec{q})$ is the coupling coefficient between the ions on the sublattice κ and monopole distributions on the sublattice t via screened bare-ion potential $v^\kappa_b(\vec{q} + \vec{G})/\epsilon_0(\vec{q} + \vec{G})$. Similarly $Y^\kappa_{l\alpha}(\vec{q})$ is the coupling coefficient between the ions on the sublattice κ and the dipole distributions on the sublattice l via screened bare-ion potential. In general $W^\kappa_t(\vec{q})$ and $Y^\kappa_l(\vec{q})$ may not be Hermitian. All the terms in Eq.(4.158) in the square brackets contain the bond bending or noncentral forces because these contributions couple the two ions together via the polarization produced on the third ion. Equation (4.158) is the general expression for dynamical matrix and it is similar to that for the breathing shell model equation [58], [59] in the sense that in addition to the terms for dipole polarization, there is also a term for monopole polarization.

The formalism described above is quite rigorous. However, its practical applications are difficult. Inclusion of even the nearest neighbour overlap integrals makes the dimension of dielectric matrix quite large.

4.4   Applications of Microscopic Theories

It is difficult to calculate the effective masses and Fermi momenta of s and d(f)-bands from the exact band structure calculations. Moreover heavy computations are required to evaluate the dielectric matrix using realistic band structure. Prakash and Joshi [55] suggested isotropic non-interacting band model which has been used by several authors [42, 43, 57, 60, 61] to calculate phonon frequencies. The isotropic band model is described here in brief.

4.4.1   Isotropic Non-Interacting Band Model

The band structure of TMs consists of hybridized s and d-subbands as shown in Fig.4.2. The s and d-characters of crystal wave functions vary as a function of electron wave vector $\vec{k}$. In a particular example of paramagnetic Ni [62], as shown in Fig.4.3, the s-character dominatas at the symmetry point Γ_1 (center of the BZ) and d- character dominates in the vicinity of X , L and K symmetry points (near BZ faces). The non-interacting bands are constructed along the principal symmetry directions $[\xi 00]$, $[\xi\xi 0]$ and $[\xi\xi\xi]$ using the compatibility relations [63]-[65]. The non-interacting bands along the principal symmetry directions $[\xi 00]$, $[\xi\xi 0]$ and $[\xi\xi\xi]$ for paramagnetic Ni are shown in Fig.4.3. These bands are similar to those obtained in the modified tight-binding approximation [50]. The d- band is five fold degenerate and each d-subband is assigned a magnetic quantum number m by examining the d-components of basis functions of representations Γ, X, L and K [55]. However this assignment of magnetic quantum numbers is assumed to be valid throughout the BZ [66]. The joining of the bands and the assignment of m values is given in Table 4.1.

It is evident from Fig. 4.3 that the s-band and one d-subband with $m = 1$ in the $[\xi 00]$ direction are partially filled and all other d-subbands are completely filled. The charges Z_s and Z_d in the s and d-subbands are calculated from the chosen electronic configuration of the metal. For $3d^9 4s^1$ configuration of Ni, 8 electrons are accomodated in the four filled d-subbands, one in the partially filled d-subband and one in the s-band i.e. $Z_s = Z_{d1} = 1$ and $Z = Z_s + Z_{d1} = 2$. In the parabolic band approximation, Fermi momenta of partially filled bands are

$$k_{FS} = (3\pi^2 Z_s/\Omega_0)^{1/3};$$
$$k_{Fd1} = (3\pi^2 Z_{d1}/\Omega_0)^{1/3}. \tag{4.161}$$

The effective masses of the electrons in the partially filled bands $\Gamma_1 A$ and $\Gamma_{25'} B$ are given by the relation

$$m_s = \frac{\hbar^2 k_{FS}^2}{2[E_F - E(\Gamma_1)]};$$
$$m_{d1} = \frac{\hbar^2 k_{Fd1}^2}{2[E_F - E(\Gamma_{25'})]}. \tag{4.162}$$

$E(\Gamma_1)$ and $E(\Gamma_{25'})$ are the energies at the symmetry points Γ_1 and $\Gamma_{25'}$ respectively. The energy scale is set to zero at Γ_1 point. To estimate the average effective masses associated with the filled d-subbands, the effective masses for the subbands are calculated along different symmetry directions using tabulated energy eigenvalues [62] but with a shift of zero as mentioned earlier. Using these effective masses, the energy eigenvalues at $k = k_B$ (radius of Brillouin sphere) are calculated for all the filled d-subbands and in all the three principal symmetry directions under consideration. Then the Houston's method is used to average the eigenvalues for the three directions. The final values of effective masses are calculated from the averaged eigenvalues. The isotropic non-interacting band model obtained from these effective masses is shown in Fig.4.4 for paramagnetic Ni.

The electronic band structure of bcc TMs [67] -[70] is more complicated as compared to the electronic structure of fcc TMs. The number of partially filled d-subbands is more than one and their number is different in different symmetry directions. Therefore, to consturct a reliable isotropic non-interacting band model, the low symmetry directions along with the principal symmetry directions should also be included. Singh et al. [60] constructed the non-interacting band model for bcc TMs Cr and V considering three principal symmetry directions $[\xi 00]\Gamma H, [\xi\xi 0]\Gamma N, [\xi\xi\xi]\Gamma P$ and three lower symmetry directions $[3\xi, \xi, 0]\Gamma G, [3\xi, \xi, \xi]\Gamma F$ and $[2\xi, 2\xi, \xi]\Gamma D$. The compatibility relations for bcc lattice are used for all the six directions [64]. The isotropic non-interacting band model is constructed in the same way as for fcc TMs. The six directional Houston's formula [71] is used in obtaining the averaged energy eigenvalues for the filled d-subbands. The number of d-electrons among the partially filled d-subbands are distributed in the ratio of the volumes occupied by these partially filled d-subbands.

Singh and Prakash [57] also sugggested a simplified non-interacting band scheme for hcp TMs. The non-interacting bands are constructed along the principal symmetry directions $\Gamma K, \Gamma M$ and ΓA using the compatibility relations for hcp structure [65],[72]. Only those subbands are considered which are either below the Fermi energy or intersect with the Fermi energy. Isotripic non-interacting bands are constructed in the same way as for fcc and bcc TMs. Equal weightage is given to all the three principal symmetry directions.

4.4.2 Dielectric Matrix and Phonon Frequencies

In the general formalism of dynamical matrix, the rigid ion approximation implies that only the conduction electrons are responsible for the dynamical interatomic forces while the core electrons move rigidly with nuclei. Therefore one has to make a separation between the core and conduction electrons per atom. The TMs and REMs exhibit variable valency in the chemical reactions. Therefore the problem arises as how to decide about the number of conduction electrons per atom in the metallic phase [73]. In the isotropic non-interacting band model, the electrons in the partially filled s and d-subbands are considered as dynamical (or conduction) electrons while other electrons are rigidly attached to the nuclei. For example, the number of dynamical electrons per Cr atom (valency) is two [64] for the electronic configuration $3d^5\,4s^1$ and for Sc it is three for the configuration $3d^2 4s^1$ [57].

To evaluate $\epsilon(\vec{q}+\vec{G}, \vec{q}+\vec{G}')$, one needs the functions $\epsilon_0(\vec{q}+\vec{G})$, $A_t(\vec{q}+\vec{G})$, $B_l(\vec{q}+\vec{G})$, $f_{tt'}(\vec{q})$ and $F_{ll'}(\vec{q})$ given in Eqs.(4.121) and (4.146). The s-band contribution $\epsilon_0(\vec{q}+\vec{G})$ given in Eq.(4.138) becomes

$$
\begin{aligned}
\epsilon(\vec{q}+\vec{G}) \;=\; & 1 + v(\vec{q}+\vec{G})\frac{\Omega_0 m_s k_{FS}}{2\pi^2\hbar^2} \\
& \times \left[1 + \frac{4k_{FS}^2 - \mid \vec{q}+\vec{G}\mid^2}{4k_{FS}\mid \vec{q}+\vec{G}\mid}\ln\left|\frac{2k_{FS}+\mid \vec{q}+\vec{G}\mid}{2\kappa_{FS}-\mid \vec{q}+\vec{G}\mid}\right|\right].
\end{aligned}
$$

$$(4.163)$$

$A_t(\vec{q}+\vec{G})$ is given in Eq.(4.142) and $f_{tt'}(\vec{q})$ in Eq.(4.141). Integrating Eq.(4.141) over $\vec{k}$ we get

$$
f_{tt'}(\vec{q}) = -\frac{\Omega_0}{2\pi^2 e^2}m_{di}k_{Fdi}\left[1 + \frac{4k_{Fdi}^2 - q^2}{4k_{Fdi}+q}\ln\left|\frac{2k_{Fdi}+q}{2\kappa_{Fdi}-q}\right|\right]e^{-iq\cdot(R_t - R_{t'})}
$$

$$(4.164)$$

where m_{di} and k_{Fdi} are the effective mass and Fermi momentum of ith partially filled d-subband. The functions $B_l(\vec{q}+\vec{G})$ and $F_{ll'}(\vec{q})$ are defined in Eqs.(4.144) and (4.145). The functions $B(\vec{q}+\vec{G})$ and $F(\vec{q})$ are obtained by Singh et al [60] by parameterizing the total interband contribution (arising from ϵ_{ds}, ϵ_{sd} and ϵ_{dd} for $m \neq m'$) in the form

$$
F(\vec{q}) \;=\; -C_1 m_s k_{FS}\left[1 + \frac{4k_{FS}^2 - q^2}{4k_{FS}q}\ln\left|\frac{2k_{FS}+q}{2\kappa_{FS}-q}\right|\right] \qquad (4.165)
$$

and

$$
\begin{aligned}
B(\vec{q}+\vec{G}) \;=\; & \lambda_1\mid \vec{q}+\vec{G}\mid^2 \exp\left(\mid \vec{q}+\vec{G}\mid^{1/2}\right) \quad \text{for} \quad \mid \vec{q}+\vec{G}\mid < p_1 \\
\;=\; & \mu_1\mid \vec{q}+\vec{G}\mid^2 \exp\left(-\delta_1\mid \vec{q}+\vec{G}\mid + \gamma_1\mid \vec{q}+\vec{G}\mid^2\right) \\
& \qquad\qquad \text{for} \quad p_1 < \mid \vec{q}+\vec{G}\mid < p_2 \\
\;=\; & \mu_2 \exp\left(-\delta_2\mid \vec{q}+\vec{G}\mid^{1/2}\right) \quad \text{for} \quad \mid \vec{q}+\vec{G}\mid > p_2.
\end{aligned}
$$

$$(4.166)$$

The parameters $C_1, \lambda_1, \mu_1, \delta_1, \gamma_1, \mu_2, \delta_2, p_1$ and p_2 are obtained by the least square fit. At $\mid \vec{q}+\vec{G} \mid = 0$, $B(\vec{q}+\vec{G}) = 0$ which is the required limit for interband contribution to the polarizability function. To evaluate $A(\vec{q}+\vec{G})$, given in Eq. (4.142), the radial wave function $R_{nd}(r)$ is adopted in the analytical form [74],

$$R_{nd}(r) = \sum_i a_i r^2 \exp(-\alpha_i r) \tag{4.167}$$

where a_i and α_i are the constants obtained by the least square fit of the exact wave function into above analytic expresion.

Lastly, the bare-ion potential is needed to calculate the phonon frequencies. Prakash and Joshi [55] found that the bare-ion potential calculated in the self-consistent Hartree-Fock-Slater scheme gives imaginary phonon frequencies. It is due to the fact that the s and d-wave functions, used in the calculations, were neither mutually orthogonal nor orthogonal to th core wave functions. The effect of orthogonality can be included by replacing the bare-ion potential by a pseudopotential for s and d-electrons. The pseudopotential for s-electrons consists of an attractive part and a repulsive part. But the pseudopotential for d-electrons will only be attractive as the d-wave functions are orthogonal to the core states. Therefore one should use different pseudopotentials for the s and d-electrons. However, it is found that the use of different pseudopotentials for s and d-electrons does not yield significant effect on the phonon frequencies [43].

(a) Transition Metals

Prakash and Joshi [55] calculated the phonon frequencies of paramagnetic Ni in the non-interacting band scheme taking into account only diagonal parts of the intraband and interband contributions to the dielectric matrix. These authors used Harrison's simple metal model potential (SMMP) for the bare-ion potential. The calculated phonon frequencies are shown by dashed lines in Fig.4.5. The difference in the phonon frequencies for the configurations $3d^9\ 4s^1$ and $3d^{9.4}\ 4s^{0.6}$ is found about 3%. Hanke and Bilz [42] and Hanke [43] also calculated the phonon frequencies of Ni and Pd in the non-interacting band scheme by including the local field effects. The phonon frequencies of paramagnetic Ni due to Hanke and Bilz [42] are also shown by solid lines in Fig. 4.5. It is found that the diagonal and off-diagonal contributions, arising due to the presence of d-electrons, sum up in the longitudinal modes to a maximum of about 20% contribution while in the transverse modes these contributions cancel due to symmetry requirements. Thus the d-band contribution is needed in the description of elastic constants C_{11} and C_{12}. Hanke [43] showed that the exchange and correlation interactions lower the phonon frequencies by about 15% in the longitudinal modes, while the transverse modes are not affected.

Singh et al. [60], [61] also calculated the phonon frequencies of paramagnetic and ferromagnetic Ni, Pd and Pt in the non-interacting band scheme including the local field effects arising from both the intraband and interband contributions. These authors used Animalu transition metal model potential (TMMP) for the bare-ions. The exchange and correlation interactions for s-electrons due to Singwi et al [31] and exchange interactions for d-electrons due to Lindgren and Schwarz [33] are used. It is found that the interband contribution to the polarizability function increases in magnitude and in oscillatory nature as one moves from Ni to Pt. Therfore the long range forces increase as one goes from Ni to Pt through Pd. The nature of the polarizability function is consistent with the nature of 3d, 4d and 5d radial wave functions. The anomalous behaviour in the $[\xi\xi0]T_1$ branch in the phonon dispersion curves of Pd and Pt is found in the calculated results also. However the anomalous behaviour in the phonon dispersion of Ni is too weak to be observed because the long range forces are weak in Ni. The calculated results predict that the phonon anomaly in the $[\xi\xi0]T_1$ mode is due to the long range forces in Pd and Pt and the anomaly may not be identified as Kohn anomaly because in the calculations the spherical Fermi surfaces for the s and d-subbands are used. The local field corrections in Ni, Pd and Pt are found 20%, 22% and 26%, respectively. Thus the local field corrections increase with the increase of core size. The disagreement between the calculated and experimental results of phonon frequencies in the T-modes indicates the significance of core-core exchange overlap interactions which are neglected in these calculations.

Singh et al [60] also studied the phonon frequencies of paramagnetic Cr and V including the local field effects in the non-interacting band scheme. The results for paramagnetic Cr are shown in Fig.4.6. Although there are number of anomalies in the phonon spectrum of Cr (Ref.56 to 58 in Chapter 3), only a few are reproduced in the theoretical calculations. The calculated results show the softening of longitudinal modes in the $[\xi 00]$ and $[\xi\xi\xi]$ directions for the same values of wave vectors as found in the experimental results. The calculated dip in the $[\xi\xi\xi]L$ mode is more pronounced as compared to the experimental results. But there are other anomalies also which arise due to electron and hole pockets and their interfaces on the Fermi surface [75], [76] which occur at the extremal portions of Fermi surface or at the wave vectors separating the nesting portions of Fermi surface. These anomalies are not present in the calculated results due to the spherical nature of Fermi surfaces for s and d-subbands. There is a need to incorporate the exact electronic energy band structure in the the calculations of phonon frequencies for these metals.

The phonon frequencies of a crystal with hcp structure are obtained along the principal symmetry directions $[000\xi]$ and $[0\xi\bar{\xi}0]$ by factorizing the secular equation given in Eqs. (2.173 to 2.175)[77]. Singh and Prakash [57] calculated the phonon frequencies of Sc and Y for the atomic configurations $3d^2\ 4s^1$ and $4d^2\ 5s^1$ respectively. These configurations explain the Knight shift [78] and electronic specific heat data of these metals [79]. The calculated phonon dispersion curves for Sc, using Animalu TMMP, are shown in Fig. 4.7. The results for Y are similar to those of Sc. The calculated phonon frequencies of Sc and Y show considerable anisotropy. The local field effects add to an over all better agreement with the experimental results. However, the Kohn anomalies in the LA mode for Sc and in the LA and LO modes for Y in the $[000\xi]$ direction are yet to be explained. For other TMs with hcp structure, the electronic band structure is more complex and hence the construction of non-interacting bands becomes difficult.

(b) Transition Metal Carbides

Hanke et al [80] used the microscopic theory of dielectric screening in the Wannier representation to analyze the phonon anomalies in TM carbides. The last term of Eq.(4.127) shows that the electronic contribution to the dynamical matrix increases as $S'^{-1} = (N^{-1} - A^\dagger v \epsilon_0^{-1} A)$ increases and hence leads to the phonon dips. Therefore, in the Wannier representation too, the phonon anomalies are related to the resonance like increase in the dielectric matrix [81].

The band structure calculations of transition metal carbides [82] show strong hybridization between the metal d-states and nonmetal p-states (i.e. p-d hybridization) near E_F in accoradance with the X-ray emission spectra [83]. The hybridization can be described in the form of covalent ionic bonding where the most important contribution comes from the planar interactions. Therefore Hanke et al [80] calculated the phonon frequencies of NbC by constructing the linear combinations of bonding and antibonding hybrids considering p and d-band complexes at the Γ_{15} and $\Gamma_{25'}$ symmetry points. The polarizability due to p-d hybridization is treated in the nonlocal form $ANA^\dagger$ (see Eq.4.118) confining the p-d overlap upto the first nearest neighbours. All other contributions to polarizability are in the diagonal form χ_{diog} with d-d metal and p-p nonmetal overlaps upto the first nearest neighbours. The calculated phonon frequencies of NbC obtained by fitting the elastic constants C_{11} and C_{44} are shown in Fig.4.8. It is found that the over all features of the dispersion curves are determined by the scalar function χ_{diog} (i.e. p-p and d-d screening) but the phonon dips in the longitudinal and transverse modes are the consequence of nonlocal p-d screening [80], [84].

The strength of phonon anomalies is determined by the degree of p-d hybridization. The plasmons are calculated by the condition

$$det \mid N^{-1} - A^\dagger v \epsilon_0^{-1} A \mid = 0. \tag{4.168}$$

However, the coupling between the phonon and the plasmon modes can cause the lattice instabilty. If one puts the local field factor equal to zero (i.e. $A^\dagger v \epsilon_0^{-1} A = 0$) in the screening matrix, the phonon dips disappear. It can be shown from the last term of Eq.(4.127) that due to local fields the phonon frequency at $\mid \vec{q} \mid = 0$ remains finite which corresponds to the energy required to make a rigid translation of localized electrons. In order to explain the phonon anomalies quantitatively Hanke et al [80] emphasized the need to include many body interactions beyond RPA in the d-d interactions.

4.5   Transition Metal Pseudopotentials

In the lattice dynamical study of metals, the crystal potential $V(\vec{r}) = V_b(\vec{r}, \vec{R}^0) + V_{ee}(\vec{r})$, is of primary interest. $V(\vec{r})$ enters the Schrödinger equation as

$$H_e^0 \mid \psi >= [T + V(\vec{r})] \mid \psi >= E \mid \psi > \qquad (4.169)$$

where $\mid \psi >$ are the electronic eigenstates of Hamiltonian H_e^0 which is used in the energy band calculations (0^0K static lattice). Exact calculation of $V(\vec{r})$ is difficult due to many body nature of the problem. The matrix elements of $V(\vec{r})$ are required to evaluate the atomic form factors and the electron-phonon interaction. Therefore, some simplifications are needed to solve the Eq.(4.169). In general, the following approximations are adopted:

1. The core and conduction states are separated. The core states are the same as in the free atom. This implies that the ion-core is very small. Further the core states $\mid \alpha >$ with energy E_α are the eigenfunctions of H_e^0 i.e.

$$H_e^0 \mid \alpha >= E_\alpha \mid \alpha > . \qquad (4.170)$$

   Here the subscript α denotes the quantum numbers (nlm) and

$$\mid \alpha >=\mid \phi_\alpha(\vec{r} - R_l) > .$$

2. The many body problem is reduced to one body problem by using self-consistent field approximation. In this approximation $V(\vec{r})$ is an averaged potential seen by an electron due to all the ions and all other electrons. It also includes the electron-electron exchange and correlation interactions.

In these approximations $| \psi >$ is the true eigenfunction and Eq.(4.169) can be written as

$$\left[-\frac{\hbar^2}{2m} \nabla^2 + V(\vec{r}) \right] | \psi >= E | \psi > . \qquad (4.171)$$

The electron wavefunction $\psi(\vec{r})$ can be regarded as plane wave like in the weak potential region (i.e. away from the ions) while it exhibits large oscillatory nature (atomic wave function like) near the ion-cores due to strong attractive potential. Therefore, the wave function $\psi(\vec{r})$ can be writtten as the linear combination of plane waves and core states i.e.

$$| \psi >= \sum_{\vec{k}} a_{\vec{k}} | \vec{k} > + \sum_{\alpha} a_\alpha | \alpha >, \qquad (4.172)$$

where $a_{\vec{k}}$ and a_α are the constants and these are to be determined. The set which consist of conduction band states $| \vec{k} >$ and core states $| \alpha >$ is an overcomplete set of basis functions.

Herring [85] pointed out that the conduction band states must be orthogonal to the core states, therefore, he defined an orthogonalized plane wave (OPW) as

$$| \vec{k}_{OPW} >=| \vec{k} > - \sum_{\alpha} | \alpha >< \alpha | \vec{k} >= (1 - P) | \vec{k} > \qquad (4.173)$$

where $P = \sum_\alpha | \alpha >< \alpha |$ is a projection operator which projects any function on to the core states. From Eq.(4.173), one can show that $< \beta | \vec{k}_{OPW} >= 0$ where $| \beta >$ is any core state. The eigenfunction ψ can be written as a linear combination of OPW's as

$$| \psi >= \sum_{\vec{k}} a_{\vec{k}} | \vec{k}_{OPW} >= (1 - P) | \varphi > \qquad (4.174)$$

where

$$| \varphi >= \sum_{\vec{k}} a_{\vec{k}} | \vec{k} > . \qquad (4.175)$$

The function φ is smooth every where in space and it is called pseudowave function as it differs from the real wave function. From Eq.(4.174), one can show that $< \beta \mid \psi >= 0$. Using Eq.(4.174) in Eq.(4.171) and rearranging the terms, one gets

$$\left[-\frac{\hbar^2}{2m} \nabla^2 + W(\vec{r}) \right] \mid \varphi >= E \mid \varphi > \tag{4.176}$$

where

$$W(\vec{r}) = V(\vec{r}) + \sum_{\alpha}(E - E_\alpha) \mid \alpha >< \alpha \mid . \tag{4.177}$$

Equation (4.176) is the wave equation satisfied by the pseudowave function and it is called pseudowave equation. The eigenvalues of Eqs.(4.169) and (4.176) are the same. The second term in Eq.(4.177) is always positive (repulsive) as E_α is negative. In writing the Eq.(4.176), the effect of orthogonality of core states is transferred into the potential as a fictitious additive repulsive potential [2] given by the second term of Eq. (4.177). Therefore the net effect of the ion cores upon the conduction band states may be represented by a weak effective potential $W(\vec{r})$ which is known as pseudopotential. The repulsive term is an operator, therefore the pseudopotential is nonlocal in nature and the energy eigenvalues are appearing in it. The smallness of pseudopotential allows, for many purposes, to treat it as a perturbation term. This facilitate the calculation of various properties of solids.

Nonuniqueness has been found in both the pseudowave function and pseudopotential [86]. Let any linear combination of core states $\mid \beta >$ be added up to $\mid \varphi >$ to give a new pseudowave function φ' as

$$\mid \varphi' >=\mid \varphi > + \sum_{\beta} a_\beta \mid \beta > . \tag{4.178}$$

Using Eq.(4.178) in (4.176) it is found that $\mid \varphi' >$ is also an eigenfunction of Eq.(4.176) with the same energy eigenvalue. To see the nonuniqueness of $W(\vec{r})$, one writes the general linear pseudopotential $W'(\vec{r})$ as

$$W'(\vec{r}) \mid \varphi >= V(r) \mid \varphi > + \sum_{\alpha} \mid \alpha >< f(\vec{r}, \alpha) \mid \varphi >, \tag{4.179}$$

where $f(\vec{r}, \alpha)$ is an arbitrary function of position and core state index. Using $W'(\vec{r})$ in Eq.(4.176) in place of $W(\vec{r})$ we write

$$\left[-\frac{\hbar^2}{2m} \nabla^2 + W'(\vec{r}) \right] \mid \varphi' >= E' \mid \varphi' > \tag{4.180}$$

where E' is the energy eigenvalue corresponding to $\mid \varphi' >$. Operating on Eq.(4.180) from the left hand side by $< \psi \mid$ one gets,

$$E < \psi \mid \varphi' >= E' < \psi \mid \varphi' > . \qquad (4.181)$$

As $\mid \psi >$ and $\mid \varphi >$ are not orthogonal therefore $E = E'$. It shows that for an arbitrary function $f(\vec{r}, \alpha)$, the pseudowave equation gives the same energy eigenvalues. The function $f(\vec{r}, \alpha)$ can be expanded in terms of plane waves in the convenient form (Fourier expansion) as

$$f(\vec{r}, \alpha) = \sum_{\vec{k}} f^*(\vec{k}, \alpha) < k \mid \alpha >\mid \vec{k} > . \qquad (4.182)$$

Substituting Eq.(4.182) in Eq.(4.179) and using Eq.(4.175) one finds

$$W'(\vec{r}) \mid \vec{k} >= V(\vec{r}) \mid \vec{k} > + \sum_{\alpha} \mid \alpha > f(\vec{k}, \alpha) < \alpha \mid k > . \qquad (4.183)$$

As $f(\vec{k}, \alpha)$, and hence $f(\vec{r}, \alpha)$, is any positive function of $\vec{k}$ and α , Eq.(4.183) shows nonuniqueness in the repulsive part of the pseudopotential. Thus the pseudopotential is nonlocal, nonunique and non-Hermitian too. These unpleasent properties of $W(\vec{r})$ allow to make various approximations for $W(\vec{r})$.

In general the pseudopotentials are classified into two categories. In the first category, wave function are included explicitly and this involves the first principle calculations. This is very well documented by Harrison [86]. These pseudopotentials are complicated and of limited accuracy since it is too difficult to go beyond Hartree-Fock approximation. This is true, in particular, for the materials which have large core sizes and partially filled shells e.g., TMs and REMs. In the second category of pseudopotentials, the nonuniqueness of the repulsive part of potential is made use of. In fact, the pseudopotential is replaced by a model potential having a simple analytical form and consisting of a few parameters which are determined by reproducing some experimental results such as liquid resistivity, phonon frequencies or a band gap in the electron spectrum on the Brillouin zone boundary etc. In this way the requirement of orthogonal core states and the exact form of true potential is bypassed. A number of model potentials are given for simple metals [86]-[88]. Following the same procedure the pseudopotentials are developed for d and f- band metals and these are used to investigate the phonon frequencies.

4.5.1   First Principle Pseudopotentials For Transition Metals

In the d-band metals, there are d-band states in addition to core states and conduction band states. The quasilocalized d-states get polarized (or distorted) by the crystal field or any other external field. As a result, the d-band states contribute towards the electronic properties of TMs. Therefore, the d-band states no more remain the eigenstates of crystal Hamiltonian. The d-band has finite width, therefore, these states cannot be included in the core states. At the same time the d-band states retain enough of their atomic character and therefore cannot be included in the conduction band states. If the d-states are included in the conduction states, the pseudowave function is no more smooth and the pseudopotential becomes strong, and hence the perturbation theory will not be applicable. To circumvent these difficulties, Harrison [14] reformulated the pseudopotential theory for TMs by expanding the conduction band states $| \psi >$ in an overcomplete set which consist of plane waves $| \vec{k} >$, core states $| \alpha >$ and atomic d-states $| d >$ [89] i.e.,

$$| \psi > = \sum_{\vec{k}} a_{\vec{k}} | \vec{k} > + \sum_{\alpha} a_{\alpha} | \alpha > + \sum_{d} a_d | d >, \qquad (4.184)$$

where a_d is a constant which can be determined with the help of orthogonality condition. The conduction band states satisfy the self-consistent Schrödinger Eq.(4.171). The core states are assumed to be the same in the metal as in the free atom and satisfy Eq.(4.170). The atomic d-states are the eigenfunctions of Hamiltonian of free ion (or atom) and satisfy the equation

$$\left[-\frac{\hbar^2}{2m} \nabla^2 + V^a(\vec{r}) \right] | d > = E_d^a | d >, \qquad (4.185)$$

where $V^a(\vec{r})$ is the atomic potential and E_d^a is the d-state energy. To define a unified pseudopotential one needs to know the eigenvalues of crystal Hamiltonian while operating on the atomic d-states. This involves the difference between the atomic potential and the self-consistent crystal potential in the neighbourhood of the ion i.e.

$$V^a(\vec{r}) - V(\vec{r}) = \delta V(\vec{r}). \tag{4.186}$$

Use of Eq.(4.186) in (4.185) gives

$$\left[-\frac{\hbar^2}{2m}\nabla^2 + V(\vec{r})\right] \mid d> = E_d^a \mid d> - \delta V(\vec{r}) \mid d>. \tag{4.187}$$

The expectation value of crystal Hamiltonian between atomic d-states is obtained as

$$\begin{aligned}
E_d &= <d\left|-\frac{\hbar^2}{2m}\nabla^2 + V(\vec{r})\right|d> \\
&= E_d^a \mid d> - <d|\delta V(\vec{r})|d>. \tag{4.188}
\end{aligned}$$

Using E_d^a from Eq.(4.188) in Eq.(4.187) one gets

$$\begin{aligned}
\left[-\frac{\hbar^2}{2m}\nabla^2 + V(\vec{r})\right] \mid d> &= E_d \mid d> - \Delta \mid d> \\
&= (E_d - \Delta) \mid d>, \tag{4.189}
\end{aligned}$$

where

$$\Delta \mid d> = \delta V(\vec{r}) \mid d> - \mid d><d \mid \delta V(\vec{r}) \mid d>. \tag{4.190}$$

Thus the d-band state energy E_d is lower by $<d \mid \delta V(\vec{r}) \mid d>$ with respect to E_d^a. Here Δ is the s-d hybridization contribution which will become more clear later.

With these definitions the pseudowave equation and the pseudopotential for TMs can be obtained. Using Eq.(4.184) in Eq.(4.171) and collecting all the terms which consist of core and d-states, on the left, one gets

$$\begin{aligned}
&\left[-\frac{\hbar^2}{2m}\nabla^2 + V(\vec{r})\right] \mid \varphi> + \sum_\alpha (E_\alpha - E)a_\alpha \mid \alpha> \\
&+ \sum_d (E_d - E - \Delta)a_d \mid d> = E \mid \varphi>, \tag{4.191}
\end{aligned}$$

where $| \varphi >$ is the same as defined in Eq.(4.175). The constant a_α can be evaluated by operating on the left side of Eq.(4.191) with a core state $< \alpha |$ and using the Hermitian property of Hamiltonian $< \alpha | H | \varphi >= E_\alpha < \alpha | \varphi >$. It is further assumed that $\delta V(\vec{r})$ and hence $\Delta(\vec{r})$ is almost constant in the region of appreciable spread up of core states $< \alpha |$ i.e.$< \alpha | \Delta | d >=< \alpha | d > \Delta = 0$, as $| \alpha >$ and $| d >$ states are orthogonal. Therefore one gets

$$a_\alpha = - < \alpha | \varphi > \tag{4.192}$$

which is the same as in the simple metal pseudopotential.

The coefficient a_d can be obtained by operating from the left by an atomic d-state $< d |$ on Eq.(4.191). Here again the two approximations are made. First, the overlap of atomic d-states on the adjacent atomic sites is neglected and second $\delta V(\vec{r})$ is assumed to be spherically symmetric so that $< d' | \delta V | d >= 0$. The first approximation is in general true and if it is not, one can truncate the d-states [89] and modify the potential $\delta V(\vec{r})$. With these approximations

$$a_d = - < d | \varphi > +\frac{< d | \Delta | \varphi >}{E_d - E}. \tag{4.193}$$

The second term arises due to the shift Δ defined in Eq. (4.190).

Using a_α and a_d in Eq.(4.191), the TM pseudopotential wave equation becomes

$$\left[-\frac{\hbar^2}{2m}\nabla^2 + W \right] | \varphi > - \sum_d \frac{\Delta | d >< d | \Delta | \varphi >}{E_d - E} = E | \varphi >, \tag{4.194}$$

where TM pseudopotential W is

$$\begin{aligned}
W | \varphi > \; = \; & V | \varphi > +\sum_\alpha (E - E_\alpha) | \alpha >< \alpha | \varphi > \\
& +\sum_d (E - E_d) | d >< d | \varphi > \\
& +\sum_d [| d >< d | \Delta | \varphi > +\Delta | d >< d | \varphi >].
\end{aligned}$$

$$\tag{4.195}$$

In Eq.(4.195), the first two terms on the right side are the same as in Eq.(4.177) but the presence of d-states has added three more repulsive terms (last three terms): One through the d-band energy E_d and other two through $\Delta(\vec{r})$. The pseudowave equation (4.194) in the presence of $\Delta(\vec{r})$ gives resonance at $E = E_d$ and it is called hybridization term where $\Delta(\vec{r})$ is the hybridization potential. If $\Delta(\vec{r}) = 0$, Eqs.(4.194) and (4.195) reduce to simple metal pseudopotential wave equation with the atomic d-states as the core states.

Equation (4.194) is used to calculate the total energy of crystal. The second order perturbation theory is used [14]. Moriarty [90] also included the overlap interaction $\Phi^r(R_l - R_{l'})$ for the nearest neighbour ion sites. The total energy E_T per ion is expressed as

$$E_T = E_{es} + E_{0l} + E_{fe} + E_{bs}. \tag{4.196}$$

Here E_{es} is electrostatic energy per ion of a system of point ions of charge Z^*e immersed in a compensating uniform background. E_{es} is also known as Ewald energy. E_{0l} consist of two contributions: First contribution is due to overlap of d-wave functions centered on the neighbouring ion sites and second contribution is due to interactions between the perturbed charge densities of the d-electrons $\Delta n_d(\vec{r})$ on different ion sites and the interaction of $\Delta n_d(\vec{r})$ with the point ions. E_{fe} is free electron energy which does not depend upon the ionic configuration but only on the atomic volume. E_{bs} is band structure energy which also includes the contribution due to screening charge density n_{sc}, orthogonalization hole charge density n_{oh} and exchange and correlation interactions.

In the calculations of phonon dispersion relations, the elements of dynamical matrix $\bar{D}^c_{\alpha\beta}(\vec{q}, \kappa\kappa')$, given in Eq.(4.92), arise from E_{es} which remains the same. $\bar{D}^R_{\alpha\beta}(\vec{q}, \kappa\kappa')$, given in Eq.(4.97), arise due to overlap potential which is derived from the detailed calculations of E_{0l} with the help of the relation

$$E_{0l} = \frac{1}{2N} \sum_{ll', l \neq l'} \Phi^r(R_l - R_{l'}). \tag{4.197}$$

290   *S. PRAKASH*

The overlap potential $\Phi^r(r)$ is expressed in the analytic form as

$$\Phi^r(r) = \alpha_1' \left\{ 1 + \alpha_2' \left(\frac{r}{R_{nn}} - 1 \right) + \alpha_3' \alpha_2'^2 \left(\frac{r}{R_{nn}} - 1 \right)^2 \right\}$$
$$\times \exp\left(-\alpha_4' \left(\frac{r}{R_{nn}} - 1 \right) \right).$$

$$(4.198)$$

The constants α_1', α_2', α_3' and α_4' are determined by calculating $\Phi^r(r)$ and its first and second derivatives at the first nearest neighbour distance R_{nn}. Equation (4.198) can directly be used in Eq.(4.95), to obtain $\bar{D}^R_{\alpha\beta}(\vec{q}, \kappa\kappa')$ with the help of Eq.(2.63).

The band structure energy E_{bs} is written as

$$E_{bs} = \sum_q |S(\vec{q})|^2 F_N(\vec{q}) \tag{4.199}$$

where $S(\vec{q})$ is structure factor and $F_N(\vec{q})$ is energy wave number characteristic function given in Eq. (4.100). The electronic contribution to dynamical matrix $\bar{D}^E_{\alpha\beta}(\vec{q}, \kappa\kappa')$ is related to $F_N^{\kappa\kappa'}(\vec{q} + \vec{G}, \vec{q} + \vec{G}')$ in a more genral form of Eq. (4.99) as

$$\bar{D}^E_{\alpha\beta}(\vec{q}, \kappa\kappa') = -\omega_{p\kappa}^2 \left(\frac{M_\kappa}{M_{\kappa'}} \right)^{1/2} \sum_{\vec{G}\vec{G}'} \frac{(\vec{q} + \vec{G})_\alpha (\vec{q} + \vec{G}')_\beta}{|\vec{q} + \vec{G}|^2}$$
$$\times F_N^{\kappa\kappa'}(\vec{q} + \vec{G}, \vec{q} + \vec{G}') \exp\left[i(\vec{G} \cdot \vec{R}_\kappa - \vec{G}' \cdot \vec{R}_{\kappa'}) \right]$$

$$(4.200)$$

where

$$F_N^{\kappa\kappa'}(\vec{q} + \vec{G}, \vec{q} + \vec{G}') = \left(\frac{4\pi Z_\kappa e^2}{\Omega_0 |\vec{q} + \vec{G}|^2} \right)^{-2} v^\kappa(\vec{q} + \vec{G}) v^{\kappa'}(\vec{q} + \vec{G}')$$
$$\times \left[\delta_{\vec{G}\vec{G}'} - \epsilon^{-1}(\vec{q} + \vec{G}, \vec{q} + \vec{G}') \right] \tag{4.201}$$

and $\omega_{p\kappa}$ is plasma frequency of the κth type of atoms which is given as

$$\omega_{p\kappa} = (4\pi Z_\kappa^2 e^2 / M_\kappa \Omega_0)^{1/2}. \tag{4.202}$$

Equation (4.200) can be obtained from Eq.(4.93) by substituting $\chi(\vec{q} + \vec{G}, \vec{q} + \vec{G}')$ from Eq.(4.15). In the present formalism of pseudopotential theory if $F_N^{\kappa\kappa'}(\vec{Q}, \vec{Q}')$ is diagonal, it can be simplified for a monoatomic lattice as [90],

$$F(\vec{q}) \;=\; \frac{2\Omega_0}{(2\pi)^3} \left[\int_{k<k_F} d\vec{k} \frac{|\,\alpha_{\vec{q}}\,|^2}{(\Delta k)^2} - \int_{k>k_F} \frac{d\vec{k}}{(\Delta k)^2} \left(\sum_\alpha \frac{\alpha_d \alpha_{dq} \alpha_{\vec{q}} + c.c.}{k^2 - E_\alpha} \right. \right.$$

$$\left. \left. + \,|\, \sum_d \frac{\alpha_d \alpha_{dq}}{k^2 - E_\alpha} \,|^2 \right) \right] - \frac{4\pi\Omega_0}{q^2} \left[(1 - f_{xc}(\vec{q})) \,|\, n_{sc}(\vec{q}) \,|^2 \right.$$

$$\left. + f_{xc} \,|\, n_{0h}(\vec{q}) \,|^2 \right] \tag{4.203}$$

where

$$\alpha_{\vec{q}} = <\vec{k} + \vec{q}\,|\, W_0 \,|\, \vec{k} >$$

$$\alpha_d = <\vec{k}\,|\, \Delta \,|\, d >$$

$$\alpha_{dq} = <d\,|\, \Delta \,|\, \vec{k} + \vec{q}\,|>$$

$$(\Delta k)^2 = k^2 - |\,\vec{k} + \vec{q}\,|^2 \,. \tag{4.204}$$

Here $F(\vec{q})$ is not the normalized energy wavenumber characteristic function [86]. $n_{sc}(\vec{q})$ and $n_{0h}(\vec{q})$ are the Fourier transforms of screening and orthogonalization hole charge densities respectively. W_0 is the optimized pseudopotential which consist of the contributions from the self-consistent potential associated with the single ion site and the orthogonalization corrections for both the core and d-band states. In Eq.(4.203) if $\Delta = 0$, $F(\vec{q})$ simplifies to that for a simple metal [86] given in Eq. (4.100).

Moriarty [90] evaluated the form factors $<\vec{k} + \vec{q}\,|\, W_0 \,|\, \vec{k} >$ and normalized $F(\vec{q})$ for noble metals. The calculated phonon frequencies show the anomalous behaviour in the transverse modes due to s-d hybridization term which is proportional to $(E - E_d)^{-1}$. Such anomalies are not observed in the experimental results for noble metals. Moriarty developed the density-functional version of generalized pseudopotential theory for empty, filled and partially filled d-band TMs [90]. The total energy is calculated component wise as given in Eq.(4.196). The contributions are furhter written in terms of two and three body potentials in the parameterized form as given in Eq.(4.197). Similar attempts are also made by Harrison and co-workers [91]-[93] which will be discussed later.

4.5.2   Model Potentials for Transition Metals and Rare Earth Metals

Equation (4.194) can be rewritten as

$$\left[-\frac{\hbar^2}{2m}\nabla^2 + v_t \right] | \varphi > \; = \; E \, | \varphi >, \tag{4.205}$$

where

$$v_t \, | \varphi > \; = \; \left[W - \sum_d \frac{\Delta \, | d > < d \, | \, \Delta}{E_d - E} \right] | \varphi > . \tag{4.206}$$

Here v_t is an energy dependent nonlocal operator which is nonunique. Therefore the calculation of its matrix elements is difficult in practice. In most of the physical properties of metals, the electronic states in the vicinity of Fermi energy E_F contribute maximum. Therefore attempts have been made to parameterize $v_t(\vec{r}, E)$ such that Eq.(4.205) maps out Fermi surface as exactly as possible.

Here v_t represents a periodic potential and it must be the sum of bare ion pseudopotential v_M (electron-ion interaction) and the Hartree self-consistent potential v_H (electron-electron interaction). Thus the self-consistent pseudopotential

$$v_t \; = \; v_M + v_H. \tag{4.207}$$

According to Hartree self-consistent calculations of section (4.1.2) and following Eq.(4.40), one can write the matrix elements of $v_t(\vec{r}, E)$ with respect to plane wave states $| \vec{k} >$ as

$$< \vec{k} + \vec{q} \, | \, v_t(\vec{r}, E) \, | \, \vec{k} > \; = \; \frac{< \vec{k} + \vec{q} \, | \, v_M(\vec{r}, E) \, | \, \vec{k} >}{\epsilon(\vec{q})} \tag{4.208}$$

where $\epsilon(\vec{q})$ is the same as defined in Eqs.(4.49) and (4.51). The equivalent local pseudopotential in $\vec{r}$-space is

$$v_t(\vec{r}) \; = \; v_M(\vec{r})/\epsilon(\vec{r}). \tag{4.209}$$

Since $\epsilon(\vec{r})$ with various exchange and correlation corrections for free electron gas has been known, the main emphasis has been on proposing the model potentials for $v_M(\vec{r}, E)$.

Heine and Abarenkov (HA) [94] proposed one of the most general form of model potential which is defined as

$$
\begin{aligned}
v_M(\vec{r}, E) &= -\sum_{l=0}^{\infty} A_l(E) P_l \quad \text{for} \quad r \leq R_M \\
&= -\frac{Ze^2}{r} \quad\quad\quad \text{for} \quad r > R_M.
\end{aligned}
\tag{4.210}
$$

Here $A_l(E)$ are the energy dependent potential well-depths, P_l is projection operator for the orbital quantum number l and R_M is the model radius. Here the true potential is replaced by the square well potentials for each orbital quantum number.

Animalu [95] extended the quantum defect method for TMs and solved Eq.(4.210) by assuming $A_l(E) = Ze^2/R_M = C$, a constant, for $l \geq 3$. With these assumptions, HA model potential for transition metal becomes

$$
\begin{aligned}
v_m(\vec{r}, E) &= -C - (A_0 - C)P_0 - (A_1 - C)P_1 - (A_2 - C)P_2 \\
&\quad\quad\quad \text{for} \quad r \leq R_M \\
&= -Ze^2/r \quad\quad\quad \text{for} \quad r > R_M.
\end{aligned}
\tag{4.211}
$$

Parameters $A_l(E)$ are obtained by solving the Schrödinger equation, for model potential $v_M(\vec{r}, E)$, inside and outside the potential well and by matching the wavefunction and its derivatives at the potential well boundary. It is found that $A_l(E)$ and E are proportional to Z^2 if Z is chemical valency. $A_l(E)$ is also found proportional to E, therefore $A_l(E)$ are extrapolated linearly upto $E = E_F$. All the relavent model potential parameters are tabulated in Table 4.2. The parameter $A_2(E)$ is found proportional to $(E_d - E)^{-1}$ and thus exhibits a resonant behaviour at $E = E_d$. Therefore the last term of the first part of Eq.(4.211) is equivalent to the second term of Eq.(4.206) and can be regarded as an s-d hybridization potential $v_{sd}(\vec{r}, E)$ i.e.

$$
\begin{aligned}
v_{sd}(\vec{r}, E) &= -(A_2 - C)P_2 \quad\quad \text{for} \quad r \leq R_M, \\
&= 0 \quad\quad\quad\quad\quad\quad \text{for} \quad r > R_M.
\end{aligned}
\tag{4.212}
$$

Animalu [95] evaluated the matrix element $< k + q \mid v_M \mid \vec{q} >$ in the semilocal approximation and this is given in Appendix D.

Upadhyaya and Animalu [96] extended Eq.(4.210) for REMs by assuming $A_l(E) = Ze^2/R_M = C$ (a constant) for $l \geq 4$. With this approximation, Eq.(4.210) becomes

$$\begin{aligned} v_M(\vec{r}, E) &= -C - (A_0 - C)P_0 - (A_1 - C)P_1 - (A_2 - C)P_2 \\ &\quad -(A_3 - C)P_3 \quad \text{for} \quad r \leq R_M, \\ &= -Ze^2/r \quad\quad\quad \text{for} \quad r > R_M. \end{aligned} \tag{4.213}$$

$A_3(E)$ is found proportional to $\frac{1}{2}W_f/(E_f - E)$ where E_f is the energy of f-state in the metal and W_f is the width of f-band resonance. Thus the last term of the first part of Eq.(4.213) exhibits resonance at $E = E_f$ and represents s-f hybridization interaction which can be written as

$$\begin{aligned} v_{sf}(\vec{r}, E) &= -(A_3 - C)P_3 \quad\quad \text{for} \quad r \leq R_M, \\ &= 0 \quad\quad\quad\quad\quad \text{for} \quad r > R_M. \end{aligned} \tag{4.214}$$

The parameters $A_l(E)$ are determined in the same way as for $v_M(\vec{r}, E)$ for TMs in Eq. (4.211).

In the calculation of phonon frequencies, one needs the matrix elements $< \vec{k}+\vec{q} \mid v_M \mid \vec{k} >$ which are found proportional to $[1/(E_{k+q} - E_d)]$ where $E_{\vec{k}+\vec{q}} \neq E_d$. Thus the resonance term is found to be small. Therefore Animalu used second order perturbation theory to calculate phonon frequencies using $< \vec{k} + \vec{q} \mid v_M \mid \vec{k} >$ given in Appendix D for fcc and bcc TMs. It is found that the calculated transverse phonon modes are too low and longitudinal phonon modes get softened even before reaching the zone boundary. This emphasized the need to include explicitly the nonlocal part of TMMP [97].

The calculation of phonon frequencies in the pseudopotential theory, including the local field effects, involves the calculation of energy wave number characteristic function $F_N^{\kappa\kappa'}(\vec{q} + \vec{G}, \vec{q} + \vec{G}')$ which again involves the inverse dielectric matrix (see Eqs.(4.200) and (4.201)). The dielectric matrix, from Eqs.(4.45) and (4.46), can be written as

$$\begin{aligned} \epsilon(\vec{q} + \vec{G}, \vec{q} + \vec{G}') &= \delta_{\vec{G}\vec{G}'} - v(\vec{q} + \vec{G}) \sum_{\vec{k}, \vec{G}''} \frac{f(E_{\vec{k}}) - f(E_{\vec{k}+\vec{q}+\vec{G}''})}{E_{\vec{k}} - E_{\vec{k}+\vec{q}+\vec{G}''}} \\ &\quad \times M_{\vec{k}, \vec{G}''}(\vec{q} + \vec{G}, \vec{q} + \vec{G}') \end{aligned} \tag{4.215}$$

where

$$M_{\vec{k},\vec{G}''}(\vec{q}+\vec{G},\vec{q}+\vec{G}') \;=\; <\psi_{\vec{k}}\,|\,\exp(-i(\vec{q}+\vec{G})\cdot\vec{r})\,|\,\psi_{\vec{k}+\vec{q}+\vec{G}''}>$$
$$\times <\psi_{\vec{k}+\vec{q}+\vec{G}''}\,|\,\exp(i(\vec{q}+\vec{G}')\cdot\vec{r})\,|\,\psi_{\vec{k}}>\,. \tag{4.216}$$

Equation (4.215) is for single energy band only, therefore the band index μ is dropped. To evaluate the inverse dielectric matrix, the overlap matrix elements have to be calculated for model potential.

In the model potential theory, the Bloch function $\psi_{\vec{k}}$ can be transformed into pseudowave function $\varphi_{\vec{k}}$ using Shaw and Harrison [98],[99] model wave function transformation

$$|\,\psi_{\vec{k}}> \;=\; \left[1-\frac{\partial v_M}{\partial E}\right]|\,\varphi> \tag{4.217}$$

provided the pole in $v_M(\vec{r},E)$ at $E=E_d$ (or E_f) is treated carefully. The quantity $(\partial v_M/\partial E)$ satisfies the properties of a projection operator i.e.

$$\left(\frac{\partial v_M}{\partial E}\right)^2 \;=\; \frac{\partial v_M}{\partial E}. \tag{4.218}$$

In the lowest order approximation 'one can replace $|\,\varphi>$ by a plane wave $|\,\vec{k}>$ giving

$$|\,\psi_{\vec{k}}> \;=\; \left(1-\frac{\partial v_M}{\partial E}\right)|\,\vec{k}>\,. \tag{4.219}$$

Using Eqs.(4.218) and (4.219) it can be easily proved that

$$<\psi_{\vec{k}}\,|\,\exp-i(\vec{q}+\vec{G})\cdot\vec{r}\,|\,\psi_{\vec{k}+\vec{q}+\vec{G}''}>$$
$$= \delta_{\vec{G},\vec{G}''}- <\vec{k}+\vec{q}+\vec{G}\,|\,\frac{\partial v_M}{\partial E}\,|\,\vec{k}+\vec{q}+\vec{G}''>\,. \tag{4.220}$$

Similarly one can calculate the other matrix element of Eq.(4.216) which finally gives

$$M_{\vec{k},\vec{G}''}(\vec{q}+\vec{G},\vec{q}+\vec{G}')$$

$$
= \delta_{\vec{G},\vec{G}''}\delta_{\vec{G}'',\vec{G}'}
$$
$$
- < \vec{k}+\vec{q}+\vec{G} \mid \frac{\partial v_M}{\partial E} \mid \vec{k}+\vec{q}+\vec{G}'' > \delta_{\vec{G}'',\vec{G}'}
$$
$$
- < \vec{k}+\vec{q}+\vec{G}'' \mid \frac{\partial v_M}{\partial E} \mid \vec{k}+\vec{q}+\vec{G}' > \delta_{\vec{G},\vec{G}''}
$$
$$
+ < \vec{k}+\vec{q}+\vec{G} \mid \frac{\partial v_M}{\partial E} \mid \vec{k}+\vec{q}+\vec{G}'' >
$$
$$
\times < \vec{k}+\vec{q}+\vec{G}'' \mid \frac{\partial v_M}{\partial E} \mid \vec{k}+\vec{q}+\vec{G}' > .
$$

$$(4.221)$$

Using Eq.(4.221) in Eq.(4.215) one gets

$$
\epsilon(\vec{q}+\vec{G},\vec{q}+\vec{G}')
$$
$$
= \epsilon_0(\vec{q}+\vec{G})\delta_{\vec{G},\vec{G}'} + 2v(\vec{q}+\vec{G}) \sum_{\vec{k}} \left[\frac{f(E_{\vec{k}}) - f(E_{\vec{k}+\vec{q}+\vec{G}})}{E_{\vec{k}} - E_{\vec{k}+\vec{q}+\vec{G}}} \right]
$$
$$
\times < \vec{k}+\vec{q}+\vec{G} \mid \frac{\partial v_M}{\partial E} \mid \vec{k}+\vec{q}+\vec{G}' >
$$
$$
- v(\vec{q}+\vec{G}) \sum_{\vec{k},\vec{G}''} \left[\frac{f(E_{\vec{k}}) - f(E_{\vec{k}+\vec{q}+\vec{G}''})}{E_{\vec{k}} - E_{\vec{k}+\vec{q}+\vec{G}''}} \right]
$$
$$
\times < \vec{k}+\vec{q}+\vec{G} \mid \frac{\partial v_M}{\partial E} \mid \vec{k}+\vec{q}+\vec{G}'' >
$$
$$
\times < \vec{k}+\vec{q}+\vec{G}'' \mid \frac{\partial v_M}{\partial E} \mid \vec{k}+\vec{q}+\vec{G}' > .
$$

$$(4.222)$$

The inversion of dielectric matrix given in Eq.(4.222) is difficult.

Oli and Animalu [97] neglected $\vec{G}''$ dependence of the band structure part in the last term of Eq.(4.222) and used the closure relation

$$
\sum_{\vec{G}''} \mid \vec{k}+\vec{q}+\vec{G}'' > < \vec{k}+\vec{q}+\vec{G}'' \mid = 1 \qquad (4.223)
$$

in the overlap part. Further using the property (4.218), the second order contribution given in the last term of Eq.(4.222), is reduced to the first order. Thus one gets

$$
\epsilon(\vec{q}+\vec{G},\vec{q}+\vec{G}') = \epsilon_0(\vec{q}+\vec{G})\delta_{\vec{G},\vec{G}'} + v(\vec{q}+\vec{G}) \sum_{\vec{k}} \frac{f(E_{\vec{k}}) - f(E_{\vec{k}+\vec{q}+\vec{G}})}{E_{\vec{k}} - E_{\vec{k}+\vec{q}+\vec{G}}}
$$

$$\times <\vec{k}+\vec{q}+\vec{G}\mid \frac{\partial v_M}{\partial E}\mid \vec{k}+\vec{q}+\vec{G}'> . \qquad (4.224)$$

The second term in Eq.(4.224) is called depletion hole contribution and is associated with the nonlocality of the model potential. If one uses HA type model potential (given in Eq.(4.210)), the $l = 2$ term is a resonance term which gives the major contribution to the depletion hole [97], [100]. The depletion hole in the model potential theory is analogous to the orthogonalization hole in the orthogonalized plane wave theory [86]. The form of Eq.(4.224) is similar to that obtained by Hanke [43] and Singh et al [60] and $\epsilon^{-1}(\vec{q}+\vec{G}, \vec{q}+\vec{G}')$ can be calculated as described earlier. Using the averaged value (averaged over wave vectors $\vec{k}$) of matrix elements which are written as

$$A_{\vec{G},\vec{G}'}(\vec{q}) \;=\; <\vec{k}+\vec{q}+\vec{G}\mid \frac{\partial v_M}{\partial E}\mid \vec{k}+\vec{q}+\vec{G}'>_{av}, \qquad (4.225)$$

equation (4.224) can be written in the matrix form as

$$\begin{aligned}
\epsilon \;&=\; \epsilon_0 \mathbf{I} + \mathbf{v}\chi^0\mathbf{A} \\
&=\; \left[I + \frac{v}{\epsilon_0}\chi^0 A\right]\epsilon_0,
\end{aligned} \qquad (4.226)$$

where $\chi^0(\vec{q})$ is the diagonal polarizability [see Eq.(4.56)]. The inverse dielectric matrix can be evaluated by matrix expansion method which gives

$$\epsilon^{-1} \;=\; \epsilon_0^{-1}\left[I - \frac{v}{\epsilon_0}\left\{\chi^{0-1} + \frac{v}{\epsilon_0}A\right\}^{-1}A\right]. \qquad (4.227)$$

Equation (4.227) can be written explicitly as

$$\epsilon^{-1}(\vec{q}+\vec{G}, \vec{q}+\vec{G}') \;=\; \frac{1}{\epsilon_0(\vec{q}+\vec{G})}\left[\delta_{\vec{G},\vec{G}'} - \frac{A_{dp}}{B_{dp}}\right] \qquad (4.228)$$

where

$$A_{dp} = \frac{v(\vec{q}+\vec{G})}{\epsilon_0(\vec{q}+\vec{G})}A_{\vec{G}'\vec{G}}(\vec{q}),$$

$$B_{dp} = \chi^{0-1}(\vec{q}+\vec{G}) + \sum_{\vec{G}''} \frac{v(\vec{q}+\vec{G}'')}{\epsilon_0(\vec{q}+\vec{G}'')} A_{\vec{G}''\vec{G}'}(\vec{q})$$

and

$$\chi^0(\vec{q}+\vec{G}) = \sum_{\vec{k}} \frac{f(E_{\vec{k}}) - f(E_{\vec{k}+\vec{q}+\vec{G}})}{E_{\vec{k}} - E_{\vec{k}+\vec{q}+\vec{G}}}, \tag{4.229}$$

is free electron polarizability function. Equation (4.228) is equivalent to Eq.(4.121) if the site dependence in Eq.(4.121) is ignored and $A_s(\vec{q}+\vec{G})A_{s'}^*(\vec{q}+\vec{G}')$ is taken equivalent to $< \vec{k}+\vec{q}+\vec{G} \mid \frac{\partial v_M}{\partial E} \mid \vec{k}+\vec{q}+\vec{G}' >_{av}$. Thus the local field effects in Eq.(4.228) are represented by depletion hole contribution and these are nonlocal in nature.

Using Eq.(4.228) in Eq.(4.201) one finds for a monatomic lattice

$$F_N(\vec{q}+\vec{G}, \vec{q}+\vec{G}') = F_N^0(\vec{q}+\vec{G}) + F_{dp}(\vec{q}+\vec{G}, \vec{q}+\vec{G}') \tag{4.230}$$

where

$$F_{dp}(\vec{q}+\vec{G}, \vec{q}+\vec{G}') = \left[\frac{4\pi Z e^2}{\Omega_0 \mid \vec{q}+\vec{G} \mid^2} \right]^2 \frac{A_{dp} v_M(\vec{q}+\vec{G}) v_M(\vec{q}+\vec{G}')}{B_{dp}}$$

$$\tag{4.231}$$

and

$$F_N^0(\vec{q}+\vec{G}) = \frac{4\pi Z e^2}{\Omega_0 \mid \vec{q}+\vec{G} \mid^2} \left[1 - \frac{1}{\epsilon_0(\vec{q}+\vec{G})} \right] \mid v_M(\vec{q}+\vec{G}) \mid^2 .$$

$$\tag{4.232}$$

$F_N^0(\vec{q}+\vec{G})$ is simple metal like contribution given in Eq.(4.98) and $F_{dp}(\vec{q}+\vec{G}, \vec{q}+\vec{G}')$ is the depletion hole contribution. The electronic contribution to the phonon frequencies can be calculated with the help of Eqs.(4.199) and (4.229) provided the matrix elements $< \vec{k}+\vec{q}+\vec{G} \mid \frac{\partial v_M}{\partial E} \mid \vec{k}+\vec{q}+\vec{G}' >_{av}$ are known. The averaged matrix elements can be calculated in the same way as for simple metals [98] - [100]. Using the HA model potential the averaged depletion hole matrix elements can be written as [97]

$$< \vec{k} + \vec{q} + \vec{G} \mid \frac{\partial v_M}{\partial E} \mid \vec{k} + \vec{q} + \vec{G}' >_{av}$$

$$= \frac{1}{Z} \sum_{\vec{k}\vec{G}\vec{G}'} f(E_{\vec{k}}) < \vec{k} + \vec{q} + \vec{G} \mid \frac{\partial v_M}{\partial E} \mid \vec{k} + \vec{q} + \vec{G}' >$$

$$= \sum_{\vec{G}\vec{G}'} \frac{4}{\pi Z} \int_0^{k_F} dk k^2 \left\{ 2 \sum_{l=0}^{\infty} \frac{dA_l}{dE} P_l(\cos\theta) \right.$$

$$\left. \times \int_0^{R_M} j_l \left(\mid \vec{k} + \vec{q} + \vec{G} \mid r \right) j_l \left(\mid \vec{k} + \vec{q} + \vec{G}' \mid r \right) r^2 dr \right\}$$

$$(4.233)$$

where dA_l/dE are energy derivatives of potential well depth A_l for various values of l. Equation (4.233) is applicable to both the d and f-band metals by retaining the terms upto $l = 2$ and $l = 3$, respectively.

Oli and Animalu [97] calculated the electronic contribution to phonon frequencies for paramagnetic vanadium using Eqs.(4.200), (4.230) and (4.233). Equation (4.233) involves the sum over $\vec{G}$, $\vec{G}'$ and $\vec{k}$ in the averaging process which is difficult to evaluate. Therefore Oli and Animalu parameterized Eq.(4.233) as

$$< \vec{k} + \vec{q} + \vec{G} \mid \frac{\partial v_M}{\partial E} \mid \vec{k} + \vec{q} + \vec{G}' >_{av} = \alpha_t K_l(\vec{Q}, \vec{Q}') \quad (4.234)$$

where α_t is the value of Eq.(4.233) at $(\vec{q} + \vec{G}) = (\vec{q} + \vec{G}') = 0$ and $K_l(\vec{Q}, \vec{Q}')$ are constructed from the cubic harmonics to satisfy the cubic symmetry. The depletion hole contribution cancels in the transverse modes but this contribution is added up in the longitudianl modes which is the usual behaviour of local field effects. The Coulomb part is calculated by Ewald's method and the overlap part is calculated using Born-Mayer potential given in Eq.(4.94). The calculated results are shown in Fig.4.9. It is found that the depletion hole contribution softens the longitudinal modes in the $[\xi\xi 0]$ and $[\xi\xi\xi]$ directions but the longitudinal mode in the $[\xi 00]$ direction remains unaffected. It is worth mentioning here that the d-band effects contribute the maximum to the depletion hole.

The approximation (4.223) when applied to second order contribution to $\epsilon(\vec{q}+\vec{G}, \vec{q}+\vec{G}')$, it reduces to the first order contribution thereby the depletion hole contribution is underestimated. On the other hand if we neglect the second order contribution, assuming it to be small, the first order contribution to the depletion hole becomes twice as given in Eq.(4.224) [100]. Singh and Prakash [57] and Singh et al [101] used, for the first time, the Animalu TMMP to calculate the phonon frequencies of d-band metals (Sc, Y, Zr, Tl) and f-band metals (Tb and Ho) with hcp structure. These authors used the free electron dielectric function (Eq.4.53) modified by exchange and correlation corrections due to Singwi et al [31]. Upadhyaya and Animalu [96] used the model potential theory to calculate the phonon frequencies of Tb and Ho. However, the depletion hole contribution is underestimated in their calculations. Their results do not give any Kohn anomaly. Very little evidence is found experimentally for the Kohn anomaly in the $[00\xi]$LA mode but as yet, it has not been confirmed [102]-[104].

4.6   Empirical Potentials

The most important physical quantity, in the theoretical study of phonon frequencies of TMs and REMs, is the ion-ion interaction potential. There have been consistent efforts to find an empirical ion-ion potential for these metals. Harrison and co-workers [91]-[93] made an attempt to obtain a general ion-ion potential for TMs and REMs. The expressions for the different contributions to the total energy, [Eq.(4.196)], of a TM are written upto second order in $\Delta(r)$ defined in Eq.(4.190). It is found that the Coulomb terms cancel exactly. The remaining terms are calculated using atomic data such that the bulk properties are reproduced reasonably. From this total energy calculations, the ion-ion interaction potential is estimated which is written as

$$v_{ii}(r) \;=\; v_{fe}(r) + v_{dd}(r) + v_d(r). \tag{4.235}$$

Here $v_{fe}(r)$ is the effective two body interaction between the ions screened by conduction electrons. If one uses Thomas-Fermi approximation for the screening and Ashcroft empty core model potential for the bare-ions, $v_{fe}(r)$ is given as

$$v_{fe}(r) \;=\; \frac{Z_s^2 e^2}{r}\cosh^2(k_{TF}r_c)\frac{e^{-k_{TF}r}}{r}, \qquad (4.236)$$

where $k_{TF}(= (4k_F/\pi a_0)^{1/2})$ is Thomas-Fermi inverse screening length, r_c is effective core radius and it is taken as the model potential parameter and a_0 is Bohr radius.

The potential $v_{dd}(r)$ arises due to broadening of d-states into the bands. The d-states on different ion sites are taken to be non-overlapping. However the d-states overlap through free electron states and the coupling between them is given as

$$v_{dd'} \;=\; \sum_{\vec{k}} \frac{< d \mid \Delta \mid \vec{k} >< \vec{k} \mid \Delta \mid d' >}{E_d - E_{\vec{k}}}. \qquad (4.237)$$

Using atomic d-state wave functions and summing over $\vec{k}$, the interionic separation dependence of $v_{dd'}(r)$ is given as $v_{ddm} = \eta_{ddm}\left[\hbar^2 r_d^3/mr^5\right]$ where r_d is radius of d-state wave function and $\eta_{ddm} = (-45/\pi,\, 30/\pi,\, -15/2\pi)$ for $m = (0, 1, 2)$ respectively. Further the effect of d-bands is included through rectangular model of density of d-band states as suggested by Friedel [66]. Retaining only the one central integrals within the Friedel model for the d-bands, the averaged d-d overlap interaction is given as

$$v_{dd}(r) \;=\; -Z_d\left(1 - \frac{Z_d}{10}\right)\left(\frac{12}{n}\right)^{1/2}\frac{28.1}{\pi}\frac{\hbar^2 r_d^3}{mr^5} \qquad (4.238)$$

where n is number of nearest neighbours.

There is also a contribution to the interionic potential due to s-d hybridization between the occupied and unoccupied d-states and the free electron states $\mid \vec{k} >$. The s-d hybridization modifies the d-charge in the occupied $\mid d >$ states which gives a change in the d-state occupation say by ΔZ_d. This ΔZ_d is accompanied by an equal and opposite change in the s-band charge ΔZ_s so as to keep the atom neutral. Wills and Harrison [91] accounted for s-d hybridization by modifying the values of Z_s which have been calculated self-consistently by Pettifor [105] and

Moriarty [106]. Since the $\mid d >$ states overlap through $\vec{k} >$ states, therefore $\mid d >$ states are perturbed and are rendered nonorthogonal. An additional interaction arises due to overlap of perturbed $\mid d >$ states on the nearest neighbours. The first order calculations yield averaged overlap interaction as

$$v_d(r) \;=\; Z_d \frac{225.0}{\pi^2} \frac{\hbar^2 r_d^6}{mr^8}. \tag{4.239}$$

$v_d(r)$ has the effect of shifting the d-band center.

The interionic potential given in Eq.(4.235) is further extended for REMs by including the effect of quasilocalized $\mid f >$ states. The effective ion-ion interaction potential for REMs is written as [92]

$$v_{ii}(r) \;=\; v_{fe}(r) + v_{dd}(r) + v_d(r) + v_f(r) + v_{ff}(r). \tag{4.240}$$

Here $v_{fe}(r)$ is the same as given in Eq.(4.236). The contributions $v_{dd}(r)$ and $v_d(r)$ are quite small as the d-states are well below the Fermi energy and therefore these can be neglected. $v_f(r)$ arises due to shift in the f-band center and is given as

$$v_f(r) \;=\; 2Z_f \frac{(3.11r_f)^{10}\hbar^2}{mr^{12}} \tag{4.241}$$

where r_f is the radius of $\mid f >$ state and Z_f is the number of f-electrons related to the valency $Z = Z_s + Z_f$. The interionic potential which areises due to overlap of $\mid f >$ states via $\mid k >$ states is written as $v_{ff}(r)$. The analytical expression is given as

$$v_{ff}(r) \;=\; \frac{Z_f}{\sqrt{n}} \left(1 - \frac{Z_f}{14}\right) \frac{(5.06r_f)^5\hbar^2}{mr^7}. \tag{4.242}$$

The effect of s-f hybridization is included by modifying the values of Z_s and Z_f.

McDonald and Taylor [107] also calculated the ion-ion interaction for TMs using a model d-band structure and found that the potential is not very sensitive to ionic configuration. These authors parametrized the ion-ion interaction for fcc structure as

$$v_{ii}(r) \;=\; A \exp\left(-Br/a_{fcc}\right) \frac{a_f cc}{r} - \frac{Cf_d}{(1 + (r/r_t)^2 f_d)} \left(\frac{a_{fcc}}{r}\right)^5 \tag{4.243}$$

where A,B,C and r_t are adjustable parameters. a_{fcc} is lattice constant for an ideal fcc lattice and

$$f_d = \frac{N_d(10 - N_d)}{(1 + N_d/50)}.$$ (4.244)

Here N_d is the number of occupied d-states per spin and it is defined as

$$\sum_d \theta(E_F - E_d) = N_d.$$ (4.245)

In Eq.(4.243), the first term is screened Coulomb repulsive interaction and second term is attractive d-band bonding interaction. However the contribution of the shift in the d-band center is not included in this expression.

Hartree dielectric function $\epsilon(\vec{q})$ given in Eq.(4.53) exhibits logrithmic singularity at $q = 2k_F$. Therefore it becomes almost impossible to evaluate analyticaliy the band structure contribution to the phonon frequencies. Singh et al [108] expressed $\epsilon^{-1}(\vec{q})$ by a rational function, proposed by Pettifor [109] which is written as

$$\epsilon^{-1}(\vec{q}) = \sum_{n=1}^{6} D_n \frac{q^2}{q^2 - q_n^2}$$ (4.246)

where the coefficients D_n and the poles q_n are determined by reporducing $\epsilon^{-1}(\vec{q})$ as exactly as possible except the logrithmic singularity. Use of Eq.(4.246) augments the free electron part of interionic potential which can be obtained by taking the Fourier transform of $v_M(\vec{q})\epsilon^{-1}(q)$ where $v_M(\vec{q})$ is the Fourier transform of HA model potential. Other contributions of Eqs.(4.235) and (4.240) remain unaltered. Singh et al [108],[110] also used the temperature dependent damping factor $exp(-\pi k_B Tr/\hbar v_F)$ where k_B is Boltzmann constant, T is temperature and v_F is Fermi velocity. Equations (4.235) and (4.240) are used to obtain the dynamical matrices and hence the phonon frequencies of d and f-band metals [108], [110], [111]. The anomalous structure in the phonon dispersion relations is not found in these calculations due to use of rational dielectric function. The exchange and correlation interactions in the d-band metals are important and these should be included [112].

4.7   Remarks

In this chapter an attempt has been made to present a unified microscopic theory of dielectric screening and lattice dynamics of TMs and REMs. The Wannier representation allows for an exact inversion of infinite dimensional dielectric matrix within RPA. However, this scheme becomes impractical due to lack of sound knowledge of Wannier functions. On the other hand, the non-interacting band scheme is practically applicable to TMs, REMs and TMCs. It also allows the exact inversion of dielectric matrix. The non-interacting band scheme incorporates the band structure effects in a simplified manner but it contains the essential features of detailed band structure effects. The model potential approach also makes possible the exact inversion of dielectric matrix. However, in this scheme the d and f-band effects are represented by a single parameter.

The intercomparison of different representations shows that in the Wannier and non-interacting band representations the local field effects arise purely due to the presence of d(f)-electrons in the TMs(REMs). In the pseudopotential and model potential representations, these effects arise from the depletion hole which is associated with the nonlocality of model potential. The major contribution to the depletion hole arises from the s-d and s-f hybridization effects. However in all these representations, the local fields produce the softening of phonons modes and some of the anomalies are also related to the resonance like increase in the dielectric function. In some of the systems, the dips in the phonon dispersion relations due to resonant screening seem to indicate strong electron-phonon coupling, higher superconducting transition temperature and lattice instabilities. The above mentioned approaches do not give phonon anomalies arising due to the nesting of Fermi surface.

The calculations for the phonon spectrum of d and f-band materials can be improved quantitatively by taking into account the following effects:
(i) One should use the nonlocal electron-ion interaction with suitable exchange and correlation interactions for d and f-electrons. The third order perturbation of electron-ion interaction gives the contribution similar to the local field effects and should be included. For the loosely

packed bcc lattice, the anharmonic contribution may also be significant. (ii) In all the calculations described above, the inner d and f-shells are assumed to be bound and static. In a realistic formalism, the movement of bound d or f-shells around the nuclei should be incroporated. (iii) The suitable ion-ion overlap contribution to dynamical matrix and hence to the phonon frequencies should be taken into account. (iv) To reproduce all the phonon anomalies, the exact electronic band structure should be included in the calculations of phonon frequencies of the d and f-band materials.

Some of the possible improvements mentioned above will be discussed in the next chapter.

Appendix A
Lindhard Function

The dielectric function in the free electron approximation is given in Eq.(4.52) as

$$\epsilon(\vec{q}) = 1 - v(\vec{q})\chi^0(\vec{q}) \tag{A1}$$

where the polarizability function

$$\chi^0(\vec{q}) = \sum_{k\mu} \frac{f_{\vec{k}\mu} - f_{\vec{k}+\vec{q}\mu}}{E_{\vec{k}\mu} - E_{\vec{k}+\vec{q}\mu}}. \tag{A2}$$

We evaluate the polarizability function $\chi^0(\vec{q})$ for a single band and hence the subscript μ is dropped. At absolute zero temperature, $f_{\vec{k}\mu} = 1$ or 0 for an occupied or unoccupied state respectively. This simplification gives

$$\chi^0(\vec{q}) = \sum_{\vec{k}} \left[\frac{1}{E_{\vec{k}} - E_{\vec{k}+\vec{q}}} + \frac{1}{E_{\vec{k}} - E_{\vec{k}-\vec{q}}} \right]. \tag{A3}$$

Replacing the sum over $\vec{k}$ by integration by the relation

$$\sum_{\vec{k}} \rightarrow \frac{2\Omega_0}{(2\pi)^3} \int d\vec{k}, \tag{A4}$$

and using $(E_{\vec{k}} = \hbar^2 k^2/2m)$, Eq.(A3) simplifies as,

$$\chi^0(\vec{q}) = -\frac{2m}{q\hbar^2} \frac{2\Omega_0}{(2\pi)^3} \int_0^{k_F} dk(2\pi k^2)$$

$$\int_0^{\pi} \left[\frac{1}{q + 2k\cos\theta} + \frac{1}{q - 2k\cos\theta} \right] \sin\theta d\theta. \tag{A5}$$

In Eq.(A4), the factor of 2 accounts for spin degeneracy and Ω_0 is primitive cell (or atomic) volume. The angular integral in Eq.(A5)

$$\int_0^{\pi} \left[\frac{1}{q + 2k\cos\theta} + \frac{1}{q - 2k\cos\theta} \right] \sin\theta d\theta = \frac{1}{k} \ln \left| \frac{q + 2k}{q - 2k} \right|. \tag{A6}$$

Using Eq.(A6) in (A5), one finds

$$\chi^0(\vec{q}) = -\frac{m\Omega_0}{\pi^2\hbar^2 q} \int_0^{k_F} dk\, k \ln \mid \frac{q+2k}{q-2k} \mid .$$
(A7)

The integral in Eq.(A7) is solved by parts to get

$$\chi^0(q) = -\frac{m\Omega_0}{\pi^2\hbar^2 q} \left[\frac{k_F^2}{2} \ln \mid \frac{q+2k_F}{q-2k_F} \mid -2q \int_0^{k_F} \frac{k^2 dk}{q^2 - 4k^2} \right] .$$
(A8)

where

$$-2q \int_0^{k_F} \frac{k^2 dk}{q^2 - 4k^2} = \frac{qk_F}{2} - \frac{q^2}{8} \ln \mid \frac{q+2k_F}{q-k_F} \mid .$$
(A9)

Substituting (A9) in (A8) and rearranging the terms, one gets

$$\chi^0(q) = -\frac{mk_F\Omega_0}{2\pi^2\hbar^2} \left[1 + \frac{1-\eta}{2\eta} \ln \mid \frac{1+\eta}{1+\eta} \mid \right] ,$$
(A10)

where $\eta = q/2k_F$. In the small q limit

$$\lim_{q\to 0} \chi^0(q) = -\frac{mk_F\Omega_0}{\pi^2\hbar^2} = -D(E_F)$$
(A11)

where $D(E_F)$ is electron density of states per atom at the Fermi surface. From Eqs.(A1) and (A10) one gets Lindhard dielectric function given in Eq. (4.53).

Appendix B
Ion-Ion Coulomb Contribution to the Dynamical Matrix

The ion-ion Coulomb interaction contribution to the dynamical matrix $\bar{D}^c_{\alpha\beta}(\vec{q}, \kappa\kappa')$ given in Eq.(4.92) is not suitable as such for numerical calculations because it diverges as $\mid \vec{q} + \vec{G} \mid \to 0$. It is evaluated by Ewald's θ-function transformation [Refs. 1, 3 and 17 of Chapter 2] The philosophy of this method is to devide the slowly convergent lattice sum for Coulomb interaction into two parts. The first part which is short ranged, is evaluated in the direct (or crystal) space and the second part which is long ranged, is evaluated in the reciprocal space. The two sums are combined together to get $\bar{D}^c_{\alpha\beta}(\vec{q}, \kappa\kappa')$ [Ref.17 of Chapter 2]. The details of Ewald's method are given as follows:

In Eq.(2.63),

$$\bar{D}^c_{\alpha\beta}(\vec{q}, \kappa\kappa') = \frac{1}{(M_\kappa M_{\kappa'})^{1/2}} \sum_l \bar{\Phi}^c_{\alpha\beta}(o\kappa, l\kappa') \exp\left[i\vec{q}\cdot(\vec{R}^0_l - \vec{R}^0_{\kappa\kappa'})\right] \quad (B1)$$

where

$$\vec{R}^0_{\kappa'\kappa} = \vec{R}^0_{\kappa'} - \vec{R}^0_{\kappa} = -\vec{R}^0_{\kappa\kappa'}. \quad (B2)$$

$\Phi^c_{\alpha\beta}(o\kappa, l\kappa')$ is the ion-ion Coulomb force constant. The Coulomb potential between the two ions with charges Z_κ and $Z_{\kappa'}(e = 1)$ is

$$\Phi^c(o\kappa, l\kappa') = \frac{Z_\kappa Z_{\kappa'}}{\mid \vec{r} - \vec{R}^0_l \mid} \mid_{\vec{r}=\vec{R}^0_{\kappa\kappa'}} . \quad (B3)$$

Therefore the force constant

$$\Phi^c_{\alpha\beta}(o\kappa, l\kappa') = Z_\kappa Z_{\kappa'} \frac{\partial^2}{\partial R'_\alpha \partial R'_\beta} \left(\frac{1}{\mid \vec{r} - \vec{R}^0_l \mid}\right) \mid_{\vec{r}=\vec{R}^0_{\kappa\kappa'}} \quad (B4)$$

where

$$\vec{R}' = \vec{r} - \vec{R}^0_l. \quad (B5)$$

Using Eq.(B4) in Eq.(B1) we get

$$\bar{D}^c_{\alpha\beta}(\vec{q}, \kappa\kappa') = \frac{Z_\kappa Z_{\kappa'}}{(M_\kappa M_{\kappa'})^{1/2}} \sum_l \frac{\partial^2}{\partial R'_\alpha \partial R'_\beta} \left(\frac{1}{\mid \vec{r} - \vec{R}^0_l \mid}\right) \mid_{\vec{r}=\vec{R}^0_{\kappa\kappa'}}$$

$$\times \exp\left[i\vec{q}\cdot(\vec{R}_l^0 + \vec{R}_{\kappa\kappa'}^0)\right]. \qquad (B6)$$

Equation (B6) is solved with the help of Kellerman method [1]. The basic quantity in Eq.(B6) is the sum over all the unit cells in the crystal which we write as

$$A = \sum_l \frac{\partial^2}{\partial R'_\alpha \partial R'_\beta}\left(\frac{1}{|\,\vec{r} - \vec{R}_l^0\,|}\right) e^{i\vec{q}\cdot\vec{R}_l^0}. \qquad (B7)$$

We use the identity

$$\frac{1}{|\vec{r}|} = \frac{2}{\sqrt{\pi}}\int_0^\infty e^{-\xi^2 r^2}\,d\xi$$

$$= \frac{2}{\sqrt{\pi}}\left[\int_0^\eta e^{-\xi^2 r^2}\,d\xi + \int_\eta^\infty e^{-\xi^2 r^2}\,d\xi\right]. \qquad (B8)$$

Here ξ is a continuous variable and η is a suitably chosen number. Using Eq.(B8) in (B7) we get

$$A = A_1 + A_2 \qquad (B9)$$

where

$$A_1 = \frac{2}{\sqrt{\pi}}\sum_l\left\{\frac{\partial^2}{\partial R'_\alpha \partial R'_\beta}\int_0^\eta e^{-\xi^2|\vec{r}-\vec{R}_l^0|^2}\,d\xi\right\} e^{i\vec{q}\cdot\vec{R}_l^0} \qquad (B10)$$

and

$$A_2 = \frac{2}{\sqrt{\pi}}\sum_l\left\{\frac{\partial^2}{\partial R'_\alpha \partial R'_\beta}\int_\eta^\infty e^{-\xi^2|\vec{r}-\vec{R}_l^0|^2}\,d\xi\right\} e^{i\vec{q}\cdot\vec{R}_l^0}. \qquad (B11)$$

Let us first solve for A_2. The differentiation in Eq. (B11) gives

$$A_2 = \frac{2}{\sqrt{\pi}}\sum_l e^{i\vec{q}\cdot\vec{R}_l^0}\int_\eta^\infty\left[4\xi^4\left(\vec{r}-\vec{R}_l^0\right)_\alpha\left(\vec{r}-\vec{R}_l^0\right)_\beta - 2\xi^2\delta_{\alpha\beta}\right]$$

$$\times \exp(-\xi^2\,|\,\vec{r}-\vec{R}_l^0\,|^2)d\xi. \qquad (B12)$$

Substituting

$$\chi = \xi\,|\,\vec{r}-\vec{R}_l^0\,| \qquad (B13)$$

in Eq.(B12) we get

$$A_2 = \sum_l e^{i\vec{q}\cdot\vec{R}_l^0}\frac{4\left(\vec{r}-\vec{R}_l^0\right)_\alpha\left(\vec{r}-\vec{R}_l^0\right)_\beta}{|\,\vec{r}-\vec{R}_l^0\,|^5}\frac{2}{\sqrt{\pi}}\int_{\eta|\vec{r}-\vec{R}_l^0|}^\infty \chi^4 e^{-\chi^2}\,d\chi$$

$$-\sum_l e^{i\vec{q}\cdot\vec{R}_l^0}\frac{2\delta_{\alpha\beta}}{\mid\vec{r}-\vec{R}_l^0\mid^3}\frac{2}{\sqrt{\pi}}\int_{\eta\mid\vec{r}-\vec{R}_l^0\mid}^{\infty}\chi^2 e^{-\chi^2}d\chi. \qquad (B14)$$

We rewrite the integral

$$\frac{2}{\sqrt{\pi}}\int_{\eta\mid\vec{r}-\vec{R}_l^0\mid}^{\infty}\chi^4 e^{-\chi^2}d\chi=\frac{2}{\sqrt{\pi}}\int_0^{\infty}\chi^4 e^{-\chi^2}d\chi-\frac{2}{\sqrt{\pi}}\int_0^{\eta\mid\vec{r}-\vec{R}_l^0\mid}\chi^4 e^{-\chi^2}d\chi.$$

The first term is Gamma-function integral and second term is integrated by parts. This gives

$$\frac{2}{\sqrt{\pi}}\int_{\eta\mid\vec{r}-\vec{R}_l^0\mid}^{\infty}\chi^4 e^{-\chi^2}d\chi=\frac{3}{4}+\left[\frac{\eta^3\mid\vec{r}-\vec{R}_l^0\mid^3}{\sqrt{\pi}}+\frac{3\eta\mid\vec{r}-\vec{R}_l^0\mid}{2\sqrt{\pi}}\right]e^{-\eta^2\mid\vec{r}-\vec{R}_l^0\mid^2}$$

$$-\frac{3}{2\sqrt{\pi}}\int_0^{\eta\mid\vec{r}-\vec{R}_l^0\mid}e^{-\chi^2}d\chi. \qquad (B15)$$

Similarly one can get

$$\frac{2}{\sqrt{\pi}}\int_{\eta\mid\vec{r}-\vec{R}_l^0\mid}^{\infty}\chi^2 e^{-\chi^2}d\chi=\frac{1}{2}+\left(\frac{\eta\mid\vec{r}-\vec{R}_l^0\mid}{\sqrt{\pi}}\right)e^{-\eta^2\mid\vec{r}-\vec{R}_l^0\mid^2}$$

$$-\frac{1}{\sqrt{\pi}}\int_0^{\eta\mid\vec{r}-\vec{R}_l^0\mid}e^{-\chi^2}d\chi. \qquad (B16)$$

Substituting Eqs.(B15) and (B16) in Eq.(B14) we get

$$A_2=\frac{2}{\sqrt{\pi}}\sum_l e^{i\vec{q}\cdot\vec{R}_l^0}\frac{\left(\vec{r}-\vec{R}_l^0\right)_{\alpha}\left(\vec{r}-\vec{R}_l^0\right)_{\beta}}{\mid\vec{r}-\vec{R}_l^0\mid^5}$$

$$\times\left[3+\left\{\frac{4\eta^3\mid\vec{r}-\vec{R}_l^0\mid^3}{\sqrt{\pi}}+\frac{6\eta\mid\vec{r}-\vec{R}_l^0\mid}{\sqrt{\pi}}\right\}\right.$$

$$\left.\times e^{-\eta^2\mid\vec{r}-\vec{R}_l^0\mid^2}-\frac{6}{\sqrt{\pi}}\int_0^{\eta\mid\vec{r}-\vec{R}_l^0\mid}e^{-\chi^2}d\chi\right]-\sum_l e^{i\vec{q}\cdot\vec{R}_l^0}\frac{\delta_{\alpha\beta}}{\mid\vec{r}-\vec{R}_l^0\mid^3}$$

$$\times\left[1+\frac{2\eta\mid\vec{r}-\vec{R}_l^0\mid}{\sqrt{\pi}}e^{-\eta^2\mid\vec{r}-\vec{R}_l^0\mid^2}-\frac{2}{\sqrt{\pi}}\int_0^{\eta\mid\vec{r}-\vec{R}_l^0\mid}e^{-\chi^2}d\chi\right]. \qquad (B17)$$

To evalute A_1, Eq.(B10) is rearranged as

$$A_1 = \frac{2}{\sqrt{\pi}} \frac{\partial^2}{\partial R'_\alpha \partial R'_\beta} \left\{ e^{i\vec{q}\cdot\vec{r}} \int_0^\eta \sum_l e^{-\xi^2 |\vec{r} - \vec{R}_l^0|^2 - i\vec{q}\cdot(\vec{r} - \vec{R}_l^0)} d\xi \right\}. \qquad (B18)$$

The integrand in Eq. (B18)

$$f(\vec{r}) = \sum_l e^{-\xi^2 |\vec{r} - \vec{R}_l^0|^2 - i\vec{q}\cdot(\vec{r} - \vec{R}_l^0)} \qquad (B19)$$

must satisfy the periodicity condition

$$f(\vec{r}) = f\left(\vec{r} + \vec{R}_l^0\right) \qquad (B20)$$

and therefore its Fourier expansion can be written as

$$f(\vec{r}) = \frac{1}{\sqrt{\Omega_0}} \sum_{\vec{G}} e^{i\vec{G}\cdot\vec{r}} f(\vec{G}), \qquad (B21)$$

where the Fourier coefficient

$$f(\vec{G}) = \frac{1}{\sqrt{\Omega_0}} \int_{\Omega_0} f(\vec{r}) e^{-i\vec{q}\cdot\vec{G}} d\vec{r}. \qquad (B22)$$

Substituting $f(\vec{r})$ from Eq. (B19) in Eq. (B22) one gets

$$f(\vec{G}) = \frac{1}{\sqrt{\Omega_0}} \int_{\Omega_0} \sum_l e^{-\xi^2 |\vec{r} - \vec{R}_l^0|^2 - i(\vec{q} + \vec{G})\cdot(\vec{r} - \vec{R}_l^0)} d\vec{r}. \qquad (B23)$$

The sum over l is equivalent to an integration in each cell of the whole crystal space. Writing $(\vec{r} - \vec{R}_l^0) = \vec{r}'$ in Eq. (B23) one gets

$$f(\vec{G}) = \frac{1}{\sqrt{\Omega_0}} \int e^{-\xi^2 \vec{r}'^2 - i(\vec{q} + \vec{G})\cdot\vec{r}'} d\vec{r}' \qquad (B24)$$

where the integration is over the total crystal volume. The integral in Eq. (B24) is written in the Cartesian co-ordinates and the standard integral

$$\int_{-\infty}^{\infty} e^{-\left(\xi r'_\alpha + \frac{i(\vec{q} + \vec{G})_\alpha}{2\xi}\right)^2} d\left(\xi r'_\alpha\right) = \sqrt{\pi}, \qquad (B25)$$

is used (r'_α is $\alpha-$ Cartesian component of $\vec{r}'$) to get

$$f(\vec{G}) = \frac{1}{\sqrt{\Omega_0}} \left(\frac{\sqrt{\pi}}{\xi}\right)^3 e^{-|\vec{q}+\vec{G}|^2/4\xi^2}. \tag{B26}$$

Use of Eq. (B26) in Eq. (B21) gives

$$f(\vec{r}) = \frac{1}{\Omega_0} \sum_{\vec{G}} \left(\frac{\sqrt{\pi}}{\xi}\right)^3 e^{-|\vec{q}+\vec{G}|^2/4\xi^2} e^{i\vec{G}\cdot\vec{r}}. \tag{B27}$$

From Eqs. (B18), (B19) and (B27)

$$A_1 = \frac{2\pi}{\Omega_0} \frac{\partial^2}{\partial r_\alpha \partial r_\beta} \left(\sum_{\vec{G}} e^{i(\vec{q}+\vec{G})\cdot\vec{r}} \int_0^\eta e^{-|\vec{q}+\vec{G}|^2/4\xi^2} \frac{d\xi}{\xi^3}\right)$$

$$= \frac{4\pi}{\Omega_0} \frac{\partial^2}{\partial r_\alpha \partial r_\beta} \sum_{\vec{G}} \frac{e^{i(\vec{q}+\vec{G})\cdot\vec{r}}}{|\vec{q}+\vec{G}|^2} e^{-|\vec{q}+\vec{G}|^2/4\eta^2}. \tag{B28}$$

Differentiating the above equation

$$A_1 = -\frac{4\pi}{\Omega_0} \sum_{\vec{G}} \frac{(\vec{q}+\vec{G})_\alpha (\vec{q}+\vec{G})_\beta}{|\vec{q}+\vec{G}|^2} e^{i(\vec{q}+\vec{G})\cdot\vec{r}} e^{-|\vec{q}+\vec{G}|^2/4\eta^2}. \tag{B29}$$

From Eqs. (B7), (B9), (B17) and (B29) one gets A given in Eq.(B9). Further when A is substituted in Eq. (B9) and $r' = R^0_{\kappa\kappa'}$ is used, one gets the dynamical matrix $\bar{D}^c_{\alpha\beta}(\vec{q},\kappa\kappa')$ as

$$\bar{D}^c_{\alpha\beta}(\vec{q},\kappa\kappa') = \frac{Z_\kappa Z_{\kappa'}}{(M_\kappa M_{\kappa'})^{1/2}} \left[-\frac{4\pi}{\Omega_0} \sum_{\vec{G}} \frac{(\vec{q}+\vec{G})_\alpha (\vec{q}+\vec{G})_\beta}{|\vec{q}+\vec{G}|^2} e^{-|\vec{q}+\vec{G}|^2/4\eta^2} \right.$$

$$\times e^{i\vec{G}\cdot\vec{R}^0_{\kappa\kappa'}} + \sum_l e^{i\vec{q}\cdot(\vec{R}^0-\vec{R}^0_{\kappa\kappa'})} \frac{\left(\vec{R}^0_l - \vec{R}^0_{\kappa\kappa'}\right)_\alpha \left(\vec{R}^0_l - \vec{R}^0_{\kappa\kappa'}\right)_\beta}{|\vec{R}^0_l - \vec{R}^0_{\kappa\kappa'}|^5}$$

$$\times \left\{ 3 + \left(\frac{4\eta^3 |\vec{R}^0_l - \vec{R}^0_{\kappa\kappa'}|^3}{\sqrt{\pi}} + \frac{6\eta |\vec{R}^0_l - \vec{R}^0_{\kappa\kappa'}|}{\sqrt{\pi}} \right) e^{-\eta^2 |\vec{R}^0_l - \vec{R}^0_{\kappa\kappa'}|^2} \right.$$

$$-\frac{6}{\sqrt{\pi}}\int_0^{\eta|\vec{R}_l^0-\vec{R}_{\kappa\kappa'}^0|}e^{-\chi^2}d\chi\Bigg\}$$

$$-\sum_l e^{-i\vec{q}\cdot(\vec{R}_l^0-\vec{R}_{\kappa\kappa'}^0)}\frac{\delta_{\alpha\beta}}{|\vec{R}_l^0-\vec{R}_{\kappa\kappa'}^0|^3}\Bigg\{1+\frac{2\eta\,|\,\vec{R}_l^0-\vec{R}_{\kappa\kappa'}^0\,|}{\sqrt{\pi}}e^{-\eta^2|\vec{R}_l^0-\vec{R}_{\kappa\kappa'}^0|^2}$$

$$-\frac{2}{\sqrt{\pi}}\int_0^{\eta|\vec{R}_l^0-\vec{R}_{\kappa\kappa'}^0|}e^{-\chi^2}d\chi\Bigg\}\Bigg].\qquad(B30)$$

Appendix C
Dynamical Matrix

The force constants $\Phi_{\alpha\beta}(l\kappa, l'\kappa')$ in Eqs.(2.25b) and (4.74) can be written as

$$\Phi_{\alpha\beta}(l'\kappa') = \Phi^i_{\alpha\beta}(l\kappa, l'\kappa') + \Phi^e_{\alpha\beta}(l\kappa, l'\kappa') \qquad (C1)$$

where

$$\Phi^i_{\alpha\beta}(l\kappa, l'\kappa') = \nabla_{l\kappa\alpha}\nabla_{l'\kappa'\beta}V_{II}(\vec{R}) \qquad (C2)$$

and

$$\Phi^e_{\alpha\beta}(l\kappa, l'\kappa') = \nabla_{l\kappa\alpha}\nabla_{l'\kappa'\beta}E^e(\vec{R}). \qquad (C3)$$

$\Phi^i_{\alpha\beta}(l\kappa, l'\kappa')$ and $\Phi^e_{\alpha\beta}(l\kappa, l'\kappa')$ are the ionic and electronic contibutions to the force constants. Here

$$V_{II}(\vec{R}) = \Phi^c(\vec{R}) + \Phi^r(\vec{R}).$$

Using Eq.(C1) in Eq. (2.25b) we can calculate the second order change in crystal energy as

$$E^{(2)}(\vec{R}) = E^{i(2)}(\vec{R}) + E^{e(2)}(\vec{R}), \qquad (C4)$$

where

$$E^{i(2)}(\vec{R}) = \frac{1}{2}\sum_{l\kappa\alpha}\sum_{l'\kappa'\beta} \Phi^i_{\alpha\beta}(l\kappa, l'\kappa')u_\alpha(l\kappa)u_\beta(l'\kappa') \qquad (C5)$$

and

$$E^{e(2)}(\vec{R}) = \frac{1}{2}\sum_{l\kappa\alpha}\sum_{l'\kappa'\beta} \Phi^e_{\alpha\beta}(l\kappa, l'\kappa')u_\alpha(l\kappa)u_\beta(l'\kappa'). \qquad (C6)$$

The ionic force constants $\Phi^i_{\alpha\beta}(l\kappa, l'\kappa')$ are obtained with the help of ion-ion interaction potential defined as

$$V_{II}(\vec{R}) = \frac{1}{2}\sum_{l\kappa, l'\kappa'(l\kappa \neq l'\kappa')} v_{ii}\left(\vec{R}(l\kappa) - \vec{R}(l'\kappa')\right). \qquad (C7)$$

v_{ii} is separated in the Coulomb and non-Coulomb interactions and the corresponding contributions to the dynamical matrix, namely $\bar{D}^c_{\alpha\beta}$ and $\bar{D}^R_{\alpha\beta}$, are obtained as discussed in Eqs.(B.30) and (4.97) respectively.

The electronic force constants $\Phi^e_{\alpha\beta}(l\kappa, l'\kappa')$ can be evaluated with the help of electron-ion interaction energy operator

$$v^e(\vec{R}) = \sum_{l\kappa} \int d\vec{r}\,\hat{\rho}(\vec{r})v\left(\vec{r} - \vec{R}(l\kappa)\right), \tag{C8}$$

where $\hat{\rho}(\vec{r})$ is the electron density operator and $v(\vec{r} - \vec{R}(l\kappa))$ is self-consistent electron-ion potential which includes the exchange and correlation interactions. $v^e(\vec{R})$ is expanded in power series of ionic displacements as follows:

$$v^e(\vec{R}) = v^{e(0)}(\vec{R}) + v^{e(1)}(\vec{R}) + v^{e(2)}(\vec{R}) + \ldots, \tag{C9}$$

where

$$v^{e(0)}(\vec{R}) = \sum_{l\kappa} \int d\vec{r}\,\hat{\rho}(\vec{r})v\left(\vec{r} - \vec{R}^0(l\kappa)\right), \tag{C10}$$

$$v^{e(1)}(\vec{R}) = \sum_{l\kappa\alpha} \int d\vec{r}\,\hat{\rho}(\vec{r})u_\alpha(l\kappa)\frac{\partial v(\vec{r} - \vec{R}^0(l\kappa))}{\partial r_\alpha}$$

$$\equiv \int d\vec{r}\,\hat{\rho}(\vec{r})v^{(1)}(\vec{r}), \tag{C11}$$

and

$$v^{e(2)}(\vec{R}) = \frac{1}{2}\sum_{l\kappa\alpha}\sum_{l'\kappa'\beta} \int d\vec{r}\,\hat{\rho}(\vec{r})u_\alpha(l\kappa)u_\beta(l'\kappa')\frac{\partial^2 v(\vec{r} - \vec{R}^0(l\kappa))}{\partial r_\alpha \partial r_\beta}$$

$$\equiv \int d\vec{r}\,\hat{\rho}(\vec{r})v^{(2)}(\vec{r}). \tag{C12}$$

$v^{e(0)}(\vec{R})$ is the electron-ion interaction energy in the perfect lattice, $v^{e(1)}(\vec{R})$ is the first order correction to the electron-ion energy and is linear in $\vec{u}(l\kappa)$ and $v^{e(2)}(\vec{R})$ is the second order correction and is quadratic in $\vec{u}(l\kappa)$ and so on.

The lattice vibrations at a finite temperature produce perturbation in $v(\vec{r} - \vec{R}(l\kappa)$. Consequently there is a change in the electron density in the crystal. If the change in electron density is assumed to be linearly proportional to the change in electron-ion potential, the first order change in electron density $\rho^{(1)}(\vec{r})$ can be written as

$$\rho^{(1)}(\vec{r}) = \int d\vec{r}'\chi(\vec{r}, \vec{r}')v^{(1)}(\vec{r}') \tag{C13}$$

316   S. PRAKASH

where $\chi(\vec{r}, \vec{r}')$ is electron density response function which is the measure of $\rho^{(1)}(\vec{r})$ when the crystal is subjected to a unit change in potential at $\vec{r}'$ which is defined in Eq.(C11). Further $\chi(\vec{r}, \vec{r}')$ is the property of a perfect crystal and follows the translational symmetry i.e.

$$\chi(\vec{r} + \vec{R}_l, \vec{r}' + \vec{R}_l) = \chi(\vec{r}, \vec{r}') \tag{C14}.$$

The eletronic contribution to the dynamical matrix is obtained with the help of $E^{e(2)(\vec{R})}$ (Eq. (C6)) which is the second order perturbation term. Since $v^e(\vec{R})$ involves both $\hat{\rho}(\vec{r})$ and $v(\vec{r} - \vec{R}(l\kappa))$, the second order contribution to electronic energy consists of two parts: The first contribution comes from the first order change in electron density $\rho^{(1)}(\vec{r})$ and the first order change in potential $v^{(1)}(\vec{r})$. The second contribution comes from the unperturbed charge density $\rho^0(\vec{r})$ and the second order change in potential$v^{(2)}(\vec{r})$ defined in Eq.(C12). The second contribution to $E^{e(2)}(\vec{R})$ is the expectation value of $v^{(2)}(\vec{r})$ with respect to ground state of the electron system and it is given as

$$E^{e(22)}(\vec{R}) = \int d\vec{r} \rho^{(0)}(\vec{r}) v^{(2)}(\vec{r})$$

$$= \frac{1}{2} \sum_{l\kappa\alpha} \sum_{l'\kappa'\beta} u_\alpha(l\kappa) u_\beta(l'\kappa') \delta_{ll'} \delta_{\kappa\kappa'} \int d\vec{r} \rho^{(0)}(\vec{r}) \frac{\partial^2 v(\vec{r} - \vec{R}^0(l\kappa))}{\partial r_\alpha \partial r_\beta}. \tag{C15}$$

The first contribution to $E^{e(2)}(\vec{R})$ is

$$E^{e(21)}(\vec{R}) = \frac{1}{2} \int d\vec{r} \rho^{(1)}(\vec{r}) v^{(1)}(\vec{r})$$

$$= \frac{1}{2} \int d\vec{r} \int d\vec{r}' v^1(\vec{r}) \chi(\vec{r}, \vec{r}') v^{(1)}(\vec{r}'). \tag{C16}$$

Using $v^{(1)}(\vec{r})$ from Eq. (C11) in Eq. (C16), we get

$$E^{e(21)} = \frac{1}{2} \sum_{l\kappa\alpha} \sum_{l'\kappa'\beta} u_\alpha(l\kappa) u_\beta(l'\kappa') \int d\vec{r} \int d\vec{r}' \frac{\partial v(\vec{r} - \vec{R}^0(l\kappa)}{\partial r_\alpha}$$

$$\times \chi(\vec{r}, \vec{r}') \frac{\partial v(\vec{r}' - \vec{R}^0(l'\kappa'))}{\partial r'_\beta}. \tag{C17}$$

Adding both the contributions

$$E^{e(2)}(\vec{R}) = E^{e(21)}(\vec{R}) + E^{e(22)}(\vec{R}). \qquad (C18)$$

From Eqs.(C6), (C15), (C17) and (C18) one can find $\Phi^e_{\alpha\beta}(l\kappa, l'\kappa')$ which will not be translationally invariant. The traslational invariance is achieved by considering the uniform infinitesimal translation of the whole crystal through a constant displacement $\vec{u}_c$. Then the frist order change in electron-ion potential becomes

$$v^{(1)}(\vec{r}) = \vec{u}_c \cdot \frac{\partial V_p(\vec{r})}{\partial \vec{r}}$$

$$= \vec{u}_c \cdot \frac{\partial}{\partial \vec{r}} \sum_{l\kappa} v(\vec{r} - \vec{R}^0(l\kappa)), \qquad (C19)$$

where $V_p(\vec{r})$ is the potential of perfect crystal. The corresponding first order change in electron density, in analogy with Eq. (C13), is

$$\rho^{(1)}(\vec{r}) = \vec{u}_c \cdot \frac{\partial}{\partial \vec{r}}\rho^{(0)}(\vec{r})$$

$$= \vec{u}_c \cdot \int d\vec{r}' \chi(\vec{r}, \vec{r}') \frac{\partial V_p(\vec{r}')}{\partial \vec{r}'}. \qquad (C20)$$

Comparison of both the sides of Eq. (C20) gives

$$\frac{\partial \rho^{(0)}(\vec{r})}{\partial \vec{r}} = \int d\vec{r}' \chi(\vec{r}, \vec{r}') \frac{\partial V_p(\vec{r}')}{\partial \vec{r}'}. \qquad (C21)$$

Integrating Eq. (C15) by parts and using Eq. (C21), one finds

$$E^{e(22)}(\vec{R}) = -\frac{1}{2} \sum_{l\kappa\alpha} \sum_{l'\kappa'\beta} u_\alpha(l\kappa) u_\beta(l'\kappa') \delta_{ll'} \delta_{\kappa\kappa'} \int d\vec{r} \int d\vec{r}'$$

$$\times \left\{ \frac{\partial v(\vec{r} - \vec{R}^0(l\kappa))}{\partial r_\alpha} \right\} \chi(\vec{r}, \vec{r}') \left\{ \frac{\partial v(\vec{r}' - \vec{R}^0(l'\kappa'))}{\partial r'_\beta} \right\}. \qquad (C22)$$

Using Eqs. (C17) and (C22) in Eq. (C18) and comparing the results so obtained with Eq. (C6) one finds the electronic contribution to the force constant as

$$\Phi^e_{\alpha\beta}(l\kappa, l'\kappa') = \int d\vec{r} \int d\vec{r}' \left[\frac{\partial v(\vec{r} - \vec{R}^0(l\kappa))}{\partial r_\alpha} \chi(\vec{r}, \vec{r}') \frac{\partial v(\vec{r}' - \vec{R}^0(l'\kappa'))}{\partial r'_\beta} \right.$$

$$-\delta_{ll'}\delta_{\kappa\kappa'}\frac{\partial v(\vec{r}-\vec{R}^0(l\kappa))}{\partial r_\alpha}\chi(\vec{r},\vec{r}\,')\frac{\partial v(\vec{r}\,'-\vec{R}^0(l'\kappa'))}{\partial r'_\beta}\Bigg]. \tag{C23}$$

Equation (C23) can be written in a convenient form as

$$\Phi^e_{\alpha\beta}(l\kappa,l'\kappa')=\bar{\Phi}^e_{\alpha\beta}(l\kappa,l'\kappa')-\delta_{ll'}\delta_{\kappa\kappa'}\bar{\Phi}^e_{\alpha\beta}(l\kappa,l'\kappa') \tag{C24}$$

where

$$\bar{\Phi}^e_{\alpha\beta}(l\kappa,l'\kappa')=\int d\vec{r}\int d\vec{r}\,'\frac{\partial v(\vec{r}-\vec{R}^0(l\kappa))}{\partial r_\alpha}\chi(\vec{r},\vec{r}\,')\frac{\partial v(\vec{r}\,'-\vec{R}^0(l'\kappa'))}{\partial r'_\beta}.$$
$$\tag{C25}$$

Equation (C24) satisfies the infinitesimal translational invariance condition (2.35).

We introduce the Fourier transform of electron density response function and the force as

$$\chi(\vec{Q},\vec{Q}')=\int d\vec{r}\int d\vec{r}\,'e^{-i(\vec{Q})\cdot\vec{r}}\chi(\vec{r},\vec{r}\,')e^{i(\vec{Q}')\cdot\vec{r}\,'} \tag{C26}$$

and

$$\frac{\partial v(\vec{r}-\vec{R}^0(l\kappa))}{\partial r_\alpha}=\sum_{\vec{k}}\sum_{\vec{Q}}M^\alpha_{\vec{k},\vec{k}+\vec{Q}}e^{-i(\vec{Q})\cdot(\vec{r}-\vec{R}^0(l\kappa'))}, \tag{C27}$$

where

$$M^\alpha_{\vec{k},\vec{k}+\vec{Q}}=<\vec{k}\mid\frac{\partial v(\vec{r}-\vec{R}^0(l\kappa))}{\partial r_\alpha}\mid\vec{k}+\vec{Q}>. \tag{C28}$$

$M^\alpha_{\vec{k},\vec{k}+\vec{Q}}$ are called electron-phonon matrix elements. Using Eqs. (C26) and (C27) in Eq. (C25) and keeping $\vec{Q}=\vec{q}+\vec{G}$ and $\vec{Q}'=\vec{q}+\vec{G}'$, one gets

$$\bar{\Phi}^e_{\alpha\beta}(l\kappa,l'\kappa')=\sum_{\vec{k}\vec{q}}\sum_{\vec{G},\vec{G}'}M^\alpha_{\vec{k},\vec{k}+\vec{q}+\vec{G}}M^\beta_{\vec{k}+\vec{q}+\vec{G}',\vec{k}}\chi(\vec{q}+\vec{G},\vec{q}+\vec{G}')$$

$$\times e^{i\vec{q}\cdot(\vec{R}^0(l\kappa)-\vec{R}^0(l'\kappa'))}e^{i(\vec{G}\cdot\vec{R}^0_\kappa-\vec{G}'\cdot\vec{R}^0_{\kappa'})}. \tag{C29}$$

Using Eq. (C29) in Eq. (2.63) we get the electronic contribution to the dynamical matrix as

$$\bar{D}^E_{\alpha\beta}(\vec{q},\kappa\kappa')=\frac{1}{(M_\kappa M'_\kappa)^{1/2}}\sum_{\vec{k}}\sum_{\vec{G},\vec{G}'}M^\alpha_{\vec{k},\vec{k}+\vec{q}+\vec{G}}M^\beta_{\vec{k}+\vec{q}+\vec{G}',\vec{k}}\chi(\vec{q}+\vec{G},\vec{q}+\vec{G}')$$

$$\times e^{i(\vec{G}\cdot\vec{R}^0_\kappa - \vec{G}'\cdot\vec{R}^0_{\kappa'})}. \tag{C30}$$

This is the expression in terms of electron-phonon matrix elements $M^\alpha_{\vec{k},\vec{k}+\vec{q}+\vec{G}}$ and the response function $\chi(\vec{q}+\vec{G}, \vec{q}+\vec{G}')$.

Equation (C30) can also be written in terms of Fourier transform of bare electron-ion potential $v_b(\vec{q}+\vec{G})$ if we define

$$v\left(\vec{r} - \vec{R}^0(l\kappa)\right) = \sum_{\vec{q}\vec{G}} v_b(\vec{q}+\vec{G})e^{i(\vec{q}+\vec{G})\cdot(\vec{r}-\vec{R}^0(l\kappa))}. \tag{C31}$$

Use of Eq. (C26) and (C31) in Eq. (C25) gives

$$\bar{\Phi}^e_{\alpha\beta}(l\kappa, l'\kappa') = \sum_{\vec{q}\vec{G}\vec{G}'} (\vec{q}+\vec{G})_\alpha v_b(\vec{q}+\vec{G})\chi(\vec{q}+\vec{G}, \vec{q}+\vec{G}')v_b(\vec{q}+\vec{G}')(\vec{q}+\vec{G}')_\beta$$

$$\times e^{i\vec{q}\cdot\left(\vec{R}^0(l\kappa)-\vec{R}^0(l'\kappa')\right)}e^{i\left(\vec{G}\cdot\vec{R}^0_\kappa - \vec{G}'\vec{R}^0_{\kappa'}\right)}. \tag{C32}$$

Substituting Eq. (C32) in Eq. (2.14) one gets

$$\bar{D}^E_{\alpha\beta}(\vec{q}, \kappa\kappa') = \frac{1}{(M_\kappa M_{\kappa'})^{1/2}} \sum_{\vec{G},\vec{G}'} (\vec{q}+\vec{G})_\alpha v_b(\vec{q}+\vec{G})\chi(\vec{q}+\vec{G}, \vec{q}+\vec{G}')$$

$$\times v_b(\vec{q}+\vec{G}')(\vec{q}+\vec{G}')_\beta e^{i\left(\vec{G}\cdot\vec{R}^0_\kappa - \vec{G}'\cdot\vec{R}^0_{\kappa'}\right)}. \tag{C33}$$

Appendix D
Transition Metal Model Potential Form Factor

In the semilocal approximation the Fourier transform of model potential defined in Eq.(4.210) is $< \vec{k}_F + \vec{q} \mid v_M(\vec{r}, \vec{E}) \mid \vec{k}_F >$. Animalu [95] added the orthogonalization effects and correlation corrections given by the potentials v_{oe} and v_{cc} [94] in defining the Fourier transform of bare electron-ion potential $v_b(\vec{q})$. Therefore, in the semilocal approximation

$$v_b(\vec{q}) = < \vec{k}_F + \vec{q} \mid v_M + v_{oe} + v_{cc} \mid \vec{k}_F >$$

$$= v_M^{(1)}(\vec{k}_F, \vec{k}_F + \vec{q}) + v_M^{(2)}(\vec{q}) \qquad (D1)$$

where

$$v_M^{(2)} = -\frac{8\pi C}{\Omega_0 q^3}[\sin(qR_M) - qR_M \cos(qR_M)] - \frac{8\pi Z}{\Omega_0 q^2}\cos(qR_M)$$

$$+ \left[\frac{4\pi \mid E_c \mid}{\Omega_0 q^3} - \frac{24\pi Z \alpha_{eff}}{\Omega_0 q^2 (qR_c)^3}\right][\sin(qR_c) - qR_c \cos(qR_c)]. \qquad (D2)$$

Here E_c is correlation energy, α_{eff} is the parameter which takes care of orthogonalization effects and $e = 1$.

For $\mid \vec{k}_F + \vec{q} \mid = \mid \vec{k}_F \mid$,

$$v_M^{(1)}(\vec{k}_F, \vec{k}_F + \vec{q}) = -\frac{4\pi}{\Omega_0}R_M^3(A_0 - C)\{j_0^2(x) - x^{-1}\cos(x)j_1(x)\}$$

$$-\frac{12\pi}{\Omega_0}R_M^3(A_1 - C)\{j_1^2(x) - j_0(x)j_2(x)\}P_1(\cos\theta) - \frac{20\pi R_M^3}{\Omega_0}(A_2 - C)$$

$$\times \left\{j_2^2(x) - j_1(x)j_3(x)\right\}P_2(\cos\theta) \qquad (D3)$$

with

$$x = k_F R_M,$$

$$\cos\theta = \left(1 - \frac{q^2}{2k_F^2}\right), \qquad (D4)$$

$$j_0(x) = [x^{-1}\sin x], j_1(x) = [x^{-2}\sin x - x^{-1}\cos x]$$

$$j_2(x) = [(3x^{-3} - x^{-1})\sin x - 3x^{-2}\cos x], \, j_3(x) = [5x^{-1}j_2(x) - j_1(x)], \tag{D5}$$

and

$$P_1(\cos\theta) = \cos\theta, \, P_2(\cos\theta) = \frac{1}{2}\left[3\cos^2\theta - 1\right]. \tag{D6}$$

For

$$|\,\vec{k}_F + \vec{q}\,| \neq |\,\vec{k}_F\,|,$$

$$v_M^{(1)}(\vec{k}_F, \vec{k}_F + \vec{q}) = -\frac{8\pi R_M^3(A_0 - C)}{\Omega_0(x^2 - y^2)}\left[xj_1(x)j_0(y) - yj_1(y)j_0(x)\right]$$

$$-\frac{24\pi R_M^3(A_1 - C)}{\Omega_0(x^2 - y^2)}\left[xj_2(x)j_1(y) - yj_2(y)j_1(x)\right]P_1(\cos\theta')$$

$$-\frac{40\pi R_M^3(A_2 - C)}{\Omega_0(x^2 - y^2)}\left[xj_3(x)j_2(y) - yj_3(y)j_2(x)\right]P_2(\cos\theta') \tag{D7}$$

where

$$x = k_F R_M, \, y = |\,\vec{k}_F + \vec{q}\,|\,R_M, \, E_F = \hbar^2 k_F^2 / 2m^* \tag{D8}$$

and

$$\cos\theta' = \left[x^2 + y^2 - (qR_M)^2\right]/2xy \tag{D9}.$$

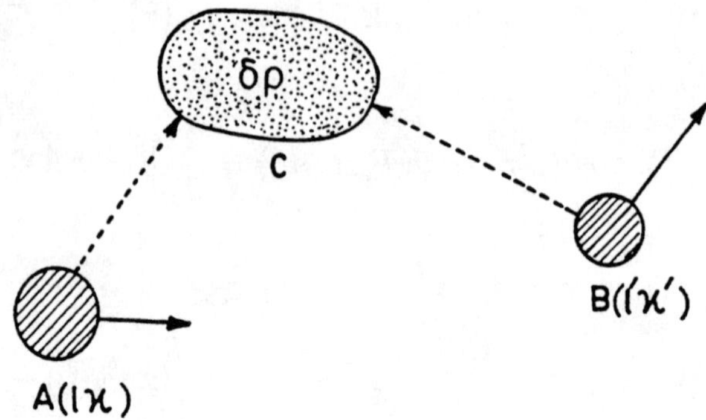

Figure 4.1: Ion-electron-ion interaction between ions $A(l, \kappa)$ and $B(l', \kappa')$.

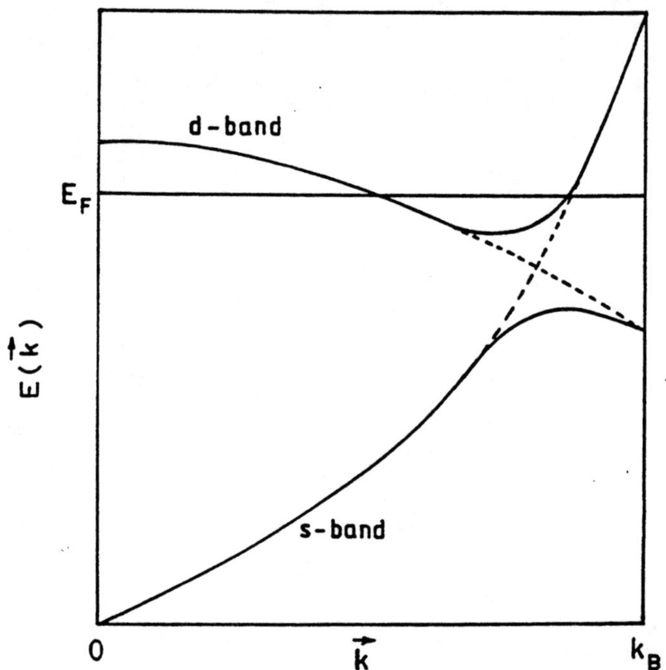

Figure 4.2: $E(\vec{k})$ versus $\vec{k}$ for s and d-bands in a transition metal. Continuous lines show hybridized (interacting) s and d-bands and dashed lines show unhybridized (non-interacting) s and d-bands. k_B is Brillouin zone radius and E_F is Fermi energy.

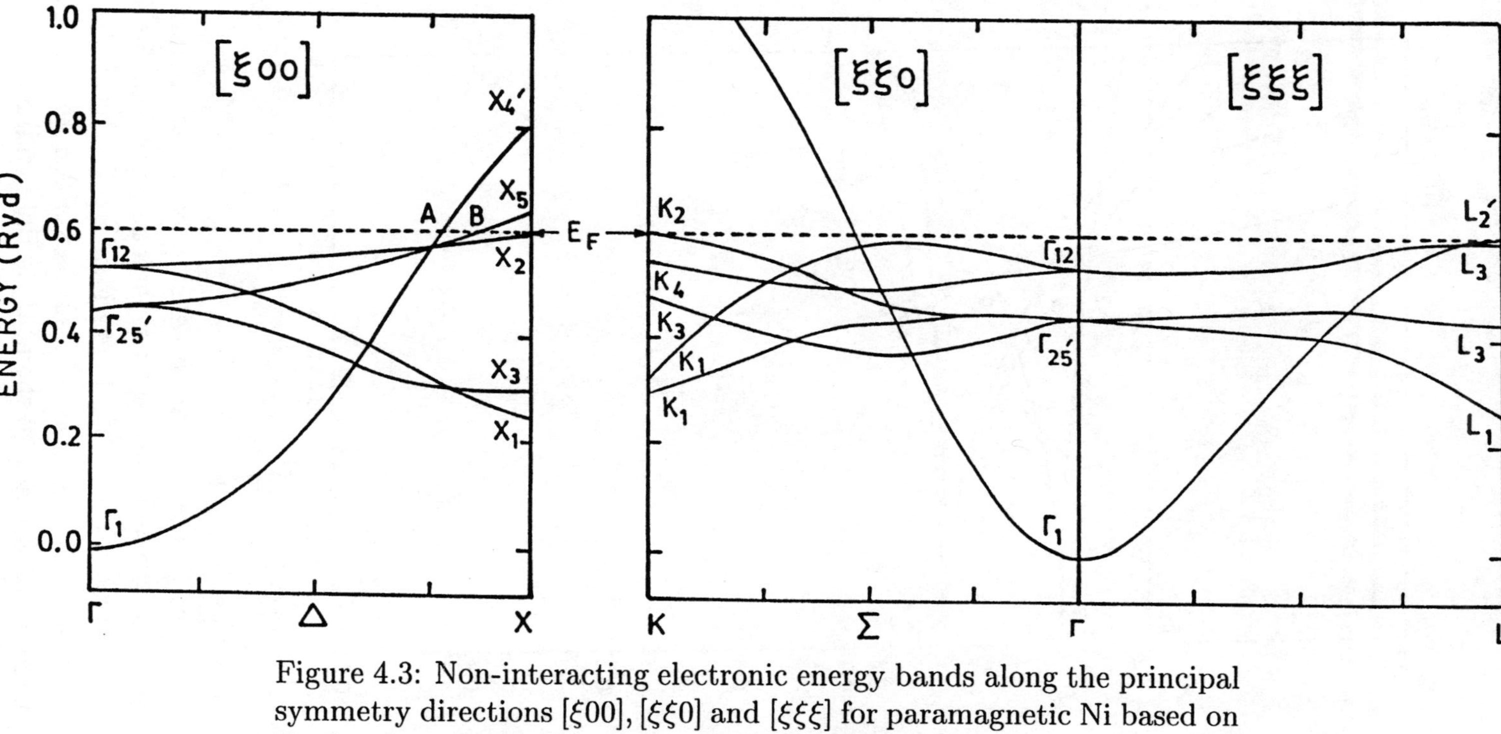

Figure 4.3: Non-interacting electronic energy bands along the principal symmetry directions [ξ00], [ξξ0] and [ξξξ] for paramagnetic Ni based on the band structure calculations of Hanus [62].

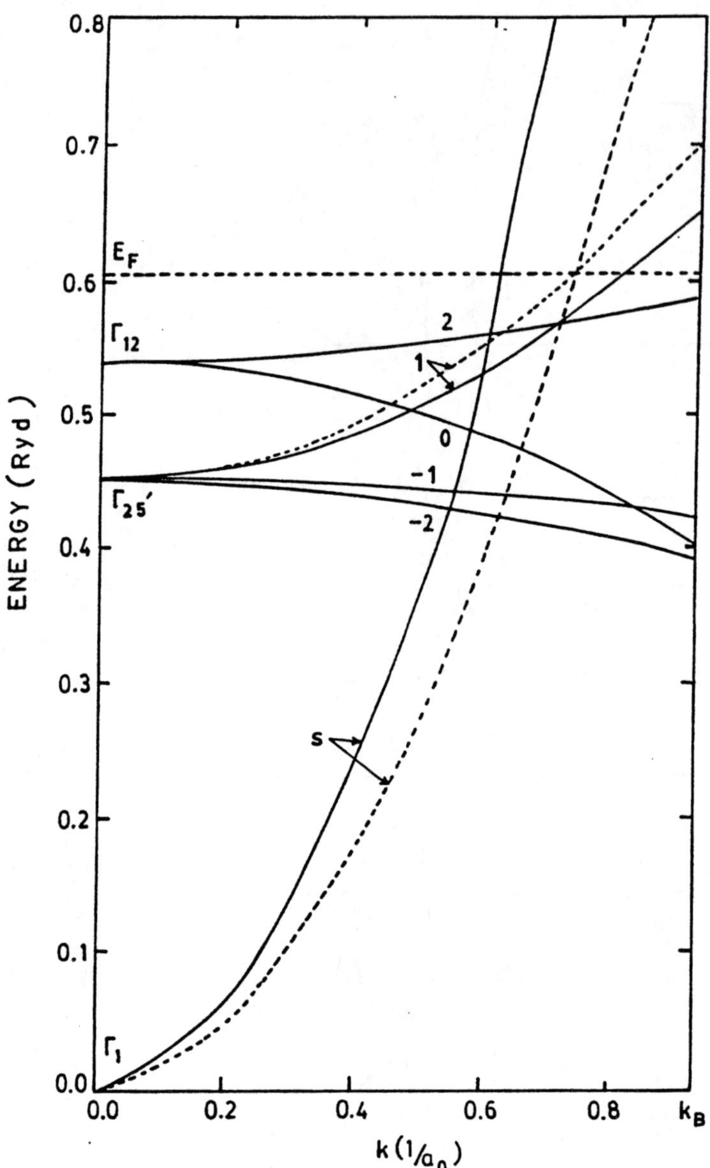

Figure 4.4: Isotoropic non-interacting energy bands for paramagnetic Ni. The solid and dashed lines are for the configurations $3d^9 4s^1$ and $3d^{9.4} 4s^{0.6}$ respectively. Filled bands are identical for both the configurations. Numbers by the side of d-subbands denote the magnetic quantum member m assigned to them. a_0 is Bohr radius.

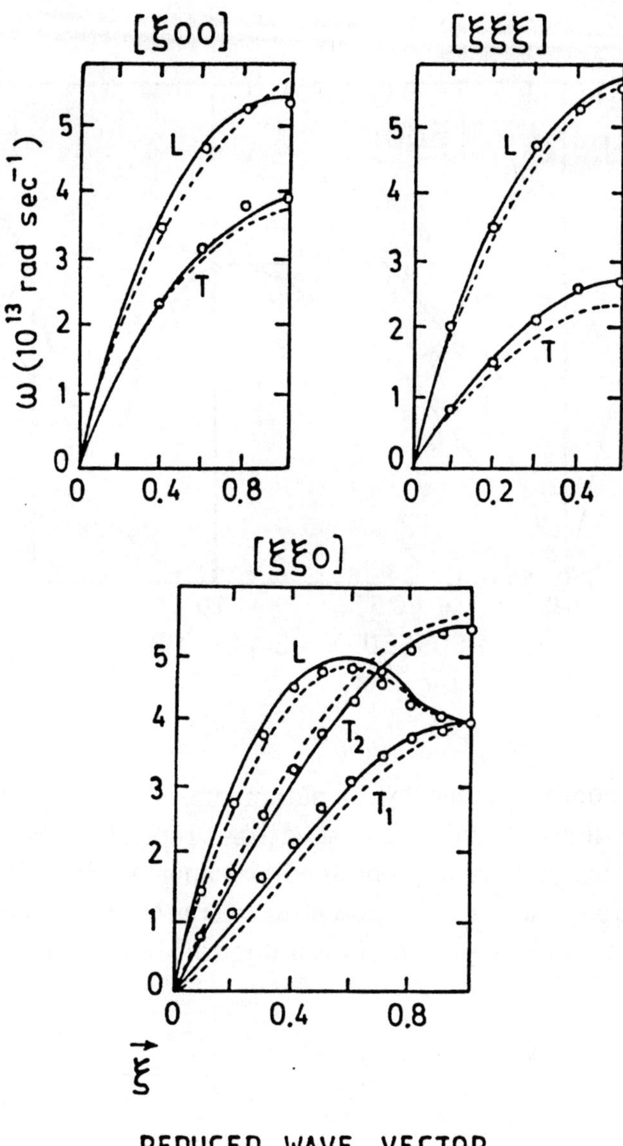

Figure 4.5: Phonon dispersion curves for paramagnetic Ni along three principal symmetry directions $[\xi00], [\xi\xi\xi], [\xi\xi0]$. Solid lines represent the theoretical results due to Hanke and Bilz [42] while the dashed lines are due to Prakash and Joshi [55]. Circles represent the experimental results due to Birgeneau et al [Ref.50 Chapter 3].

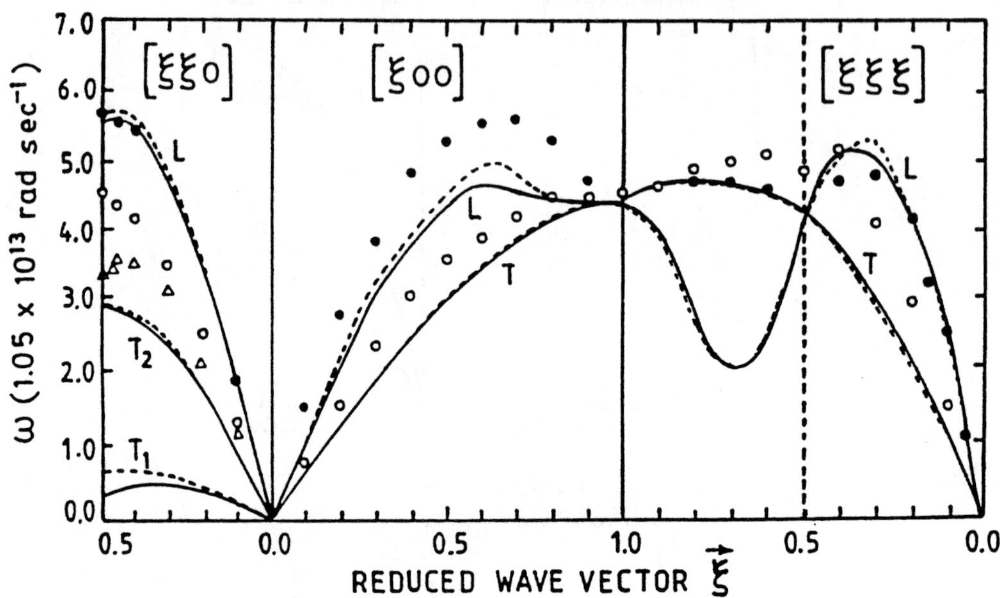

Figure 4.6: Phonon frequencies of paramagnetic Cr along $[\xi00]$, $[\xi\xi0]$ and $[\xi\xi\xi]$ directions. Solid and dashed lines represent the theoretical results due to Singh et al [60] obtained by using Animalu TMMP and Harrison's simple metal model potential, respectively. Solid and open circles represent the experimental data due to Mühlestein et al [Ref.58 Chapter 3].

Table 4.1: Assignment of magnetic quantum number m to different d-subbands along three principal symmetry directions.

$[\xi 00]$	$[\xi\xi 0]$	$[\xi\xi\xi]$	m
$\Gamma_{12} \to X_2$	$\Gamma_{12} \to K_4$	$\Gamma_{12} \to L_3$	2
$\Gamma_{12} \to X_1$	$\Gamma_{12} \to K_1$	$\Gamma_{12} \to L_3$	0
$\Gamma_{25'} \to X_5$	$\Gamma_{25'} \to K_2$	$\Gamma_{25'} \to L_3$	1
$\Gamma_{25'} \to X_3$	$\Gamma_{25'} \to K_3$	$\Gamma_{25'} \to L_3$	-1
$\Gamma_{25'} \to X_5$	$\Gamma_{25'} \to K_1$	$\Gamma_{25'} \to L_1$	-2

Table 4.2: Parameters of transition and rare earth metal model potential. All the quantities are in atomic units except $\mid E_c \mid$ which is in Rydberg units.

	A_0	A_1	A_2	R_m	Ω_0	Z	m^*	R_c	α_{eff}	$\mid E_c \mid$
Cu	0.25	0.40	0.215	2.2	79.4	1	1.0	1.814	0.157	0.086
Ag	0.223	0.40	0.218	2.6	115.4	1	1.0	2.381	0.245	0.082
Au	0.15	0.50	0.212	2.6	114.6	1	1.0	2.589	0.317	0.082
Zn	0.99	1.14	0.98	2.2	102.0	2	1.1	1.570	0.079	0.091
Cd	0.89	0.98	0.87	2.6	144.8	2	1.0	1.950	0.107	0.087
Hg	0.97	1.11	0.85	2.6	157.8	2	1.0	2.120	0.126	0.086
Sc	1.60	1.65	1.40	2.0	168.7	3	1.0	1.531	0.045	0.090
Y	0.75	1.30	1.10	2.0	223.1	3	1.0	1.739	0.049	0.087
La	0.90	1.40	0.85	2.0	252.2	3	1.0	2.154	0.083	0.086
Ti	2.30	2.50	2.10	2.0	119.0	4	1.0	1.285	0.037	0.096
Zr	1.15	1.70	1.50	2.0	157.0	4	1.0	1.493	0.044	0.095
Hf	1.30	1.80	1.35	2.0	150.2	4	1.0	1.474	0.045	0.095
V	3.25	3.50	2.90	1.6	93.9	5	1.0	1.115	0.031	0.101
Nb	1.70	2.30	2.25	2.0	121.3	5	1.0	1.304	0.038	0.100
Ta	1.75	2.35	2.25	2.0	121.3	5	1.0	1.285	0.038	0.100
Cr	1.60	1.47	1.40	2.5	80.6	3	1.0	1.191	0.044	0.102
Mo	2.30	2.93	2.50	2.0	105.5	6	1.0	1.323	0.046	0.100
W	2.30	2.85	2.50	2.0	106.5	6	1.0	1.172	0.032	0.101
Mn	0.89	0.98	0.87	2.2	81.9	2	1.0	1.512	0.088	0.095
Tc	3.10	3.20	3.30	2.0	96.5	7	1.0	1.058	0.026	0.102
Re	2.95	3.55	3.30	2.0	99.3	7	1.0	1.058	0.025	0.102
Fe	1.60	1.65	1.40	2.0	79.8	3	1.0	1.400	0.072	0.090
Ru	1.15	1.70	1.50	2.0	91.9	4	1.0	1.266	0.046	0.098
Os	1.30	1.80	1.35	2.0	94.8	4	1.0	1.304	0.049	0.098
Co	0.99	1.05	0.98	2.2	74.9	2	1.0	1.360	0.070	0.094
Rh	0.75	1.30	1.10	2.0	92.6	3	1.0	1.285	0.048	0.096
Ir	1.30	1.80	1.35	2.0	95.5	4	1.0	1.285	0.047	0.098
Ni	0.99	1.05	0.98	2.2	73.6	2	1.0	1.304	0.063	0.093
Pd	0.89	0.75	0.87	2.6	99.3	2	1.0	1.512	0.073	0.091
Pt	0.97	1.11	0.85	2.6	101.6	2	1.0	1.512	0.071	0.091

	A_0	A_1	A_2	A_3	R_M	Ω_0	Z	m^*	R_c	α_{eff}	$\mid E_c \mid$
La	0.90	1.40	0.85	2.30	2.0	252.20	3	1.0	2.154	0.083	0.085
Tb	0.90	1.40	0.85	2.30	2.0	214.54	3	1.0	1.894	0.066	0.085
Ho	0.90	1.40	0.85	2.30	2.0	210.03	3	1.0	2.000	0.066	0.086

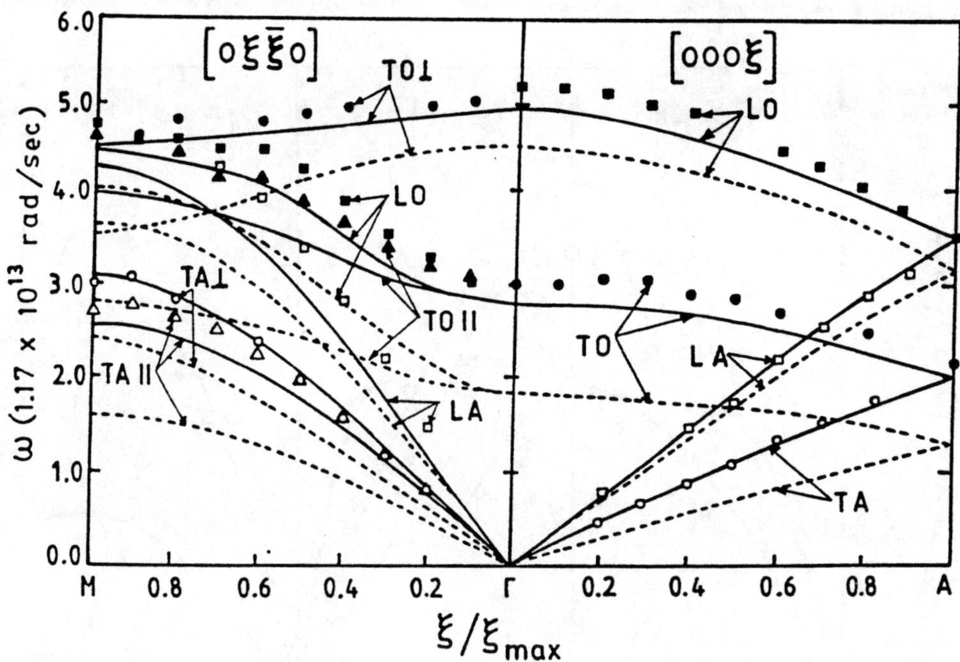

Figure 4.7: Phonon dispersion curves for Sc due to Singh and Prakash [57]. Solid lines represent the results obtained by renormalizing the parameters of Animalu TMMP to obtain a reasonable fit of the experimental data while the dashed lines give the results obtained by using the parameters tabulated by Animalu (see Table 2). Experimental results, represented by circles, triangles and squares, are taken from Wakabayashi et al [Ref.87 Chapter 3].

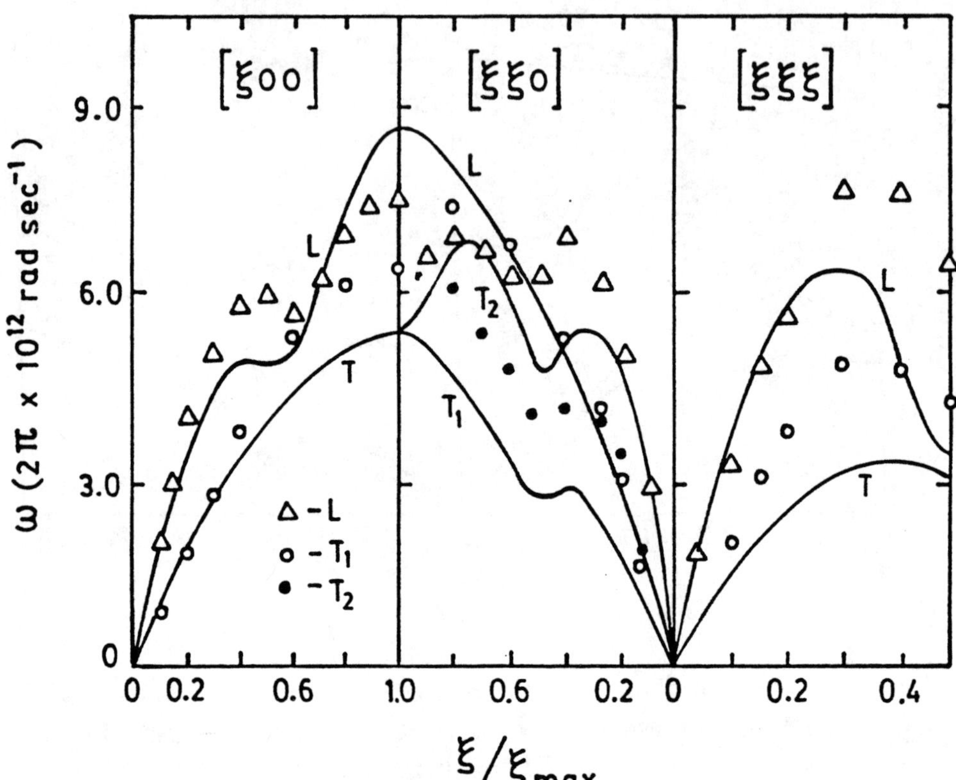

Figure 4.8: Phonon dispersion curves for acoustic phonons in NbC due to Hanke et al [80]. Circles and triangles represent the experimental results of Smith and Gläser [Ref.160 Chapter 3].

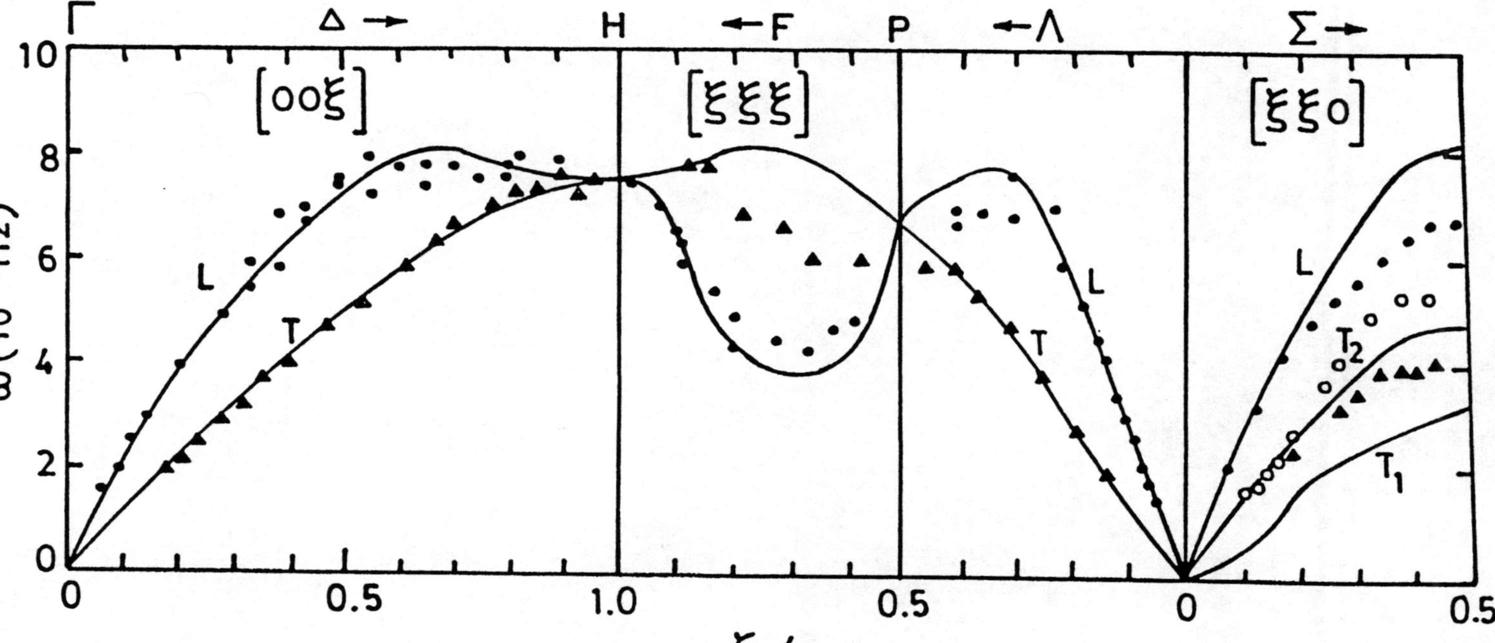

Figure 4.9: Phonon dispersion curves for paramagnetic Vanadium. Continuous curves represent the calculated values of phonon frequencies for depletion hole parameter $\alpha = 0.206$. Experimental values, represented by circles and traingles and are taken from Collela and Batterman [Ref.77 Chapter 3].

Bibliography

[1] M. Born and J.R. Oppenheimer, Ann. Physik 84, 457(1927); E.W. Kellerman, Phil. Trans. Roy. Soc.(London) A238, 513(1940).

[2] G.Baym, Ann. Physik, 14, 1(1961).

[3] W. Cochran, Proc. Roy. Soc.(London) A276, 308(1963).

[4] S.H.Vosko, R.Taylor and G.H.Keech, Can. J. Phys. 43, 1187(1965).

[5] A. Sjölander and R. Johnson, in "Inelastic Scattering of Neutrons" (IAEA, Vienna, 1965) Vol.1, p.61.

[6] L.J. Sham, Proc. Roy. Soc.(London) A238, 33(1965).

[7] P.N. Keating, Phys. Rev. 176, 1171(1968).

[8] S.K. Joshi and A.K. Rajagopal, Solid State Physics, 22, 159(1968).

[9] L.J. Sham, in "Modern Solid State Physics" Eds. R.H. Enns and R.R. Haering (Gordon and Beach, 1969) Vol.2, p.143.

[10] L.J. Sham, Phys. Rev. 188, 1431(1969).

[11] R.M. Pick, Adv. Phys. 19, 269(1970).

[12] R.M. Pick, M.H. Cohen and R.M. Martin, Phys. Rev. B1, 910(1970); in " Neutron Inelastic Scattering" (IAEA, Vienna, 1968) Vol.1, p.119.

[13] L.J. Sham, in "Dynamical Properties of Solids" Eds. G.K. Horton and A.A. Maradudin (North-Holland, Amsterdam 1974) Vol.1, p.301.

[14] W.A. Harrison, Phys. Rev. 181, 1036(1969).

[15] G. Venkataraman, L.A. Feldkamp and V.C. Sahni, "Dynamics of Perfect Crystals" (MIT Press, Cambridge, 1975).

[16] V. Heine and D.Waeire, Solid State Physics 24, 249(1970).

[17] V. Heine, P. Nozieres and J.W. Wilkins, Phil. Mag.13, 741(1966).

[18] L.J. Sham and J.M. Ziman, Solid State Physics 15 221(1965).

[19] S.K. Sinha, in "Dynamical Theory of Solids" Ed. G.K. Horton and A.A. Maradudin (North-Holland, Amsterdam 1980) Vol.3, p.1.

[20] P. Nozieres and D. Pines, Phys. Rev. 109, 741, 762, 1062(1958); ibid 111, 442(1958); ibid 113, 1254(1959).

[21] H. Ehrenreich and M.H. Cohen, Phys. Rev. 115, 786(1959).

[22] S. Adler, Phys. Rev. 126, 413(1962).

[23] N. Wiser, Phys. Rev. 129, 62(1963).

[24] T. Toya, J. Res. Inst. Catal.6, 161(1958); ibid 6, 183(1958); ibid 7, 60(1959); Prog. Theor. Phys.(Kyota) 20, 974(1958); in "Lattice Dynamics" Ed. R.F. Wallis (Pergamon Press,1964) p. 250.

[25] D. Pines, "Elementary Excitations in Solids"(Benjamin Inc., New York 1963).

[26] W. Kohn and L.J. Sham, Phys. Rev. 140, A1133(1965).

[27] J.E. Robinson, F. Bassani, R.S. Knox and J.R. Schrieffer, Phys. Rev. B1, 1044(1970).

[28] D.J.W. Geldart and S.H. Vosko, Cand. J. Phys. 44, 2137(1966).

[29] D.J.W. Geldart and R. Taylor, Cand. J. Phys. 48, 155, 167(1970).

[30] F. Toigo and T.O. Woodruff, Phys. Rev. B2, 3959(1970).

[31] K.S. Singwi, A. Sjölander, M.P. Tosi and R.H. Land, Phys. Rev. B1, 1044(1970).

[32] P. Vashishta and K.S. Singwi, Phys. Rev. B6, 875, 4883(1972).

[33] I. Lindgren and K. Schwarz, Phys. Rev. A5, 542(1972).

[34] A.O.E. Animalu, Phil. Mag. 11, 379(1965).

[35] J. Hubbard, Proc. Roy. Soc.(London) A240, 539(1957); ibid A243, 336(1958).

[36] L.J. Sham, Proc. Phys. Soc. (London) 78, 895(1961).

[37] D. Pines and P. Nozieres, "The theory of Interacting Fermi Systems" (Benjamin Inc., New York, 1964).

[38] S.K. Sinha, Phys. Rev. 169, 477(1968).

[39] D.C. Golibersuch, Phys. Rev. 157, 532(1964).

[40] P.B. Allen, in "Dynamical Properties of Solids" Eds. G.K. Horton and A.A. Maradudin (North-Holland, Amsterdam, 1980) Vol.3, p.95.

[41] L.G. Ferreira, J. Phys. C4, 3(1971); L.G. Ferreira and N.J. Parada, J. Phys. C4, 15(1971); L.G. Ferreira and G.W. Pratt, J. Phys. C4, 20 (1971).

[42] W. Hanke and H. Bilz, in "Inelastic Scattering of Neutrons" (IAEA, Vienna, 1972) p.3.

[43] W. Hanke, Phys. Rev. B8, 4585, 4591(1973).

[44] J. Callaway, "Quantum Theory of the Solid State" (Acad. Press, New York, 1974) p.371.

[45] J.C. Slater and G.F. Koster, Phys. Rev. 94, 1498(1954).

[46] L.J. Sham, Phys. Rev. Lett. 27, 1725(1971); Phys. Rev. B6, 3584(1972).

[47] R. Pick, in "Proceedings of International Conference on Phonons" Ed. M.A. Nusimovice (Flammarion, Paris, 1971) p.20.

[48] W. Hanke, in "Proceedings of International Conference on Phonons" Ed. M.A. Nusimovice (Flammarion, Paris, 1971) p.294.

[49] F.M. Müller, Phys. Rev. 153, 659(1967).

[50] L.F. Mattheiss, Phys. Rev. 134, A970(1964); J. Yamashita, M. Fukuchi and S. Wakoh, J. Phys. Soc. Japan 18, 999(1963); ibid 19, 1342(1964); L. Hodges, H. Ehrenreich and N.D. Lang, Phys. Rev. 152, 505(1966); J.W.D. Connolly, Phys. Rev. 159, 415(1967); J. Callaway and C.S. Wang, Phys. Rev. B7, 1096(1973).

[51] S.K. Sinha, Phys. Rev. 177, 1256(1969).

[52] S.K. Sinha, R.P. Gupta and D.L. Price, Phys. Rev. Lett. 26, 1324(1971).

[53] S.K. Sinha, R.P. Gupta and D.L. Price, Phys. Rev. B9, 2564, 2573(1974).

[54] W. Hanke and H. Bilz, Z. Naturforsch, A26, 585(1971).

[55] S. Prakash and S.K. Joshi, Phys. Rev. B2, 915(1970); ibid B4, 1770(1971).

[56] J.S. Brown, J. Phys. F1, 63(1971); ibid F2, 115(1972).

[57] J. Singh and S. Prakash, Nuovo Cimento 37B, 131(1977).

[58] U. Schröder, Solid State Commun. 4, 347(1966).

[59] U. Schröder and V. Nusslein, Phys. Stat. Sol. 21(b), 309(1967).

[60] J. Singh, N. Singh and S. Prakash, Phys. Rev. B12, 3159, 3166(1975); ibid B18, 2954(1975).

[61] N. Singh, J. Singh and S. Prakash, Phys. Rev. B12, 1076, 5415(1975).

[62] J. Hanus, "MIT Solid State and Molecular Theory Group, Quarterly Progress Report No.44"(1962) p.62, (unpublished).

[63] J. Callaway, "Energy Band Theory" (Acad.Press, New York,1964).

[64] J.F. Cornwell, "Group Theory and Electronic Energy Bands" (North-Holland, Amsterdam, 1969).

[65] C. Herring, J. Franklin Inst. 233, 525(1942).

[66] V. Heine, in "Physics of Metals" Ed. J.M. Ziman (Combridge Univ. Press, London, 1969) p.25; ibid J.Friedel p.340.

[67] M. Asdente and J. Friedel, Phys. Rev. 124, 384(1961).

[68] Y. Yasui, E. Hayashi and M. Shimizu, J. Phys. Soc. Japan. 29, 1446(1970).

[69] R.P. Gupta and S.K. Sinha, Phys. Rev. B3, 2401(1971).

[70] D.A. Papaconstantopoulos, J.R. Anderson and J.W. McCaffrey, Phys. Rev. B5, 1214(1972).

[71] D.D. Bett, A.B. Bhatia and M. Wyman, Phys. Rev. 104, 37(1965).

[72] S.L. Altmann and C.J. Bradley, Proc. Phys. Soc.(London) 92, 764(1967); Rev. Mod. Phys. 37, 33(1965).

[73] S. Prakash and P. Lucasson, Phys. Stat. Sol. 91(b), 339(1979); J. Phys. F11, 2515(1981).

[74] R.E. Watson, "MIT Solid State and Molecular Theory Group, Technical Report No.12" (1952) (unpublished).

[75] W.H. Lomer, Proc. Phys. Soc. (London)80, 489(1962); ibid 84, 327(1964).

[76] J.E. Graebner and J.A. Marcus, Phys. Rev. 175, 659(1968).

[77] A. Czachor, in "Inelastic Scattering of Neutrons in Solids and Liquids" (IAEA, Vienna, 1965) Vol.1, p.181.

[78] V. Jaccarino, in "Theory of Magnetism in Transition Metals" Ed. W. Marshall (Acad. Press, New York, 1967) p.335.

[79] T.L. Loucks, Phys. Rev. 144, 504(1966); ibid 159, 444(1967).

[80] W. Hanke, J. Hafner and H. Bilz, Phys. Rev. Lett. 37, 1560(1976).

[81] W. Weber, Phys. Rev. B8, 5082, 5093(1973).

[82] A. Neckel, P. Raste, R. Eibler, P. Weinberger and K.H. Schwarz, J. Phys. C9, 579(1976).

[83] L. Ramqvist, B. Ekstig, E. Källne, E. Noreland and R. Manne, J. Phys. Chem. Solids 32, 149(1971).

[84] W. Hanke, J. Hafner and H. Bilz, Ferroelectrics 16, 161(1977).

[85] C. Herring, Phys. Rev. 57, 1169(1940).

[86] W. A. Harrison, "Pseudopotentials in the Theory of Metals" (Banjamin Inc., New York, 1966).

[87] W.A. Harrison, "Solid State Theory" (McGraw-Hill, 1970).

[88] V. Heine, Solid State Physics, 24, 1(1970).

[89] R.A. Deegan and W.D. Twose, Phys. Rev. 164, 993(1967).

[90] J.A. Moriarty, Phys. Rev. B1,1363(1970); ibid B5, 2066 (1972); ibid B6, 1239, 4445(1972); ibid B26, 1754 (1982); ibid B38, 3199(1988).

[91] J.M. Wills and W. A. Harrison, Phys. Rev. B28, 4363(1983).

[92] W.A. Harrison, Phys. Rev. B28, 550(1983); ibid B29, 2917(1984).

[93] W.A. Harrison and G.K. Straub, Phys. Rev. B36, 2695 (1987).

[94] V. Heine and I.Abarenkov, Phil. Mag. 9, 451(1964).

[95] A.O.E Animalu, Phys. Rev. B8, 3542, 3555 (1973).

[96] J.C. Upadhyaya and A.O.E. Animalu, Phys. Rev. B15, 1867(1977).

[97] B.A. Oli and A.O.E. Animalu, Phys. Rev. B13, 2398(1976).

[98] R.W. Shaw and W.A. Harrison, Phys. Rev. 163, 604(1967).

[99] R.W. Shaw, Phys. Rev. 176, 769(1968).

[100] J. Singh and S. Prakash, J.Phys. F11, 2409(1981).

[101] J. Singh, R. Singh and S. Prakash, Physica 90B, 223(1977).

[102] O.W. Deitrich and J. Als-Nielsen, Phys. Rev. 162, 315(1967).

[103] S.C. Keeton and T.L. Loucks, Phys. Rev. 168, 672(1968).

[104] W.E. Evenson and S.H. Liu, Phys. Rev. 178, 783(1969).

[105] D.G. Pettifor, J. Phys. F7, 613(1977).

[106] J.A. Moriarty, Phys. Rev. B26, 1754(1982).

[107] A.H. MacDonald and R. Taylor, Can. J. Phys. 62, 796(1984).

[108] N. Singh, Phys. Lett. 145A, 363(1990); N. Singh, N.S. Banger
and S.P. Singh, Phys. Rev. B39, 3097(1989).

[109] D.G. Pettifor, Phys. Scr. T1, 26(1982); D.G. Pettifor and M.A.
Ward, Solid State Commun. 49, 291(1984).

[110] N. Singh and B. Kumar, Physica 176B, 222(1992).

[111] N. Singh, Physica, 160B, 313(1990).

[112] G. Treglia, F. Ducastelle and D. Spanjaerd, J. de Physique 41,
281(1980).

Chapter 5

MICROSCOPIC MODELS

Contents

List of Figures

List of Tables

Chapter 5

MICROSCOPIC MODELS

It has been pointed out in Chapter 4 that the exact calculations of band structure effects on phonon frequencies are too difficult. Even then, one may not be sure where one is lost in the huge numerical data. Therefore microscopic models are developed to understand the dynamics of transition metals, their compounds and alloys. These models have been used to study the phonon anomalies and interatomic forces in these materials. We present these models systematically in this chapter.

5.1   Charge Fluctuation Model

5.1.1   General Theory

The phonon frequencies for the mode $(\vec{q}j)$ are obtained by solving the secular Eq.(2.57) where one gets

$$\omega_j^2(\vec{q})e_\alpha(\vec{q}j;\kappa) = \sum_{\kappa'\beta} D_{\alpha\beta}(\vec{q},\kappa\kappa')e_\beta(\vec{q}j;\kappa').$$

The elements of dynamical matrix are given in Eq.(2.62) as

$$D_{\alpha\beta}(\vec{q},\kappa\kappa') = \bar{D}_{\alpha\beta}(\vec{q},\kappa\kappa') - \bar{D}_{\alpha\beta}(0,\kappa\kappa').$$

In the microscopic theories (Eq.(4.91))

$$\bar{D}_{\alpha\beta}(\vec{q},\kappa\kappa') = \bar{D}_{\alpha\beta}^c(\vec{q},\kappa\kappa') + \bar{D}_{\alpha\beta}^E(\vec{q},\kappa\kappa')$$

where $\bar{D}^c_{\alpha\beta}(\vec{q}, \kappa\kappa')$ arises due to ion-ion Coulomb interaction and $\bar{D}^E_{\alpha\beta}(\vec{q}, \kappa\kappa')$ due to ion-electron-ion interaction i.e. the interaction between the ions via conduction electrons. Thus in conjunction with Eqs.(2.58), (4.92) and (4.93), we can write the phonon frequencies as the sum of two contributions for a monatomic lattice as

$$\omega_j^2(\vec{q}) \;=\; \omega_j^{c2}(\vec{q}) + \frac{1}{M} \sum_{\vec{G}\vec{G}'} \Delta v_b(\vec{q}+\vec{G})\chi(\vec{q}+\vec{G}, \vec{q}+\vec{G}')\Delta v_b^*(\vec{q}+\vec{G}')$$

$$\;=\; \omega_j^{c2}(\vec{q}) - \omega_j^{E2}(\vec{q}). \tag{5.1}$$

Here $\omega_j^c(\vec{q})$ is ion plasma frequency which arises due to $D^c_{\alpha\beta}(\vec{q})$. The second term is electronic contribution which arises due to $D^E_{\alpha\beta}(\vec{q})$ and the corresponding frequency is $\omega_j^E(\vec{q})$. $\Delta v_b(\vec{q}+\vec{G})$ is the change in ionic potential due to displacement of ions in the mode $(\vec{q}j)$ and $\chi(\vec{q}+\vec{G}, \vec{q}+\vec{G}')$ is the electron density response function given in Eq.(4.14).

The first term in Eq.(5.1) arises due to repulsive interaction between the ions. The second term arises due to attractive interaction between the ions via conduction electrons. These two contributions are large and tend to cancel each other. The first term depends only on the geometry of the lattice while the second term involves all the many-body interactions of the electron-ion system. Hence the second term depends on both the geometry and the volume of the system. Therefore the second term must be evaluated as accurately as possible. But the bare ion potential $v_b(\vec{q}+\vec{G})$ is strong and rapidly varying in the vicinity of ion cores and $\chi(\vec{q}+\vec{G}, \vec{q}+\vec{G}')$ involves the inverse of infinite dimensional dielectric matrix as given in Eq.(4.15). Therefore the accurate evaluation of the second term of Eq.(5.1), as it stands, is a formidable task.

Sinha and co-workers [1] [2] have shown that a large part of $\Delta v(\vec{q}+\vec{G})$, in particular the strong and rapidly varying part near the ion core, had the same effect as if the atomic like part of the electron wave function near the ion core is moving rigidly with the ion core. This implies that this large part of the second term may be separated and this may explicitly cancel $D^c_{\alpha\beta}(\vec{q})$, the Coulomb coupling coefficient. This view point can be achieved if we begin with a problem of rigid motion of some part of conduction electrons (say the electronic charge in the muffin-tin volume v_{ol}) around the ion core . This charge together with the ion core will constitute a "pseudoatom" which will have much reduced charge as compared to the bare ion charge. This part may be taken to contribute towards the first term $D^c_{\alpha\beta}(\vec{q})$ or $\omega^c_j(\vec{q})$ and thus renormalizing it to $D'^c_{\alpha\beta}(\vec{q})$ or $\bar{\omega}^c_j(\vec{q})$. Now the second term of Eq.(5.1) will involve the remaining distortions of the electron density in and outside the volume v_{ol}. These remaining distortions can be described by a rapidly converging Fourier expansion. Consequently, the size of the dielectric matrix which is to be inverted to obtain $\chi(\vec{q}+\vec{G}, \vec{q}+\vec{G}')$ will be reduced. Sinha and co-workers [2], [3] formulated this concept using the method of linear combination of atomic orbitals (LCAO).

By generalizing Eq.(4.116), we can write the Bloch function in terms of atomic orbiatls as

$$\psi_{\vec{k}\mu}(\vec{r}) = |\,\vec{k}\mu >$$

$$= \frac{1}{\sqrt{N}} \sum_{l\nu} e^{\nu}_{\vec{k}\mu} \exp(i\vec{k}\cdot\vec{R}_l)\phi_\nu(\vec{r}-\vec{R}_l) \tag{5.2}$$

where $\phi_\nu(\vec{r}-\vec{R}_l)(\equiv|\,\nu l >)$ is the atomic orbital of type ν which is centered at the site $\vec{R}_l$ and $e^{\nu}_{\vec{k}\mu}$ is the eigenvector coefficient. The frequency dependent density response function can readily be written in terms of thermal retarted Green's function as [4]

$$\chi(\vec{q}+\vec{G}, \vec{q}+\vec{G}',\omega) = -2\pi \sum_{\vec{k}_1\vec{k}_2} \sum_{\mu_1\mu_2\mu_3\mu_4} \sum_{\sigma}$$
$$\times < \vec{k}_1\mu_1 \mid \exp[-i(\vec{q}+\vec{G})\cdot\vec{r}] \mid \vec{k}_1+\vec{q}\mu_2 >$$
$$\times < \vec{k}_2+\vec{q}\mu_3 \mid \exp[i(\vec{q}+\vec{G}')\cdot\vec{r}] \mid \vec{k}_2\mu_4 >$$
$$\times G(\vec{k}_1\mu_1\sigma, \vec{k}_1+\vec{q}\mu_2\sigma, \vec{k}_2+\vec{q}\mu_3\sigma, \vec{k}_4\mu_4\sigma, \omega)$$
$$\tag{5.3}$$

where $G(\vec{k}_1\mu_1\sigma, \vec{k}_1+\vec{q}\mu_2\sigma, \vec{k}_2+\vec{q}\mu_3\sigma, \vec{k}_4\mu_4\sigma, \omega)$ is the time Fourier transform of two particle Green's function. One can obtain $G(\omega)$ by solving the equation of motion for $G(\omega)$ and by reducing the four particle operators into two particle operators. Finally the use of Eq.(5.2) for the Bloch states simplify Eq.(5.3) as

$$\chi(\vec{q}+\vec{G}, \vec{q}+\vec{G}',\omega) = \sum_{\nu_1\nu_2\nu_3\nu_4} \sum_{ll'} < \nu_1 \mid e^{-i(\vec{q}+\vec{G})\cdot\vec{r}} \mid \nu_4 l' >$$

352   *S. PRAKASH*

$$\times < \nu_2 \mid e^{i(\vec{q}+\vec{G}')\cdot\vec{r}} \mid \nu_3 l >$$

$$\times \chi_{\nu_1\nu_2;\nu_4 l'\nu_3 l}(\vec{q},\omega) \tag{5.4}$$

where the density response function in the atomic orbital representation is given as

$$\chi_{\nu_1\nu_2;\nu_4 l'\nu_3 l}(\vec{q},\omega) = \chi^0_{\nu_1\nu_2;\nu_4 l'\nu_3 l}(\vec{q},\omega) + \sum_{\nu_5\nu_6\nu_7\nu_8}\sum_{tt'}\chi^0_{\nu_1\nu_5;\nu_4 l'\nu_6 t}(\vec{q},\omega)$$

$$\times \bar{v}_{\nu_5\nu_7;\nu_6 t\nu_8 t'}(\vec{q})\chi_{\nu_7\nu_2;\nu_8 t'\nu_3 l}(\vec{q},\omega). \tag{5.5}$$

Here the polarizability function

$$\chi^0_{\nu_1\nu_2;\nu_4 l'\nu_3 l}(\vec{q},\omega) = \sum_{\vec{k}\mu_1\mu_2\sigma}\frac{f_{\vec{k}\mu_1\sigma} - f_{\vec{k}+\vec{q}\mu_2\sigma}}{E_{\vec{k}\mu_1\sigma} - E_{\vec{k}+\vec{q}\mu_2\sigma} + i\omega}e^{\nu_1*}_{\vec{k}\mu_1}e^{\nu_2}_{\vec{k}+\vec{q}\mu_2}$$

$$\times e^{\nu_4}_{\vec{k}+\vec{q}\mu_2}e^{i(\vec{k}+\vec{q})\cdot\vec{R}_{l'}}e^{\nu_3}_{\vec{k}\mu_1}e^{i\vec{k}\cdot\vec{R}_l}, \tag{5.6}$$

$$\bar{v}_{\nu_1\nu_2;\nu_4 l'\nu_3 l}(\vec{q}) = \left[v^c(\vec{q}) - (1/2)v^x(\vec{q}) - (1/2)v^0(\vec{q})\right]_{\nu_1\nu_2;\nu_4 l'\nu_3 l} \tag{5.7}$$

$$v^c_{\nu_1\nu_2;\nu_4 l'\nu_3 l}(\vec{q}) = \sum_j e^{i\vec{q}\cdot\vec{R}_j}\int\int d\vec{r}\,d\vec{r}'\,\phi^*_{\nu_1}(\vec{r})\phi^*_{\nu_2}(\vec{r}'-\vec{R}_j)$$

$$\times v(\vec{r}-\vec{r}')\phi_{\nu_3}(\vec{r}'-\vec{R}_l)\phi_{\nu_4}(\vec{r}-\vec{R}_j-\vec{R}_{l'}) \tag{5.8}$$

and

$$v^x_{\nu_1\nu_2;\nu_4 l'\nu_3 l}(\vec{q}) = \sum_j{}'e^{i\vec{q}\cdot\vec{R}_j}\int\int d\vec{r}\,d\vec{r}'\,\phi^*_{\nu_1}(\vec{r}')\phi^*_{\nu_2}(\vec{r}-\vec{R}_j)$$

$$\times v(\vec{r}-\vec{r}')\phi_{\nu_3}(\vec{r}'-\vec{R}_l)\phi_{\nu_4}(\vec{r}-\vec{R}_j-\vec{R}_{l'}). \tag{5.9}$$

The prime on summation in Eq.(5.9) denotes that the term $\vec{R}_j = 0$ is excluded. $v^c(\vec{q})$ and $v^x(\vec{q})$ are the Fourier transforms of Coulomb and exchange interactions between the electrons respectively and $v^0(\vec{q})$ is $v^c(\vec{q})$ for $\vec{R}_j = 0$.

If the pair indices $(\nu_1, \nu_4 l')$ and $(\nu_2, \nu_3 l)$ are represented by s_1 and s_2 respectively, Eq.(5.5) can be written as

$$\chi_{s_1 s_2}(\vec{q},\omega) = \chi^0_{s_1 s_2}(\vec{q},\omega) + \sum_{s_3 s_4}\chi^0_{s_1 s_3}(\vec{q},\omega)\bar{v}_{s_3 s_4}(\vec{q})\chi_{s_4 s_2}(\vec{q},\omega)$$

$$\tag{5.10}$$

which has the solution

$$\chi_{s_1 s_2}(\vec{q}, \omega) \;=\; \left[\left(1 - \chi^0 \bar{v}\right)^{-1} \chi^0 \right]_{s_1 s_2} \tag{5.11}$$

where $s_3 (= \nu_5 \to \nu_6 t)$ and $s_4 (= \nu_7 \to \nu_8 t')$. One can calculate $\chi(\vec{q} + \vec{G}, \vec{q} + \vec{G}', \omega)$ by using Eq.(5.11) in Eq.(5.4). In Eq.(5.10), $s_1 (= \nu_1 0 \to \nu_4 l')$ represents the phonon induced electronic charge fluctuation corresponding to a virtual transition from an orbital ν_1 centered at some lattice site to another orbital ν_4 centered on the neighbouring site $\vec{R}_{l'}$. Similarly $s_2 (= \nu_2 0 \to \nu_3 l)$ reprsents the phonon induced charge fluctuation corresponding to a virtual transition from an orbital ν_2 centered at some lattice site to another orbital ν_3 centered on the neighbouring site $\vec{R}_l$ and so on. The intra-atomic and the first nearest neighbour inter-atomic virtual transitions are shown in Fig.5.1. The charge fluctuation due to intra-atomic virtual transitions i.e. $s_1 (= \nu_1 0 \to \nu_1 0)$ has monopole symmetry while the other transitions $s_1 (= \nu_1 0 \to \nu_2 0)$ may have dipole, quadrupole and other lower order symmetries. These are shown in Figs. 5.1a, 5.1b and 5.1c. Similarly the charge fluctuations due to inter-atomic and inter-orbital virtual transitions $s_1 (= \nu_1 0 \to \nu_2 1)$ and so on will have dipole, quadrupole and other lower order symmetries. One of the virtual transitions which give rise to charge fluctuation of dipole symmetry, is shown in Fig. 5.1d.

The first term of Eq.(5.10) $\chi^0_{s_1 s_2}(\vec{q}, \omega)$ is the polarizability of non-interacting electrons and $\bar{v}_{s_1 s_2}(\vec{q})$ can be interpreted as the $\vec{q}$-dependent interaction coefficient between the unit amplitude of charge fluctuations of type s_1 and s_2 which are propagating through the crystal with periodicity $\vec{q}$. Therefore $\chi_{s_1 s_2}(\vec{q}, \omega)$ is the measure of the amount of charge fluctuation of type s_1 which is induced by an external field of frequency ω and wave vector $\vec{q}$ and which is coupled to the charge fluctuation of type s_2. The poles of $\chi_{s_1 s_2}(\vec{q}, \omega)$, given in Eq.(5.11), describe the electronic excitation spectrum which can be obtained by solving the determinental equation

$$\left| \, \bar{v}(\vec{q}) - \chi^{0-1}(\vec{q}, \omega) \, \right| \;=\; 0. \tag{5.12}$$

For $\omega = 0$, Eq.(5.12) gives the electronic instabilities against the charge fluctuations. The simplest monopole type charge fluctuations, as shown in Fig.5.1a, give ionic type instabilities. The transitions between different orbitals on the same site, as shown in Fig. 5.1b, give dipolar instabilities and the overlapping orbitals on different sites, as shown in Fig.5.1d, give covalent type instabilities and so on. Before $\chi(\vec{q}, \omega)$ actually blows up, the incipient electronic instability makes the lattice unstable by deriving a particular phonon mode soft.

Prakash and co-workers [5] - [7] evaluated the static density response function $\chi(\vec{q}+\vec{G}, \vec{q}+\vec{G'})$ for transition metals Ni, Pt, Pd, V, Sc and Y in the non-interacting band scheme as discussed in Chapter 4. The simple tight-binding wave function is used for d-electrons. In these calculations all the d-d intra-site transitions and d-d inter-site transitions upto first nearest neighbours are considered. Therefore the charge fluctuations of monopole and dipole symmetries are included in these calculations.

To calculate the phonon frequencies in the charge fluctuation model, the crystal Hamiltonian is written as

$$H = H_e + \sum_{\vec{q}j} \omega_j^c(\vec{q}) a_{\vec{q}j}^\dagger a_{\vec{q}j} + H_{ep}. \tag{5.13}$$

The first term H_e is the electronic part of Hamiltonian given in Eq.(2.3) for $\vec{R}(l\kappa) = \vec{R}^0(l\kappa)$. The second term is the phonon part of Hamiltonian given by Eq.(2.2) and $a_{\vec{q}j}^\dagger$ and $a_{\vec{q}j}$ are phonon creation and annihilation operators respectively. In the third term H_{ep} describes the electron-phonon (e-p) interaction which arises due to displacement of ions from their mean positions and which is given as

$$\begin{aligned}
H_{ep} = & -\sum_{l\alpha} \sum_{\vec{k}_1 \vec{k}_2 \mu_1 \mu_2} \sum_{\sigma} < \vec{k}_1\mu_1 \mid \nabla_\alpha v_b(\vec{r} - \vec{R}_l) \mid \vec{k}_2\mu_2 > c_{\vec{k}_1\mu_1\sigma}^\dagger c_{\vec{k}_2\mu_2\sigma} u_\alpha(l) \\
& + (1/2) \sum_{l\alpha\beta} \sum_{\vec{k}_1 \vec{k}_2 \mu_1 \mu_2} \sum_{\sigma} < \vec{k}_1\mu_1 \mid \nabla_\alpha \nabla_\beta v_b(\vec{r} - \vec{R}_l) \mid \vec{k}_2\mu_2 > \\
& \times c_{\vec{k}_1\mu_1\sigma}^\dagger c_{\vec{k}_2\mu_2\sigma} u_\alpha(l) u_\beta(l)
\end{aligned} \tag{5.14}$$

where $c_{\vec{k}\mu\sigma}^\dagger$ and $c_{\vec{k}\mu\sigma}$ are creation and annihilation operators for the Bloch states $\mid \vec{k}\mu >$ with spin σ, $u_\alpha(l)$ is α-component of the displacement of an atom at the lth site and $v_b(\vec{r} - \vec{R}_l)$ is the bare electron-ion potential.

Now consider the canonical transformation

$$H' = e^{iS_t} H e^{-iS_t} \tag{5.15}$$

where

$$S_t = -i \sum_{\vec{k}_1 \vec{k}_2 \mu_1 \mu_2} \sum_{\sigma} < \vec{k}_1 \mu_1 \mid A \mid \vec{k}_2 \mu_2 > c^\dagger_{\vec{k}_1 \mu_1 \sigma} c_{\vec{k}_2 \mu_2 \sigma} \tag{5.16}$$

and

$$< \vec{k}_1 \mu_1 \mid A \mid \vec{k}_2 \mu_2 > = (1/2) \sum_{l\alpha} u_\alpha(l) \int_{vol} d\vec{r} \left[\psi^*_{\vec{k}_1 \mu_1} \left(\nabla_\alpha \psi_{\vec{k}_2 \mu_2} \right) \right.$$
$$\left. - (\nabla_\alpha \psi^*_{\vec{k}_1 \mu_1}) \psi_{\vec{k}_2 \mu_2} \right]. \tag{5.17}$$

In Eqs.(5.16) and (5.17), only the first order terms in $u_\alpha(l)$ are retained. Carrying out the transformation (5.15) and making the RPA like decoupling of electron operators (retaining terms upto second order in $u_\alpha(l)$), H' can be written as

$$H' = H_e + \sum_{\vec{q}j} \bar{\omega}^c_j(\vec{q}) a^\dagger_{\vec{q}j} a_{\vec{q}j} + H'_{ep}. \tag{5.18}$$

Here $\bar{\omega}^c_j(\vec{q})$ are no more the bare-ion plasma frequencies but these are the frequencies associated with the mutual interaction of pseudoatoms. The renormalized electron-phonon interaction becomes

$$H'_{ep} = \sum_{l\alpha} \sum_{\vec{k}_1 \vec{k}_2 \mu_1 \mu_2} \sum_{\sigma} \left[Q^{l\alpha}_{\vec{k}_1 \mu_1, \vec{k}_2 \mu_2} - < \vec{k}_1 \mu_1 \mid \nabla_\alpha w(\vec{r} - \vec{R}_l) \mid \vec{k}_2 \mu_2 > \right]$$
$$\times c^\dagger_{\vec{k}_1 \mu_1 \sigma} c_{\vec{k}_2 \mu_2 \sigma} u_\alpha(l) \tag{5.19}$$

where

$$Q^{l\alpha}_{\vec{k}_1 \mu_1, \vec{k}_2 \mu_2} = \int_{vol} d\vec{r} \left[(\nabla^2 \psi^*_{\vec{k}_1 \mu_1})(\nabla_\alpha \psi_{\vec{k}_2 \mu_2}) \right.$$
$$\left. - \psi^*_{\vec{k}_1 \mu_1} \nabla^2 (\nabla_\alpha \psi_{\vec{k}_2 \mu_2}) \right] + (1/2) u_\alpha(l) \left(E_{\vec{k}_1 \mu_1} - E_{\vec{k}_2 \mu_2} \right)$$
$$\times \int_{vol} d\vec{r} \nabla_\alpha (\psi^*_{\vec{k}_1 \mu_1} \psi_{\vec{k}_2 \mu_2}). \tag{5.20}$$

Here $w(\vec{r} - \vec{R}_l)$ is a local pseudopotential which can be chosen a muffin-tin potential which is defined as

$$\begin{aligned} w(\vec{r}) &= -(Z^*e^2/R_{MT}) \quad \text{for} \quad r < R_{MT} \\ &= -(Z^*e^2/r) \quad \text{for} \quad r \geq R_{MT}. \end{aligned} \tag{5.21}$$

Here R_{MT} is muffin-tin radius and Z^*e is charge on the pseudoatom inside muffin-tin volume v_{ol}. Finally $w(\vec{r})$ should also include the effect of exchange and correlation interactions due to the charge inside the volume v_{ol} but this contribution vanishes in the local density formalism.

Following Eqs.(2.85) and (2.107), we can write the components of atomic displacements in terms of phonon creation and annihilation operators as

$$u_\alpha(l) = \sum_{\vec{q}j} \left(\frac{1}{2MN\bar{\omega}_j^c(\vec{q})} \right)^{1/2} e_\alpha(\vec{q}j) e^{(i\vec{q}\cdot\vec{R}_l)} \left[a_{\vec{q}j} + a_{-\vec{q}j}^\dagger \right]. \tag{5.22}$$

Substituting Eq.(5.22) in (5.19) we get

$$\begin{aligned} H'_{ep} &= \sum_{\vec{q}j} \sum_{\vec{k}_1\vec{k}_2\mu_1\mu_2} \left(\frac{1}{2MN\bar{\omega}_j^c(\vec{q})} \right)^{1/2} \sum_\alpha P^\alpha_{\vec{k}_1\mu_1,\vec{k}_2\mu_2}(\vec{q}j) e_\alpha(\vec{q}j) \\ &\quad \times c^\dagger_{\vec{k}_1\mu_1\sigma} c_{\vec{k}_2\mu_2\sigma} \left[a_{\vec{q}j} + a_{-\vec{q}j}^\dagger \right] \end{aligned} \tag{5.23}$$

where the electron-phonon matrix element

$$\begin{aligned} P^\alpha_{\vec{k}_1\mu_1,\vec{k}_2\mu_2}(\vec{q}j) &= \sum_l \Big[Q^{l\alpha}_{\vec{k}_1\mu_1,\vec{k}_2\mu_2} \\ &\quad - < \vec{k}_1\mu_1 \mid \nabla_\alpha w(\vec{r} - \vec{R}_l) \mid \vec{k}_2\mu_2 > \Big] e^{(i\vec{q}\cdot\vec{R}_l)}. \end{aligned} \tag{5.24}$$

Use of Eq.(5.2) for $\mid \vec{k}\mu >$, simplifies Eq.(5.24) for a particular mode $(\vec{q}j)$ as

$$\begin{aligned} P^\alpha_{\vec{k}_1\mu_1,\vec{k}_2\mu_2}(\vec{q}j) &= \sum_{\nu_1\nu_2 l} e^{\nu_1*}_{\vec{k}_1\mu_1} e^{\nu_2}_{\vec{k}_2\mu_2} \exp(i\vec{k}_1 \cdot \vec{R}_l) \\ &\quad \times I^\alpha_{\nu_1,\nu_2 l}(\vec{q}j) \delta_{\vec{k}_1,\vec{k}_2+\vec{q}} \end{aligned} \tag{5.25}$$

where

$$
\begin{aligned}
I^\alpha_{\nu_1,\nu_2 l}(\vec{q}j) \;=\; & (1/2) \int_{s_0} d\vec{S} \left[\left\{ \nabla_n \phi^*_{\nu_1}(\vec{r}) \right\} \left\{ \nabla_\alpha \phi_{\nu_2}(\vec{r}-\vec{R}_l) \right\} \right. \\
& - \phi^*_{\nu_1}(\vec{r}) \nabla_n \left\{ \nabla_\alpha \phi_{\nu_2}(\vec{r}-\vec{R}_l) \right\} + \left\{ \nabla_n \phi_{\nu_2}(\vec{r}-\vec{R}_l) \right\} \\
& \times \left\{ \nabla_\alpha \phi^*_{\nu_1}(\vec{r}) \right\} - \phi_{\nu_2}(\vec{r}-\vec{R}_l) \nabla_n \left\{ \nabla_\alpha \phi^*_{\nu_1}(\vec{r}) \right\} \Big] \\
& - i \sum_{\vec{G}} (\vec{q}+\vec{G})_\alpha w(\vec{q}+\vec{G}) \int_{v_{0l}} d\vec{r}\, \phi^*_{\nu_1}(\vec{r}) \\
& \times \exp[i(\vec{q}+\vec{G})\cdot\vec{r}]\phi_{\nu_2}(\vec{r}-\vec{R}_l).
\end{aligned}
\tag{5.26}
$$

Here $\nabla_n \phi_{\nu_1}(\vec{r})$ is the derivative of $\phi_{\nu_1}(\vec{r})$ normal to the surface s_0 which encloses the volume v_{ol} in the unit cell at the origin. $w(\vec{q}+\vec{G})$ is the Fourier transform of $w(\vec{r}-\vec{R}_l)$. $I^\alpha_{\nu_1,\nu_2 l}(\vec{q}j)$ involves the integral over the surface s_0 and plays the role of electron-phonon matrix element in the localized orbital representation.

The expressions for the phonon frequencies are obtained by defining the phonon Green's function which is obtained from the transformed Hamiltonian H' by the method of equation of motion [8]. Let the phonon Green's function is written as

$$
G_{\vec{q}j}(t-t') \;=\; i\theta(t-t') \left\langle \left[B_{\vec{q}j}(t), B^\dagger_{\vec{q}j}(t') \right] \right\rangle
\tag{5.27}
$$

where $B_{\vec{q}j} = \left(a_{\vec{q}j} + a^\dagger_{-\vec{q}j} \right)$ and $\theta(t)$ is Heaviside unit step function. One can evaluate the commutation relations of $a_{\vec{q}j}$ and $a^\dagger_{\vec{q}j}$ with H' with the help of Eqs. (5.16) and (5.17). These are given as [9]

$$
\begin{aligned}
[a_{\vec{q}j}, H'] \;=\; & \bar{\omega}^c_j(\vec{q}) a_{\vec{q}j} + \sum_{\vec{k}_1 \alpha \mu_1 \mu_2} \sum_\sigma \left(\frac{1}{2MN\bar{\omega}^c_j(\vec{q})} \right)^{1/2} P^\alpha_{\vec{k}_1\mu_1, \vec{k}_1+\vec{q}\mu_2}(\vec{q}j) \\
& \times e^*_\alpha(\vec{q}j) c^\dagger_{\vec{k}_1\mu_1\sigma} c_{\vec{k}_1+\vec{q}\mu_2\sigma}
\end{aligned}
\tag{5.28}
$$

and

$$
\begin{aligned}
\left[a^\dagger_{\vec{q}j}, H' \right] \;=\; & -\bar{\omega}^c_j(\vec{q}) a^\dagger_{-\vec{q}j} - \sum_{\vec{k}_1 \alpha \mu_1 \mu_2} \sum_\sigma \left(\frac{1}{2MN\bar{\omega}^c_j(\vec{q})} \right)^{1/2} \\
& \times P^\alpha_{\vec{k}_1\mu_1, \vec{k}_1+\vec{q}\mu_2}(-\vec{q}j) e^*_\alpha(\vec{q}j) c^\dagger_{\vec{k}_1\mu_1\sigma} c_{\vec{k}_1+\vec{q}\mu_2\sigma}.
\end{aligned}
\tag{5.29}
$$

358   *S. PRAKASH*

Introducing the equation of motion for $G_{\bar{q}j}(\omega)$ which is the time Fourier transform of $G_{\bar{q}j}(t)$, one gets [4]

$$\left[\omega^2 - \omega_j^2(\vec{q})\right] G_{\bar{q}j}(\omega) \;=\; -(1/\pi)\bar{\omega}_j^c(\vec{q}) + 2\bar{\omega}_j^c(\vec{q}) \sum_{\vec{k}_1\alpha\mu_1\mu_2} \sum_{\sigma}$$

$$P^\alpha_{\vec{k}_1\mu_1,\vec{k}_1+\vec{q}\mu_2}(-\vec{q}j)e^*_\alpha(\vec{q}j)G_{\vec{k}_1\mu_1\sigma;\vec{k}_1+\vec{q}\mu_2\sigma;\bar{q}j}(\omega)$$

$$(5.30)$$

where $G_{\vec{k}_1\mu_1\sigma;\vec{k}_1+\vec{q}\mu_2\sigma;\bar{q}j}(\omega)$ is Fourier transform of Green's function

$$G_{\vec{k}_1\mu_1\sigma;\vec{k}_1+\vec{q}\mu_2\sigma;\bar{q}j}(t-t') = i\theta(t-t')$$

$$\times \left\langle \left[c^\dagger_{\vec{k}_1\mu_1\sigma}(t)c_{\vec{k}_1+\vec{q}\mu_2\sigma}(t), B^\dagger_{\bar{q}j}(t') \right] \right\rangle. \qquad (5.31)$$

Again using the equation of motion for $G_{\vec{k}_1\mu_1\sigma;\vec{k}_1+\vec{q}\mu_2\sigma;\bar{q}j}(\omega)$ in Eq. (5.30) and decoupling the electron and phonon operators as $\left\langle c^\dagger_{\vec{k}_1\mu_1\sigma}c_{\vec{k}_2\mu_2\sigma}B_{\bar{q}j} \right\rangle \cong \left\langle c^\dagger_{\vec{k}_1\mu_1\sigma}c_{\vec{k}_2\mu_2\sigma} \right\rangle B_{\bar{q}j}$, one finds the solution

$$G_{\vec{k}_1\mu_1\sigma;\vec{k}_1+\vec{q}\mu_2\sigma;\bar{q}j}(\omega) \;=\; \sum_{\nu_1\nu_2\nu_3\nu_4}\sum_{l_1l_2} e^{\nu_1*}_{\vec{k}_1\mu_1} e^{\nu_2}_{\vec{k}_1+\vec{q}\mu_2} \exp(i\vec{k}_1\cdot\vec{R}_{l_1})$$

$$\times \frac{f_{\vec{k}_1\mu_1\sigma} - f_{\vec{k}_1+\vec{q}\mu_2\sigma}}{E_{\vec{k}_1\mu_1\sigma} - E_{\vec{k}_1+\vec{q}\mu_2\sigma}} \left[1 - \bar{v}\chi^0\right]^{-1}_{\nu_1\nu_3;\nu_2l_1\nu_4l_2}$$

$$\times \left(\frac{1}{2MN\bar{\omega}_j^c(\vec{q})}\right)^{1/2} \sum_\alpha I^\alpha_{\nu_3,\nu_4l_2}(\vec{q}j)e_\alpha(\vec{q}j)G_{\bar{q}j}(\omega).$$

$$(5.32)$$

Use of Eqs.(5.25) and (5.32) in Eq.(5.30) gives

$$G_{\bar{q}j}(\omega) \;=\; -(1/\pi)\bar{\omega}_j^c(\vec{q})\left[\omega^2 - \bar{\omega}_j^{c2}(\vec{q}) + 2\bar{\omega}_j^c(\vec{q})\Sigma(\vec{q}j\omega)\right]^{-1} \quad (5.33)$$

where

$$\Sigma(\vec{q}j\omega) \;=\; \frac{1}{2M\bar{\omega}_j^c(\vec{q})} \sum_{\alpha\beta}\sum_{s_1s_2s_3} e^*_\alpha(\vec{q}j)I^\alpha_{s_1}(-\vec{q})\chi^0_{s_1s_2}(\vec{q}\omega)$$

$$\times \left[1 - \bar{v}\chi^0\right]^{-1}_{s_2s_3} I^\beta_{s_3}(\vec{q})e_\beta(\vec{q}j). \qquad (5.34)$$

The poles of $G_{\vec{q}j}(\omega)$, given in Eq.(5.33), give the phonon frequencies. The final expression for the phonon frequencies with the help of Eq.(5.34) can be written as

$$\omega_j^2(\vec{q}) \;=\; \bar{\omega}_j^{c2}(\vec{q}) - (1/M)\sum_{\alpha\beta} e_\alpha^*(\vec{q}j) \sum_{s_1 s_2} I_{s_1}^\alpha(-\vec{q})\chi_{s_1 s_2}(\vec{q})$$
$$\times I_{s_2}^\beta(\vec{q})e_\beta(\vec{q}j) \tag{5.35}$$

where $\chi = \chi^0(1 - \bar{v}\chi^0)^{-1}$.

It is evident from Eq.(5.35) that $\omega_j(\vec{q})$ are the eigenvalues of the dynamical matrix

$$D_{\alpha\beta}(\vec{q}) \;=\; D'^c_{\alpha\beta}(\vec{q}) - (1/M)\sum_{s_1 s_2} I_{s_1}^\alpha(-\vec{q})\chi_{s_1 s_2}(\vec{q})I_{s_2}^\beta(\vec{q}) \tag{5.36}$$

where $D'^c_{\alpha\beta}(\vec{q})$ arises due to Coulomb interaction between the pseudoatoms and its eigenvalues are $\bar{\omega}_j^{c2}(\vec{q})$. In Eq.(5.36) $\chi_{s_1 s_2}(\vec{q}) = \chi_{s_1 s_2}(\vec{q}\omega)$ at $\omega = 0$ and $I_{s_1}^\alpha(\vec{q})$ represent the coupling of lattice with the deformation part of electron charge density fluctuations or "the dynamic charge density waves of type s_1 (apart from the rigid motion of the pseudoatoms)". Comparison of Eq.(5.36) with Eq. (4.91) shows that $D_{\alpha\beta}^c(\vec{q})$ is replaced by $D'^c_{\alpha\beta}(\vec{q})$ and the electronic term $D_{\alpha\beta}^E(\vec{q})$ has been transformed in the charge fluctuation representation.

In order to have a more physical insight in Eq.(5.35), we introduce $\rho_s(\vec{q}j)$ as the amplitude of charge fluctuation s in response to lattice vibrational mode $(\vec{q}j)$. Then with the help of Eq.(2.67) and Eq.(5.11) one can write the equation of motion (5.35) as (as in shell model)

$$\omega_j^2(\vec{q})e_\alpha(\vec{q}j) \;=\; \sum_\beta D'^c_{\alpha\beta}(\vec{q})e_\beta(\vec{q}j) + \sum_{s_2} I_{s_2}^\alpha(\vec{q})\rho_{s_2}(\vec{q}j) \tag{5.37}$$
$$0 \;=\; \sum_\beta I_{s_2}^\beta(\vec{q})e_\beta(\vec{q}j) + \sum_{s_2}(\bar{v} + \chi^{0-1})_{s_2}\rho_{s_2}(\vec{q}j). \tag{5.38}$$

Equation (5.38) is the self-consistency equation which determines the amplitude of charge fluctuations. Equations (5.37) and (5.38) are the general lattice dynamical equations of motion in the charge fluctuation model. This formalism can be related to other microscopic theories [5] - [7], [10] which describe a particular type of charge fluctuation.

To know the small $\vec{q}$ behaviour of second term in Eq.(5.36), we consider, for simplicity, only one orbital in the bands near the Fermi level. At larger r, $w(r)$ varies as $-Z^*e^2/r$ (Eq.(5.21)). The dominent term in Eq.(5.36), as $\vec{q} \to 0$, will arise from the monopole charge fluctuations, i.e. $(s_1 \to s_1)$ will correspond to transitions from one orbital to the same orbital on the same site. From Eq.(5.20) $Q^{l\alpha}_{\vec{k}_1\mu_1,\vec{k}_2\mu_2}$ is finite as $\vec{q} \to 0$, therefore the dominent term in $I^\alpha_{s_1}(\vec{q})$, as $\vec{q} \to 0$, arises from the term involving $\nabla_\alpha w(\vec{r}-\vec{R}_l)$ and it is $(4\pi Z^*e^2/\Omega_0 q^2)q_\alpha$ where Ω_0 is the volume of unit cell. The monopole term in the response function $\chi_{s_1 s_2}(\vec{q})$ approaches to $(4\pi e^2/\Omega_0 q^2)^{-1}$ as $\vec{q} \to 0$. Therefore the second term in Eq.(5.36) goes as $(-1/M)(4\pi Z^{*2}e^2/\Omega_0)(q_\alpha q_\beta/q^2)$ which is exactly cancelled by the singular behaviour of $D'^c_{\alpha\beta}(\vec{q})$ as $\vec{q} \to 0$, since the pseudoatoms have the charge Z^*e.

5.1.2 Applications

To stydy the phonon anomalies and lattice instabilities, we consider a simple system which consists of a single d-band at the Fermi level and the overlap between the d-orbitals on the neighbouring atoms is negligible. In this system, the charge fluctuations will constitute only monopoles due to charge fluctuations in d-shells on the same site (Fig.5.1a). The other transitions which involve the s-electrons, will only have the effect of further renormalizing $\bar{\omega}^c_j(\vec{q})$ and the screening of electron-ion interaction $w(\vec{q}+\vec{G})$. The first term of Eq.(5.26) for such a system vanishes as it satisfies the selection rule $l \to l+1$. These approximations simplify Eq.(5.35) as

$$\omega^2_j(\vec{q}) \;=\; \bar{\omega}^{c2}_j(\vec{q}) + \frac{1}{M}\frac{\mid W_{\vec{q}j}\mid^2}{\bar{v}(\vec{q})-[1/\chi^0(\vec{q})]} \tag{5.39}$$

where

$$W_{\vec{q}j} \;=\; \sum_\alpha e_\alpha(\vec{q}j)\sum_{\vec{G}}(\vec{q}+\vec{G})_\alpha\frac{w(\vec{q}+\vec{G})}{\epsilon_s(\vec{q}+\vec{G})}$$

$$\times \left\langle \phi_d \mid e^{i(\vec{q}+\vec{G})\cdot\vec{r}} \mid \phi_d \right\rangle. \tag{5.40}$$

Here $\epsilon_s(\vec{q}+\vec{G})$ is the screening function due to s-electrons which is given in Eq. (4.163) and $\chi^0(\vec{q})$ is generalized susceptibility in the constant matrix element approximation given by the first term of Eq. (4.157).

For a single tight-binding band which involves orbital ν, we can write $\bar{v}(\vec{q})$ as a direct lattice sum

$$\bar{v}(\vec{q}) \;=\; \sum_l [v^c(l) + v^x(l)]\exp(i\vec{q}\cdot\vec{R}_l) \tag{5.41}$$

where $v^c(l)$ is the Coulomb interaction between charge distributions $(\phi_\nu^*\phi_\nu)$ centered at the origin and at the site l and $v^x(l)$ is exchange and correlation interaction between these two charge distributions. The introduction of free electron like s-band (neglecting its hybridization with ν band) produces screening and this reduces $v^c(l)$ to a short ranged potential (except for possible Friedel oscillations). Note that $v^x(l)$ which is short ranged in any case, tends to cancel $v^c(l)$ for $l = 0$ i.e. to remove the " self-interaction" of ν-orbital charge distribution on a particular site. Thus we may write

$$\bar{v}(\vec{q}) \;=\; v^{ef} + \sum_l{}' v^{sl}\exp(i\vec{q}\cdot\vec{R}_l) \tag{5.42}$$

where v^{ef} is the effective intra-atomic Coulomb interaction which is reduced by exchange and correlation interactions and the screening effects due to s-electrons. v^{sl} is the screened inter-atomic Coulomb interaction which is also reduced by exchange and correlation interactions and by the screening effects due to s-electrons.

The $\vec{q}$ dependence of $\bar{v}(\vec{q})$, $\chi^0(\vec{q})$ and $I^\alpha(\vec{q})$ for a model band structure due to Sinha and co-workers [2]- [4] is shown in Fig. 5.2. $\bar{v}(\vec{q})$ measures the increase in the potential energy due to interaction of unit amplitudes of charge fluctuations on the d-shells of pseudoatoms. $\bar{v}(\vec{q})$ is maximum at $\vec{q} = 0$ and it is minimum at the zone boundary where the charge fluctuation has ionic character. The order of magnitude of $\chi^0(\vec{q})$ is the density of d-states at the Fermi energy $n_d(E_F)$ and it has fine structure due to the nesting of Fermi surface. The detailed calculations of $\chi^0(\vec{q})$ and its relation to the phonon anomalies will be discussed in the next chapter.

The vanishing of $\{\bar{v}(\vec{q}) - [1/\chi^0(\vec{q})]\}$ in Eq. (5.39) implies an instability against the charge density wave (CDW) on the d-shell. If $[\chi^0(\vec{q})]^{-1}$ is large, particularly in the regions of $\vec{q}$ where $\bar{v}(\vec{q})$ is most negative (near zone boundary), the system tends to become unstable with regard to such charge fluctuations. Thus Eq.(5.39) shows that the mode $\omega_j(\vec{q})$ is driven soft before the CDW instability occures.

With the above simplifications, one can write $I^\alpha(\vec{q})$ given in Eq. (5.26) in the real space as

$$I^\alpha(\vec{q}) \;=\; -\sum_l \int d\vec{r}\, \phi_\nu^*(\vec{r})\phi_\nu(\vec{r})\nabla_\alpha w_{sc}(\vec{r} - \vec{R}_l)\exp(i\vec{q}\cdot\vec{R}_l) \quad (5.43)$$

where $w_{sc}(\vec{r})$ is the Fourier transform of $w(\vec{q})/\epsilon_s(\vec{q})$. $I^\alpha(\vec{q})$ measures the q-dependent coupling coefficient between the charge fluctuation

$$\delta n(\vec{r}) \;=\; A'(\vec{q}j)\sum_l [\phi_\nu^*(\vec{r} - \vec{R}_l)\phi_\nu(\vec{r} - \vec{R}_l)]\exp(i\vec{q}\cdot\vec{R}_l) \quad (5.44)$$

and the displaced pseudoatoms. For α along the direction of $\vec{q}$, the schematic variation of $I^\alpha(\vec{q})$ with $\vec{q}$ is shown in Fig.5.2. $I^\alpha(\vec{q})$ vanishes at $\vec{q} = 0$ and at the zone boundary. As $I^\alpha(\vec{q}) \to 0$, the pseudoatoms move together and thus their effect on the charge fluctuation at the atom on the origin get cancelled. The position of phonon anomalies depends on $I^\alpha(\vec{q})$. Thus the phonon anomalies and hence the instabilities occur whenever the second term of Eq. (5.35) is maximum which is in general half way to the zone boundary. However the explicit $\vec{q}$ dependence of $\chi_{s_1 s_2}(\vec{q})$ may further alter the position and magnitude of phonon anomalies.

The above mentioned charge fluctuation theory has been used to calculate the phonon dispersion relations and lattice instabilities for a large number of transition metals:

(1) The model is used to calculate the phonon frequencies of niobium [2]. The characteristic dips in the phonon dispersion relations in the longitudinal(L) modes in the $[\xi 00]$ and $[\xi\xi\xi]$ directions are reproduced. In the $[\xi\xi\xi]$ direction, the phonon dip is very sensitive to the magnitude of $\chi^0(\vec{q})$ and it occures at $\vec{q} = 0.8(2\pi/a)$. This dip may be related to the central peak observed in Nb.

(2) The measurements for $Nb_{0.8}Zr_{0.2}$ alloys [11] for which the density

of states $n(E_F)$ at Fermi energy is larger than in pure Nb in the rigid band model, show that there is softening of longitudinal acoustic(LA) modes relative to those in pure Nb. However, Stassis et al [12] showed a dramatic anomaly in the LA mode in bcc Zr. On alloying Mo into Nb [13], $n_d(E_F)$ decreases and consequently anomalies disappear continuously as expected on the basis of this model. The same trend is found in the phonon spectra of non-stoichiometric compounds NbH_x and NbD_x [14]. The band structure calculations show that the addition of interstitial H to Nb reduces $n_d(E_F)$.

(3) In the calculations for NbC, d-orbitals of t_{2g}-symmetry are taken on Nb sites and p-orbitals on the C sites. The overlap between the orbitals on Nb and C sites is neglected. $\chi^0(\vec{q})$ is calculated separately for Nb and C sites and it is taken to be equal to partial density of states at E_F for Nb and C respectively. The calculated phonon frequencies of NbC due to Sinha and Hermon [2], as shown in Fig. 5.3, explain the observed anomalies in the LA modes. If the partial contributions to $\chi^0(\vec{q})$ i.e. density of states at Nb and C sites are decreased, the anomalies rapidly tend to vanish. This characteristic also explains the absence of anomalies in the phonon spectra of ZrC and HfC [3.15].

$I^\alpha(\vec{q})$ given in Eq. (5.26) vanishes for transverse modes if the overlap between the $d-d$ and $s-d$ orbitals is neglected. Therefore the monopole model of charge fluctuations is unable to explain the anomalous features in the transverse modes. In Eq. (5.26), the first term does not vanish if there is an overlap between the orbitals and it gives the coupling to the transverse modes. For NbC there is such an overlap as the conduction band consists of hybridized p and d-bands [15] [16]. As $I^\alpha(\vec{q}) = 0$ at the zone boundary, the phonon anomalies at the symmetry point H in Mo are not explained.

5.1.3 Phenomenological Charge Fluctuation Model

As discussed earlier, the electronic transitions between different orbitals of ions produce the charge fluctuations and these are accounted for in the second term of Eq. (5.36). In other words, one may say that these charge fluctuations produce fluctuations in the effective charge on the ions. Wakabayashi [17] used this concept and proposed a phenomenological charge fluctuation model to study the dynamics of solids. In this model, the total energy for a monoatomic crystal is given as

$$E_T \;=\; \frac{1}{2}M\sum_{l\alpha}\dot{R}_{l\alpha}^2 + \Phi^r(\vec{R}) + \sum_{ll'}Z_l v_s(|\,\vec{R}_l - \vec{R}_{l'}\,|)Z_{l'} \quad (5.45)$$

where

$$Z_l \;=\; Z + \Delta Z_l \qquad\qquad (5.46)$$

and $\vec{R} = (\vec{R}_1, \vec{R}_2 \ldots \vec{R}_N)$ is the collective atomic coordinate. The first term of Eq.(5.45) is the kinetic energy of pseudoatoms and the last two terms are the total potential energy. $\Phi^r(\vec{R})$ is short ranged atomic overlap potential and $v_s(|\,\vec{R}_l - \vec{R}_{l'}\,|)$ is screened Coulomb interaction between the pseudoatoms at the sites $\vec{R}_l$ and $\vec{R}'_l$ with fluctuating charges Z_l and $Z_{l'}$ respectively. Z is the charge in the equilibrium position and ΔZ_l is the instantaneous fluctuation in the charge of the lth pseudoatom. Thus in this model the dynamics of pseudoatoms is coupled to the redistribution of their charges.

Expanding Eq.(5.45) in power series of $u_\alpha(l)$ and ΔZ_l, we get

$$\begin{aligned}
E_T \;=\; & \frac{1}{2}M\sum_{l\alpha}\dot{u}_\alpha^2(l) + \frac{1}{2}\sum_{ll'}\sum_{\alpha\beta}\left[\Phi^r_{\alpha\beta}(l,l') + Zv^{(2)}_{s\alpha\beta}(l,l')Z\right] \\
& \times u_\alpha(l)u_\beta(l') + \sum_{ll'}\sum_{\alpha}Zv^{(1)}_{s\alpha}(l,l')(\Delta Z_l + \Delta Z_{l'}) \\
& \times (u_\alpha(l) - u_\alpha(l')) + \sum_{ll'}\Delta Z_l v^{(0)}_s(l,l')\Delta Z_{l'} + \ldots. \quad (5.47)
\end{aligned}$$

Using the equation of motion and adiabatic condition $\partial E_T/\partial Z_l = 0$, one gets the dynamical equation

$$\begin{aligned}
M\ddot{u}_\alpha(l) \;=\; & -\sum_{l'\beta}\left[\Phi^r_{\alpha\beta}(l,l') + Zv^{(2)}_{s\alpha\beta}(l,l')Z\right]u_\beta(l') \\
& -2\sum_{l'}Zv^{(1)}_{s\alpha}(l,l')\Delta Z_{l'} \\
0 \;=\; & -2Z\sum_{l'\alpha}v^{(1)}_{s\alpha}(l,l')u_\alpha(l') + 4\sum_{l'}v^{(0)}_s(l,l')\Delta Z_{l'}. \quad (5.48)
\end{aligned}$$

Equations (5.48) are transformed in q-space as

$$M\omega^2 u_\alpha(\vec{q}) = \sum_\beta \left[\Phi^r_{\alpha\beta}(\vec{q}) + Z v^{(2)}_{s\alpha\beta}(\vec{q}) Z \right] u_\beta(\vec{q})$$

$$+2Z v^{(1)}_{s\alpha}(\vec{q}) \Delta Z(\vec{q}),$$

$$0 = -2Z \sum_\alpha v^{(1)}_{s\alpha}(\vec{q}) u_\alpha(\vec{q}) + 4 v^{(0)}_s(\vec{q}) \Delta Z(\vec{q}). \quad (5.49)$$

Combining the two parts of Eq.(5.49), the expression governing the phonon frequencies becomes

$$M\omega^2 u_\alpha(\vec{q}) = \sum_\beta D^{eff}_{\alpha\beta}(\vec{q}) u_\beta(\vec{q}) \quad (5.50)$$

where

$$D^{eff}_{\alpha\beta}(\vec{q}) = D^0_{\alpha\beta}(\vec{q}) + \frac{Z v^{(1)}_{s\alpha}(\vec{q}) v^{(1)}_{s\beta}(\vec{q}) Z}{v^{(0)}_s(\vec{q})} \quad (5.51)$$

$$D^0_{\alpha\beta}(\vec{q}) = \Phi^r_{\alpha\beta}(\vec{q}) + Z v^{(2)}_{s\alpha\beta}(\vec{q}) Z \quad (5.52)$$

$$v^{(0)}_s(\vec{q}) = \sum_{ll'} v_s(l, l') \exp[i\vec{q} \cdot (\vec{R}_l - \vec{R}_{l'})] \quad (5.53)$$

$$v^{(1)}_\alpha(\vec{q}) = \sum_{ll'} \frac{\partial v_s(l, l')}{\partial R_{l\alpha}} \exp[i\vec{q} \cdot (\vec{R}_l - \vec{R}_{l'})] \quad (5.54)$$

$$v^{(2)}_{\alpha\beta}(\vec{q}) = \sum_{ll'} \frac{\partial^2 v_s(l, l')}{\partial R_{l\alpha} \partial R_{l'\beta}} \exp[i\vec{q} \cdot (\vec{R}_l - \vec{R}_{l'})]. \quad (5.55)$$

The first term of Eq.(5.51) $D^0_{\alpha\beta}(\vec{q})$ contains the conventional contributions from the short range overlap and screened Coulomb interactions. The second term is due to monopolar charge fluctuations, therefore this term does not alter the transverse phonon modes. Equations (5.50) and (5.51) can be generalized for the crystals with more number of atoms in the unit cell by replacing $v^{(0)}_s(\vec{q}), v^{(1)}_{s\alpha}(\vec{q}), v^{(2)}_{s\alpha\beta}(\vec{q})$ with matrix elements $v^{(0)}_s(\vec{q}, \kappa\kappa'), v^{(1)}_{s\alpha}(\vec{q}, \kappa\kappa'), v^{(2)}_{s\alpha\beta}(\vec{q}, \kappa\kappa')$ respectively. If one accounts for the nearest neighbour interactions only in the calculations of $v^{(0)}_s(\vec{q})$ and $v^{(1)}_{s\alpha}(\vec{q})$, the charge fluctuation model becomes equivalent to the breathing shell model for ionic crystals which was proposed by Schröder [18].

As the second term of Eq. (5.51) does not affect the transverse modes, the parameters (inter-atomic force constants) of the first term of Eq. (5.51) are obtained by fitting the experimental phonon frequencies of transverse modes at the zone boundary points. Keeping these parameters fixed, the phonon frequencies of longitudinal modes are fitted to determine other parameters (intra and inter shell force constants) of the second term which describes the charge fluctuations. The phonon frequencies of Nb and Mo are calculated including the inter-atomic interactions upto third nearest neighbours in both the first and second term of Eq. (5.51). The calculated results for Nb and Mo due to Wakabayashi et al [20] are shown in Fig.5.4. It is evident from the figure that the charge fluctuation term modifies the frequencies and re-produces anomalies in the longitudinal modes. This contribution also produces the flattening in the longitudinal modes near the symmetry point H and a sharp rise in the longitudinal mode frequencies near the point N. In Mo, the charge fluctuation term neither produces any significant improvement nor any anomaly in the longitudinal modes of phonon frequencies. This is due to the fact that in these calculations the long range forces in Mo and the dipolar charge fluctuations are neglected which may be prominent in Mo.

Smith and Wakabayashi [19] and Wakabayashi et al [20] used the charge fluctuation model, with monopole symmetry, to calculate the phonon frequencies of hcp TMs Sc, Y, Ti, Zr, Hf and Tc. It is found that the charge fluctuation term is not needed to fit the experimental data of phonon frequencies of Sc and Y. This indicates that the electron-phonon interaction in these transition metals is not very strong. These model calculations are further improved by including the charge fluctuations of dipole symmetry. It is found that the phonon anomalies in these hcp transition metals are explained by including the charge fluctuation of dipole symmetry. For example, the flattening of $[00\xi]$ longitudinal optical mode near Γ point in Zr and Hf and the sharp decrease in the phonon frequencies of Tc near the Γ point are reproduced. The softening of longitudinal phonon modes in the $[00\xi]$ direction near Γ point in the hcp transition metals is closely related to an incipient instability in the electron system which leads towards the formation of a dipolar charge density wave. Still many anomalies in the phonon spectra of hcp TMs remain unexplained. Perhaps the inclusion of coupling beyond the $3NN$ may explain the anomalies in the transverse modes in Tc [19], [20]. The charge fluctuation model is yet to be tested for other transition metal compounds.

5.2 Charge Density Distortion Model

In TM's, as the ions move about their equilibrium positions, the charge density distortion (CDD) is produced in the partially localized d-electron charge distribution. The CDDs possess monopole, dipole and quadrupole symmetries etc. and these cannot be represented by localized (or scalar) parameters. Allen [21] showed that the monopole (having Γ_1 symmetry) type CDD is needed to explain the phonon anomalies in the longitudinal modes in bcc TMs while the phonon anomalies in the transverse modes are associated with the quadrupole (having $\Gamma_{25'}$ symmetries) type CDDs. On the phenomenonlogical basis Allen [21] assumed let $\rho_{L\alpha}(l)$ be the $\alpha-$ component of amplitude of CDD of L-symmetry on the lth atom. For $L = 15$, the dipole moment vector transforms according to Γ_{15} irreducible representation of the point group of the crystal.

In accordance with the description of Eq.(5.47), the crystal Hamiltonian can be written as

$$H = H_I + H_e \tag{5.56}$$

where the ionic part of the Hamiltonian

$$H_I = \frac{1}{2}M\sum_{l\alpha}\dot{R}_{l\alpha}^{\,2} + \frac{1}{2}\sum_{\alpha\beta}\sum_{lR}u_\alpha(l)K_{\alpha\beta}(R)u_\beta(l+R) \tag{5.57}$$

and the electronic part

$$H_e = \sum_{lLR}\sum_{\alpha\beta}\rho_{L\alpha}(l)A_{\alpha\beta}^L u_\beta(l+R)$$
$$+\frac{1}{2}\sum_{\alpha\beta}\sum_{LL'lR}\rho_{L\alpha}(l)B_{\alpha\beta}^{LL'}(\vec{R})\rho_{L'\beta}(l+R). \tag{5.58}$$

In Eq.(5.57), the first term is kinetic energy of ions and in the second term Coulomb interaction between the ions is combined with the ion-electron and electron-electron interactions. Thus the second term represents the screened interaction between the ions via conduction electrons and $K_{\alpha\beta}$ are Born-von-Karman force constants. In Eq.(5.58), the matrix **A** describes the CDD-displacement interaction, **B** describes the CDD-CDD interactions. Here the non-dioganal components of CDD-CDD coupling are included in **B** explicitly. There is no term in the theory corresponding to momentum which is conjugate to $\rho_{L\alpha}(l)$.

The dynamics of a system with CDDs is specified by the constraints

$$\partial H_e / \partial \rho_{L\alpha}(l) \;=\; 0 \tag{5.59}$$

which means that the CDD amplitudes $\rho_{L\alpha}(l)$ adjust instantaneously corresponding to any lattice displacement $u_\alpha(l)$ to minimize the electronic energy. Equations (5.56) to (5.59) are used to obtain the effective dynamical matrix which is written as

$$D_{\alpha\beta}^{eff}(\vec{q}) \;=\; \mathbf{K}_{\alpha\beta}^{\gamma}(\vec{q}) - \sum_{\mathbf{LL'}} \sum_{\gamma,\delta} \left[\mathbf{A}_{\alpha\gamma}^{\mathbf{L}}(\vec{q})\right]^{\dagger}$$

$$\times \left[\{\mathbf{B}(\vec{q})\}^{-1}\right]_{\gamma\delta}^{LL'} \mathbf{A}_{\delta\beta}^{\mathbf{L'}}(\vec{q}) \tag{5.60}$$

where $\mathbf{K}(\vec{q}), \mathbf{A}(\vec{q})$ and $\mathbf{B}(\vec{q})$ are the Fourier transforms of $\mathbf{K}(\vec{R}), \mathbf{A}(\vec{R})$ and $\mathbf{B}(\vec{R})$ respectively. Equation (5.60) consists of $\{\mathbf{B}(\vec{q})\}^{-1}$ which is a multidimensional matrix and therefore the second term of Eq. (5.60) accounts for CDDs of all symmetries. Thus the theory describes the long range forces even though the coupling constants $\mathbf{K}(\vec{R}), \mathbf{A}(\vec{R})$ and $\mathbf{B}(\vec{R})$ are short ranged. Allen [21] has shown that the microscopic formalism for TMs [22], [23], [5], [5] can be achieved if one expands the change in charge density in a complete set of orthonormal functions $f_L(\vec{r} - \vec{R})$ as

$$\delta n(\vec{r}) = \sum_{lL\alpha} \rho_{L\alpha}(l) f_{L\alpha}(\vec{r} - \vec{R}) \tag{5.61}$$

where $\rho_{L\alpha}(l)$ are the expansion coefficients.

Let S stands for the genral element of a point group and $\bar{S}$ be the corresponding three dimensional rotation matrix with $\bar{S}\vec{R} = \vec{R'}$ i.e.a lattice vector $\vec{R_l}$ is transformed into another lattice vector $\vec{R_l'}$.

The representation matrices for Lth irreducible representation will be denoted as $\Gamma^L_{\mu\nu}(S)$. Now under the rotation $\bar{S}$ of the point group, the displacement $\vec{u}(l)$ and CDD transform as

$$u_\alpha(\bar{S}, l) = \Gamma^{15}_{\alpha\beta}(S)u_\beta(l)$$
$$\rho_{L\alpha}(\bar{S}, l) = \Gamma^L_{\alpha\beta}(S)\rho_{L\beta}(l). \tag{5.62}$$

Thus the rotational invariance of Hamiltonian gives the following laws for coupling constants:

$$\mathbf{K}^\gamma_{\alpha\beta}(\bar{\mathbf{S}}, \vec{\mathbf{R}}) = \Gamma^{15}_{\alpha\alpha'}(S)\mathbf{K}^\gamma_{\alpha'\beta'}(\vec{\mathbf{R}})\Gamma^{15}_{\beta'\beta}(\mathbf{S}^{-1}),$$
$$A^L_{\alpha\beta}(\bar{S}, \vec{R}) = \Gamma^L_{\alpha\alpha'}(S)\mathbf{A}^{\mathbf{L}}_{\alpha'\beta'}(\vec{\mathbf{R}})\Gamma^{15}_{\beta'\beta}(\mathbf{S}^{-1}),$$

and

$$\mathbf{B}^{\mathbf{LL'}}_{\alpha\beta}(\bar{\mathbf{S}}\vec{\mathbf{R}}) = \Gamma^L_{\alpha\alpha'}(S)\mathbf{B}^{\mathbf{LL'}}_{\alpha'\beta'}\Gamma^{\mathbf{L'}}_{\beta'\beta}(\mathbf{S}^{-1}). \tag{5.63}$$

The first equation is the transformation law for the force-constants. The second and third equations are the generalization of the first equation for CDD-dispalcement and CDD-CDD coupling constants.

The force constants given in Eq.(5.63) are block-diagonal along the symmetry directions. For example, in the bcc structure along the principal symmetry directions $[\xi 00]$, $[\xi\xi 0]$ and $[\xi\xi\xi]$, there are Δ, Σ and Λ irreducible representations respectively (see Fig. 5.5). The first basic distortion is the flow of charge from one atom to another. Therfore the net charge on the ion may be regarded as a first electronic variable. The Γ_1-symmetry affects only the longitudinal modes [2], [3], [4], [17]. For $l = 1, \Gamma_{15}$-symmetry has been studied extensively in the shell model [18], [24],[25], [26]. However these symmetries of CDD did not explain the phonon spectrum of pure TMs and their alloys although some success has been obtained in the explanation of phonon frequencies of transition metal carbides [24]. Therfore $l = 2$ with $(\Gamma_{12} + \Gamma_{25'})$ symmetries are considered. The CDD with $\Gamma_{25'}$ symmetry has the quadrupole degree of freedom. $\Gamma_{25'}$ representation couples every transverse acoustic branch except the Σ_4 branch and also couples to longitudinal acoustic branches everywhere except the Δ_1 branch at N and P symmetry points in the bcc crystal. Therefore one expects that the phonon anomalies in the transverse acoustic modes can be explained by including the CDDs of quadrupole symmetry.

5.2.1   Applications

The experimental results for phonon frequencies of Nb (Fig 5.5) show that the transverse branches except the Σ_4 branch, have anomalous inflection points. This suggests the importance of including the CDD of $\Gamma_{25'}$- symmetry. The charge density calculations by Ho et al [27] also show a noticeable buildup of bonding valence charge between nearest neighbour atoms from the states near the Fermi surface. These charge distributions are susceptable to distortions when the atoms move about their equilibrium positions. Therefore the amplitude of bond charge distortions can be taken as a fluctuating variable in Nb. Allen [21] introduced four electronic parameters $\rho_i(l)$ per atom of Nb. These variables form a basis for a four dimensional representation of the cubic point group which reduces to $(\Gamma_1 + \Gamma_{25'})$ symmetry. Therefore the CDD with Γ_{15}- symmetry is ignored. In choosing the coupling constant matrices $\mathbf{A}$ and $\mathbf{B}$, the parameters are allowed to be as general as the point group symmetries permit. However it is assumed that these couplings are short ranged in nature. The zeroth order coupling constant $\mathbf{B}_{\alpha\beta}^{\mathbf{LL'}}(0) = \mathbf{B}_0 \delta_{\alpha\beta}\delta_{\mathbf{LL'}}$ and $\mathbf{A}_{\alpha\beta}^{\mathbf{L}}(0) = \mathbf{A}_0\delta_{\alpha\beta}$. As only even L representations are present in this model, therefore $\mathbf{A}_{\alpha\beta}^{\mathbf{L}}(\vec{\mathbf{R}})$ is an odd function of $\vec{R}$. Using the translational invariance of coupling matrices one can write

$$K_{\alpha\beta}(\vec{q}) \; = \; -\sum_{\vec{R}\neq 0} K_{\alpha\beta}(\vec{R})\left[\cos(\vec{q}\cdot\vec{R}) - 1\right],$$

$$\mathbf{A}_{\alpha\beta}^{\mathbf{L}}(\vec{\mathbf{q}}) \; = \; \sum_{\vec{R}}\mathbf{A}_{\alpha\beta}^{\mathbf{L}}(\vec{\mathbf{R}})\sin(\vec{\mathbf{q}}\cdot\vec{\mathbf{R}}),$$

and

$$\mathbf{B}_{\alpha\beta}^{\mathbf{LL'}}(\vec{\mathbf{q}}) \; = \; \sum_{\vec{R}}\mathbf{B}_{\alpha\beta}^{\mathbf{LL'}}(\vec{\mathbf{R}})\cos(\vec{\mathbf{q}}\cdot\vec{\mathbf{R}}). \tag{5.64}$$

The parameters $\mathbf{A}_{\alpha\beta}^{\mathbf{L}}(\vec{\mathbf{q}})$ and $\mathbf{B}_{\alpha\beta}^{\mathbf{LL'}}(\vec{\mathbf{q}})$ are scaled to unity in the calculations of phonon frequencies without affecting the results. The parameters $K_{\alpha\beta}(\vec{R})$, $\mathbf{A}_{\alpha\beta}^{\mathbf{L}}(\vec{\mathbf{R}})$ and $\mathbf{B}_{\alpha\beta}^{\mathbf{LL'}}(\vec{\mathbf{R}})$ are extended upto second nearest neighbours. Because the Γ_1 and $\Gamma_{25'}$ symmetries do not couple at the H and N symmetry points, the third nearest neighbour central force constants are also included. In all 18 parameters are introduced and the values of these parameters are obtained from the least square fit of the measured phonon frequencies by the following expressions:

$$M\omega^2(\Delta_1) = 16K_{1a}\sin^2 z + (4K_{2b} + 16K_{3a} + 16K_{3b})\sin^2 2z$$
$$-(8A_1\sin 2z + 2A_2\sin 4z)^2/[1 + 8b_1\cos 2z$$
$$+b_2(4 + 2\cos 4z)]$$

and

$$M\omega^2(\Delta_5) = 16K_{1a}\sin^2 z + (4K_{2a} + 16K_{3a} + 8K_{3b})\sin^2 2z$$
$$-(8B_{1b}\sin 2z + 2B_2\sin 4z)^2/[1 + 8C_{1a}\cos 2z$$
$$+C_{2a}(4 + 2\cos 4z) + 2C_{2b}]$$

where

$$z = \pi\xi/2 \quad \text{and} \quad q = (2\pi/a)(\xi 00).$$

These parameters are given in Tables 5.1 and 5.2. The calculated results for Nb are shown in Fig.5.5.

The calculated phonon frequencies yield overall good agreement with the experimental data and the phonon anomalies are reproduced except at the few points. For example the CDD model does not give satisfactory fit at the Λ_1 maximum and on the Δ_5-branch near Γ and H symmetry points. Because Γ_{15} (dipolar) degree of freedom couples the H and N symmetry points, the results may be improved by including Γ_{15} degree of freedom. The calculated results further suggest that $\mathbf{B}_{\alpha\beta}^{\mathbf{LL'}}(\vec{q})$ may be neglected for $L \neq L'$ and this parameter is extremly short ranged for $L = L' = 25$. The matrices $\mathbf{A^L}(\vec{R})$ are long ranged to some extent and possibly only nearest neighbour coupling is needed except for $L = 1$. The calculations show that the resonance of $\mathbf{B}^{-1}(\vec{q})$ arising from the charge transfer gives a natural explanation for the longitudinal acoustic mode anomalies while the anomalies in the transverse acoustic modes are associated with the dispersionless quadrupolar fluctuations.

It is to be pointed out that the charge fluctuation and CDD models may also be useful in understanding the phonon anomalies in A-15 compounds.

5.3   Non-orthogonal Tight-Binding Theory

We have described the calculations of phonon frequencies in the Wannier representation in Chapter 4. For the calculational purposes, Wanneir functions are replaced by time independent orthogonal tight-binding wave functions. These calculations do not account for the time variation of charge fluctuations on the ions in the lattice. There exist many calculations of electron-phonon interaction in the literature in the augmented plane wave representation (APW)[28]-[31]. However, none of these calculations have been used to investigate the phonon frequencies of transition metals and their compounds and alloys. Varma and co-workers [32]- [35] calculated the electron-phonon interaction in the time dependent non-orthogonal tight-binding (NTB) scheme and used these calculations to investigate the origin of phonon anomalies in the phonon spectrum of transition metals and transition metal carbides. In the following, we discuss these calculations.

5.3.1   Electron-Phonon Matrix Elements

We assume that it is possible to find loclized orbitals $\phi_\nu(\vec{r} - \vec{R}_l^0)(= \phi_{\nu l})$ centered on the site $\vec{R}_l^0$ with quantum number ν which bears one-to-one correspondence with the wave function of an isolated atom. The atomic orbitals at the same site may be taken as orthogonal but the atomic orbitals on the different sites, in genral, are not orthogonal i.e.

$$< \phi_{\nu l}|\phi_{\nu' l'} > \; = \; S_{\nu l, \nu' l'} =< \nu l|\nu' l' > \tag{5.65}$$

where $S_{\nu l, \nu' l'}$ is overlap integral and

$$S_{\nu l, \nu' l'} = \delta_{\nu \nu'} \qquad \text{for} \;\; l = l'. \tag{5.66}$$

If the crystal is perturbed on the time scale of phonon energy, one can construct the time dependent Bloch function as [33], [34]

$$\Psi(\vec{r}, t) \; = \; \sum_{\vec{k}\mu} C_{\vec{k}\mu}(t)\psi_{\vec{k}\mu}(\vec{r}, t) \tag{5.67}$$

where μ is band index and the coefficients $C_{\vec{k}\mu}(t)$ explicitly depend on time. The electronic wave function

$$\psi_{\vec{k}\mu}(\vec{r},t) \;=\; N^{-1/2}\sum_{\nu l} e^{\nu}_{\vec{k}\mu}\exp(i\vec{k}\cdot\vec{R}_l)\phi_{\nu l}(\vec{r},t)\exp(-E_{\vec{k}\mu}t). \quad (5.68)$$

Here $\phi_{\nu l}(\vec{r},t) = \phi_{\nu l}(\vec{r}-\vec{R}_l^0,t)$ varies with time on the perturbation scale of phonon energy. Equation (5.68) is the generalization of Eq. (5.2) to include the time dependence of orbital wave function. Equation (5.67) can also be written as the linear combination of atomic orbitals $\phi_{\nu}(\vec{r}-\vec{R}_l^0)$ as

$$\Psi(\vec{r},t) \;=\; \sum_{\nu l} C_{\nu l}(t)\phi_{\nu l}. \quad (5.69)$$

where the coefficients $C_{\nu l}(t)$ are related to $C_{\vec{k}\nu}(t)$ by the relation

$$C_{\nu l}(t) \;=\; \sum_{\vec{k}\mu} e^{\nu}_{\vec{k}\mu}C_{\vec{k}\mu}(t)\exp(i\vec{k}\cdot\vec{R}_l)\exp(-E_{\vec{k}\mu}t). \quad (5.70)$$

As a result of perturbation, due to the displacement of ions, the matrix elements of Hamiltonian of the perturbed lattice can be written as

$$H'_{\nu l,\nu' l'} \;=\; H_{\nu l,\nu' l'} + \delta H_{\nu l,\nu' l'}. \quad (5.71)$$

Similarly the overlap integral for the perturbed lattice is expressed as

$$S'_{\nu l,\nu' l'} \;=\; S_{\nu l,\nu' l'} + \delta S_{\nu l,\nu' l'}. \quad (5.72)$$

Here

$$\begin{aligned}
H_{\nu l,\nu' l'} &= \; < \phi_{\nu l}|H|\phi_{\nu' l'} > \\
&= \; < \nu l|H|\nu' l' >
\end{aligned} \quad (5.73)$$

and

$$\begin{aligned}
\delta S_{\nu l,\nu' l'} &= \; < \delta\phi_{\nu l}|\phi_{\nu' l'} > + < \phi_{\nu l}|\delta\phi_{\nu' l'} > \\
&= \; < \delta(\nu l)|(\nu' l') > + < (\nu l)|\delta(\nu' l') > .
\end{aligned} \quad (5.74)$$

The expressions for the matrix elements $H_{\nu l,\nu' l'}$ and $\delta H'_{\nu l,\nu' l'}$ can also be written as in Eqs.(5.73) and (5.74). Here δH and $\delta(\nu l) = \delta\phi_\nu(\vec{r} - \vec{R}_l^0, t)$ are changes in H and $\phi_\nu(\vec{r}, t)$ respectively due to thermal perturbation.

The equation of motion for $C_{\vec{k}\mu}(t)$ is

$$
i\dot{C}_{\vec{k}\mu} = \sum_{\nu\nu'} \sum_{\vec{k}'\mu'} e_{\vec{k}\mu}^{\nu*} \left[\delta H_{\nu\nu'}(\vec{k}) e_{k'\mu'}^{\nu'} \right.
$$
$$
\left. - \delta S_{\nu\nu'}(\vec{k}) e_{k'\mu'}^{\nu'} E_{\vec{k}\mu} \right] C_{\vec{k}'\mu'} \tag{5.75}
$$

where

$$
\delta H_{\nu\nu'}(\vec{k}) = N^{-1/2} \sum_{ll'} \delta H_{\nu l,\nu' l'} \exp(i\vec{k}\cdot\vec{R}_{ll'}^0) \tag{5.76}
$$

and

$$
\delta S_{\nu\nu'}(\vec{k}) = N^{-1/2} \sum_{ll'} \delta S_{\nu l,\nu' l'} \exp(i\vec{k}\cdot\vec{R}_{ll'}^0) \tag{5.77}
$$

with $\vec{R}_{ll'}^0 = \vec{R}_l^0 - \vec{R}_{l'}^0$. Equation (5.75) is exact if the quantities δH and δS are interpreted as renormalized quantities which include the effects of screening, exchange and correlation interactions etc.

We can write the matrix elements of H as the sum of one, two, and three central integrals as

$$
H_{\nu\nu'}(\vec{k}) = H_{\nu\nu'}^{(1)}(\vec{k}) + H_{\nu\nu'}^{(2)}(\vec{k}) + H_{\nu\nu'}^{(3)}(\vec{k}) \tag{5.78}
$$

where the first, second and third terms represent the Fourier transforms of one, two, and three central integral contributions respectively. For a monatomic lattice, one central integral $H_{\nu\nu'}^{(1)}(\vec{k}) = H_{\nu l,\nu' l}^{(1)}$ where

$$
H_{\nu l,\nu' l}^{(1)} = \,<\phi_{\nu l}|T_e + v(\vec{r} - \vec{R}_l^0)|\phi_{\nu' l}> + \sum_{l'} <\phi_{\nu l}|v(\vec{r} - \vec{R}_{l'}^0)|\phi_{\nu' l}.
$$
$$
\tag{5.79}
$$

Similarly

$$
H_{\nu\nu'}^{(2)}(\vec{k}) = N^{-1/2} \sum_{ll'} H_{\nu l,\nu' l'}^{(2)} \exp(i\vec{k}\cdot\vec{R}_{ll'}^0) \tag{5.80}
$$

and

$$H^{(3)}_{\nu\nu'}(\vec{k}) \;=\; N^{-1/2}\sum_{ll'} H^{(3)}_{\nu l,\nu'l'}\exp(i\vec{k}\cdot\vec{R}^0_{ll'}) \tag{5.81}$$

where

$$H^{(2)}_{\nu l,\nu'l'} \;=\; <\phi_{\nu l}|T_e + v(\vec{r}-\vec{R}^0_l) + v(\vec{r}-\vec{R}^0_{l'})|\phi_{\nu'l'}> \tag{5.82}$$

and

$$H^{(3)}_{\nu l,\nu'l'} \;=\; \sum_{l''\neq ll'} <\phi_{\nu l}|T_e + v(\vec{r}-\vec{R}^0_{l''})|\phi_{\nu'l'}> . \tag{5.83}$$

The energy eigenvalues $E_{\vec{k}\mu}$ in the NTB scheme are obtained by solving the determinental equation

$$\left|H_{\nu\nu'}(\vec{k}) - S_{\nu\nu'}(\vec{k})E_{\vec{k}\mu}\right| \;=\; 0. \tag{5.84}$$

The stationary states are obtained by substituting $t=0$ in Eq. (5.68). This gives

$$\psi_{\vec{k}\mu}(\vec{r}) \;=\; N^{-1/2}\sum_{\nu l} e^{\nu}_{\vec{k}\mu}\exp(i\vec{k}\cdot\vec{R}^0_l)\phi_{\nu l}(\vec{r}) \tag{5.85}$$

which is the same as given in Eq. (5.2). The eigenvector matrix $\mathbf{e}$ is defined by the matrix equation

$$\mathbf{He} \;=\; \mathbf{SeE} \tag{5.86}$$

with the normalization condition

$$\mathbf{e}^{\dagger}\mathbf{Se} = \mathbf{I}. \tag{5.87}$$

Here $\mathbf{H}, \mathbf{S}$ and $\mathbf{I}$ are Hamiltonian, overlap and identity matrices respectively. Thus the matrix $\mathbf{e}$ defined in Eq.(5.86) is not a unitary matrix. However to get the unitary matrix for eigenvectors, Eqs. (5.86) and (5.87) are written as

$$\mathbf{S}^{-1/2}\mathbf{H}\mathbf{S}^{1/2}\mathbf{e'} \;=\; \mathbf{e'E} \tag{5.88}$$

where $\mathbf{e}'$ is unitary matrix which is related to $\mathbf{e}$ as

$$\begin{aligned}
\mathbf{e} &= \mathbf{S}^{-1/2}\mathbf{e}' \\
\mathbf{e}^{\dagger} &= \mathbf{e}'^{\dagger}\mathbf{S}^{-1/2}.
\end{aligned} \tag{5.89}$$

Using the above mentioned NTB scheme Varma et al [33], [34] calcualted the electron-phonon interaction. The variation of H and $\mathbf{S}$ with the displacement $\vec{u}(l)$ of an atom at $\vec{R}_l$ is calcualted by representing the atomic displacments, given in Eq. (2.77) in terms of normal co-ordinates $A'(\vec{q}j)$ as

$$u_{\alpha}(l) = \sum_{\vec{q}j} e_{\alpha}(\vec{q}j)A'(\vec{q}j)\exp(i\vec{q}\cdot\vec{R}_l^0). \tag{5.90}$$

It is found that the one-center term given in Eq. (5.79) is not altered as the distance between the atoms is varied. The contribution from the three-center term and other crystal field contributions are found small in comparison with the contribution of two-center term. Thus the changes in H involving two center integrals given in Eq. (5.80) are calculated. Similarly the changes in the overlap integral S involving two central integrals are calculated. Combining these terms, Eq. (5.75) simplifies as

$$i\dot{C}_{\vec{k}\mu} = \sum_{\vec{k}'\mu'}\sum_{\vec{q}j\alpha} e_{\alpha}(-\vec{q}j)M^{\alpha}_{\vec{k}\mu,\vec{k}'\mu'}C_{\vec{k}'\mu'}A'(-\vec{q}j) \tag{5.91}$$

where

$$M^{\alpha}_{\vec{k}\mu,\vec{k}'\mu'} = \sum_{\nu\nu'} e^{\nu\dagger}_{\vec{k}\mu}\left[\gamma^{\alpha}_{\nu\nu'}(\vec{k}) - \gamma^{\alpha}_{\nu\nu'}(\vec{k}')\right]e^{\nu'}_{\vec{k}'\mu'} \tag{5.92}$$

and

$$\gamma^{\alpha}_{\nu\nu'}(\vec{k}) = \sum_{ll'}\left[\nabla_{\alpha}H_{\nu l,\nu'l'} - E_{\vec{k}}\nabla_{\alpha}S_{\nu l,\nu'l'}\right]$$
$$\times \exp(i\vec{k}\cdot\vec{R}_{ll'}^0). \tag{5.93}$$

The asymmetry in the definition of $\gamma(\vec{k})$ is due to nonunitary matrix $\mathbf{e}$. $M^{\alpha}_{\vec{k}\mu,\vec{k}'\mu'}$ are the elements of electron-phonon matrix.

Now we compare the results for electron-phonon interactions in the NTB and Bloch representations [2], [4], [36]. In the Bloch representation, the localized orbitals $\phi_{\nu l}(\vec{r})$ are time independent even in the presence of perturbation which gives the equation of motion for $C_{\vec{k}\mu}$ as

$$iC_{\vec{k}\mu}^{\cdot} = \sum_{\nu\nu'}\sum_{\vec{k}'\mu'} e_{\vec{k}\mu}^{\nu\dagger}\delta H_{\nu\nu'}(\vec{k},\vec{k}')e_{\vec{k}'\mu'}^{\nu'}C_{\vec{k}'\mu'} \qquad (5.94)$$

where

$$\delta H_{\nu\nu'}(\vec{k},\vec{k}') = \sum_{ll'} <\nu l|\delta H|\nu' l'> \exp[i(\vec{k}-\vec{k}')\cdot\vec{R}_{ll'}^{0}]. \qquad (5.95)$$

In Eq. (5.75) the shift in the orbitals due to displacement of atoms from their mean positions is explicitly considered while it is not so in Eq. (5.94). But the Eqs.(5.75) and (5.94) must be equivalent as these are derived just with two different basis sets. In other words the change in the electronic wave function for a given perturbation will be the same if the scattering to all possible final states is summed up. Therefore equation (5.94) requires the sum over a larger number of higher states to build up the same change in the wave function due to perturbation as in Eq. (5.75). If one is concerned only with the energy conserving transtions i.e. $(E_{\vec{k},\mu} = E_{\vec{k}'\mu'})$ as in the transport theory or in the Fermi surface integrals, it can be shown that the two expressions are identical. Therefore the arguments given by Birnboim and Gütfreund [29] that the Bloch representation is not applicable to TMCs does not seem to be valid. It is, however, true that the Bloch formalism is not as convenient for quasilocalized d-electrons as it is for nearly free electrons.

5.3.2   Lattice Dynamics

The study of phonon spectra involves the evaluation of total electronic energy $E_e(\vec{R})$ for the instantaneous configuration $\{\vec{R}\}$ of the lattice. The wave functions $\psi_{\vec{k}\mu}(\vec{r},\vec{R})$ and the energy eigenvalues $E_{\vec{k}\mu}(\vec{R})$ of the electrons are determined in the adiabatic approximation using one electron Hamiltonian, given in Eq. (2.3), for a particular configuration $\{\vec{R}\}$. In Eq.(2.3) the electron- electron interaction $V_{ee}(\vec{r},\vec{R})$ can be written as the sum of Hartree and exchange-correlation potential i.e.

$$V_{ee}(\vec{r},\vec{R}) \;=\; V_H(\vec{r},\vec{R}) + V_{xc}(\vec{r},\vec{R}) \tag{5.96}$$

where $V_{xc}(\vec{r},\vec{R})$ is an approximate self-consistant one-body potential. Using Eq.(5.96) in (2.3), the energy eigenvalues $E_{\vec{k}\mu}(\vec{R})$ satisfy the equation

$$(T_e + V_b + V_{ee})\psi_{\vec{k}\mu}(\vec{r},\vec{R}) \;=\; E_{\vec{k}\mu}(\vec{R})\psi_{\vec{k}\mu}(\vec{r},\vec{R}). \tag{5.97}$$

Thus the total electronic energy

$$\begin{aligned}
E_e(\vec{R}) &\;=\; \sum_{\vec{k}\mu} f_{\vec{k}\mu}E_{\vec{k}\mu}(\vec{R}) - <V_{ee}(\vec{r},\vec{R})> \\
&\;=\; E_T(\vec{R}) - <V_{ee}(\vec{r},\vec{R})>
\end{aligned} \tag{5.98}$$

where $f_{\vec{k}\mu}$ is the occupation probability and $<\ldots>$ denotes the expectation value. The second term on the right side is subtracted as it has been counted twice in the first term. Using Eq.(5.96) in (5.98) we get

$$\begin{aligned}
E_e(\vec{R}) &\;=\; \sum_{\vec{k}\mu} f_{\vec{k}\mu}E_{\vec{k}\mu}(\vec{R}) - <V_H(\vec{r},\vec{R}) + V_{xc}(\vec{r},\vec{R})> \\
&\;=\; \sum_{\vec{k}\mu} f_{\vec{k}\mu}E_{\vec{k}\mu}(\vec{R}) - V_H(\vec{R}) - V_{xc}(\vec{R}).
\end{aligned} \tag{5.99}$$

Writing $E_e(\vec{R})$ in the above form, the total potential energy $\Phi(\vec{R})$ becomes

$$\Phi(\vec{R}) \;=\; \Phi_0(\vec{R}) + \sum_{\vec{k}\mu} f_{\vec{k}\mu}E_{\vec{k}\mu}(\vec{R}) \tag{5.100}$$

where

$$\Phi_0(\vec{R}) \;=\; \Phi^c(\vec{R}) - \left[V_H(\vec{R}) + V_{xc}(\vec{R})\right] \tag{5.101}$$

and $\Phi^c(\vec{R})$ is Coulomb interaction between the ions.

Using Eqs. (5.100) and (5.101) in Eq. (2.54), the dynamical matrix for a monatomic lattice becomes

$$D_{\alpha\beta}(\vec{q}) = D_{\alpha\beta}^{(0)}(\vec{q}) + D_{\alpha\beta}^{RC}(\vec{q}) \tag{5.102a}$$

where

$$D_{\alpha\beta}^{(0)}(\vec{q}) = \frac{1}{M} \sum_{ll'} \Phi_{\alpha\beta}^{(0)}(l, l') \exp[-i\vec{q} \cdot (R_l^0 - R_{l'}^0)], \qquad (5.102b)$$

$$D_{\alpha\beta}^{RC}(\vec{q}) = \frac{1}{M} \sum_{ll'} \Phi_{\alpha\beta}^{RC}(l, l') \exp[-i\vec{q} \cdot (R_l^0 - R_{l'}^0)], \qquad (5.102c)$$

$$\Phi_{\alpha\beta}^{(0)}(l, l') = \nabla_{l\alpha} \nabla_{l'\beta} \Phi_0(l, \vec{R}) \qquad (5.102d)$$

and

$$\Phi_{\alpha\beta}^{RC}(l, l') = \nabla_{l\alpha} \nabla_{l'\beta} \left(\sum_{\vec{k}\mu} f_{\vec{k}\mu} E_{\vec{k}\mu}(\vec{R}) \right). \qquad (5.102e)$$

$$(5.102)$$

In Eq.(5.101), $V_{xc}(\vec{R})$ is short range potential and the long range Coulomb potentials $\Phi^c(\vec{R})$ and $V_H(\vec{R})$ may get cancelled in the limit $\vec{R} \to 0$. Therefore $(\Phi^c(\vec{R}) - V_H(\vec{R}))$ is much smaller as compared to $\Phi^c(\vec{R})$ or $V_H(\vec{R})$. The cancellation between $\Phi^c(\vec{R})$ and $V_H(\vec{R})$ for more tightly bound orbitals is more closer. Thus $\Phi_0(\vec{R})$ becomes a short ranged weak potential. Therefore the major problem of cancellation of ion-ion interaction and ion-electron-ion interaction is resolved in the NTB scheme.

The elements of dynamical matrix $D_{\alpha\beta}^{RC}(\vec{q})$ arise from $E_T(\vec{R})$. As the ions move from their equilibrium positions, the energy eigenvalues $E_{\vec{k}\mu}$ are shifted. This shift $\delta E_{\vec{k}\mu}$ can be expressed in terms of $\delta H_{\nu l, \nu' l'}$ and $\delta S_{\nu l, \nu' l'}$ given in Eqs. (5.71)-(5.74). We write $\delta E_{\vec{k}\mu}$ as the sum of two parts:

$$\delta E_{\vec{k}\mu} = \delta E_{\vec{k}\mu}^{(1)} + \delta E_{\vec{k}\mu}^{(2)} \qquad (5.103a)$$

where

$$\delta E_{\vec{k}\mu}^{(1)} = \sum_{\nu\nu'} e_{\vec{k}\mu}^{\nu\dagger} [\delta H_{\nu\nu'}(\vec{k}) - E_{\vec{k}\mu} \delta S_{\nu\nu'}(\vec{k})] e_{\vec{k}\mu}^{\nu'} \qquad (5.103b)$$

and

$$\delta E_{\vec{k}\mu}^{(2)} = - \sum_{\vec{k}'\mu'} \sum_{\nu\nu'} e_{\vec{k}\mu}^{\nu\dagger} [\delta H_{\nu\nu'}(\vec{k}) - E_{\vec{k}\mu} \delta S_{\nu\nu'}(\vec{k})] e_{\vec{k}'\mu'}^{\nu'} e_{\vec{k}'\mu'}^{\nu'\dagger}$$

$$\times [\delta H_{\nu\nu'}(\vec{k}') - E_{\vec{k}\mu}\delta S_{\nu\nu'}(\vec{k}')]e^{\nu}_{\vec{k}\mu}\left(E_{\vec{k}\mu} - E_{\vec{k}'\mu'}\right)^{-1}\left(1 - f_{\vec{k}'\mu'}\right).$$

$$(5.103c)$$
$$(5.103)$$

Here $\delta E^{(1)}_{\vec{k}\mu}$ is the first order correction to $E_{\vec{k}\mu}$ which arises from the second order displacements and $\delta E^{(2)}_{\vec{k}\mu}$ is the second order correction to $E_{\vec{k}\mu}$ which arises from the first order displacements.

As discussed earlier, the crystal-field and three-center integrals in the Hamiltonian are ignored. With these approximations, the change in Hamiltonian and change in overlap integrals given in Eqs. (5.71) and (5.74) respectively simplify as

$$\delta H_{\nu l,\nu' l'} = \vec{\nabla}H_{\nu l,\nu' l'} \cdot (\vec{u}(l) - \vec{u}(l')) + \frac{1}{2}(\vec{u}(l) - \vec{u}(l'))$$
$$\times \nabla^2 H_{\nu l,\nu' l'} \cdot (\vec{u}(l) - \vec{u}(l')) \qquad (5.104a)$$

and

$$\delta S_{\nu l,\nu' l'} = \vec{\nabla}S_{\nu l,\nu' l'} \cdot (\vec{u}(l) - \vec{u}(l')) + \frac{1}{2}(\vec{u}(l) - \vec{u}(l'))$$
$$\times \nabla^2 S_{\nu l,\nu' l'} \cdot (\vec{u}(l) - \vec{u}(l')). \qquad (5.104b)$$

$$(5.104)$$

Using Eqs. (5.103) and (5.104) in Eq. (5.102) one gets

$$D^{RC}_{\alpha\beta}(\vec{q}) = D^{(1)}_{\alpha\beta}(\vec{q}) + D^{(2)}_{\alpha\beta}(\vec{q}) \qquad (5.105)$$

where

$$D^{(1)}_{\alpha\beta}(\vec{q}) = \frac{1}{2M}\sum_{\vec{k}\mu}f_{\vec{k}\mu}\sum_{\nu l\nu' l'}e^{\nu\dagger}_{\vec{k}\mu}\left[\nabla_{l\alpha}\nabla_{l'\beta}H_{\nu l,\nu' l'} - E_{\vec{k}\mu}\nabla_{l\alpha}\nabla_{l'\beta}S_{\nu l,\nu' l'}\right]$$
$$\times e^{\nu'}_{\vec{k}\mu}\exp\left[-i\vec{q}\cdot(\vec{R}_l - \vec{R}_{l'})\right] \qquad (5.106)$$

and

$$D^{(2)}_{\alpha\beta}(\vec{q}) = -\frac{1}{M}\sum_{\vec{k}\mu\mu'}\frac{M^{\alpha*}_{\vec{k}\mu,\vec{k}+\vec{q}\mu'}M^{\beta}_{\vec{k}+\vec{q}\mu',\vec{k}\mu}\left[f_{\vec{k}\mu} - f_{\vec{k}+\vec{q}\mu'}\right]}{E_{\vec{k}\mu} - E_{\vec{k}+\vec{q}\mu'}}. \qquad (5.107)$$

The above equations can easily be generalized for the lattice with a basis.

In the conventional method of calculating the phonon frequencies one calculates the bare ion-ion dynamical matrix instead of $D_{\alpha\beta}^{(0)}(\vec{q})$ and then one uses the bare electron-ion matrix elements instead of $M_{\vec{k}\mu,\vec{k}'\mu'}^{\alpha}$. However it is extremely hard to compute accurately the fully renormalized susceptibilty for materials with non-uniform charge densities. It is shown by Varma and Weber [35] that both the conventional and NTB methods are formally equivalent in RPA. However, the NTB method avoids the difficulties involved in the grouping of terms for the convergence purposes [36], [37]. The total dynamical matrix in both the approaches is the same. The conceptual advantage in the NTB method is that the features of electronic structure are accounted for more explicitly in the calculation of electron-phonon matrix elements.

$D_{\alpha\beta}^{(0)}(\vec{q})$ is derived from $\Phi^c(\vec{R})$ and therefore the long range forces in $D_{\alpha\beta}^{(0)}(\vec{q})$ arising from the scattering near the Fermi surface [35], are much smaller than those in $D_{\alpha\beta}^{(2)}(\vec{q})$. The long range forces in $D_{\alpha\beta}^{(1)}(\vec{q})$ have the same range as that in the matrix elements $H_{\nu l,\nu'l'}$ and $S_{\nu l,\nu'l'}$. Therefore $D_{\alpha\beta}^{(0)}(\vec{q})$ and $D_{\alpha\beta}^{(1)}(\vec{q})$ are combined as $(D_{\alpha\beta}^{0}(\vec{q}) + D_{\alpha\beta}^{(1)}(\vec{q}))$ and this quantity is parameterized using nearest and next nearest neighbour force constants such that the eigenvalues obtained by the diagonalization of $\left(\mathbf{D}^{(0)} + \mathbf{D}^{(1)} + \mathbf{D}^{(2)}\right)$ are in the best agreement with the experimental data of phonon frequencies.

$D_{\alpha\beta}^{(2)}(\vec{q})$ is calculated from the first principles utilizing the knowledge of electronic band structure, Fermi surface, eigenvectors $e_{\vec{k}\mu}^{\nu}$ and electron-phonon matrix elements $M_{\vec{k}\mu,\vec{k}'\mu'}^{\alpha}$. Varma et al [34] calcualted these quantities as function of electron-atom (e/a) ratio for 4d bcc transition metals and their alloys. The electronic band structure calculations of Nb due to Anderson et al [38] and Mattheiss [39] are fitted in the NTB band scheme using nine s-p-d orbitals as the basis. The NTB matrix elements are calculated upto second NNs using one and two-center integrals. It is found that in the NTB scheme the actual calcualtions of energy band structure are well represented just within few percent. The calculated results are shown in Fig. 5.6. The derivatives $\vec{\nabla}H$ and $\vec{\nabla}S$ are obtained by calculating H and S for different lattice constants using Herman-Skillman atomic wave functions and atomic potentials and by fitting the self-consistent APW band structure calculations. The calculated phonon frequencies of Nb, Nb-Mo alloys and for Mo are compared with the experimental data in Fig.5.7. For Nb, the dip in the $[00\xi]$ longitudinal branch at $q = 0.7$, the deep minimum and the kink in the $[\xi\xi\xi]$ longitudinal branch near $q = 0.7$ and $q = 0.4$, the negative curvature in the $[\xi00]$ transverse and $[\xi\xi0]$ transverse branches and the crossing of two $[\xi\xi0]$ transverse branches are reproduced. With increasing concentration of Mo the anomalous features decrease and disappear for $Nb_{0.25}Mo_{0.75}$ configuration. With further increase of Mo concentration new features appear. The most important features are the softening of phonon modes around **H** point, crossing of longitudinal and transverse branches in the $[\xi\xi\xi]$ direction near $q = 0.9$ and decrease in the frequencies in the $[\xi\xi0]$ transverse mode. The experimental results of Mo in the transverse branches in the $[\xi\xi0]$ direction are not reproduced well.

5.3.3   Analysis of Phonon Anomalies

a. Transition Metals and their Alloys

Most of the anomalies in the phonon dispersion relations arise due to the scattering processes which are involved in $D^{(2)}_{\alpha\beta}(\vec{q})$ (given in Eq. (5.107)) when the energies $E_{\vec{k}\mu}$ and $E_{\vec{k}'\mu'}$ are in the vicinity of Fermi energy E_F. Let this contribution be denoted as $D^{(2)*}_{\alpha\beta}(\vec{q})$. If $D^{(2)*}_{\alpha\beta}(\vec{q})$ is omitted from $D^{(2)}_{\alpha\beta}(\vec{q})$, all the anomalies except the negative curvature of the transverse modes, disappear in the phonon dispersion relations of Nb and Mo as shown in Fig.5.7. From the band structure calculations of Nb and Nb-rich alloys, it is inferred that $D^{(2)*}_{\alpha\beta}(\vec{q})$ is equivalent to intraband scattering of band 3 (counted from the bottom in Fig. 5.6). In Mo, $D^{(2)*}_{\alpha\beta}(\vec{q})$ arises from the inter-band scattering between bands 3 and 4. It is further noted that if the scattering within the lowest six bands, which are mainly d-like, is ignored, the negative curvature also disappears. The contribution to $D^{(2)}_{\alpha\beta}(\vec{q})$ due to scattering from the d-like bottom to p-like top of the band complex is found structureless. Therefore all the structures in the phonon dispersion relations originate from d-d scattering contribution to $D^{(2)}_{\alpha\beta}(\vec{q})$. It is found in the analysis of various coupling gradients $(\vec{\nabla} H)$ that those involving d-d orbitals contribute upto 90% to $D^{(2)}_{\alpha\beta}(\vec{q})$. The variation of magnitude of d-d gradients makes changes in $D^{(2)}_{\alpha\beta}(\vec{q})$, however, its q-dependence remains unaltered. Similarly the omission of $(\vec{\nabla} S)$ term in $M^{\alpha}_{\vec{k}\mu,\vec{k}'\mu'}$ leads to an increase in the magnitude of $D^{(2)}_{\alpha\beta}(\vec{q})$ but does not change its $\vec{q}$-dependence.

The phonon anomalies are related with the $\vec{q}$-dependene of electron-phonon matrix elements $M^{\alpha}_{\vec{k}\mu,\vec{k}'\mu'}$. The physical origin of these anomalies and their comparison with other calculations can be discussed if $M^{\alpha}_{\vec{k}\mu,k'\mu'}$ is written in the form [34], [35]

$$M^{\alpha}_{\vec{k}\mu,\vec{k}'\mu'} \simeq \left(v^{\alpha}_{\vec{k}\mu} - v^{\alpha}_{\vec{k}'\mu'}\right) \sum_{\nu} e^{\nu*}_{\vec{k}\mu} e^{\nu}_{\vec{k}\mu'}, \qquad (5.108)$$

where $v^{\alpha}_{\vec{k}\mu}$ is the group velocity $\partial E_{\vec{k}\mu}/\partial k_{\alpha}$. Substituting Eq.(5.108) in Eq.(5.107), one gets

$$\begin{aligned}
D^{(2)}_{\alpha\beta}(\vec{q}) \;\simeq\; & \frac{1}{M} \sum_{\vec{k}\mu\mu'} \sum_{\nu\nu'} \frac{(v^{\alpha}_{\vec{k}\mu} - v^{\alpha}_{\vec{k}+\vec{q}\mu'})(v^{\beta}_{\vec{k}\mu} - v^{\beta}_{\vec{k}+\vec{q}\mu'})}{E_{\vec{k}\mu} - E_{\vec{k}+\vec{q}\mu'}} \\
& \times (f_{\vec{k}\mu} - f_{\vec{k}+\vec{q}\mu'}) e^{\nu}_{\vec{k}\mu} e^{\nu*}_{\vec{k}\mu'} e^{\nu'*}_{\vec{k}\mu'} e^{\nu'}_{\vec{k}\mu}.
\end{aligned} \qquad (5.109)$$

If the state $\mid \vec{k}'\mu' >$ is almost at the Fermi surface, $\mid v^{\alpha}_{\vec{k}\mu} - v^{\alpha}_{\vec{k}'\mu'} \mid$ will be large only when the energy of the state $\mid \vec{k}\mu >$ decreases rapidly in the vicinity of Fermi surface i.e. for the large and negative $v^{\alpha}_{\vec{k}\mu}$. Therefore, for the occurrence of an anomaly, the phonon wave vector $\vec{q} = (\vec{k} - \vec{k}')$ will reduce the energy of the state $\mid \vec{k}\mu >$ significantly. Thus for the nesting of the bands near the Fermi surface, the NTB scheme predicts phonon anomalies for large $\mid v^{\alpha}_{\vec{k}\mu} \mid$ whereas Kohn-Overhauser theory [40] predicts the phonon anomalies for small $\mid v^{\alpha}_{\vec{k}\mu} \mid$. On the other hand the energy denominator in Eq.(5.109) should be small to get the strong anomalies and for this reason the initial and final states in the scattering process may be nearly parallel. This is, in general, likely if the energy bands are flat in these directions. Therefore it is concluded that the materials with sharp saddle points in the band structure near the Fermi surface give rise to more prominent phonon anomalies.

It is found in Nb, $Nb_{1-x}Mo_x$ alloys and Mo that the sharpness of the saddle point (with large $|\, v^{\alpha}_{\vec{k}\mu} - v^{\alpha}_{\vec{k}'\mu'} \,|$) washes away the q-dependence of $\chi^0(\vec{q})$ but it makes q-dependence very predominant in $\mathbf{D}^{(2)}(\vec{q})$. This need not be true always. The phonon dispersion curves of $Ta_{1-x}W_x$ are similar to those of $Nb_{1-x}Mo_x$ and Varma-Weber calculations display some features observed in the experimental data for $Ta_{1-x}W_x$ [41]. In Hf, Zr, Ti and Tc metals, the phonon anomaly in the longitudinal optical mode in the $[00\xi]$ direction is due to electronic structure of these metals and it is explained on the basis of Varma-Weber theory [42]. Using NTB scheme, Simons and Varma [43] have shown that the ω-phase anomaly in $Nb_{1-x}Zr_x$ alloy is electronically driven, which is caused by changing features of Fermi surface. The phonon frequencies of $Nb_{1-x}Zr_x$ decrease with the increase of x. But the phonon frequencies decrease more rapidly in the longitudinal branch near $\xi = \left(\frac{2}{3}\frac{2}{3}\frac{2}{3}\right)$ which is in agreement with the experimental data. The NTB scheme also predicts correctly the wave vector and the composition of alloy at which the ω-phase anomalies occur.

b. Transition Metal Compounds

Weber et al [44] applied the NTB scheme to calculate the phonon spectrum of ideal superconducting vanadium nitride VN_x. The band structure calculations due to Neckel et al [15] are fitted in the NTB band scheme using nine orbital basis set (s and p orbitals at nitrogen sites and d orbitals at vanadium sites). The contribution $(\mathbf{D}^{(0)} + \mathbf{D}^{(1)})$ is parameterized using two first nearest neighbour vanadium-nitrogen force constants and three first nearest neighbour vanadium-vanadium force constants. The calculations reproduce the phonon anomalies in the longitudinal acoustic modes at the X symmetry point. However, the minimum is too sharp to be compared with the experimental observations. It is perhaps due to the nonstoichiometry of $VN_{0.86}$ sample which is used in the experiment. To study the effect of nitrogen vacancies on $\mathbf{D}^{(2)}(\vec{q})$, the number of valence electrons (VEs) per atom is varied from 9 to 10 by shifting E_F in accordance with the rigid band scheme. It is found that $\mathbf{D}^{(2)}(\vec{q})$ is very sensitive to the number of VEs and for $VN_{0.86}$ the best agreement with the experimental data is found for 9.65 VEs.

The origin of phonon anomalies in the 10VE compounds can be understood by considering the effect of vacancies on the band filling in the two extreme limits: (i) In the free electron limit, nitrogen vacancies will lower E_F in accordance with the rigid band model and (ii) in the case of non-interacting s and p-bands at nitrogen site and d-bands at vanadium site, the nitrogen vacancies will remove the states from the s and p-bands but will not affect the Fermi energy E_F. In 9VE materials such as TiN, E_F lies within the lowest t_{2g} band which forms jungle-gym Fermi surface [45], [46]. Therefore the intra-band contributions are dominant. In VN_x, E_F lies within the t_{2g} bands. Therefore the inter-band transitions in the t_{2g} bands, which are most important for $\mathbf{D}^{(2)}(\vec{q})$, occur between the states near the symmetry points W(Fig. 5.8) where the p admixing is largest. The electron density of states increases with increase in x in VN_x. Also $\mathbf{D}^{(2)}(\vec{q})$ contribution increases as one moves towards 10 VE system. Therefore VN_x leads to softer phonon modes with increasing x. However the large phonon line widths in VN_x suggest a considerable number of vacancies in both the vanadium and nitrogen sublattices as observed in the nitrides and oxides [47].

In transition metal carbides, the peaks in the polarizability function $\chi^0(\vec{q})$ are found at the positions of phonon anomalies [45], [46] but these peaks do not explain the strength of phonon anomalies. In the NTB scheme the electron-phonon matrix element enhances the magnitude of these peaks by an order of magnitude [35]. Therefore in the TM carbides the anomalies are explained by the combined effect of structure in the electron-phonon matrix element and peaks in the polarizability function $\chi^0(\vec{q})$. Weber's double shell model was used to study the effect of carbon vacancies on the phonon spectrum of NbC_x. The disappearance of the phonon anomalies in NbC_x at $x = 0.8$ was related to the stabilization of second neighbour Nb-Nb interactions upon the removal of carbon atoms [47],[48]. Klein et al [49] studied the effect of carbon vancancies on the substoichiometric NbC_x, TaC_x and HfC_x using the linear combination of atomic orbitals method in the coherent potential approximation. They have shown that there is a charge transfer from

TM atom to carbon atom which coupled with the nesting Fermi surface features, drive the phonon instability. In order to establish this conjecture ab initio calculations like that of Varma and Weber are needed.

5.4   Discussion

From the preceding sections we observe that there are two group of thoughts to explain the phonon anomalies in the d and f-band materials. Sinha and co-workers [2], [3], [4], Singh et al [5],[6], [7] and Hanke et al [10], [50] believe that the local orbital symmetry is more important. However due to intricacies of the formalism, the ab initio calculations are formidable. Therefore their attempts resulted in the simplified models which are the variations of phenomenological shell model for the transition metal carbides. Allen [21] and Wakabayashi [17] attempted to generalize these models. Sinha and Harmon [2] represented the d-d intra-band contribution to polarizability function in q-independent parameters and these authors explained the phonon anomalies in the longitudinal acoustic branches of NbC. These authors further emphasized the importance of local field corrections in the interpretation of phonon anomalies. However, Hanke et al [50], [51] recall that exchange and correlation interactions among d-electrons also contribute towards the phonon anomalies in NbC.

Pickett and Gyorffy [36] and Varma and Weber [32], [33], [34], [35] believe that the bands near E_F are crucial in the explanation of phonon anomalies and a careful calculation of polarizability function $\chi^0(\vec{q})$ is required. A similar conclusion is also drawn by Freeman and co-workers [45], [52], [53], [54]. They have shown in a variety of materials that if only those bands which intersect E_F are included in the calculation of $\chi^0(\vec{q})$, then strong structures appear in $\chi^0(\vec{q})$ and these structures are directly related to phonon anomalies in the phonon dispersion relations. Varma and co-workers [32], [33], [34], [35] have shown that the bands near E_F are responsible for the structure appearing in both the electron-phonon matrix element and $\chi^0(\vec{q})$ and both the physical quantities are required to explain the position and magnitude of phonon anomalies in transition metals and their compounds.

The reason of contradiction between the two interpretations is that in the band theory, the chemical bonds come from the valence electron charge density which can seldom be adequately described by examining a narrow set of states near E_F. But the reconciliation between the two interpretations is possible if the relevant bond orbitals are carefully specified in the band theoretic calculations. This procedure will bring the results nearer to the calculations of Varma and co-workers [32], [33], [34], [35]. It may be interesting to reparameterize the calculations of Sinha and Harmon [2], [4], [36] taking into account the $\vec{q}$-dependence of $\chi^0(\vec{q})$. This would lead to a different set of parameters to explain the phonon anomalies. In this context it is important to note that Kamitakahara et al [55] succeeded in reproducing the phonon dispersion relations of tungston bronzes, using only the structure of $\chi^0(\vec{q})$ and a smoothly varying electron-ion pseudopotential. However the relative importance of (i) the number of bands to be summed up in the calculation of $\chi^0(\vec{q})$, (ii) the effect of overlap matrix elements and exchange and correlation interactions in $\chi^0(\vec{q})$ and (iii) the correctness of electron-phonon matrix elements in the interpretation of phonon spectrum of these materials is not very clear.

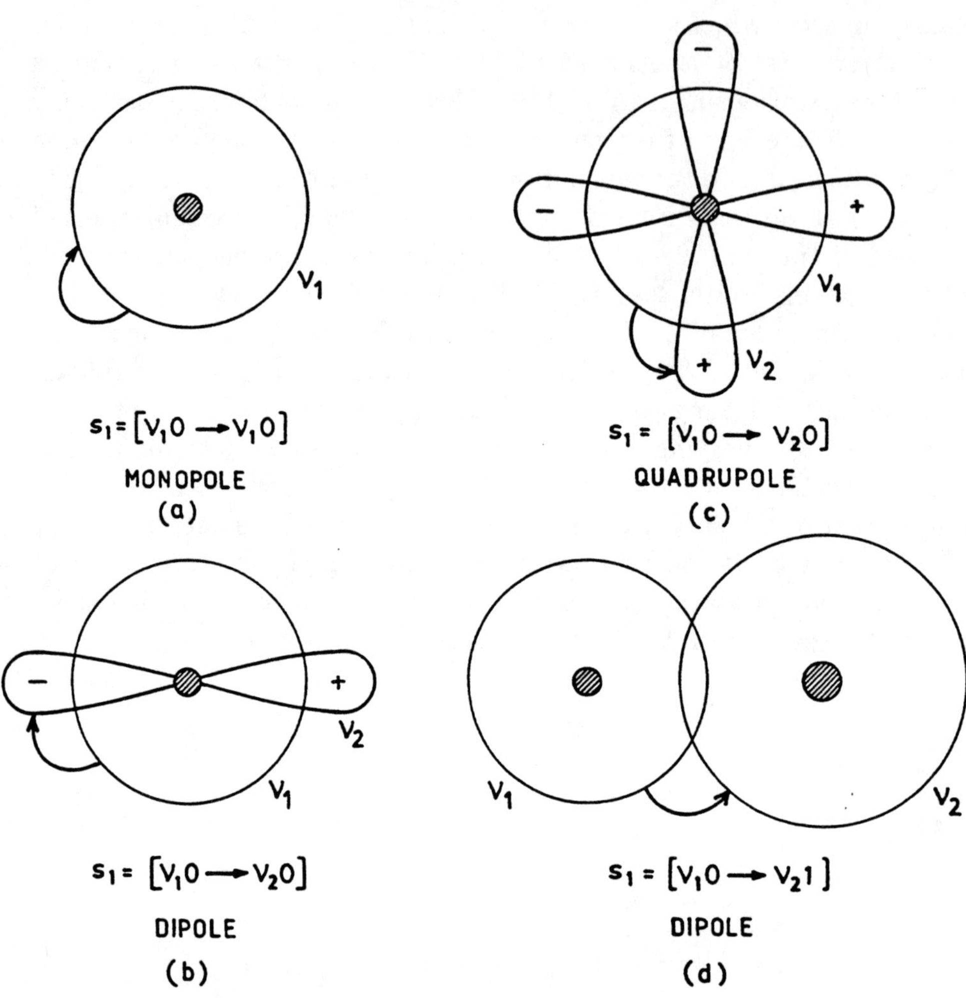

Figure 5.1: Virtual transitions between orbitals corresponding to charge fluctuations of various symmetries [4]: (a) Charge fluctuations of monopole symmetry which arise due ($\nu_1 0 \rightarrow \nu_1 0$) intra-atomic and intra-orbital virtual transitions. (b) Charge fluctuations of dipole symmetry which arise due to ($\nu_1 0 \rightarrow \nu_2 0$) intra-atomic and inter-orbital virtual transitions. (c) Charge fluctuations of quadrupole symmetry which arise due to manifold ($\nu_1 0 \rightarrow \nu_2 0$) virtual transitions. (d) Charge fluctuations of dipole symmetry which arise due to ($\nu_1 0 \rightarrow \nu_2 1$) intra-atomic and inter-orbital virtual transitions.

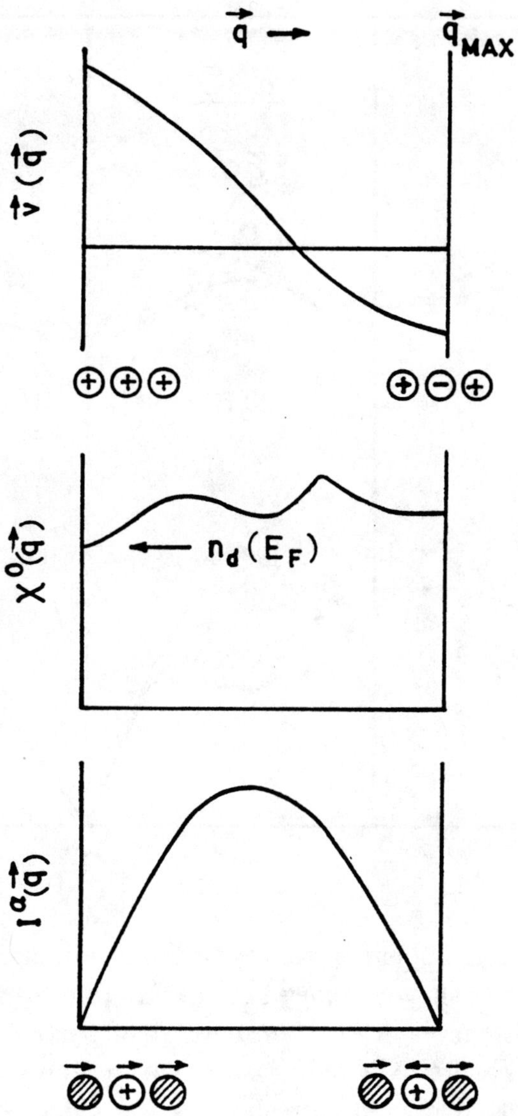

Figure 5.2: Schematic variation of effective electron-electron interaction $\bar{v}(\vec{q})$, polarizability function $\chi^0(q)$ and the electron-phonon interaction in the localized representation $I^\alpha(\vec{q})$ with $\vec{q}$ [4]. The behaviour of ions and ionic displacements at the extreme limits of $\vec{q}$ for $\bar{v}(\vec{q})$ and $I^\alpha(\vec{q})$ is also shown there.

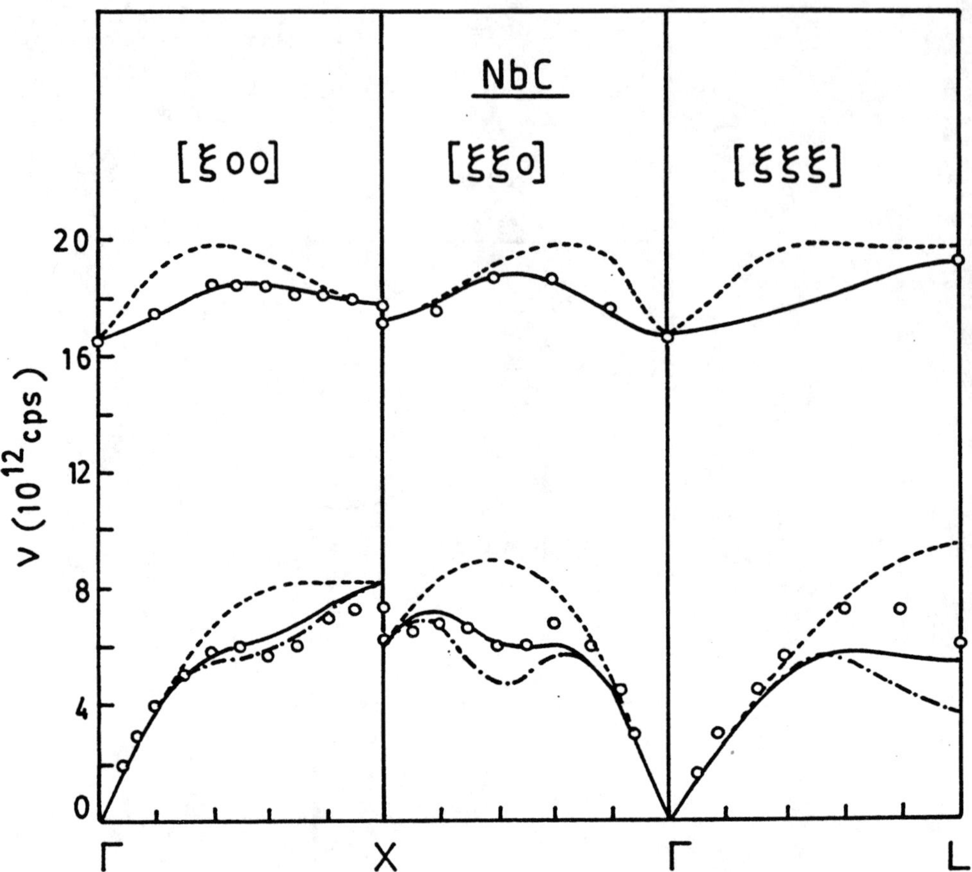

Figure 5.3: Phonon dispersion curves for longitudinal modes of NbC [2]. The measurements shown by open circles are due to Smith and Gläser [3.160]. Smooth curves are for the best fit of partial density of states parameters at E_F for Nb and C sites within the charge fluctuation model. The dotted curves correspod to the calculations when these parameters are increased by 10% . The dashed curves represent the calculations when these parameters are set to zero i.e., there is no charge fluctuation effect at all. For the optical modes, the smooth and dotted curves coincide.

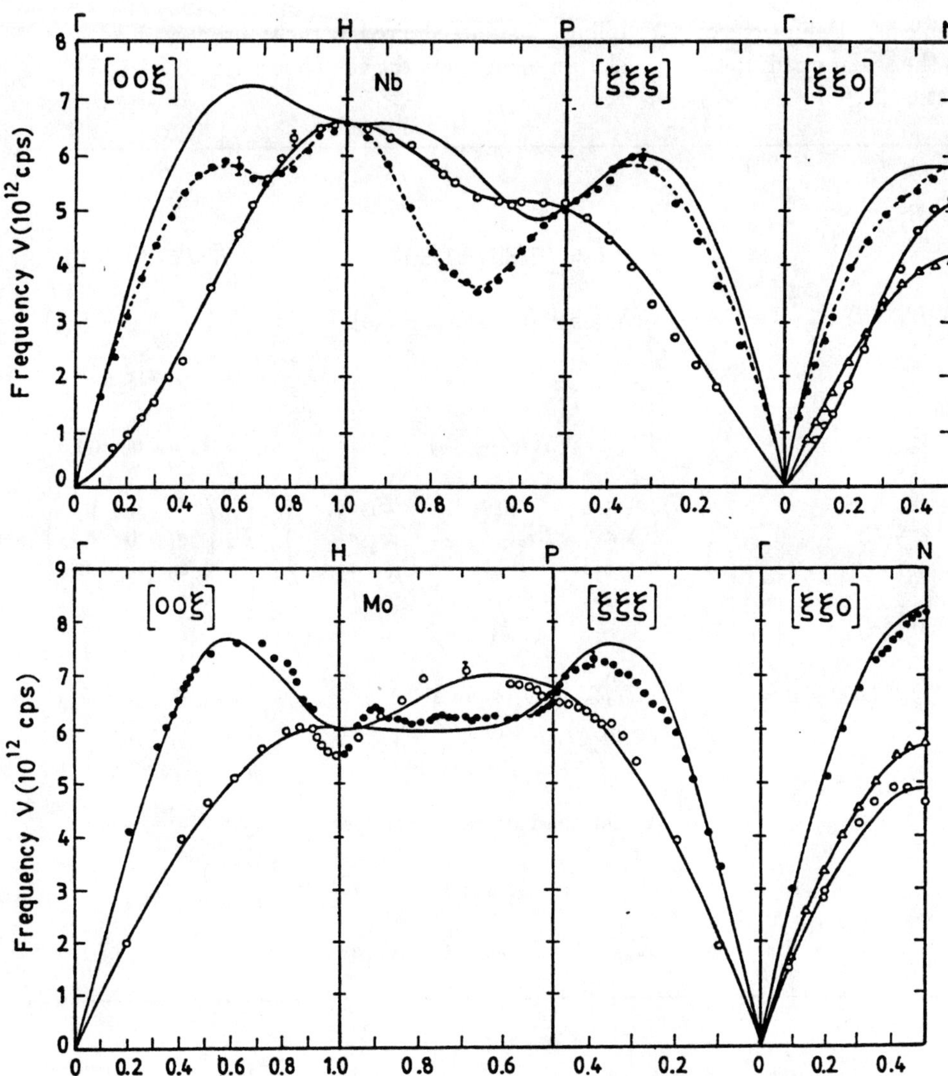

Figure 5.4: The calculated and the experimental results of phonon dispersion relations of Nb and Mo due to Wakabayashi et al [20]. The solid lines reprsent the calculations for 3NN BvK model. The dashed lines are the calculation of charge fluctuation model. The tranverse modes are not affected by the charge fluctuations. There is no significant change in the results for Mo due to charge fluctuations.

Table 5.1: Definitions of 16 coupling constants allowed by the symmetry in the scalar quadrupolar model upto second nearest neighbours for bcc lattice [21]

	Zeroth Order	1NN	2NN
		$\vec{R} = \frac{1}{2}\vec{a}\{\pm 1, \pm 1, \pm 1\}$	$\vec{R} = \vec{a}\{\pm 1, 0, 0\}$
$-K_{\alpha\beta}(R)$	0	$K_{1a}\delta_{\alpha\beta} + K_{1b}a_\alpha a_\beta(1 - \delta_{\alpha\beta})$	$[K_{2a} + (K_{2b} - K_{2a}) \times a_\alpha^2]\delta_{\alpha\beta}$
$A_{\alpha\alpha}^1(R)$	0	$A_1(a_x, a_y, a_z)$	$A_2(a_x, a_y, a_z)$
$A_{\alpha\beta}^{25'}(R)$	0	$\begin{pmatrix} B_{1a}a_xa_ya_z & B_{1b}a_z & B_{1b}a_y \\ B_{1b}a_z & B_{1a}a_xa_ya_z & B_{1b}a_x \\ B_{1b}a_y & B_{1b}a_x & B_{1a}a_xa_ya_z \end{pmatrix}$	$B_2\begin{pmatrix} 0 & a_z & a_y \\ a_z & 0 & a_x \\ a_y & a_x & 0 \end{pmatrix}$
$B_{\alpha\alpha}^{1,1}(R)$	1	b_1	b_2
$B_{\alpha\beta}^{25',25'}(R)$	$\delta_{\alpha\beta}$	$C_{1a}\delta_{\alpha\beta} + C_{1b}a_\alpha a_\beta(1 - \delta_{\alpha\beta})$	$(C_{2a} + C_{2b}a_\alpha^2)\delta_{\alpha\beta}$
$B_{\alpha\alpha}^{1,25'}(\vec{R})$	0	$D_1(a_ya_z, a_za_x, a_xa_y)$	0

For the third nearest neighbour

$$\vec{R} = \vec{a}(\pm 1, \pm 1, 0)$$

$$K_{\alpha\beta}(R) = K_{3a}\delta_{\alpha\beta} + K_{3b}a_\alpha a_\beta$$

Table 5.2: Values of parameters (with $\pm$ uncertainty) in the scalar-quadrupolar model obtained by the best fit of phonon frequencies of Nb [21].

Parameter				
$\left(\frac{K}{M}\right)\ (10^{12}Hz)^2$	K_{1a}	K_{1b}	K_{2a}	K_{2b}
	2.69 ± 0.24	0.98 ± 0.39	-0.48 ± 0.54	1.72 ± 1.05
	K_{3a}	K_{3b}		
	-0.59 ± 0.29	1.08 ± 0.37		
$B^{LL'}$ (dimensionless)	$b_1 = 0.138 \pm 0.12,$		$b_2 = 0.076 \pm 0.15,$	
	$C_{1a} = -0.027 \pm 0.20$		$C_{1b} = -0.057 \pm 0.13,$	
	$C_{2a} = 0.066 \pm 0.14,$		$C_{2b} = 0.214 \pm 0.45,$	
	$D_1 = -0.023 \pm 0.11,$			
$\frac{A_{\alpha\beta}^{L}}{M^{1/2}}\ (10^{12}Hz)$	$A_1 = 0.207 \pm 0.18,$		$A_2 = -0.522 \pm 0.51,$	
	$B_{1a} = 0.031 \pm 0.09,$		$B_{1b} = 0.377 \pm 0.25,$	
	$B_2 = 0.038 \pm 0.41,$			

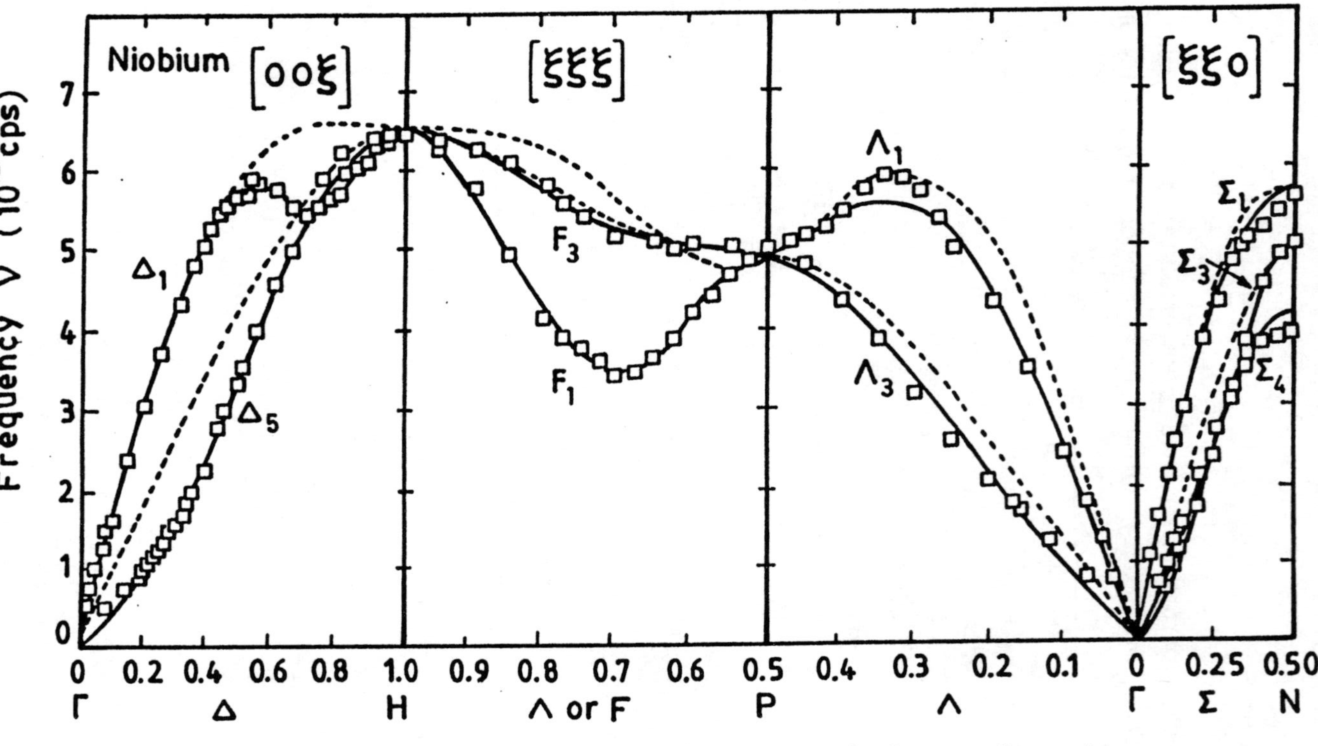

Figure 5.5: The calculated and the measured phonon dispersion relations of Nb. The solid curves are the best fit in the charge density distortion (CDD) model due to Allen [21]. The dashed curve is obtained in the same model by putting the scalar displacements and displacement-quadrupolar coupling parameters $A^L_{\alpha\beta} = 0$.

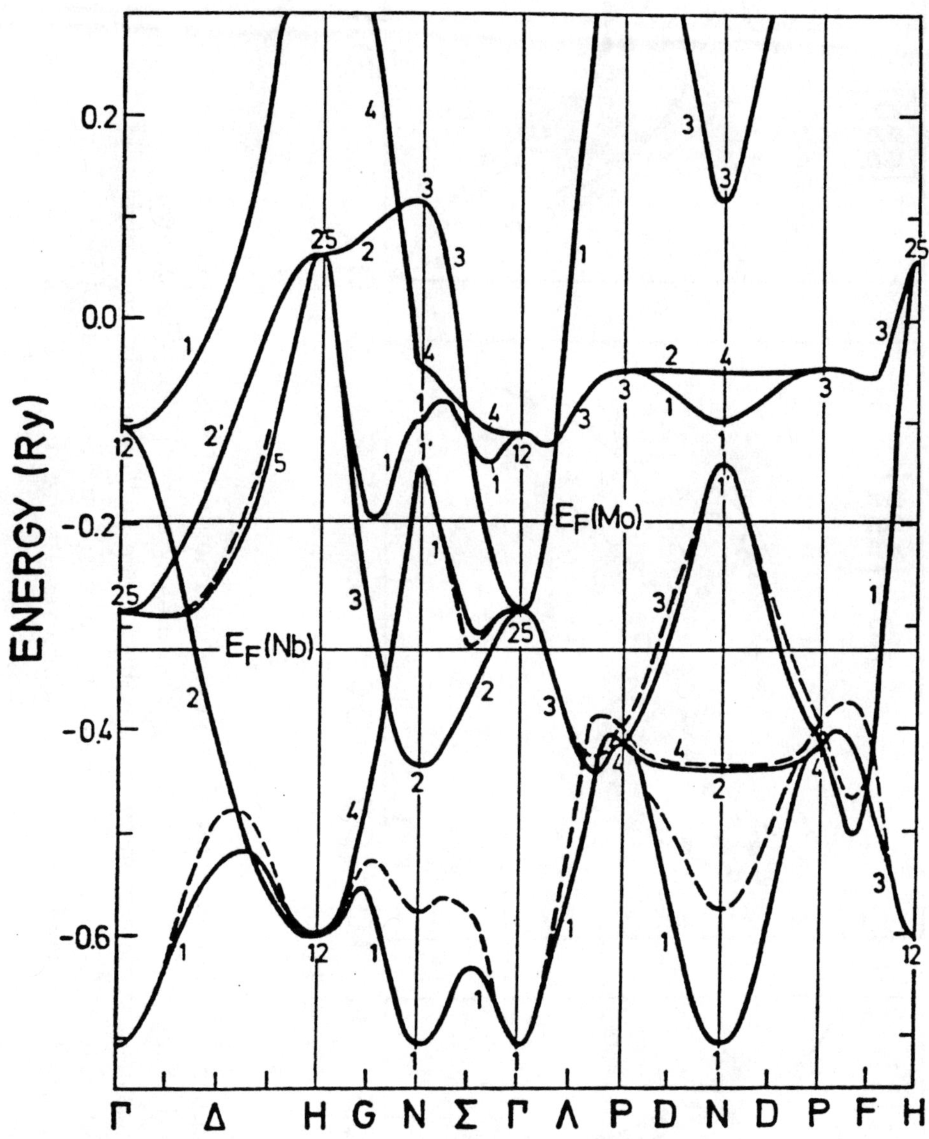

Figure 5.6: Band structure of Nb(Mo). Non-orthogonal tight-binding (NTB) band structure fit due to Varma and Weber [35] is shown by the dark lines. The first principle band structure calculations are shown by the dashed lines [40] where these calculations deviate from the NTB band structure. The Fermi energies E_F for Nb and Mo are also indicated.

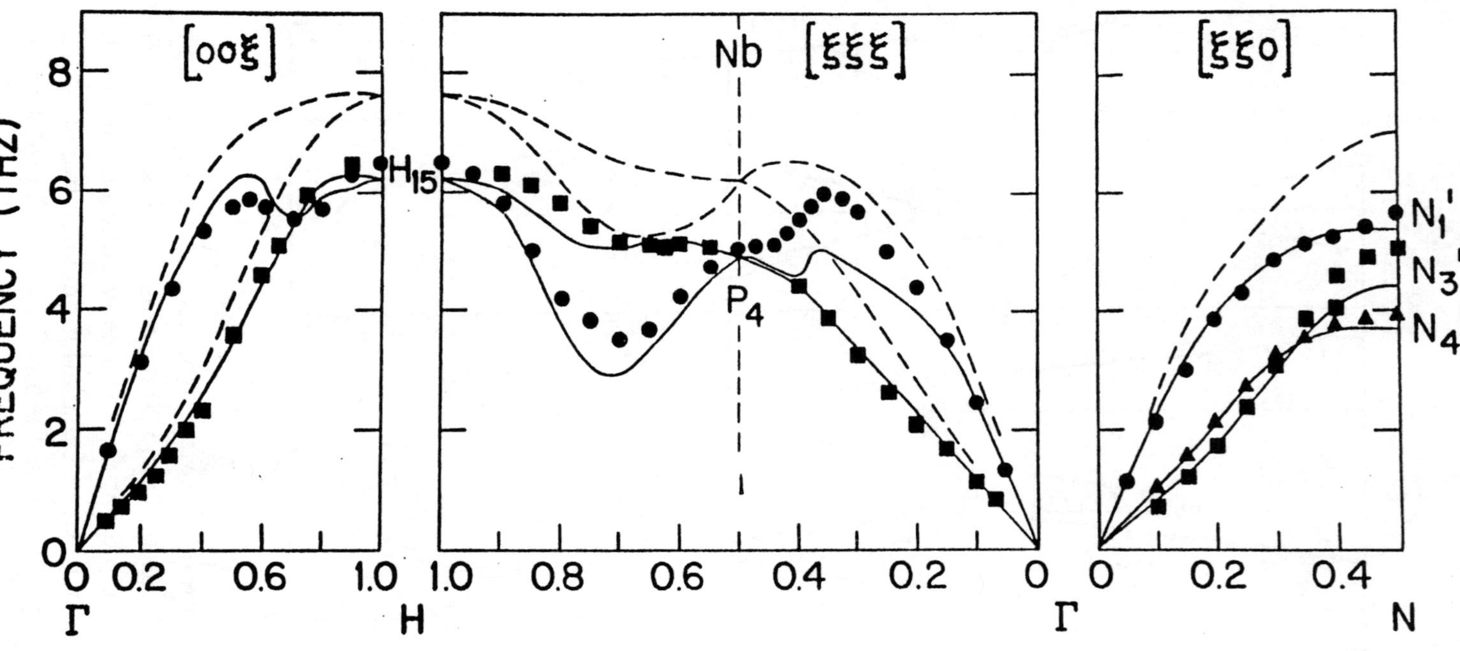

Figure 5.7: The calculated phonon dispersion relations due to Varma and Weber [35] for (a) Nb, (b) $Nb_{0.25}Mo_{0.75}$ and (c) Mo. The dashed lines are the dispersion curves obtained when $\mathbf{D}^{(2)*}$ which is the contribution to $\mathbf{D}^{(2)}_{\alpha\beta}(\vec{q})$ due to energies $E_{k\mu}$ and $E_{k'\mu'}$ in the vicinity of Fermi surface is omitted. The experimental results are due to Powell et al [56].

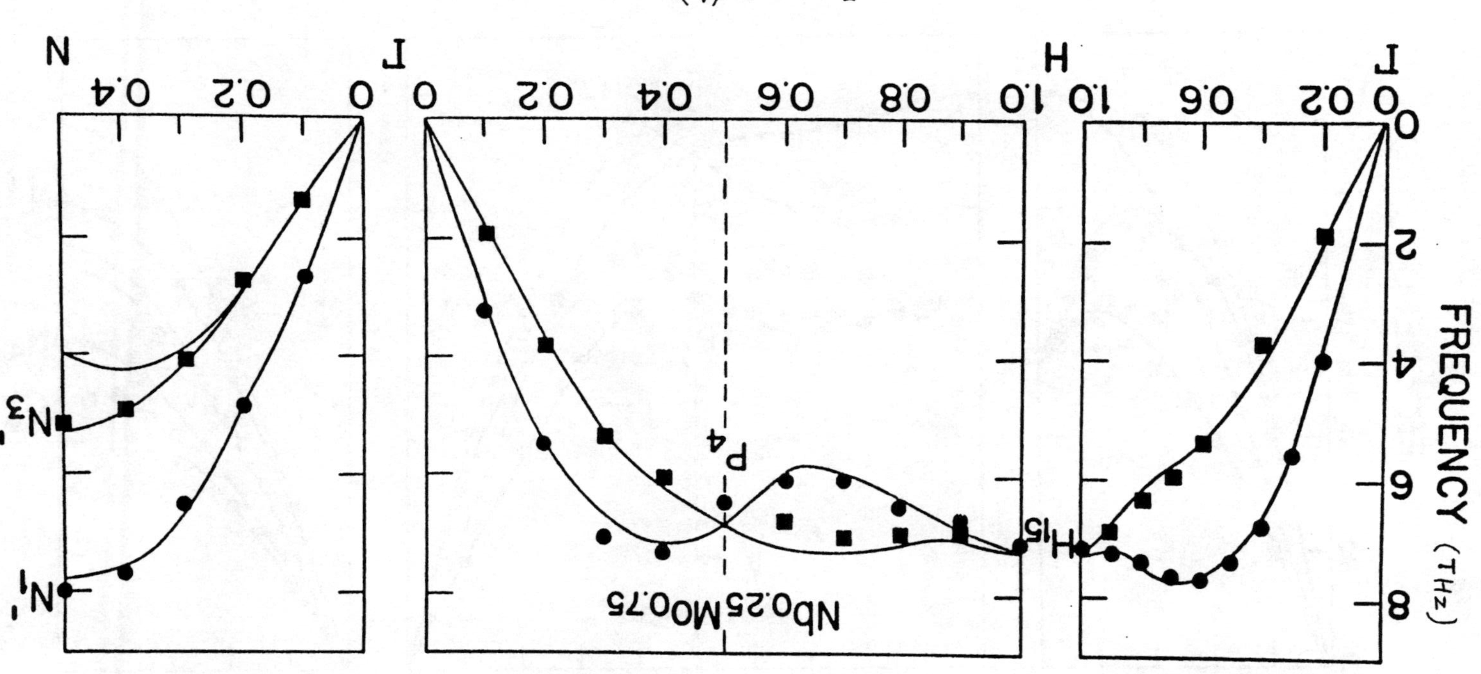

Fig. 5.7 (b)

FREQUENCY (THz)
10
8
6
4
2
0
MO
H₁₅
P₄
N₁'
N₄'
N₃'
0 0.2 0.6 1.0
Γ
H
1.0 0.8 0.6 0.4 0.2 0
Γ
0 0.2 0.4
N

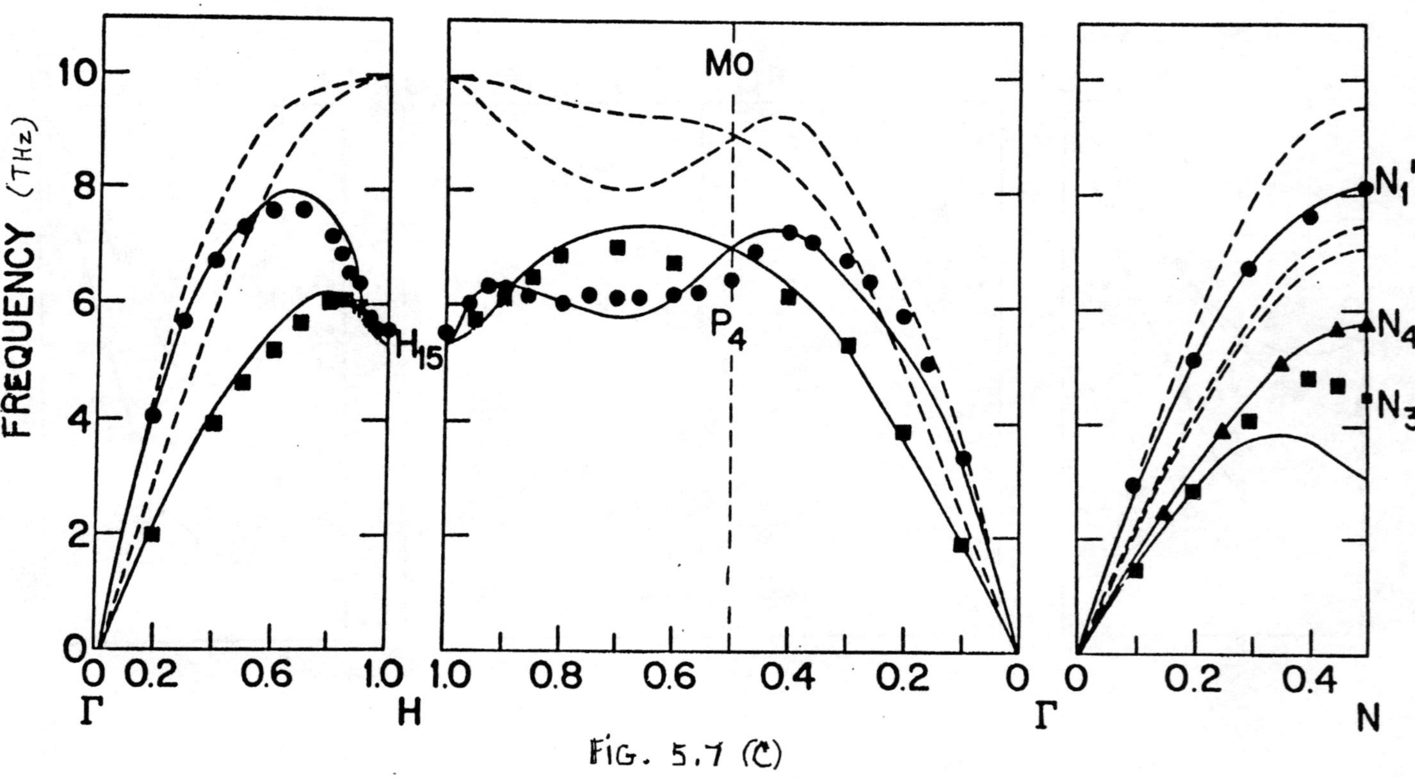

FIG. 5.7 (C)

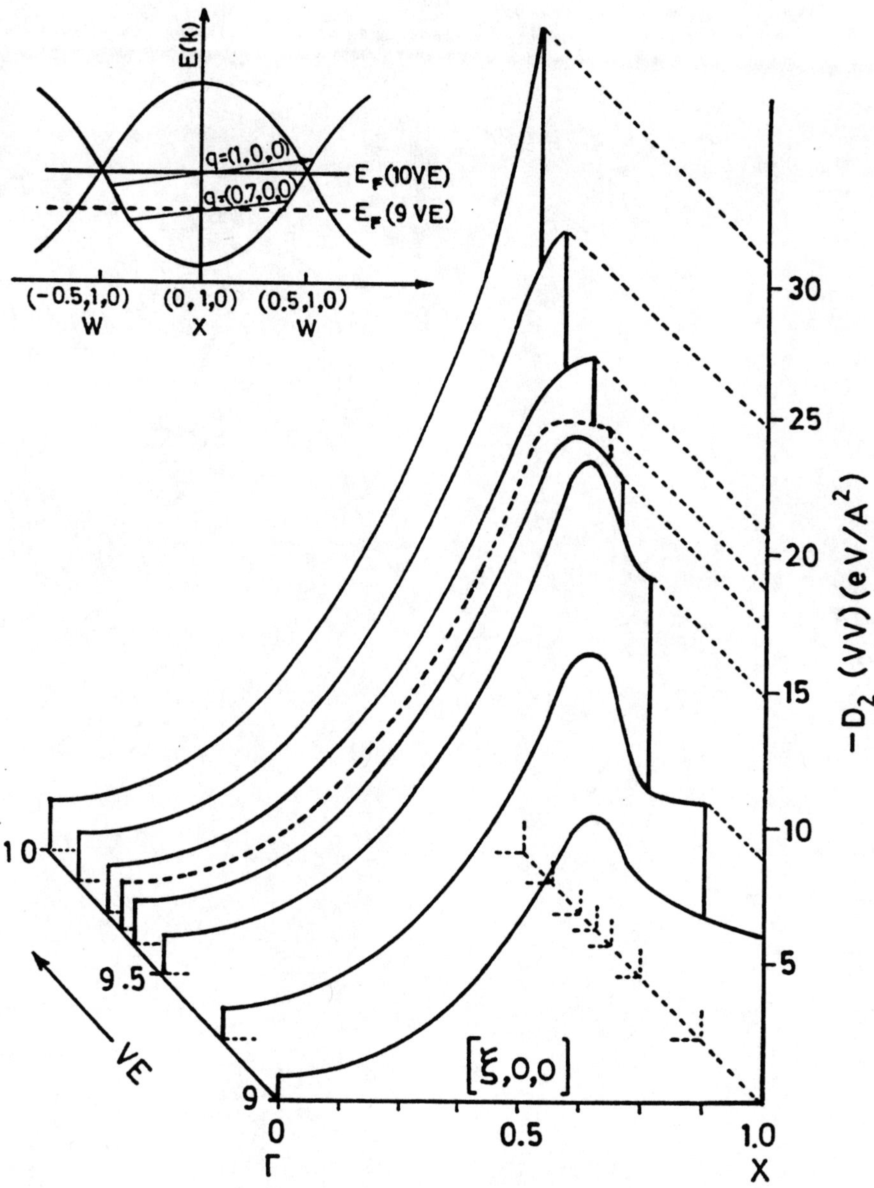

Figure 5.8: The longitudinal components of $-\mathbf{D}^{(2)}$ (vanadium-vanadium) as a function of wave vector $\vec{\xi}$ along the direction $[\xi 00]$ and VE (valence electron) concentration. The dashed curve is for 9.65 VE. The inset shows schematically the transitions most relevant to $\mathbf{D}^{(2)}$.

Bibliography

[1] S.K. Sinha, Phys. Rev. 169, 477(1968); ibid B3, 2401(1971).

[2] S.K. Sinha and B.N. Harmon, Phys. Rev. Lett. 35, 1515(1975); in "Superconductivity of d and f-band Metals", Ed. D. Doughlass (Plenum Press 1976)p.269.

[3] S.K. Sinha, in "Lattice Dynamics", Ed. M. Balkanski (Flammarion, Paris, 1977)p.7.

[4] S.K. Sinha, in "Dynamical Properties of Solids" Vol. 3(Eds. G.K. Horton and A.A. Maradudin, North Holland, 1980)p.1.

[5] J. Singh, N. Singh and S. Prakash, Phys. Rev. B12, 3159, 3166, 5415(1975).

[6] J. Singh and S. Prakash, Nuovo Cimento 37B, 131(1977).

[7] J. Singh, N. Singh and S. Prakash, Phys. Rev. B18, 2954(1978).

[8] D.N. Zubarev, Sovt. Phys. Usp. 3, 320(1960).

[9] H. Takashashi, Phys. Rev. 192, 474(1968); A.K. Rajagopal and M.H. Cohen, Collect. Phenomena 1, 9(1972).

[10] W. Hanke, Phys. Rev. B8, 4585, 4591(1973).

[11] R.L. Cappalletti, N. Wakabayashi, J.C. Traylor and A.J. Bevolo, Phys. Rev. B25, 6090(1982).

[12] C Stassis, J. Zaretsky and N. Wakabayashi, Phys. Rev. Lett. 41, 1726(1978).

[13] B.M. Powell, P. Martel and A.D.B. Woods, Phys. Rev. 171, 727(1968).

[14] V. Löttner, A. Kollma, T. Springer, W. Kress, H. Bilz and W.D. Teuchert, in "Lattice Dynamics" Ed. M. Balkanski (Flammarion, Paris, 1977)p.247.

[15] A. Neckel, P. Raste, R. Eibler, P. Weinberger and K.H. Schwarz, J. Phys. C9, 579(1976).

[16] L. Ramqvist, B. Ekstig, E. Källane, E. Noreland and R. Manne, J. Phys. Chem. Solids. 32, 149(1971).

[17] N. Wakabayashi, Solid State Commun. 23, 737(1977).

[18] U. Schröder, Solid State Commun. 4, 347(1966); U. Schröder and V. Nüsslein, Phys. Stat. Sol. 21, 309(1967).

[19] H.G. Smith and N. Wakabayashi, Solid State Commun. 39, 371(1981).

[20] N. Wakabayashi, R.H. Scherm and H.G. Smith, Phys. Rev. B25, 5122(1982).

[21] P.B. Allen, Phys. Rev. B16, 5139(1977); Bull. Am. Phys. Soc. 29, 297(1977).

[22] L.J. Sham, in " Modern Solid State Physics" Eds. R.H. Erns and R.R. Hearing (Gordon and Beach, New York, 1969)Vol.2, p.143.

[23] W. Hanke and H. Bilz, in " Inelastic Scattering of Neutrons" (IAEA, Vienna, 1972)p.3.

[24] W. Weber, Phys. Rev. B8, 5082, 5093(1973).

[25] W. Weber, H. Bilz and U. Schröder, Phys. Rev. Lett. 28, 600(1972).

[26] S.K. Sinha, Critical Review, Solid State Science 4, 273(1973).

[27] K.M. Ho, S.G. Louie, J.R. Chelikowsky and M.L. Cohen, Phys. Rev. B15, 1755(1977).

[28] S.G. Das, Phys. Rev. B7, 2238(1973).

[29] A. Birnboim and H. Gutfreund, Phys. Rev. B9, 139(1974); Phys. Rev. B12, 2682(1975).

[30] A. Birnboim, Phys. Rev. B14, 2587(1976).

[31] M. Peter, W. Klose, G. Adam, P. Entel and E. Kundla, Helv. Phys. Acta 47, 807(1974); M. Peter, J. Ashkenazi and M. Decorogna, Helv. Phys. Acta 50, 267(1977).

[32] C.M. Varma and W. Weber, Phys. Rev. Lett. 39, 1094(1977).

[33] C.M. Varma, P. Vashishta, W. Weber and E.I. Blount, Solid State Commun. 27, 919(1978).

[34] C.M. Varma, E.I. Blount, P. Vashishta and W. Weber, Phys. Rev. B19, 6130(1979).

[35] C.M. Varma and W. Weber, Phys. Rev. B19, 6142(1979).

[36] W.E. Pickett and B.L. Gyorffy, in "Superconductivity in d and f-band Metals" Ed. D.H. Doughlass (Plenum Press, New York, 1976)p.251.

[37] C.M. Varma and R.C. Dynes, in "Superconductivity in d- and f-band Metals" Ed. D.H. Doughlass (Plenum Press, New York, 1976)p.507.

[38] J.R. Anderson, D.A. Papaconstantopoulos, J.W. McCaffrey and J.E. Shirber, Phys. Rev. B7, 5115(1973).

[39] L.F. Mattheiss, Phys. Rev. B1, 373(1970).

[40] W. Kohn, Phys. Rev. Lett. 2, 393(1959); A.W. Overhauser, Phys. Rev. Lett. 4, 415(1960).

[41] B.J. Higuera, F.R. Brotzen, H.G. Smith and N. Wakabayashi, Phys. Rev. B31, 730 (1985).

[42] C. Stassis, D. Arch, J. Zarestky, O.D. McMasters and B.N. Harmon, Solid State Commun. 35, 259(1980).

[43] A.L. Simons and C. M. Varma, Solid State Commun. 35, 317(1980).

[44] W. Weber, P. Roedhammer, L. Pintschovius, W. Reichardt, F. Gompf and A.N. Christensen, Phys. Rev. Lett. 43, 868(1979).

[45] M. Gupta and A. J. Freeman, Phys. Rev. Lett. 37, 364(1976); Phys. Rev. B14, 5205(1976).

[46] B.M. Klein, D.A. Papaconstantopoulos and L.L. Boyer, Solid State Commun. 20, 937(1976).

[47] L.E. Toth, "Transition Metal Carbides and Nitrides"(Acad. Press, New York, 1971).

[48] B. Splettstösser, Z. Phys. B26, 151(1977).

[49] B.M. Klein, D.A. Papaconstantopoulos and L.L. Boyer, Phys. Rev. B22, 1946(1980).

[50] W. Hanke, J. Hafner and H. Bilz, Phys. Rev. Lett. 37, 1560(1976).

[51] W. Hanke, J. Hafner and H. Bilz, in "International Conference on Low Lying Vibrational Modes and their Relationship to Superconductivity and Ferroelectricity" (San Juan, Puerto Rico, 1975).

[52] A.J. Freeman, M. Gupta, H.W. Myron, J. Rath and T.J. Watson-Yang, in " Lattice Dynamics" Ed. M. Balkanski (Flammarion, Paris, 1977)p.204.

[53] A.J. Freeman, T.J. Watson-Yang and J. Rath, J. Mag. and Magn. Materials. 12, 140(1982).

[54] R.P. Gupta and A.J. Freeman, Phys. Rev. B13, 4376(1976).

[55] W.A. Kamitakahara, B.N. Harmon, J.C. Traylor, L. Kopp, H.R. Shanks and J. Rath, Phys. Rev. Lett. 36, 1393(1976).

[56] B.M. Powell, P. Martel and A.D.B. Woods, Phys. Rev. 171, 727(1968).

Chapter 6

PHONON ANOMALIES AND GENERALIZED SUSCEPTIBILITY

Contents

List of Figures

Chapter 6

PHONON ANOMALIES AND GENERALIZED SUSCEPTIBILITY

Microscopic theories and microscopic models of lattice dynamics, described in chapters 4 and 5, show that the characterstics of phonon spectrum of d and f-band metals are intimately related to dielectric function $\epsilon(\vec{q} + \vec{G}, \vec{q} + \vec{G}')$. Therefore various attempts have been made to compute $\epsilon(\vec{q} + \vec{G}, \vec{q} + \vec{G}')$ and to explain the anomalies and other phonon related instabilities in these materials. The dielectric function involves three factors: (i) Electronic band structure, (ii) overlap matrix elements and (iii) the exchange and correlation interactions between the conduction electrons and between the core and conduction electrons. In literature, there does not exist much discussion about the exchange and correlation interactions between the quasilocalized d and f-electrons.Therefore one uses free particle exchange and correlation interactions (chapter 4, section 4.1.3) in the calculation of phonon frequencies of these meterials. However, there exit explicit calculations of electronic energy band structure and overlap matrix elements. The positions of broad and sharp maxima and minima obtained in the calculations of dielectric function are correlated with the positions of phonon anomalies. The structures in the dielectric function become apparent in the calculations of generalized susceptibility (or polarizability) func-

413

tion.

In this chapter we present the *ab initio* calculations of generalized susceptibility function and establish a relation between the phonon anomalies and electronic and crystal structures of d and f-band materials.

6.1   Static Susceptiblity Function

The susceptiblity function defined in Eq.(4.46) can be written as

$$\chi^0(\vec{q}+\vec{G}, \vec{q}+\vec{G}') \;=\; \sum_{\mu\mu'} \chi^0_{\mu\mu'}(\vec{q}+\vec{G}, \vec{q}+\vec{G}') \tag{6.1}$$

where

$$\chi^0_{\mu\mu'}(\vec{q}+\vec{G}, \vec{q}+\vec{G}') \;=\; \sum_{\vec{k}\vec{k}'} \left[\frac{f_{\vec{k}\mu} - f_{\vec{k}'\mu'}}{E_{\vec{k}'\mu'} - E_{\vec{k}\mu}} \right] M_{\vec{k}\mu,\vec{k}'\mu'}(\vec{q}+\vec{G}, \vec{q}+\vec{G}') \tag{6.2}$$

and

$$M_{\vec{k}\mu,\vec{k}'\mu'}(\vec{q}+\vec{G}, \vec{q}+\vec{G}') \;=\; < \psi_{\vec{k}\mu} \mid e^{-i(\vec{q}+\vec{G})\cdot\vec{r}} \mid \psi_{\vec{k}'\mu'} >$$
$$\times < \psi_{\vec{k}'\mu'} \mid e^{i(\vec{q}+\vec{G}')\cdot\vec{r}} \mid \psi_{\vec{k}\mu} > \tag{6.3}$$

with $\vec{k}' = \vec{k} + \vec{q} + \vec{G}''$. Equation (6.1) is the same as Eq.(4.46) except a negative sign. In Eq.(6.2) the quantity in the square brackets is band structure part and $M_{\vec{k}\mu,\vec{k}'\mu'}(\vec{q}+\vec{G}, \vec{q}+\vec{G}')$ is overlap part. Equation (6.1) consists of both the diagonal and non-diagonal parts of susceptibility function. It is too difficult to compute the non-diagonal part of $\chi^0(\vec{q}+\vec{G}, \vec{q}+\vec{G}')$ including both the band structure and overlap parts explicitly. Therefore we present the calculations for diagonal part first and make the comments about the non-diagonal part wherever the calculations are available.

The diagonal part of susceptibility function is

$$\chi^0(\vec{Q},\vec{Q}) \;=\; \sum_{\vec{k}\vec{k}'} \sum_{\mu\mu'} \left[\frac{f_{\vec{k}\mu} - f_{\vec{k}'\mu'}}{E_{\vec{k}'\mu'} - E_{\vec{k}\mu}} \right] M_{\vec{k}\mu,\vec{k}'\mu'}(\vec{Q},\vec{Q}). \tag{6.4}$$

Further we restrict the wave vector $\vec{Q}$ in the first Brillouin zone by subtracting an appropriate reciprocal lattice vector. The electrons readjust themselves by making virtual transitions in the same energy band and between different energy bands. Thus separating the intra-band ($\mu = \mu'$) and inter-band ($\mu \neq \mu'$) parts of $\chi^0(\vec{Q}, \vec{Q})$, Eq.(6.4) simplifies as

$$\chi^0(\vec{q}, \vec{q}) \;=\; \chi^0_{intra}(\vec{q}, \vec{q}) + \chi^0_{inter}(\vec{q}, \vec{q}) \tag{6.5}$$

where

$$\chi^0_{intra}(\vec{q}, \vec{q}) \;=\; \sum_{\vec{k}\mu} \left[\frac{f_{\vec{k}\mu} - f_{\vec{k}+\vec{q}\mu}}{E_{\vec{k}+\vec{q}\mu} - E_{\vec{k}\mu}} \right] M_{\vec{k}\mu, \vec{k}+\vec{q}\mu}(\vec{q}, \vec{q}), \tag{6.6}$$

$$\chi^0_{inter}(\vec{q}, \vec{q}) \;=\; \sum_{\vec{k}} \sum_{\mu \neq \mu'} \left[\frac{f_{\vec{k}\mu} - f_{\vec{k}+\vec{q}\mu'}}{E_{\vec{k}+\vec{q}\mu'} - E_{\vec{k}\mu}} \right] M_{\vec{k}\mu, \vec{k}+\vec{q}\mu'}(\vec{q}, \vec{q})$$

$$\tag{6.7}$$

as $\vec{k}' = \vec{k} + \vec{q}$. As $| \vec{q} | \to 0$, $M_{\vec{k}\mu, \vec{k}+\vec{q}\mu'} = \delta_{\mu\mu'}$ due to orthogonality condition of Bloch functions. Therefore χ^0_{inter} vanishes in the limit $| \vec{q} | \to 0$ while χ^0_{intra} remains finite and becomes equal to electron density of states at Fermi energy $D(E_F)$ (Appendix A). Thus

$$\lim_{q \to 0} \chi^0(\vec{q}, \vec{q}) \;=\; D(E_F). \tag{6.8}$$

The intra and inter-band contributions to $\chi^0(\vec{q}, \vec{q})$ are evaluated in chapter 4 in the non-interacting energy band scheme [1]-[5]. In Eq.(6.7) the band structure part gives resonance like effect whenever $E_{\vec{k}\mu}$ and $E_{\vec{k}+q,\mu'}$ are in the close vicinity. These effects may appear in the calculation of $\chi^0(\vec{q}, \vec{q})$ in the form of peaks and kinks. The overlap part involves the square of matrix elements and its magnitude will depend upon the characteristics of wave function. For example, in the free electron approximation for a single band, $M_{\vec{k}, \vec{k}+\vec{q}}$ will be q-independent constant. In general $M_{\vec{k}, \vec{k}+\vec{q}}$ will have sinusoidal character, it may not generate new peaks and kinks but may change the magnitude of $\chi^0(\vec{q}, \vec{q})$. Thus $M_{\vec{k}, \vec{k}+\vec{q}}$ will augment the strength of peaks and kinks generated by band structure part in the calculation of $\chi^0(\vec{q}, \vec{q})$.

If the effect of overlap matrix elements is neglected by keeping $M_{\vec{k}\mu,\vec{k}+\vec{q}\mu'} = 1$, the band structure part of susceptibility function becomes

$$\chi^{0b}(\vec{q}) \;=\; \sum_{\vec{k}}\sum_{\mu\mu'} \frac{f_{\vec{k}\mu} - f_{\vec{k}+\vec{q}\mu'}}{E_{\vec{k}+\vec{q}\mu'} - E_{\vec{k}\mu}}. \qquad (6.9)$$

Thus $\chi^{0b}(\vec{q})$ of Eq.(6.9) is periodic in the reciprocal space while $\chi^0(\vec{q},\vec{q})$ given in Eq.(6.5) is not periodic due to the presence of overlap matrix elements. The $\chi^{0b}_{inter}(\vec{q})$ does not vanish at $q = 0$ due to absence of overlap part while $\chi^{0b}_{intra}(\vec{q}) = D(E_F)$. Therefore $\chi^{0b}(\vec{q})$ does not reduce exactly to $D(E_F)$ at $q = 0$. In a single band model

$$\chi^{0b}(\vec{q}) = 2 \sum_{\vec{k}} \frac{1}{E_{\vec{k}+\vec{q}} - E_{\vec{k}}}. \qquad (6.10)$$

Equation (6.10) involves only intra-band transitions, therefore

$$\lim_{q\to 0} \chi^{0b}(\vec{q}) = D(E_F). \qquad (6.11)$$

In general the phonon anomalies are explained with the help of resonant screening which arise due to electronic transitions from one state to another on the Fermi surface. It is evident from Eq.(6.2) that the large contribution to the susceptibility function is from those values of $\vec{q}$ which connect the states having nearly the same energy but different occupancies. The polarizability (dielectric) function given in Eq.(4.53) has the logarithmic singularity at $q = 2k_F$ if the Fermi surface is spherical. In general this anomalous behaviour may also occur at the extremal portions of Fermi surface [6] (Kohn anomalies) or at the large flat and parallel portions of Fermi surface which may be electron or hole sheets or electron or hole pockets [7]. The later is called the nesting of Fermi surface. We illustrate below these terms by diagrams and simple arguments.

The Fermi surface is spherical for free electron gas as shown in Fig (6.1(a)). The energy states M_1 and M_2 have the same energy E_F and these are separated by wave vector $q = 2k_F$. The tangents at M_1 and M_2 are parallel to each other. Therefore electrons get nested between these two states. Thus one expects the anomalies at $q = 2k_F$ (logarithimic singularity, see Eq.(4.53)). In some simple metals, the Fermi surface is nearly spherical. However, in d and f-band metals the Fermi surface is distortred, anisotropic and extended to more than first BZ. Even for copper, the Fermi surface consists of necks which are sticking out towards the hexagonal faces of BZ. The electron density of states below Fermi surface is not uniform. Theses factors give rise to electron pockets (or sheets) and hole pockets (or sheets). Figure (6.1(b)) shows the (001) section of Fermi surface of Cr [8]. The Fermi surface consists of electron octahedra around Γ ponit (I), electrin balls on the [100] axis (IV), hole octahedra around H points (II) and hole surface around N points (III). The electron and hole octahedra exhibit the nesting property required for antiferromagnetism of Cr. The quantity $\vec{q} = (2\pi/a)(1-\delta,0,0)$ is the nesting wave vector and $(1-\delta)$ varies from 0.976 to 0.955 in the (100) plane and from 0.974 to 0.963 in the plane displaced by the vector $(2\pi/a)(0,0,1/24)$. The experimental value is 0.963 close to Néel temperature and 0.951 at lower temperature [9].

To discuss the phonon anomalies we cinsider a hypothetical Fermi surface shown in Fig. (6.2). Figure (6.2(a)) shows the extremal portions of Fermi surface A, B and C and Fig.(6.2(b)) its cross-section. Points M_1 and M_2 on the Fermi surface are such that the planes tangential to the Fermi surface at M_1 and M_2 are parallel. The anomaly is expected to occour at the points (Fig.6.2(b)) where

$$\vec{q} = \vec{k}_1 - \vec{k}_2. \tag{6.12}$$

Here $\vec{k}_1$ and $\vec{k}_2$ define two points M_1 and M_2 on the Fermi surface with parallel tangent planes. The locus of the phonon wave vectors satisfying Eq.(6.12) defines the so called Kohn surface. The character of the phonon anomaly depends on the shape of Fermi surface. Let S_1 be the portion of Fermi surface in the neighbourhood of M_1. By translation of vector $\vec{q}$ one obtains a surface say S_1' (shown in inset) tangential to surface S_2 in the neighbourhood of M_2. Two cases may arise: (i) The two surfaces (S_1') and (S_2) just touch at one point M_2 which becomes common to both the surfaces. This is the situtation for free electron gas. (ii)The two surfaces (S_1') and (S_2) intersect and the loci of the points of intersection will define two curves say D_1 and

D_2. Then the denominator $(E_{\vec{k}+\vec{q}} - E_{\vec{k}})$ vanishes along D_1 and D_2 lines. If the two surfaces (S_1') and (S_2) intersect along the neck, the denominator will vanish along the neck. Such an anomalous behaviour gives a cusp in the susceptibility function $\chi^0(\vec{q})$.

Other important physical effects which alter the polarizability function are exchange and correlation interactions of conduction electrons. These interactions are supposed to be responsible for magnetic properties of transition metals, rare earth metals and their compounds and alloys. Let $\vec{H}(\vec{r})$ be the applied magnetic field in the z-direction which induces magnetization $\triangle \vec{M}(\vec{r})$. The z-component of magnetization $\triangle \vec{M}(\vec{r})$ is related to susceptibility function in the reciprocal space as [10]

$$\triangle M_z(\vec{q} + \vec{G}) = \sum_{\vec{G}'} \chi^0(\vec{q} + \vec{G}, \vec{q} + \vec{G}') \left[\delta_{0G'} H(\vec{q}) + I \triangle M_z(\vec{q} + \vec{G}) \right]$$

(6.13)

where the parameter I takes into account the exchange and correlation interactions. For the diagonal part of susceptibility and restricting $\vec{q}$ in the first BZ

$$\triangle M_z(\vec{q}) = \chi^{0b}(\vec{q}) \left[H(\vec{q}) + I \triangle M_z(q) \right].$$

(6.14)

The exchange enhanced susceptiblity function is defined as

$$\triangle M_z(\vec{q}) = \chi_{en}(\vec{q}) H(\vec{q}).$$

(6.15)

Comparison of Eqs.(6.14) and (6.15) gives

$$\chi_{en}(\vec{q}) = \frac{\chi^0(\vec{q})}{1 - I\chi^0(\vec{q})}.$$

(6.16)

The effect of local fields which arise due to non-diagonal part of susceptibility function is not accounted for in Eq.(6.16). To include this effect we consider the general expression given in Eq.(6.13). To make the calculations tractable for exchange enhanced susceptibility, we simplify the susceptibilty matrix for d and f-band metals given in

Eq.(4.110). In the LCAO approximation and considering only the intra-atomic overlap integrals Eq.(4.110) simplifies as

$$\chi^0(\vec{q}+\vec{G},\vec{q}+\vec{G}') \;=\; A(\vec{q}+\vec{G})A^*(\vec{q}+\vec{G}')\chi^{0b}(q). \qquad (6.17)$$

Equation (6.17) includes both the intra and inter-band contributions in the band structure part $\chi^{0b}(\vec{q})$. The factorization of overlap and band structure parts as given in Eq.(6.17) is not possible in the mixed band scheme as discussed in chapter 4 in section (4.3.1). In these circumstances the factorization ansatz due Gupta and Sinha [10, 11] can be used to represent $\chi^0(\vec{q}+G,\vec{q}+\vec{G}')$ in the analytic form given in Eq. (6.17). This procedure has been adopted by Prakash and co-workers [2]-[5] for the calculation of dielectric function in the non-interacting band scheme. To obtain the solution of Eq.(6.13) we assume that

$$\triangle M_z(\vec{q}+\vec{G}) \;=\; A(\vec{q}+\vec{G})W(\vec{q}) \qquad (6.18)$$

where $W(\vec{q})$ is determined using Eqs.(6.17) and (6.18) in Eq.(6.13). These calculations give

$$W(\vec{q}) \;=\; \frac{A^*(\vec{q})\chi^{0b}(\vec{q})H(\vec{q})}{1-I\sum_{G'}\chi^0(\vec{q}+\vec{G}',\vec{q}+\vec{G}')}. \qquad (6.19)$$

Equations (6.19) and (6.18) give the exchange enhanced diagonal part of susceptibility function

$$\chi_{en}(\vec{q},\vec{q}) \;=\; \frac{A(\vec{q})\chi^{0b}(\vec{q})A^*(\vec{q})}{1-IF(\vec{q})}$$

$$\;=\; \frac{\chi^0(\vec{q},\vec{q})}{1-IF(\vec{q})} \qquad (6.20)$$

where

$$F(\vec{q}) \;=\; \sum_G \chi^{0b}(\vec{q}+\vec{G},\vec{q}+\vec{G}). \qquad (6.21)$$

The local field effects in Eq.(6.20) are included through the factor $F(\vec{q})$ in the denominator. Thus $\chi_{en}(\vec{q},\vec{q})$ given in Eq.(6.20) is anomalous when

$$IF(\vec{q}) \;=\; 1 \qquad (6.22)$$

which is the condition for magnetic ordering i.e. for the existence of antiferromagnetic and ferromagnetic phases. The peak in $\chi_{en}(\vec{q}, \vec{q})$ occurs when $F(\vec{q})$ is maximum. This condition determines the value of spin density wave vector $\vec{q}_m$. The magnetic ordering stabilizes when

$$IF(\vec{q}) \geq 1. \tag{6.23}$$

The above facts indicate that the peaks in the susceptibility function are due to two physical effects: (i) Nesting of Fermi surface and (ii) the exchange enhancement effects. These peaks in the susceptibility function cause the electronically driven instabilities in these metallic systems.

6.2   Calculation of Susceptibility Function

The calculation of $\chi^0(\vec{q} + \vec{G}, \vec{q} + \vec{G}')$ involves the knowledge of energy eigenvalues and eigenfunctions of crystalline solid and the precise and fast method of numerical integration in the BZ. In the following, we discuss these aspects.

6.2.1   Electron Energy Bands

The Schrödinger equation for a crystalline solid in the Rydberg units $(\hbar = 2m = 1)$ is

$$\left[-\nabla^2 + V(\vec{r})\right] \psi_{\vec{k}\mu}(\vec{r}) \; = \; E_{\vec{k}\mu} \psi_{\vec{k}\mu}(\vec{r}) \tag{6.24}$$

where $\psi_{\vec{k}\mu}(\vec{r})$ and $E_{\vec{k}\mu}$ are the eigenfunctions and eigenvalues respectively for the band μ and $V(\vec{r})$ is the self-consistently calculated periodic crystal potential. A proper choice of crystal potential and self-consistency in the calculations is required to calculate $\psi_{\vec{k}\mu}$ and $E_{\vec{k}\mu}$. In most of the reliable calculations of energy bands of metallic crystals $V(\vec{r})$ is replaced by a muffin-tin potential defined as

$$
\begin{aligned}
V(\vec{r}) \; &= V(r) \quad \text{for} \quad \text{r} < \text{R}_i \\
&= V_0 \quad \text{for} \quad \text{r} \geq \text{R}_i
\end{aligned}
\tag{6.25}
$$

where $V(r)$ is spherically symmetric potential inside the sphere of radius R_i around the center of Wigner-Seitz (WS) cell of each atom. Outside these spheres, the potential has the constant value V_0. The radius R_i is arbitrary such that any two consecutive spheres do not overlap. Normally the energy scales are shifted such that V_0 is zero.

Depending upon the choice of wave function, there exist standard techniques of calculating the electronic energy band structure of crystalline solids. The augmented plane wave (APW) method [12], Korringa-Kohn-Rostoker (KKR) method [13, 14, 15] and the linear combination of atomic orbitals (LCAO) method [16]-[24] are mostly used to calculate the energy band structure of transition metals and their compounds. In the APW method, the electron wave function is written as [12]

$$\phi(\vec{k}, \vec{r}) \;=\; \exp(i\vec{k} \cdot \vec{r})\theta(\vec{r} - \vec{R_i}) + \sum_{lm} a_{lm} Y_{lm}(\theta_{\vec{k}}, \phi_{\vec{k}}) R_l(E, r)\theta(R_i - r)$$

$$(6.26)$$

where

$$a_{lm} \;=\; 4\pi i^l Y_{lm}^*(\theta_{\vec{k}}, \phi_{\vec{k}}) \left[j_l(|kR_i|)/R_l(E, R_i)\right]. \qquad (6.27)$$

Here $Y_{lm}(\theta_{\vec{k}}, \phi_{\vec{k}})$ are spherical harmonics, $R_l(E, r)$ is radial part of wave function, $\theta(R_i - r)$ is Heaviside unit step function and $j_l(kR_i)$ is spherical Bessel function of order l. Here the radial part of wave function is an explicit function of energy E. The coefficients a_{lm} are determined by the condition of continuity of $\phi(\vec{k}, \vec{r})$ across the muffin-tin spheres. The Bloch function $\psi_{\vec{k}\mu}(\vec{r})$ is generated by taking the linear combination of $\phi(\vec{k}, \vec{r})$ which is given as

$$\psi_{\vec{k}\mu}(\vec{r}) \;=\; \sum_{\vec{G}} a_\mu(\vec{k}, \vec{G})\phi(\vec{k} + \vec{G}, \vec{r}). \qquad (6.28)$$

Using Eq.(6.28) in Eq.(6.24), the Schrödinger equation is solved self-consistently to obtain $\psi_{\vec{k}\mu}(\vec{r})$ and $E_{\vec{k}\mu}$.

Another popular method for obtaining energy bands of crystalline solids is KKR Green's function method. In this method Eq.(6.24) is written as

$$\left[\nabla^2 + E\right]\psi(\vec{r}) \;=\; V(\vec{r})\psi(\vec{r}) \qquad (6.29)$$

where the subscript $\vec{k}\mu$ of E and $\psi(\vec{r})$ is dropped for simplicity. The Green's function $G(\vec{r},\vec{r}')$ for Eq.(6.29) is defined as

$$\left[\nabla^2 + E\right]G(\vec{r},\vec{r}') = \delta(\vec{r},\vec{r}'). \tag{6.30}$$

$G(\vec{r},\vec{r}')$ must satisfy the periodic boundary condition. Therefore according to Bloch theorem

$$G(\vec{r},\vec{r}' + \vec{R}_l) = \exp(i\vec{k}\cdot\vec{R}_l)G(\vec{r},\vec{r}'). \tag{6.31}$$

Once the Green's function is calculated with the help of Eqs.(6.30) and (6.31), the solution of Eq.(6.29) for the wave function is

$$\psi(\vec{r}) = \int_{\Omega_0} G(\vec{r},\vec{r}')V(\vec{r}')\psi(\vec{r}')d\vec{r}' \tag{6.32}$$

where the integral is over the atomic volume Ω_0. Kohn and Rostoker [14] used the variational method to solve the integral equation (6.32). The trial function $\psi(\vec{r})$ which is a linear combination of finite number of basis functions $\phi_n(\vec{k},\vec{r})$ with undetermined coefficients, is written as

$$\psi(\vec{r}) = \sum_{n=1}^{N} C_{n\mu}(\vec{k})\phi_n(\vec{k},\vec{r}). \tag{6.33}$$

The variational parameter Λ is defined as

$$\Lambda(\psi,\vec{k},E) = \sum_{nn',\mu\mu'} C_{n\mu}^*(\vec{k})\Lambda_{nn'}C_{n'\mu'}(\vec{k}) \tag{6.34}$$

where

$$\Lambda_{nn'} = \int_{\Omega_0} d\vec{r}\,\phi_n^*(\vec{k},\vec{r})V(\vec{r})\left[\phi_{n'}(\vec{k},\vec{r}) - \int_{\Omega_0} d\vec{r}'\,G(\vec{r},\vec{r}')V(\vec{r}')\phi_{n'}(\vec{k},\vec{r}')\right]. \tag{6.35}$$

The partial derivatives of Λ with respect to C_n^* must vanish for each n' i.e.

$$\sum_{n'=1}^{N} \Lambda_{nn'}C_{n'}(\vec{k}) = 0. \tag{6.36}$$

Equation (6.36) is a set of N linear homogeneous equations. The non-trivial solutions of determinental equation

$$\det(\Lambda_{nn'}) = 0 \qquad (6.37)$$

will give the energy eigenvalues and eigenvectors and hence the energy bands of the solid. If $\phi_n(\vec{k}, \vec{r})$ constitute a complete set, the calculated eigenvalues will approach the correct limit. These energy eigenvalues may further be used to determine the coefficients $C_{n\mu}(\vec{k})$ and hence the eigenfunctions $\psi_{\vec{k},\mu}(\vec{r})$.

The APW and KKR methods are similar and give identical results if the same muffin-tin potential is used [25]. In the KKR method the dimensions of secular determinant are smaller than in the APW method and therefore requires less computer time. However the matrix elements in the KKR method are considerably involved.

If the electron-ion potential is strongly attractive, the LCAO method, usually called the tight-binding (TB) method, is also used to study the electronic band structure of metallic solids [16]-[19]. This method gives only the narrow d and f-bands. However in the transition and rare earth metals the broad s band also exists which hybridizes with the d and f-bands. Therefore the LCAO method is not very suitable for describing the energy band structure of these materials. The combined representation of orthogonalized plane wave (OPW) and LCAO [26],[27] is also developed to compute the energy band structure of these materials. In this representation, s-electrons are represented in terms of OPWs, d-electrons are represented in the LCAO and the energy bands are calculated by solving Schrödinger equation. This is a complete subject in itself and the reader is referred to the references [12] and [21].

6.2.2   Brillouin Zone Integration

In the calculation of susceptibility function the sum is over all those values of $\vec{k}$ in the BZ for which $E_{\vec{k}\mu}$ lies below the Fermi energy E_F. However, in the energy band structure calculations, $E_{\vec{k}\mu}$ is calculated only at the high symmetry $\vec{k}$ points in the BZ. Therefore $E_{\vec{k}\mu}$ at any arbitrary $\vec{k}$ point in the BZ is generated through the interpolation scheme.

These interpolated values of $E_{\vec{k}\mu}$ are used over a given mesh of $\vec{k}$ points in the BZ in Eq.(6.4). This computational procedure requires a highly accurate and efficient computational technique. The values of $E_{\vec{k}\mu}$ possess the full rotational and translational symmetry of crystal and this property allows to divide the BZ into subzones such that all the subzones are identical and exhibit the symmetry properties of crystal lattice. Such a subzone is called Irreducible Brillouin Zone (IBZ) which are given in Figures (2.4), (2.8) and (2.11c) for fcc, bcc and hcp lattices respectively. In the literature, following methods are proposed for BZ integration:

1. Summation methods [28]-[30],
2. Quasi-analytic methods [31]-[35],
3. Quadratic (QUAD) methods [33]-[35],
4. Tetrahedron methods [36]-[37].

The summation methods adopt Simpson or equivalent sum rules. These methods are slow on computer and break down in regions where the energy denominator is small [38]. By increasing the number of mesh points in the regions of small energy denominators, computer time is substantially increased and the results remain, in general, inconsistent. This difficulty is overcome in the quasi-analytical methods of integration. In these methods the IBZ is further divided into microzones and energy is further interpolated at all the points in these microzones. The integration is done over all the microzones separately and then added to get the total contribution. Gilat and Raubenheimer [32] have chosen these microzones as cubes of equal volume. At the center of each cube $E_{\vec{k}\mu}$ and its gradient with respect to $\vec{k}$ are calculated by interpolating the given $E_{\vec{k}\mu}$. At other points in the cube, $E_{\vec{k}\mu}$ is obtained by assuming linear variation of $E_{\vec{k}\mu}$ as

$$E_{\vec{k}\mu} \;=\; E_{\vec{k}_c\mu} + (\vec{k} - \vec{k}_c) \cdot \vec{\nabla}_{\vec{k}} E_{\vec{k}\mu} \,|_{\vec{k}=\vec{k}_c} \tag{6.38}$$

where $\vec{k}_c$ is the center of the cube. This analytic interpolation makes the integral in the cube very fast. The size and hence the number of cubes in the IBZ are varied untill the convergence is achieved. Lipton and Jacob [38] used the values of $E_{\vec{k}\mu}$, given in Eq.(6.38), at the corners of the cube to interpolate the values of $E_{\vec{k}\mu}$ inside the cube. This procedure further reduces the numerical computational time to a great extent.

In the QUAD scheme of integration [33], the quadratic term is also added in the expansion given in Eq.(6.38). This interpolation scheme is more accurate, although it requires more computational time. Considering the desired accuracies in the calculations in the different regions of IBZ, the combination of linear and quadratic interpolation schemes is also used [34]- [35].

Jepsen and Anderson [36] proposed the tetrahedron method for numerical integration which has been used by many authors [37]-[39] for integration in the IBZ. In this method each cube is further splitted into six tetrahedrons. $E_{\vec{k}\mu}$ is expanded upto linear term inside each tetrahedron as given in Eq.(6.38). The known values of $E_{\vec{k}\mu}$ at the corners of tetrahedron are used in the calculations. The corner energies of a tetrahedron are arranged either in the increasing or in the decreasing order. The coefficients of expansion are determined uniquely in terms of energies and co-ordinates of the corners [40]. Let $k_i(i = 0, 1, 2, 3)$ are the co-ordinates of four corner points of a tetrahedron and $E_{\vec{k}} = E_i$ are the correspoding energies as shown in Fig.(6.3(a)). Here E_0, E_1, E_2 and E_3 are in the increasing order. The linearly interpolated energies within the tetrahedron are given as

$$E_{\vec{k}\mu} \;\; = \;\; E_0 + (\vec{k} - \vec{k}_0) \cdot \vec{\epsilon} \qquad (6.39)$$

where

$$\vec{\epsilon} \;\; = \;\; \frac{1}{\vec{a} \cdot (\vec{b} \times \vec{c})} \left[(E_1 - E_0)(\vec{b} \times \vec{c}) + (E_2 - E_0)(\vec{c} \times \vec{a}) \right.$$
$$\left. + (E_3 - E_0)(\vec{a} \times \vec{b}) \right]. \qquad (6.40)$$

Here $\vec{a}$, $\vec{b}$ and $\vec{c}$ are the primitive translation vectors of crystal lattice and $(\vec{a} \cdot (\vec{b} \times \vec{c})) = 6v_t$ where v_t is the volume of a tetrahedron. Comparing Eqs.(6.38) and (6.39) the gradient of energy can be written as

$$\vec{\nabla}_{\vec{k}} E_{\vec{k}\mu} = \;\; \vec{\epsilon} = \;\; \frac{1}{6v_t} \left[(E_1 - E_0)(\vec{b} \times \vec{c}) + (E_2 - E_0)(\vec{c} \times \vec{a}) \right.$$
$$\left. + (E_3 - E_0)(\vec{a} \times \vec{b}) \right]. \qquad (6.41)$$

The linear interpolation of energies makes the constant energy surface as a plane surface which cuts the tetrahedron in to three possible ways as shown in Fig.(6.3(b)). For $E_0 < E_F < E_1$ the plane of intersection is a triangle K_{11}, K_{12}, K_{13} which has the area

$$A_1 = \frac{(E_F - E_0)}{2(E_1 - E_0)(E_2 - E_0)(E_3 - E_0)}$$
$$\times \mid (E_1 - E_0)(\vec{b} \times \vec{c}) + (E_2 - E_0)(\vec{c} \times \vec{a}) + (E_3 - E_0)(\vec{a} \times \vec{b}) \mid.$$

$$(6.42)$$

Similar results hold when $E_1 < E_F < E_2$ and $E_2 < E_F < E_3$. This method of calculating the susceptibility function is useful when the Fermi surface is non-spherical and extends to more than one zone. In this procedure of integration over IBZ, full volume of some of the tetrahedrons and the fractional volume of other tetrahedrons contribute towards the susceptibility function. The fractional volume of tetrahedron which contributes towards $\chi^0(\vec{q} + \vec{G}, \vec{q} + \vec{G}')$ is determined by constant energy plane surfaces corresponding to $E_{\vec{k}\mu}$ and $E_{\vec{k}'\mu'}$ which are nondegenrate.

There are five different possibilities of energy ranges which may occur in the process of numerical integration. These different ranges are described below and the corresponding constant energy planes are shown in Fig.(6.3(b)).

(i) When $E_F \leq E_0$, entire tetrahedron is unoccupied and therefore does not contribute.

(ii) When $E_0 < E_F < E_1$, Fermi surface intersects the sides of tetrahedron at the points $\vec{k}_{11}$, $\vec{k}_{12}$ and $\vec{k}_{13}$. The occupied region of tetrahedron is the portion on the left side of the plane formed by $\vec{k}_{11}$, $\vec{k}_{12}$ and $\vec{k}_{13}$. This region itself forms another tetrahedron. The unoccupied portion is on the right side of this plane. This unoccupied region may be constructed as the sum of three tetrahedrons namely: $\vec{k}_3\vec{k}_1\vec{k}_{11}\vec{k}_2$ with $\vec{k}_{11}$ vortex and $\vec{k}_1\vec{k}_2\vec{k}_3$ basal plane, $\vec{k}_1\vec{k}_{11}\vec{k}_{12}\vec{k}_{13}$ with $\vec{k}_{12}$ vortex and $\vec{k}_1\vec{k}_{11}\vec{k}_{13}$ basal plane and $\vec{k}_1\vec{k}_2\vec{k}_{13}\vec{k}_{11}$ with $\vec{k}_{13}$ vortex and $\vec{k}_1\vec{k}_2\vec{k}_{11}$ basal plane.

(iii) For $E_1 < E_F < E_2$, Fermi surface intersects the two sides of tetrahedron at the points $\vec{k}_4, \vec{k}_5, \vec{k}_6$ and $\vec{k}_7$. The occupied portion of tetrahedron, on the left side of Fermi surface, is the sum of three tetrahedrons:

$(\vec{k}_0, \vec{k}_4, \vec{k}_5, \vec{k}_6)$, $(\vec{k}_0, \vec{k}_4, \vec{k}_6, \vec{k}_7)$ and $(\vec{k}_0, \vec{k}_5, \vec{k}_7, \vec{k}_4)$. The tetrahedrons can be constructed by choosing three different vortices. The unoccupied portion of tetrahedron on the right side of Fermi surface can also be taken as the sum of three tetrahedrons.

(iv) For $E_2 < E_F < E_3$, Fermi surface intersects the sides of tetrahedron at $\vec{k}_8, \vec{k}_9$ and $\vec{k}_{10}$. The occupied portion of tetrahedron is the sum of three tetrahedrons: $(\vec{k}_0, \vec{k}_1, \vec{k}_9, \vec{k}_2)$ with vortex $\vec{k}_0$ and basal plane $(\vec{k}_1, \vec{k}_9, \vec{k}_2)$, $(\vec{k}_1, \vec{k}_8, \vec{k}_9, \vec{k}_{10})$ with vortex $\vec{k}_1$ and basal plane $(\vec{k}_8, \vec{k}_9, \vec{k}_{10})$ and $(\vec{k}_2, \vec{k}_1, \vec{k}_9, \vec{k}_{10})$ with vortex $\vec{k}_2$ and basal plane $(\vec{k}_1, \vec{k}_9, \vec{k}_{10})$. The unoccupied portion is a single tetrahedron $(\vec{k}_3, \vec{k}_8, \vec{k}_9, \vec{k}_{10})$ with vortex $\vec{k}_3$ and basal plane $(\vec{k}_8, \vec{k}_9, \vec{k}_{10})$.

(v) For $E_3 \leq E_F$, the entire tetrahedron is occupied.

The above discussion shows that the occupied and unoccupied volumes are either a single tetrahedron or a sum of three tetrahedrons. Even at the finite temperature when the distortion of Fermi surface is very small, the same procedure of constructing the tetrahedrons can be adopted. Lehman and Taut [37] suggested a simpler geometry in which the occupied and unoccupied volumes are the difference of volumes of two tetrahedrons.

6.3   Results

6.3.1   Transition Metals

(a) bcc Structure

The bcc TMs are separated into two groups (Cr, Mo, W) and (V, Nb, Ta) due to their similarities in the electronic properties and phonon spectra. Evansosn et al.[41]-[42] calculated $\chi^0(\vec{q})$ using <u>abinitio</u> calculation of electron energy band structure. These authors assumed the band structure of all the six metals exactly similar to that of W [43] and adjusted E_F for each metal such that the correct number of conduction electrons per atom is reproduced. For Cr group of elements Evanson et al.[42] found number of small peaks along [00ξ] direction. However the sharp and prominent peak near H symmetry point is related to phonon softening which is observed near H point in the experimental phonon spectra of Cr [3.56, 3.57, 3.58] and Mo [3.62, 3.63, 3.64]. In the calcualtions for W, this peak is less pronounced [3.78]. In addition the exchange interaction parameter I for Cr is found large enough to satisfy the condition $I\chi^0_{max} \geq 1$ (Eq.6.16) which predicts a stable antiferromagnetic ground state of Cr.

Gupta and Sinha [10, 11] extended the calculations of susceptibility function for paramagnetic Cr. These authors used a fine mesh of points in the BZ integration and did the calculations by excluding the overlap matrix elements. A peak is found at $q = (2\pi/a)[0,0,0.88]$ which arises from the nesting of Fermi surface of Cr and can be correlated to the spin density wave vector $q_m = (2\pi/a)[0,0,0.952]$ at Nèel temperature T_N [44]. However due to neglect of overlap matrix elements $\chi^0(0)$ is found about 20 times larger than $(1/2)D(E_F)$. Figure (6.4(a)) shows the calculations of $\chi^0(\vec{q},\vec{q})$ for Cr where the overlap matrix elements are included. The intra and inter-band contributions are shown separately. The intra-band contribution is maximum at $|\vec{q}| = 0$ and decreases rapidly with increase of $\vec{q}$. The inter-band contribution is zero at $|\vec{q}| = 0$ and increases with the increase in $\vec{q}$ and shows a broad maximum. At large $\vec{q}$, the inter-band contribution dominates and intra-band contribution becomes almost negligible. It is found that the overlap matrix elements change the characteristics of susceptibility function $\chi^0(\vec{q},\vec{q})$ drastically. One gets the correct limit $(1/2)D(E_F)$ at $|\vec{q}| = 0$. However the inclusion of overlap matrix elements does not alter the position of peaks in $\chi^0(\vec{q},\vec{q})$. If a peak is at $\vec{q} = (2\pi/a)(0,0,1-\delta)$ there will be a nesting at $\vec{q} = (2\pi/a)(0,0,1+\delta)$ outside BZ, resulting in another peak although smaller in strength. The effect of temperature in $\chi^0(\vec{q},\vec{q})$ is included through Fermi-Dirac distribution function $f_{\vec{k}\mu}(T)$. These effects are found very small.

By further increasing the number of points in the mesh for numerical intigration the magnitude and shape of $\chi^0(\vec{q},\vec{q})$ in changed as shown in Fig. (6.4(b)). However the peak at the nesting wave vector continue to exist. This example illustrates the problem of inconsistency in the numerical integration due to small denominator $(E_{\vec{k}'\mu'} - E_{\vec{k}\mu})$. These peaks are yet to be correlated with the anomalies in the phonon spectra. Gupta and Sinha [10, 11] also calculated $\chi_{en}(\vec{q},\vec{q})$ using Eq. (6.20) which takes into account local field effects. The sum over $\vec{G}$ in $F(\vec{q})$

combines the two peaks at $(2\pi/a)[0,0,1\pm\delta]$ into one peak at the nesting wave vector and the susceptibility becomes periodic in the reciprocal space.

Evanson et al.[42] calculated $\chi^0(\vec{q})$ for vanadium group of transition metals in the $[\xi 00]$ direction and found two peaks: One near Γ and other near H symmetry points in the BZ. The peak near H symmetry point corresponds to the dip in the longitudinal branch of phonon spectra of Nb and NbMo alloys [3.63]. The peak near Γ symmetry point could not be correlated to any phonon anomaly. Perhaps the inclusion of overlap matrix elements may suppress this peak. Later many authors [45]-[48] calculated the static susceptibility for vanadium group of transition metals. Cooke et al. [45] calculated $\chi^0(\vec{q})$ for Nb, with and without overlap matrix elements. These authors found that the overlap matrix elements reduce $\chi^0(\vec{q})$ by an order of magnitude and introduce additional structures which contradict the predictions of Evanson et al.[42] and Gupta and Sinha [10, 11]. This allows us to say that the peak near Γ symmetry point in the calculation of Evanson et al. may be due to their integration procedure.

Figure (6.5) shows $\chi^0(\vec{q})$ for Nb and Mo due to Pickett and Allen [46] and Cooke et al [45]. Both the calculations agree reasonably well. A small hump in $\chi^0(\vec{q})$ for Nb around $\vec{q} = (2\pi/a)[0.7,0,0]$ corresponds to a dip in the longitudinal branch of phonon dispersion curves of Nb and Nb rich NbMo alloys [3.63]. However, the magnitude of hump is too small to be related to the magnitude of phonon anomaly. In the phonon spectra of Mo [3.64] there is an Kohn anomaly near the zone boundary in the $[\xi 00]$ direction. $\chi^0(\vec{q})$ for Mo shows a sharp increase of about 5% just at the position of Kohn anomaly. This increase is very sensitive to the small changes in E_F which is expected for a Kohn anomaly.

Pickett and Allen [46] found a relation between the static susceptibility $\chi^0(\vec{q})$ and the derivative of imaginary part of dynamical susceptibility $\mathrm{Im}\chi^0(\vec{q},\omega)$. The general expression for temperature dependent dynamical susceptiblity $\chi^0(\vec{q},\omega,T)$ can be written by generalizing

Eq.(6.1) for the diagonal part as

$$\chi^0(\vec{q},\omega,T) = \sum_{\vec{k}}\sum_{\mu\mu'} \frac{f_{\vec{k}\mu}(T) - f_{\vec{k}+\vec{q}\mu'}(T)}{E_{\vec{k}+\vec{q}\mu'} - E_{\vec{k}\mu} - \hbar\omega - i\delta} M_{\vec{k}\mu,\vec{k}+\vec{q}\mu'}(\vec{q}).$$

$$(6.43)$$

Here δ is an infinitismal small parameter for the adiabatic switching on of the field and ω is the frequency of the field. The temperature dependence is included through $f_{\vec{k}\mu}(T)$. $\chi^0(\vec{q},\omega,T)$ is a complex quantity and its real and imaginary parts can be separated by using the identity

$$\frac{1}{x \pm i\delta} = P(\frac{1}{x}) \mp i\pi\delta(x) \qquad (6.44)$$

where P denotes the principal value. Thus the real and imaginary parts of $\chi^0(\vec{q},\omega,T)$ can be written as

$$\mathrm{Re}\chi^0(\vec{q},\omega,T) = \sum_{\vec{k}}\sum_{\mu\mu'} \frac{f_{\vec{k}\mu}(T) - f_{\vec{k}+\vec{q}\mu'}(T)}{E_{\vec{k}+\vec{q}\mu'} - E_{\vec{k}\mu} - \hbar\omega} M_{\vec{k}\mu,\vec{k}+\vec{q}\mu'}(\vec{q}) \quad (6.45)$$

and

$$\mathrm{Im}\chi^0(\vec{q},\omega,T) = \pi\sum_{\vec{k}}\sum_{\mu\mu'} \left[f_{\vec{k}\mu}(T) - f_{\vec{k}+\vec{q}\mu'}(T) \right] M_{\vec{k}\mu,\vec{k}+\vec{q}\mu'}(\vec{q})$$
$$\times \delta(E_{\vec{k}+\vec{q}\mu'} - E_{\vec{k}\mu} - \hbar\omega). \qquad (6.46)$$

$\mathrm{Re}\chi^0(\vec{q},\omega,T)$ is related to $\mathrm{Im}\chi^0(\vec{q},\omega,T)$ through Kramers-Kronig relation [49] as

$$\mathrm{Re}\chi^0(\vec{q},\omega,T) = \frac{2}{\pi}\int_0^\infty \frac{\omega'\mathrm{Im}\chi^0(\vec{q},\omega',T)d\omega'}{\omega'^2 - \omega^2}. \qquad (6.47)$$

There have been several calculations of $\mathrm{Re}\chi^0(\vec{q},\omega,T)$ and its characteristic features are related to magnetic excitations [50]-[54].

The dynamical susceptibility for free electrons from Eqs.(6.45) and (6.46) at T=0 is

$$\mathrm{Re}\chi^0(\vec{q},\omega) = \sum_{\vec{k}}\sum_{\mu\mu'} \frac{f_{\vec{k}\mu}(0) - f_{\vec{k}+\vec{q}\mu'}(0)}{E_{\vec{k}+\vec{q}\mu'} - E_{\vec{k}\mu} - \hbar\omega} \qquad (6.48)$$

and

$$\mathrm{Im}\chi^0(\vec{q},\omega) \;=\; \pi \sum_{\vec{k}} \sum_{\mu\mu'} \left[f_{\vec{k}\mu}(0) - f_{\vec{k}+\vec{q}\mu'}(0) \right] \delta(E_{\vec{k}+\vec{q}\mu'} - E_{\vec{k}\mu} - \hbar\omega).$$

$$(6.49)$$

To establish a relation between $\chi^0(\vec{q})$ and imaginary part of dynamical susceptibility, we define a function

$$I(\vec{q}, E, E') = \sum_{\vec{k}} \sum_{\mu\mu'} \left[\delta(E_{\vec{k}\mu} - E)\delta(E_{\vec{k}+\vec{q}\mu'} - E') \right]. \qquad (6.50)$$

and rewrite Eq.(6.49) as

$$\mathrm{Im}\chi^0(\vec{q},\omega) \;=\; \pi \sum_{\vec{k}} \sum_{\mu\mu'} \left[\delta(E_{\vec{k}+\vec{q}\mu'} - E_{\vec{k}\mu} - \hbar\omega) - \delta(E_{\vec{k}\mu'} - E_{\vec{k}-\vec{q}\mu} - \hbar\omega) \right].$$

$$(6.51)$$

Taking the derivative of Eq.(6.51) with respect to ω and using the properties of Dirac-delta function, one can show that for low frequencies

$$\frac{1}{\pi}\frac{d}{d\omega}\mathrm{Im}\chi^0(\vec{q},\omega)|_{\omega=0} \;=\; I(\vec{q}, E_F, E_F). \qquad (6.52)$$

Thus for small value of ω, $\mathrm{Im}\chi^0(\vec{q},\omega)$ is linear in ω with slope $\pi I(q, E_F, E_F)$ which is a function of $\vec{q}$ and E_F and it is shown in Fig. (6.6). $I(\vec{q}, E_F, E_F)$ involves the electronic excitations only in the vicinity of E_F and can be regarded as a measure of electron-hole pair density at E_F. Near E_F, $I(\vec{q}, E_F, E_F)$ scales roughly as $[D(E_F)]^2$, however its BZ average is exactly $[D(E_F)]^2$. Thus the minimum of $I(\vec{q}, E_F, E_F)$ is related to minimum of $D(E_F)$ near E_F of Nb and Mo. The structure in $I(\vec{q}, E_F, E_F)$ versus $\vec{q}$ is correlated to Kohn anomaly in Mo near zone boundary. The calculated results show an empirical linear relation between $\chi^0(\vec{q})$ and $I(\vec{q}, E_F, E_F)$ as

$$\chi^0(\vec{q}) - \chi_c \;\cong\; 2\bar{\omega} I(\vec{q}, E_F, E_F) \qquad (6.53)$$

where the constant $\chi_c = 56(\mathrm{Ryd}^{-1}\mathrm{spin}^{-1})$ and $\bar{\omega} = 0.022$ Ryd. It is important to note that in Eq.(6.53) $\chi^o(\vec{q})$ involves the sum over all the electronic states while $I(\vec{q}, E_F, E_F)$ involves electronic states only in the vicinity of E_F.

Pickett and Allen [46] suggested that the proper inclusion of overlap matrix elements may further improve the correspondence between $\chi^0(\vec{q})$ and the phonon anomalies. This concludes that the anomalies in the phonon dispersion relation of Nb and Mo are related to that part of $\chi^0(\vec{q})$ which is due to transitions within $E_F \pm \hbar\bar{\omega} \approx E_F \pm 0.022$ Ryd. Similar conclusion may be true for other transition metals also. The calculations of Gupta and Freeman [55] for Nb and TaC show that the structures in $\chi^0(\vec{q})$ which are correlated to phonon anomalies arise due to intra-band transitions in the bands in the vicinity of E_F .

(b) fcc structure

Fradin et al [56] and Watson-Yang et al.[57] calculated energy band structure of Pt and Pd respectively using APW method. The calculated energy band structure explained reasonably well the experimental results of electrical resistivity, Fermi radius, spin lattice relaxation,temperature variation of magnetic susceptibilty, magnetic densities and neutron magnetic form factors [56]-[59]. This confirmed the validity of energy band description of these metals. The uppermost three d-subbands which intersect E_F contribute about 80-90% towards the density of states $D(E_F)$ [60, 61] and magnetic form factor. Earlier Evanson et al.[42] have found that the bands which do not intersect E_F give rise to q-independent contribution to $\chi^0(\vec{q})$.

Freeman et al. [60, 61] calculated $\chi^0(\vec{q})$ for Pd and Pt using tetrahedron method of IBZ integration. The results for Pd along $[\xi\xi 0]$, $[\xi\xi\xi]$, $[\xi 00]$ and $[\xi\xi 0]$ symmetry directions are shown in Fig (6.7). The intra-band part of $\chi^0(\vec{q})$ (shown by full diamonds) at $\vec{q} = 0$ agrees with $D(E_F)$ within 0.5%. The main peaks for Pd are along the $[\xi 00]$, $[\xi\xi 0]$ and $[\xi\xi\xi]$ directions at $\xi = 0.65(\pi/a)$. The intra-band contribution of the uppermost two subbands which intersect E_F is shown by triangles. It is noted that the contribution of these bands is maximum and contribute significantly in the characteristics of $\chi^0(\vec{q})$. In the total intra-band contribution the higest peak is in the $[\xi\xi 0]$ direction. However in the $[\xi 00]$ direction the peak height at $\xi = 0.65(\pi/a)$ is lower than the peak height at $q = 0$, therefore this peak is suppressed. Consequently the phonon anomalies are predicted in the $[\xi\xi 0]$ and $[\xi\xi\xi]$ directions. The strongest phonon anomaly is predicted along the $[\xi\xi 0]$ T_1 branch at $\xi = 0.65(\pi/a)$ which is found in agreement with the experimental observation at $\xi = 0.7(\pi/a)$ [46, 47]. The magnitude of inter-band contribution is much larger than the intra-band contribution although the peaks are less pronounced.

The results for $\chi^0(\vec{q})$ for Pt are similar to those for Pd and the peaks occour at $\xi = 0.68(\pi/a), 0.75(\pi/a)$ and $0.85(\pi/a)$ along the $[\xi 00], [\xi\xi 0]$ and $[\xi\xi\xi]$ directions respectively. As a result, as discussed earlier, the phonon anomalies are predicted along $[\xi\xi 0]$ and $[\xi\xi\xi]$ directions and not along $[\xi 00]$ direction. In the $[\xi\xi 0]T_1$ phonon branch the phonon anomaly predicted at $\xi = 0.75(\pi/a)$ is in close agreement with the experimental observation of phonon anomaly at $\xi = 0.76(\pi/a)$ to $0.80(\pi/a)$. Freeman et al.[61] predicted a weaker phonon anomaly for Pt than that for Pd along $[\xi\xi 0]$ direction which is in agreement with the experimental observation. In addition a phonon anomaly along $[\xi\xi\xi]$ direction is also predicted and the possibility of its observation is indicated by Miller [62] for Pd. Therefore a careful search for phonon anomaly in the $[\xi\xi\xi]$ direction for Pd and Pt is suggested.

(c) hcp structure

There exist many electronic band structure calculations of hcp transition metals [63]-[70] but the calculations of $\chi^0(\vec{q})$ are carried out only for Sc and Y. Although Sc and Y metals are free from the magnetic ordering effects, their dilute alloys with rare earth metals (REMs) show a spiral spin structure with magnetic wave vector $\vec{q}_m = (2\pi/c)(0, 0, 0.28)$ and $\vec{q}_m$ does not change with the small changes in concentration [71, 72]. Liu et al [73] calculated $\chi^0(\vec{q})$ for Y along $[00\xi]$ direction using only those subbands which intersect E_F. $\chi^0(\vec{q})$ for Y is shown in Fig. (6.8). It exhibits three peaks at $(2\pi/c)[00\xi]$ with $\xi = 0.375, 0.583$ and 0.75. The phonon anomalies in the $[00\xi]$ longitudinal optic branch of Y [3.94] correspond to the second and third peaks in $\chi^0(\vec{q})$ at $\xi = 0.583$ and 0.75. No anomaly is found in the phonon spectrum corresponding to the highest peak at $\xi = 0.375$. The existence of this peak may be correlated to the magnetic ordering wave vector $\vec{q}_m$ in the dilute alloys of Y with rare earth metals [71, 72].

The calculations of Gupta and Freeman [75] for $\chi^0(\vec{q})$ of Sc along $[00\xi]$ direction are shown in Fig.(6.9). The intra and inter-band contributions with (solid lines) and without (dashed lines) overlap matrix elements are shown separately. There is a qualitative change in the magnitude of $\chi^0_{intra}(\vec{q})$ and $\chi^0_{inter}(\vec{q})$ by including the APW overlap matrix elements. The intra-band contribution $\chi^0_{intra}(\vec{q})$ decreases rapidly with increase in $\vec{q}$ and exhibits a broad peak at $\xi = 0.7$. $\chi^0_{inter}(\vec{q})$ increases with increase in $\vec{q}$ and shows broad shoulders without overlap matrix elements. However by including the overlap matrix elements, $\chi^0_{inter}(\vec{q})$ starts decreasing with increase in $\vec{q}$ and the shoulder has changed to a peak at $\xi = 0.3$. While summing both the intra and inter-band contributions $\chi^0(\vec{q})$ does not show any peak at $\xi = 0.3$ due to rapid decrease in $\chi^0_{intra}(\vec{q})$. These authors also calculated the exchange enhanced susceptibility function $\chi_{en}(\vec{q})$ using Eq.(6.20). It is found that $F(\vec{q})$ exhibits a peak at $\xi = 0.25(2\pi/c)$ which is close to magnetic wave vector $\vec{q}_m$. Thus the inclusion of local field effects in $\chi_{en}(\vec{q}, \vec{q})$ gives the possibilty of existence of spin density wave in Sc metal [76].

The calculations of $\chi^0(\vec{q})$ for Sc due to Liu et al.[73] show three peaks at $(2\pi/c)[00\xi]$ with $\xi = 0.35, 0.57$ and 0.77. The calculations of $\chi^0(\vec{q})$ due to Wakabayashi et al.[74] and Roth and Freeman [39] show only two peaks in the vicinity of $\xi = 0.35$ and 0.77. The behaviour of $\chi^0(\vec{q})$ as a function of $\vec{q}$ is nearly the same in all the calculations. The first peak in $\chi^0(\vec{q})$ is related to the phonon anomaly in the $[00\xi]$ longitudinal acoustic branch at $\xi = 0.27$ and it is close to $\vec{q}_m$ for dilute alloys of Sc with rare earth metals [71, 72]. The second peak in the calculations of Liu et al. [73] is not related to the phonon spectrum. The third peak at $\xi = 0.77$ in $\chi^0(\vec{q})$ may produce an effect in the $[00\xi]$ longitudinal optic branch of phonons but no anomaly is found due to poor resolution of phonon data in this branch.

6.3.2   Rare Earth Metals and Their Alloys

Rare earth metals are a class of metals with narrow f and broad s-bands. Therefore the electronic structure of these metals is quite similar to those of transition metals. Their phonon spectra also show the phonon anomalies. Liu et al [73] calculated $\chi^0(\vec{q})$ for hcp REMs Gd, Tb, Dy, Er, Lu, Pr and Nd using a finer mesh points in the BZ integration. The calculated $\chi^0(\vec{q})$ for Gd and Dy along $[00\xi]$ direction are shown in Fig.(6.10). The results for $\chi^0(\vec{q})$ of Tb, Er and Lu are similar to those of Dy. In Fig (6.10(a)), $\chi^0(\vec{q})$ versus $\vec{q}$ for Gd shows small peaks in between Γ and A symmetry points. These peaks may not be significant as these may either be shifted towards smaller values of $\vec{q}$ or may be removed by including the overlap matrix elements. In Tb one of these peaks occours at $\vec{q} = (2\pi/c)[0,0,0.3]$ which is away from the magnetic ordering wave vector $\vec{q}_m = [0,0,0.11]$. However, this peak in Tb may be related to phonon anomaly in the $[00\xi]$ longitudinal acoustic phonon branch. In Dy, Lu and Er the peaks are sharp as shown in Fig.(6.10(b)) for Dy. However, the the first peak is not at $\vec{q} = (2\pi/c)[0,0,0.3]$ in all these metals. This conclusion agrees with the different observed values of magnetic wave vector $\vec{q}_m$.

Liu et al.[73] calculated $\chi^0(\vec{q})$ for Th and its alloys of REMs having fcc structure. Figure (6.11) shows $\chi^0(\vec{q})$ versus $\vec{q}$ for Th and its alloys along $[\xi 00]$ direction. In Fig. (6.11(a)) $\chi^0(\vec{q})$ versus $\vec{q}$ for Th exhibits peaks at $\vec{q} = (\pi/4a)[\xi,0,0]$ with $\xi = 1.6, 3.8, 4.0$ and 7.5. These peaks are related to multizone structure of Fermi surface [73]. Figures (6.11(b)) to (6.11(d)) show $\chi^0(\vec{q})$ versus $\vec{q}$ for $Th_{1-x}R_x$ for $x = 0.1, 0.3$ and 0.5 where R is any REM atom. These figures show that the shape and magnitude of $\chi^0(\vec{q})$ changes with the increase in x. The neutron diffraction studies on the powdered samples of $Th_{0.7}Tb_{0.3}$ due to Child et al. [77] reveal two humps at $\vec{q} = (\pi/4a)[4.1300]$ and $(\pi/4a)[5.0700]$ which may be due to magnetic ordering. The directions assigned to $\vec{q}_m$ are arbitrary as the powdered samples do not give any identification of direction. For $Th_{0.7}R_{0.3}$ a sharp peak at $\vec{q} = (\pi/4a)[5.100]$ and a broad peak at $\vec{q} = (\pi/4a)[3.800]$ are found theoretically and these are close to the observed values of $\vec{q}_m$. These calculations are in the rigid ion approximation and therefore a quantitatively correct information may not be expected. It is not possible to correlate the structure of $\chi^0(\vec{q})$ with the phonon spectrum anomalies as the experimental measurements of phonon spectrum of these materials are not available.

Fleming and Liu [78] studied the pressure dependece of $\chi^0(\vec{q})$ for

Gd, Tb and Dy. In Tb and Dy the peaks in $\chi^0(\vec{q})$ shift towards smaller values of $\vec{q}$ and the peak strength is reduced. This is due to the pressure induced shift in the energy bands. Most of the rare-earth metals are antiferromagnetic with a sinusoidal, spiral or a more complex arrangement of magnetic moments which are periodic along c-axis [79]. Some of the REMs such as Gd are in the ferromagnetic phase below the transition temperature.

In the rare earth metals the mechanism of magnetic ordering is believed to be Ruderman-Kittel-Kasuya-Yosida (RKKY) indirect exchange interaction [80, 81, 82] between s and quasi-localized f-electrons. The 4f-shell magnetic moment of REM ion polarizes the spins of conduction electrons in its neighbourhood through exchange interaction which depends upon two physical quantities: (i) The exchange interaction between REM ions and s-conduction electrons and (ii) the electron energy bands and Fermi surface of REMs. Therefore the exchange enhanced susceptibility $\chi_{en}(\vec{q})$ is also studied to see the effect of exchange interacton. The maximum in $\chi_{en}(\vec{q})$ fixes the minimum exchange energy and hence the magnetic wave vector $\vec{q}_m$ is determined for magnetic ordering. The peak in $\chi^0(\vec{q})$ at $\vec{q} = 0$ would mean ferromagnetism and the one at finite $\vec{q}$ would indicate a sinusoidal, a helical or a more complicated magnetic ordering. Evanson and Liu [41] and Evanson et al [42] calculated $\chi^0(\vec{q})$ for REMs and correlated the peak structure with the nesting of Fermi surface [83].

6.3.3   Transition Metal Compounds

Electronic energy band structure of transition metal compounds is studied by many authors [55],[84]-[92] using APW and KKR Green's function methods. The calculations of $\chi^0(\vec{q})$ are only a few. Gupta and Freeman [55] calculated $\chi^0(\vec{q})$ for NbC, TaC, ZrC and HfC using the results of APW band structure calculations. Their results for NbC are shown in Figs.(6.12). The main structures in $\chi^0(\vec{q})$ in NbC arise from the intra-band transitions in the bands just below the Fermi energy. The inter-band contribution is almost structureless but large in magnitude in comparison to intra-band contribution. The maxima in $\chi^0(\vec{q})$ versus $\vec{q}$ for NbC occour at $\vec{q} = (2\pi/a)[0.6, 0, 0], (2\pi/a)[0.55, 0.55, 0]$

and $(2\pi/a)[0.5, 0.5, 0.5]$ which almost exactly coincide with the dips in the longitudinal acoustic phonon branches of phonon spectra [3.160]. For TaC the maxima along $[\xi\xi 0]$ and $[\xi\xi\xi]$ directions are at the same positions as found in NbC except in the $[\xi 00]$ direction where the maximum occurs at $\vec{q} = (2\pi/a)[0.6300]$. These features coincide with the phonon anomalies observed in the phonon measurements. The small change in $\vec{q}$ values, at which the anomalies occour, in going from NbC to TaC $[\xi_{theor} = 0.60$ to $0.63]$ along $[\xi 00]$ direction is consistent with the observed experimental shift $[\xi_{exp} = 0.60$ to $0.65]$.

The results for $\chi^0(\vec{q})$ for ZrC are shown in Fig.(6.13). The total inter-band contribution to $\chi^0(\vec{q})$ decreases smoothly from its value at $\vec{q} = 0$ in all the principal symmetry directions. The intra-band contribution is also almost structureless. The magnitude of inter-band contribution is larger than the intra-band contribution by an order of magnitude. The resulting maximum at $\vec{q} = 0$ shows an overscreening at the zone center. This overscreening can be correlated with the phonon frequency measurements of Smith [3.166] where the optical mode frequencies at the zone center Γ in ZrC and HfC are significantly lower than those found in the measurements for NbC and TaC. The optical mode frequencies at Γ symmetry point are 17 THz for NbC and 13 THz for ZrC. The overall decrease of $\chi^0(\vec{q})$ away from the zone center is also consistent with the fact that the phonon anomalies are not found in these low T_c compounds.

Hanke et al.[93] have shown that there is an hybridization between the metal d-states and nonmetal p-states near E_F in transition metal carbides. Therefore these authors proposed the p-d hybridization model which leads to resonance like increase in the density response function. Based on this model Hafner and Hanke [94] calculated $\chi^0(\vec{q})$ for transition metal carbides and nitrides including full p-d band complex. The intra-band contribution to $\chi^0(\vec{q})$ in Hafner and Hanke's scheme shows structure at the positions of phonon anomalies while the inter-band contribution shows a broad maximum in the middle of BZ.

Thus the theoretical studies of $\chi^0(\vec{q})$ of 9VE (NbC, TaC) and 8VE (ZrC, HfC) transition metal carbides reveal that the phonon anomalies (dips) arise because of the overscreening of corresponding phonon modes in these materials.

6.4 Temperature Dependent Dynamical Susceptibility

Temperature dependent dynamical susceptibility is not directly related to dynamical matrix as the latter involves the temperature independent static susceptibilty function. However the dynamical susceptibility can be related indirecty to superconductivity and the phonon softening in the phonon spectra of d and f-band materials. Therefore the dynamical susceptibility is less studied both theoretically and experimentally [49],[95]. The temperature variation of dynamical susceptibility $\chi^0(\vec{q}, \omega, T)$ is described in Eq.(6.43). In fact the temperature variation of $\chi^0(\vec{q}, \omega, T)$ arises from two physical effects: The first from the Fermi-Dirac distribution function $f_{\vec{k}\mu}(T)$ and the second from the temperature dependence of electronic energies $E_{\vec{k}\mu}$. It is too difficult to include the temperature dependence of $E_{\kappa\mu}$, therefore the temperature variation of $\chi^0(q, \omega, T)$ is studied only through $f_{\vec{k}\mu}(T)$ in Eqs.(6.43) to (6.47). The temperature dependence of $f_{\vec{k}\mu}(T)$ is given as

$$f_{\vec{k}\mu}(T) \;=\; \left[1 + \exp\left(\frac{E_{\vec{k}\mu} - E_F(T)}{T}\right)\right]^{-1} \tag{6.54}$$

where $E_F(T)$ is temperature dependent chemical potential which is equal to Fermi energy at T=0 i.e. $E_F(0) = E_F$. In the constant matrix element approximation Eq.(6.43) simplifies as

$$\chi^0(\vec{q}, \omega, T) \;=\; \sum_{\vec{k}} \sum_{\mu\mu'} \frac{f_{\vec{k}\mu}(T) - f_{\vec{k}+\vec{q}\mu'}(T)}{E_{\vec{k}+\vec{q}\mu'} - E_{\vec{k}\mu} - \omega - i\delta}. \tag{6.55}$$

In the following, we discuss the calculations of Eq.(6.55) in the various model band schemes.

6.4.1 Free Electron Approximation

In the free electron (plane wave) approximation $M_{\vec{k}\mu,\vec{k}+\vec{q}\mu'}(\vec{q})$ becomes unity and hence $\chi^0(\vec{q}, \omega, T)$ is the same as given in Eq. (6.55). In this apprximation the conduction band is parabolic with $E_{\vec{k}} = (k^2/2m)$ and

($\hbar = 1$). For a single band, the imaginary part of $\chi^0(\vec{q}, \omega, T)$, given in Eq.(6.55), is

$$Im\chi^0(\vec{q}, \omega, T) = \pi \sum_{\vec{k}} \left[f_{\vec{k}}(T) - f_{\vec{k}+\vec{q}}(T) \right] \delta(E_{\vec{k}+\vec{q}} - E_{\vec{k}} - \omega).$$

$$(6.56)$$

By changing the sum over $\vec{k}$ into integration, the above equation for parabolic band simplifies as

$$Im\chi^0(\vec{q}, \omega, T) = \frac{\Omega_0}{4\pi} \int_0^\infty dk k^2 \left[f_{\vec{k}}(T) - f_{\vec{k}+\vec{q}}(T) \right]$$
$$\times \int_{-1}^{+1} \delta(\frac{1}{2m}(q^2 + 2kqz) - \omega)dz \qquad (6.57)$$

where $z = \cos\theta$. Using the property of Dirac-delta function

$$\int \delta(az - b)dz = \frac{1}{a} \int \delta(x - b)dx$$

with $x = az$, the delta function integral can be solved. Thus one gets

$$Im\chi^0(\vec{q}, \omega, T) = \frac{m\Omega_0}{4\pi q} \int_{|(m\omega/q)-(q/2)|}^\infty dk k \left[f_{\vec{k}}(T) - f_{\vec{k}+\vec{q}}(T) \right].$$

$$(6.58)$$

The integral over $f_{\vec{k}}(T)$ can be solved by using the standard integral

$$\int \frac{dx}{a + be^{mx}} = \frac{1}{am} [mx - \ln(a + be^{mx})] \qquad (6.59)$$

which finally gives

$$Im\chi^0(\vec{q}, \omega, T) = \frac{m^2\Omega_0 T}{4\pi q} \ln \left(\frac{e^{\omega/T} + e^{E(\vec{q},\omega,T)/T}}{1 + e^{E(\vec{q},\omega,T)/T}} \right) \qquad (6.60)$$

where

$$E(\vec{q}, \omega, T) = \frac{1}{2m} \left| \frac{m\omega}{q} + \frac{q}{2} \right|^2 - E_F(T). \qquad (6.61)$$

The temperature variation of $\chi^0(q, \omega, T)$ will be more apparent if we define the reduced susceptibility as

$$\bar{\chi}^0(\vec{q}, \omega, T) \;=\; \chi^0(\vec{q}, \omega, T)/\chi^0(0, 0, 0) \tag{6.62}$$

where

$$\chi^0(0, 0, 0) \;=\; (mk_F\Omega_0/2\pi^2) = \; D(E_F) \tag{6.63}$$

is the density of electronic states at E_F for one spin. From Eqs. (6.62) and (6.60) the imaginary part of reduced susceptibility becomes

$$\mathrm{Im}\bar{\chi}^0(\vec{q}, \omega, T) \;=\; \frac{\pi}{2}\frac{T}{qv_F} \ln\left(\frac{e^{\omega/T} + e^{E(\vec{q},\omega,T)/T}}{1 + e^{E(\vec{q},\omega,T)/T}}\right) \tag{6.64}$$

where $v_F(= k_F/m)$ is Fermi velocity. The real part of $\bar{\chi}^0(\vec{q}, \omega, T)$ can be obtained using Kramers-Kronig relation (6.47). It can be shown that at $T = 0^0 K$, $\mathrm{Im}\chi^0(\vec{q}, \omega, T)$ and $\mathrm{Re}\chi^0(\vec{q}, \omega, T)$ reduce to the following expressions:

$$
\begin{aligned}
\mathrm{Re}\chi^0(\vec{q}, \omega) \;=\; & \frac{\Omega_0 mk_F}{2\pi^2\hbar^2} \left[\frac{1}{2} + \frac{1}{8\eta}\left\{1 - (\frac{mW}{\hbar\eta} - \eta)^2\right\}^2 \ln\left|\frac{1 - \frac{mW}{\hbar\eta} + \eta}{1 + \frac{mW}{\hbar\eta} - \eta}\right| \right. \\
& \left. - \frac{1}{8\eta}\left\{1 - (\frac{mW}{\hbar\eta} + \eta)^2\right\}^2 \ln\left|\frac{1 - \frac{mW}{\hbar\eta} - \eta}{1 + \frac{mW}{\hbar\eta} + \eta}\right|\right]
\end{aligned}
\tag{6.65}
$$

and

$$
\begin{aligned}
\mathrm{Im}\chi^0(\vec{q}, \omega) \;=\; & \frac{\Omega_0 mk_F}{16\pi^2\hbar^2\eta} \left[\left\{1 - (\frac{mW}{\hbar\eta} + \eta)^2\right\}\theta\left(1 - (\frac{mW}{\hbar\eta} + \eta)^2\right) \right. \\
& \left. - \left\{1 - (\frac{mW}{\hbar\eta} - \eta)^2\right\}\theta\left(1 - (\frac{mW}{\hbar\eta} - \eta)^2\right)\right]
\end{aligned}
\tag{6.66}
$$

where $\eta = (q/2k_F)$, $W = (\omega/2k_F^2)$ and $\theta(x)$ is Heavyside unit step function. Doniach [50] calculated $\chi^0(\vec{q}, \omega)$ in the free electron approximation. Equations (6.65) and (6.66) are smooth functions of $\vec{q}$ and ω and may not be related to phonon anomalies directly. $E_F(T)$ which is needed to calculate $\chi^0(\vec{q}, \omega, T)$, can be calculated using number conservation condition for the electrons in a given band i.e.

$$2\sum_{\vec{k}} f_{\vec{k}}(T) \;=\; N \tag{6.67}$$

where N is the total number of electrons. The values of $E_F(T)/E_F$ are obtained with the help of Eq.(6.67) and these are used to calculate $\mathrm{Im}\bar{\chi}^0(\vec{q}, \omega, T)$ and $\mathrm{Re}\bar{\chi}^0(\vec{q}, \omega, T)$ as a function of $\omega/v_F q, q/k_F$ and T/T_F [51, 53]. Here $T_F(= E_F = k_F^2/2m)$ is Fermi temperature. As T increases, the peak in $\mathrm{Im}\chi^0(\vec{q}, \omega, T)$ gets weaken and broaden and finally disappear at higher temperature.

The exchange enhanced susceptibility $\chi_{en}(\vec{q}, \omega, T)$ can be defined in the same way as in Eq.(6.16) and is written as

$$\chi_{en}(\vec{q}, \omega, T) \;=\; \frac{\chi^0(\vec{q}, \omega, T)}{1 - I\chi^0(\vec{q}, \omega, T)}. \tag{6.68}$$

At $\vec{q} = \omega = 0$, Eq.(6.68) becomes

$$\chi_{en}(0, 0, 0) \;=\; \frac{D(E_F)}{1 - ID(E_F)}. \tag{6.69}$$

For $ID(E_F) \geq 1$, the material becomes ferromagnetic or antiferromagnetic [95] i.e. magnetic ordering is induced in the material. Therefore one can study the magnetic phase of material with the help of Eq.(6.69).

6.4.2 Non-Interacting Band Scheme

Prakash and co-workers [53] calculated $\chi^0(\vec{q}, \omega, T)$ in the non- interacting band scheme including only the leading intra-band contribution. In this scheme $\chi^0(\vec{q}, \omega, T)$ is written as the sum of contributions from the partially filled s and d-subbands as

$$\chi^0(\vec{q}, \omega, T) \;=\; \sum_{\sigma} \chi^0_{s\sigma}(\vec{q}, \omega, T) + \sum_{m\sigma} \chi^0_{dm\sigma}(\vec{q}, \omega, T) \tag{6.70}$$

where

$$\chi^0_{s\sigma}(\vec{q}, \omega, T) \;=\; \sum_{\vec{k}} \frac{f_{\vec{k}s\sigma} - f_{\vec{k}+\vec{q}s\sigma}(T)}{E_{\vec{q}+\vec{q}s\sigma} - E_{\vec{k}s\sigma} - \omega - i\delta} \tag{6.71}$$

and

$$\chi^0_{dm\sigma}(\vec{q},\omega,T) \;=\; \sum_{\vec{k}} \frac{f_{\vec{k}dm\sigma}(T) - f_{\vec{k}+\vec{q}dm\sigma}(T)}{E_{\vec{k}+\vec{q}dm\sigma} - E_{\vec{k}dm\sigma} - \omega - i\delta}\; |\, M(\vec{q})\,|^2 \,.$$

(6.72)

Here σ is spin index and $M(\vec{q})$ is overlap matrix element which is calcuated using simple tight-binding wave function (Eqs.(4.131) and (4.142)) and the hybridization between the different components of d-wave functions is included. Equation (6.71) and (6.72) can be solved analytically in the same manner as in the free electron approximation with energies $E_{\vec{k}s\sigma} = (k^2/2m_{s\sigma})$ and $E_{\vec{k}dm\sigma} = (k^2/2m_{dm\sigma})$. Thus one gets

$$\mathrm{Im}\bar{\chi}^0_{s\sigma}(\vec{q},\omega,T) \;=\; \frac{\pi}{2}\frac{T}{q v_{Fs\sigma}} \ln\left(\frac{e^{\omega/T} + e^{E_{s\sigma}(\vec{q},\omega,T)/T}}{1 + e^{E_{s\sigma}(\vec{q},\omega,T)/T}}\right)$$

(6.73)

and

$$\mathrm{Im}\bar{\chi}^0_{dm\sigma}(\vec{q},\omega,T) \;=\; \frac{\pi}{2}\frac{T}{q v_{Fdm\sigma}} \,|\, M(\vec{q})\,|^2 \ln\left(\frac{e^{\omega/T} + e^{E_{dm\sigma}(\vec{q},\omega,T)/T}}{1 + e^{E_{dm\sigma}(\vec{q},\omega,T)/T}}\right)$$

(6.74)

where

$$E_{s\sigma}(\vec{q},\omega,T) \;=\; \frac{1}{2m_{s\sigma}}\left|\frac{m_{s\sigma}\omega}{q} + \frac{q}{2}\right|^2 - E_{Fs\sigma}(T)$$

(6.75)

and

$$E_{dm\sigma}(\vec{q},\omega,T) \;=\; \frac{1}{2m_{dm\sigma}}\left|\frac{m_{dm\sigma}\omega}{q} + \frac{q}{2}\right|^2 - E_{Fdm\sigma}(T).$$

(6.76)

Here $m_{s\sigma}$ and $k_{Fs\sigma}$ are the effective mass and Fermi momentum for s-band with spin σ and $m_{dm\sigma}$ and $k_{Fdm\sigma}$ are the corresponding quantities for d-subbands with quantum number m and spin σ. $E_{Fs\sigma}(T)$ and $E_{Fdm\sigma}(T)$ are chemical potentials which can be evaluated by satisfying the number conservation conditions for s and d-electrons separately. The variation of $E_{s\sigma}(T)$ and $E_{Fdm\sigma}(T)$ with T is similar to that for $E_F(T)/E_F$.

Figure (6.14) shows $\mathrm{Im}\bar{\chi}^0(\vec{q}, \omega, T)$ as a function of $\vec{q}, \omega$ and T for ferromagnetic Ni((a) and (b)), paramagnetic Ni((c) and (d)), Pd((e) and (f)) and Pt((g) and (h)). The peaks at small values of ω arise through the contribution of partially filled d-subbands and the broad peaks at large values of ω arise from the s-band contribution. For ferromagnetic Ni there are two sharp peaks due to two partially filled minority spin d-subbands. Figure (6.14) also shows that with the increase in temperature, the structure in $\mathrm{Im}\bar{\chi}^0(\vec{q}, \omega, T)$ diminishes and ultimately disappears at very high temperatures. It means that the spin waves in the ferromagnetic phase are destroyed leading to paramagnetic phase. In the paramagnetic metals the increase in temperature destroys the paramagnons as the characteristic peaks get broadened. Figure (6.15) shows $\mathrm{Re}\bar{\chi}^0(\vec{q}, \omega, T)$ as a function of $\vec{q}, \omega$ and T for Pd metal. It exhibits a peak at the small values of $\vec{q}$ and T which shows that Pd is nearly magnetic in nature. This peak becomes more pronounced in the exchange-enhanced susceptibility functions $\chi_{en}(\vec{q}, \omega, T)$. With the increase in $\vec{q}$ and T this peak is broadened and finally disappears.

Ferromagnetism arises basically because of spatial localization of d-orbitals near the top of d-band [96] and this localization produces both large $D(E_F)$ and a relative maximum in the enhancement factor I in Eq.(6.69). The non-interacting band model of ferromagneric Ni has flat d-subbands giving a very small d-band width and large value of $D(E_F)$. It makes the product $ID(E_F) = 1.01$ for ferromagnetic Ni. But for Pd and Pt, the d-band width is comparatively large leading to smaller values of $ID(E_F)$. In other words one can say that the 4d and 5d wave functions have spatial distribution to a larger extent which implies larger overlap with the nearest neighbours. This gives rise to large d-band width and hence the smaller $D(E_F)$.

6.4.3 Model Potential Theory

The overlap matrix elements can be evaluated in the model potential approach described in chapter 4, section (4.5.2). For a single band, the overlap matrix elements can be written as

$$M_{\vec{k}, \vec{k}+\vec{q}}(\vec{q}) \;=\; |< \psi_{\vec{k}}(r)|e^{-i\vec{q}\cdot\vec{r}}|\psi_{\vec{k}+\vec{q}}(\vec{r}) >|^2 . \qquad (6.77)$$

Using the model wave function transformation given in Eq. (4.217) it can be proved that

$$< \psi_{\vec{k}}(\vec{r})|e^{-i\vec{q}\cdot\vec{r}}|\psi_{\vec{k}+\vec{q}}(\vec{r}) > \; = \; < \psi_{\vec{k}+\vec{q}}(\vec{r})|\psi_{\vec{k}+\vec{q}}(\vec{r}) >$$

$$= \; 1 + \frac{1}{2}\beta(\vec{k}+\vec{q}) \tag{6.78}$$

where

$$\beta(\vec{k}+\vec{q}) \; = \; -2 < \vec{k}+\vec{q}|\frac{\partial V_M}{\partial E}|\vec{k}+\vec{q} > . \tag{6.79}$$

Using Eq.(6.78) in (6.77) the matrix elements become

$$M_{\vec{k},\vec{k}+\vec{q}}(\vec{q}) \; = \; 1 + \gamma(\vec{k}+\vec{q})$$

$$= \; 1 + \beta(\vec{k}+\vec{q}) + \frac{1}{4}|\beta(\vec{k}+\vec{q})|^2. \tag{6.80}$$

Here $\gamma(\vec{k}+\vec{q})$ is called depletion hole contribution and it is analogous to orthogonalization hole in the OPW theory. It contains the effect of Bloch charaterstics of conduction electrons or in other words the band structure effects which will be evident latter.

Using Eq.(6.80) in Eq. (6.43) the imaginary part of $\chi^0(\vec{q},\omega,T)$ for a single band becomes

$$\mathrm{Im}\chi^0(\vec{q},\omega,T) \; = \; \mathrm{Im}\chi_f^0(\vec{q},\omega,T) + \mathrm{Im}\chi_{dp}^0(\vec{q},\omega,T) \tag{6.81}$$

where

$$\mathrm{Im}\chi_f^0(\vec{q},\omega,T) \; = \; \pi\sum_{\vec{k}}\left[f_{\vec{k}}(T) - f_{\vec{k}+\vec{q}}(T)\right]\delta(E_{\vec{k}+\vec{q}} - E_{\vec{k}} - \omega) \tag{6.82}$$

and

$$\mathrm{Im}\chi_{dp}^0(\vec{q},\omega,T) \; = \; \pi\sum_{\vec{k}}\left[f_{\vec{k}}(T) - f_{\vec{k}+\vec{q}}(T)\right]\gamma(\vec{k}+\vec{q})\delta(E_{\vec{k}+\vec{q}} - E_{\vec{k}} - \omega). \tag{6.83}$$

Here $\chi^0_f(\vec{q}, \omega, T)$ is free electron contribution and it is given in Eq.(6.60) and $\chi^0_{dp}(\vec{q}, \omega, T)$ is depletion hole contribution. The angular integration of Eq.(6.83) gives

$$\mathrm{Im}\chi^0_{dp}(\vec{q}, \omega, T) = \frac{m\Omega_0}{4\pi q} \int_{|(m\omega/q)-(q/2)|}^{\infty} dk\, k\, \left[f_{\vec{k}}(T) - f_{\vec{k}+\vec{q}}(T) \right] \gamma(\vec{k}+\vec{q}).$$

$$(6.84)$$

To compare the free electron and depletion hole contributions, we define the reduced susceptibility as

$$\bar{\chi}^0(\vec{q}, \omega, T) = \bar{\chi}^0_f(\vec{q}, \omega, T) + \bar{\chi}^0_{dp}(\vec{q}, \omega, T) \tag{6.85}$$

where

$$\bar{\chi}^0_f(\vec{q}, \omega, T) = \chi^0_f(\vec{q}, \omega, T)/\chi^0_f(0, 0, 0) \tag{6.86}$$

and

$$\bar{\chi}^0_{dp}(\vec{q}, \omega, T) = \chi^0_{dp}(\vec{q}, \omega, T)/\chi^0_f(0, 0, 0). \tag{6.87}$$

With the above definition $Im\chi^0_f(\vec{q}, \omega, T)$ is given in Eq.(6.64) and $\bar{\chi}^0_{dp}(\vec{q}, \omega, T)$ is given as

$$\mathrm{Im}\bar{\chi}^0_{dp}(\vec{q}, \omega, T) = \frac{\pi}{2qk_f} \int_{|(m\omega/q)-(q/2)|}^{\infty} dk\, k\, \left[f_{\vec{k}}(T) - f_{\vec{k}+\vec{q}}(T) \right] \gamma(\vec{k}+\vec{q}).$$

$$(6.88)$$

The real part of susceptibility can be calculated by using Kramers-Kronig relation (6.47).

$Im\bar{\chi}^0_{dp}(\vec{q}, \omega, T)$ can be calculated if $\gamma(\vec{k}+\vec{q})$ is known. $\beta(\vec{k}+\vec{q})$ and hence $\gamma(\vec{k}+\vec{q})$, can be calculated using Heine-Abarenkov model potential for $V_M(E)$ and expanding the plane wave $|\vec{k}+\vec{q}>$ in terms of spherical harmonics. It gives [97]

$$\beta(\vec{k}+\vec{q}) = \sum_{l=0}^{\infty} \beta_l(\vec{k}+\vec{q}) \tag{6.89}$$

446   *S. PRAKASH*

where

$$\beta_l(\vec{k} + \vec{q}) \;=\; \frac{8\pi}{\Omega_0}(2l+1)\frac{dA_l}{dE}\int_0^{R_M} \left[j_l(|\vec{k} + \vec{q}|r)\right]^2 r^2 dr \qquad (6.90)$$

and (dA_l/dE) is energy derivative of potential well depth $A_l(E)$ for the lth orbital. Note that $\beta(\vec{k} + \vec{q})$ and hence $\gamma(\vec{k} + \vec{q})$ is the function of both $\vec{k}$ and $\vec{q}$ as calculated by Rath and Freeman [39] and Gupta and Freeman [55]. Shaw and Harrison [98] defined averaged depletion hole by summing over all the electronic states as

$$\gamma(\vec{q}, T) \;=\; \frac{1}{Z}\sum_{\vec{k}} f_{\vec{k}}(T)\gamma(\vec{k} + \vec{q}) \qquad (6.91)$$

where Z is chemical valency. Using $\gamma(\vec{k}+\vec{q})$ from Eq.(6.80) in Eq.(6.91) and changing the sum over $\vec{k}$ into integration one gets

$$\gamma(\vec{q}, T) \;=\; \frac{\Omega_0}{\pi^2 Z}\int_0^\infty dk k^2 f_{\vec{k}}(T)\left[\beta(\vec{k} + \vec{q}) + \frac{1}{4}|\beta(\vec{k} + \vec{q})|^2\right].$$
$$(6.92)$$

Prakash and co-workers [97] calculated $\beta(\vec{k} + \vec{q})$ for vanadium as a function of $\vec{k}$ for different values of l. $\beta_0(\vec{k} + \vec{q})$ vanishes as dA_0/dE is zero for vanadium. $\beta_1(\vec{k} + \vec{q})$ exhibits the resonance behaviour which is the characteristic of a d-band metal. The averaged depletion hole $\gamma(\vec{q}, T)$ is also calculated as a function of $\vec{q}$ and T for vanadium and it is found that it decreases with the increase in $\vec{q}$ and T.

    Using $\gamma(\vec{q}, T)$ in place of $\gamma(\vec{k} + \vec{q})$ in Eq.(6.88), one can do the integration analytically to get

$$\mathrm{Im}\bar{\chi}^0_{\mathrm{dp}}(\vec{q}, \omega, \mathrm{T}) \;=\; \gamma(\vec{q}, T)\bar{\chi}^0_f(\vec{q}, \omega, T). \qquad (6.93)$$

In the averaged depletion hole approximation the total susceptibilty becomes

$$\mathrm{Im}\bar{\chi}^0(\vec{q}, \omega, \mathrm{T}) \;=\; [1 + \gamma(\vec{q}, T)]\,\mathrm{Im}\bar{\chi}^0_f(\vec{q}, \omega, \mathrm{T}) \qquad (6.94)$$

and the real part of susceptibility function using Kramers-Kronig relation becomes

$$\mathrm{Re}\bar{\chi}^0(\vec{q}, \omega, \mathrm{T}) \;=\; [1 + \gamma(\vec{q}, T)]\,\mathrm{Re}\bar{\chi}^0_f(\vec{q}, \omega, \mathrm{T}). \qquad (6.95)$$

In the temperature dependent calculations [99] $\mathrm{Im}\bar{\chi}^0_{\mathrm{dp}}(\vec{q}, \omega, \mathrm{T})$ is found about 46% of $\mathrm{Im}\bar{\chi}^0_{\mathrm{f}}(\vec{q}, \omega, \mathrm{T})$ which is quite significant. The peaks in $\mathrm{Im}\bar{\chi}^0_{\mathrm{f}}(\vec{q}, \omega, \mathrm{T})$ and $\mathrm{Im}\bar{\chi}^0_{\mathrm{dp}}(\vec{q}, \omega, \mathrm{T})$ are found at the same value of ω. The spread of $\mathrm{Im}\bar{\chi}^0_{\mathrm{dp}}(\vec{q}, \omega, \mathrm{T})$ in ω is found less than that of $\mathrm{Im}\bar{\chi}^0_{\mathrm{f}}(\vec{q}, \omega, \mathrm{T})$. However, in the non-interacting band scheme $\mathrm{Im}\bar{\chi}^0_{\mathrm{dd}}(\vec{q}, \omega, \mathrm{T})$ is more localized due to use of tight-binding approximation for d-electrons. The overall variation of $\mathrm{Im}\bar{\chi}^0(\vec{q}, \omega, \mathrm{T})$ with temperature is the same in both the model potential and non-interacting band schemes. It is interesting to note that $\mathrm{Im}\bar{\chi}^0_{\mathrm{dp}}(\vec{q}, \omega, \mathrm{T})$ obtained from the Eqs.(6.88) and (6.91) are just the same with a difference of less than even 1% at all ω values. This fact supports the decoupling procedure given by Shaw and Harrison [98] according to which the band structure and overlap matrix parts are separately averaged in k-space.

6.5   Remarks

In this chapter the calculations of generalized susceptibility function are described in the context of its relation with the phonon anomalies and the magnetic ordering effects. The <u>ab initio</u> calculations of $\chi^0(\vec{q})$ explain, to some extent, the positions and strength of some of the phonon anomalies. As regards the calculation of susceptibility function, following remarks are in order:

(i) The generalized susceptibility function defined in Eqs. (6.1) to (6.3), is based on the linear response theory. However in the d and f-band meterials the crystal potential or the pseudopotential is strong and exhibits the resonance behaviour (chapter 5). Therefore the non- linear effects are important and must be included in the calculations of susceptibility function of d and f-band materials. These effects may help in the explanation of both the positions and strength of phonon anomalies.

(ii) In Eqs.(6.16) and (6.20), the exchange enhancement parmeter I is taken as a constant. In fact $\chi^0(\vec{q})$ and I both are the functions of momentum transfer $\vec{q}$ and temperature T. Consequently Eq.(6.16) modifies as [49, 100, 101]

$$\chi_{en}(\vec{q}, T) \;=\; \frac{\chi^0(\vec{q}, T)}{1 - I(\vec{q}, T)\chi^0(\vec{q}, T)}. \tag{6.96}$$

However the available forms of $I(\vec{q}, T)$ at present are valid either at lower or higher values of $\vec{q}$ or at lower or higher values of T. The local field effects also modify the parameter $I(\vec{q}, T)$ [10, 11, 55] and $\chi^0(\vec{q}, T)$. Therefore further efforts are needed to calculate $I(\vec{q}, T)$ and hence $\chi_{en}(\vec{q}, T)$ more accurately. This may explain the detailed structure of phonon dispersion curves and magnetic ordering effects.

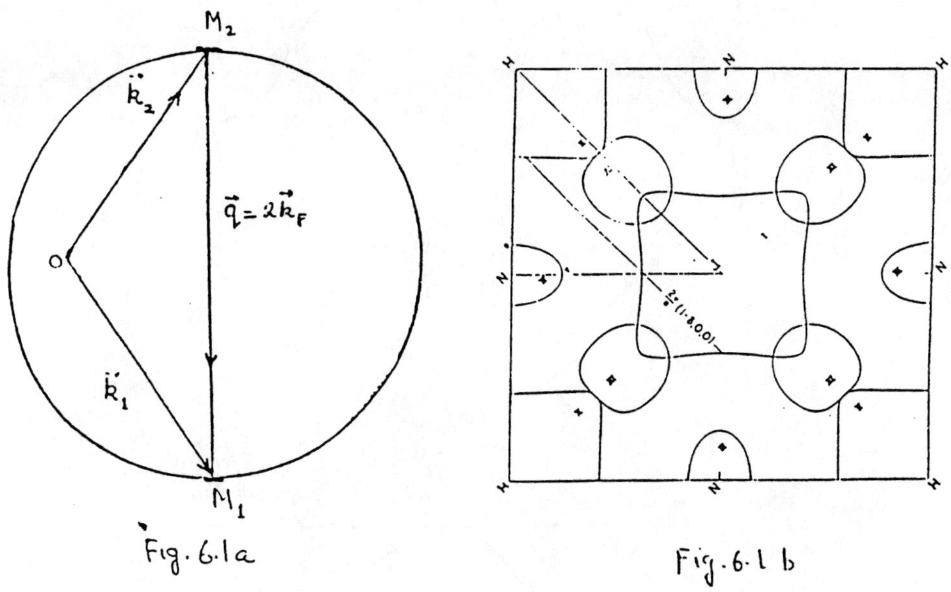

Figure 6.1: (a) Fermi surface of free electron gas. The energy states M_1 and M_2 at the Fermi surface are separated by the wave vector $\vec{k}_1 - \vec{k}_2 = \vec{q} = 2k_F$. (b) (001) cross section of BZ of bcc lattice. Γ, H and N are symmetry points. The Fermi surface consists of electron octahedra around Γ point (I), electron balls on the [100] axis (IV), hole octahedra around H points (II) and hole surface around N points (III). The electron and hole octahedra exhibit the nesting property required for antiferromagnetism of Cr. The quantity $\vec{q} = (2\pi/a)(1 - \delta, 0, 0)$ is the nesting wave vector.

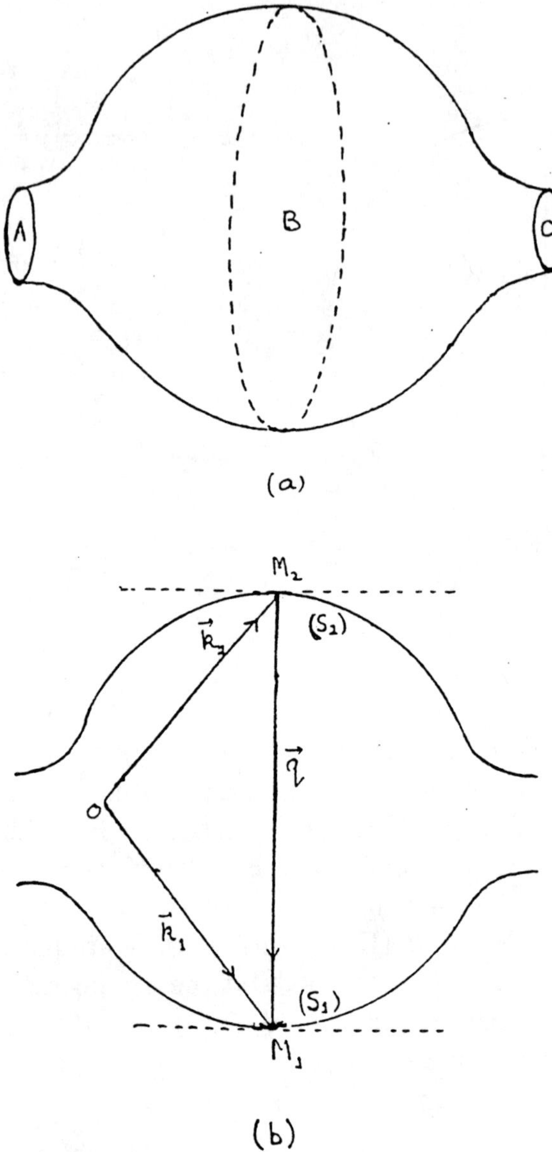

Figure 6.2: (a) Hypothetical Fermi surface of a metal exhibiting extremal portions A, B and C. (b) Cross sectin of Fermi surface showing electron $(\vec{k}_1, \vec{k}_2)$ and phonon $(\vec{q})$ wave vectors for an anomaly in the susceptibility function. S_1, S_2 and S_1' are the portions of Fermi surface and D_1 and D_2 are the loci of points of intersection.

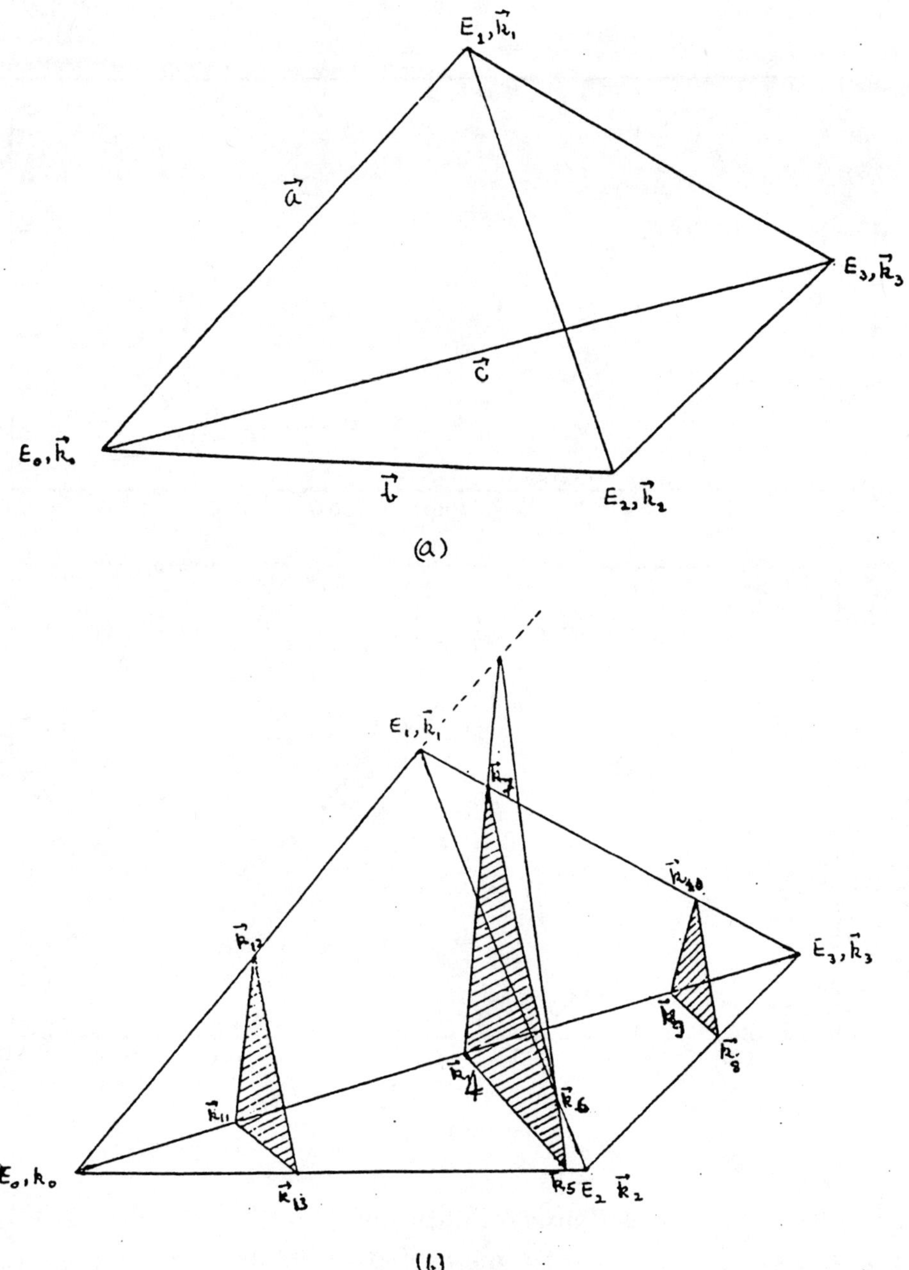

Figure 6.3: (a) A tetrahedron with primitive translation vectors $\vec{a}, \vec{b}$ and $\vec{c}$. E_0, E_1, E_2 and E_3 are the energies at the corners of tetraherdron and the corresponding wave vectors are $\vec{k}_0, \vec{k}_1, \vec{k}_2$ and $\vec{k}_3$. (b) The plane surfaces which cut the tetrahedron. The description is given in the text.

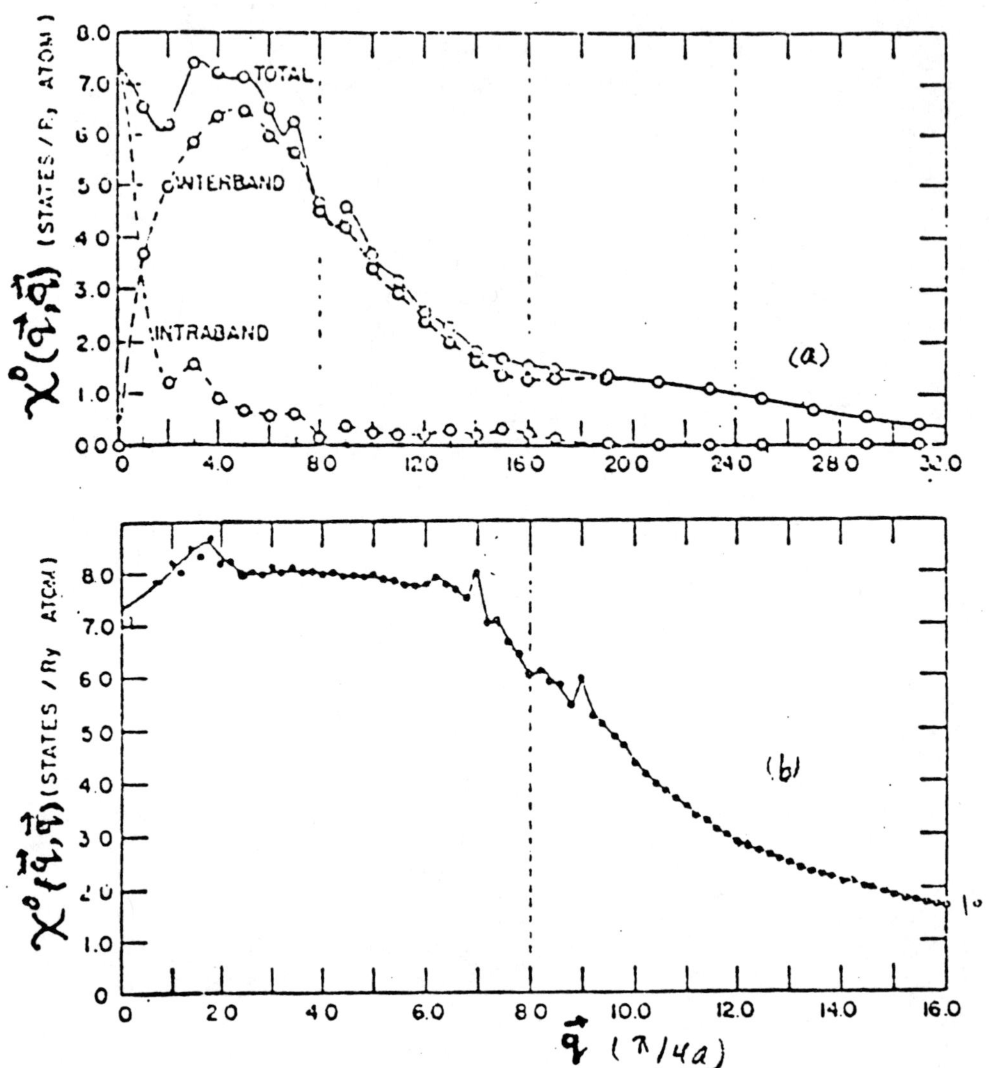

Figure 6.4: Unenhanced susceptibility function $\chi^0(\vec{q}, \vec{q})$ at $T = 0$ for paramagnetic Cr. In Fig.(a) a mesh of 1024 points is used in the BZ integration and in Fig.(b) a mesh of 12800 points is used in the BZ integration. Here $\vec{q}$ is in units of $(\pi/4a)$. In Fig.(6.4(a)), dashed lines show the intra-band and inter-band contributions separately.

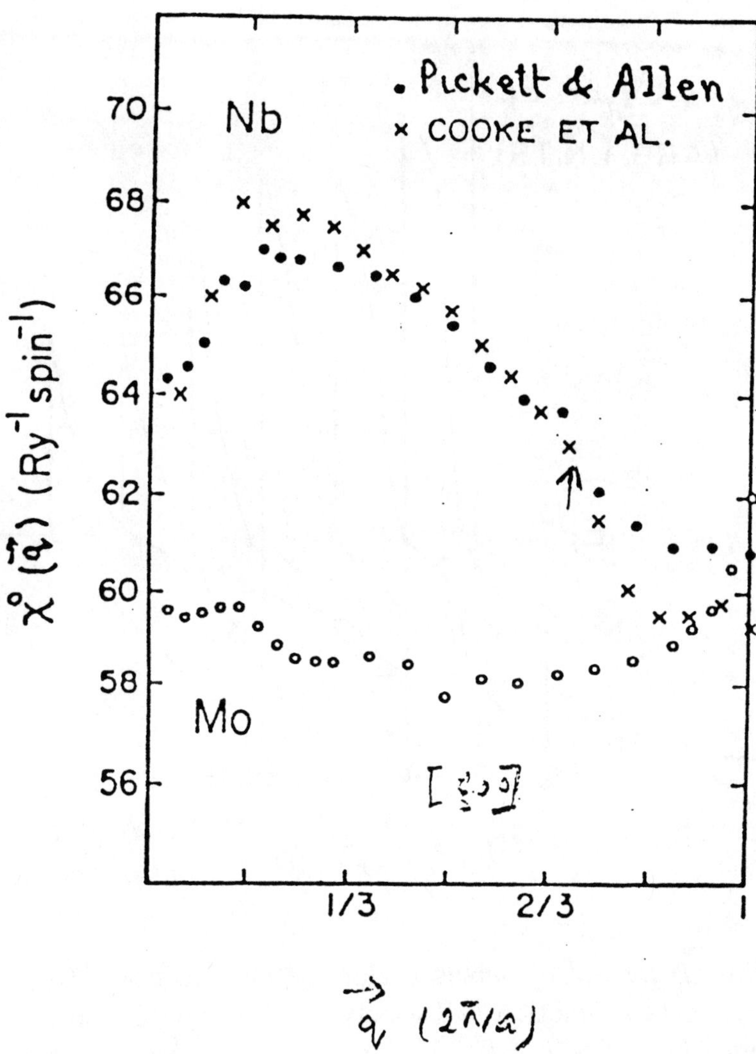

Figure 6.5: Unenhanced susceptibility function $\chi^0(\vec{q})$ along $[\xi00]$ direction for Nb (closed circles) and Mo (open circles) due to Pickett and Allen [46] obtained by using Gilat and Raubenheimer method of integration [32]. The results of Cooke et al [45] are shown by the crosses for comparison. Note that a small constant (8.0) is added in the results of Pickett and Allen to bring these results in agreement with the results of Cooke et al. This constant presumably comes from a higher energy cut off used in the calculations of Cooke et al.

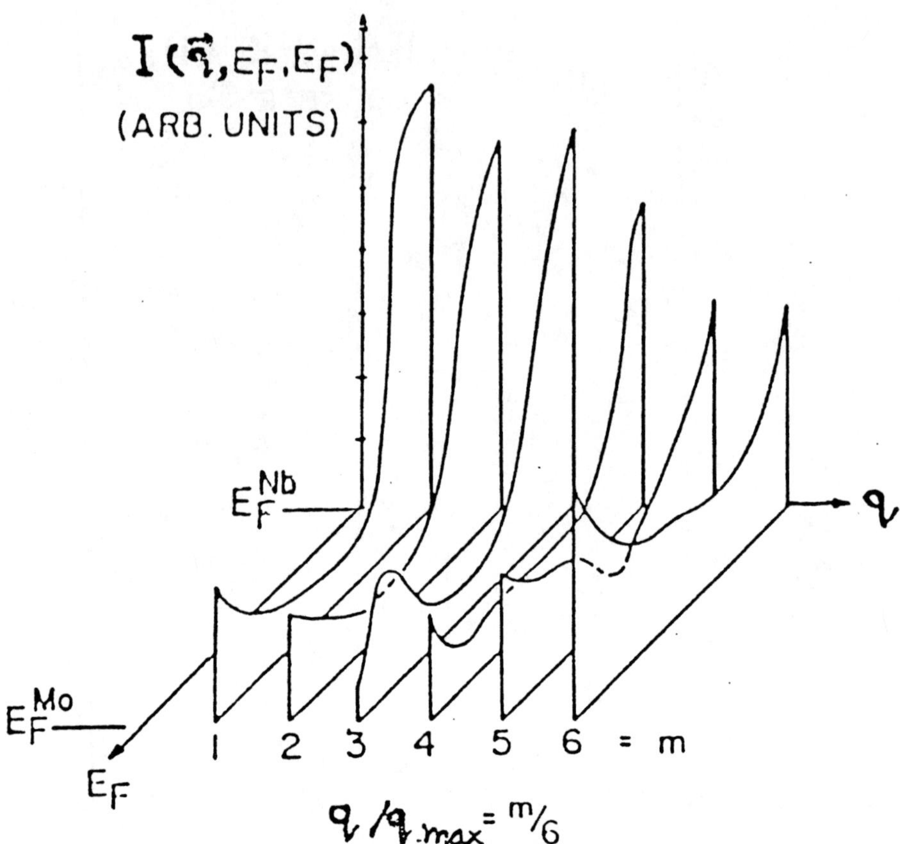

Figure 6.6: $I(\vec{q}, E_F, E_F)$ versus $\vec{q}$ along $[\xi 00]$ direction. $I(\vec{q}, E_F, E_F)$ is also shown as a function of E_F between Nb and Mo due to Pickett and Allen [46]. Here $I(\vec{q}, E_F, E_F) \approx [D(E_F)]^2$ which is a measure of electron-hole pair density in the vicinity of Fermi energy.

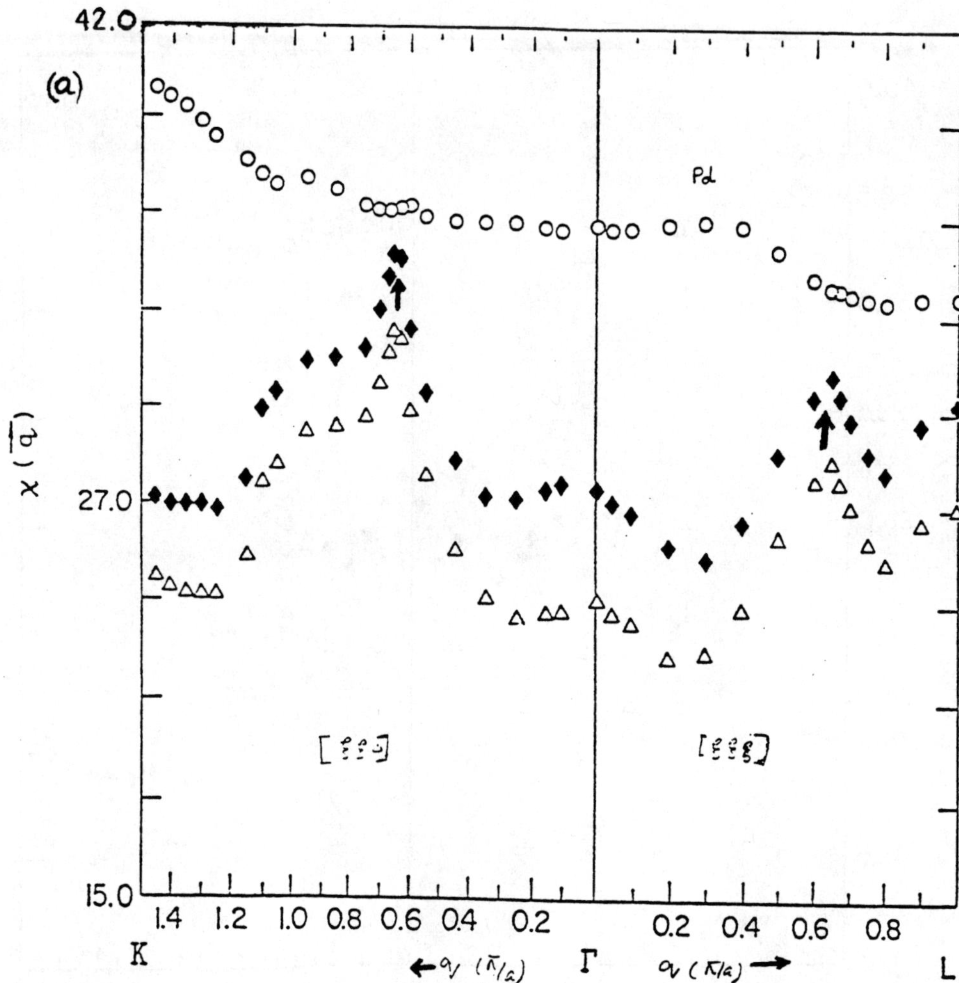

Figure 6.7: $\chi^0(\vec{q})$ versus $\vec{q}$ (π/a) for Pd (a) along $[\xi\xi 0]$ and $[\xi\xi\xi]$ directions and (b) along $[\xi 00]$ and $[\xi\xi/20]$ directions. Circles denote total inter-band contribution, full dimonds denote total intra-band contribution and triangles denote the intra-band contribution of the uppermost bands which intersect Fermi energy [60, 61].

Figure 6.7(b) continued

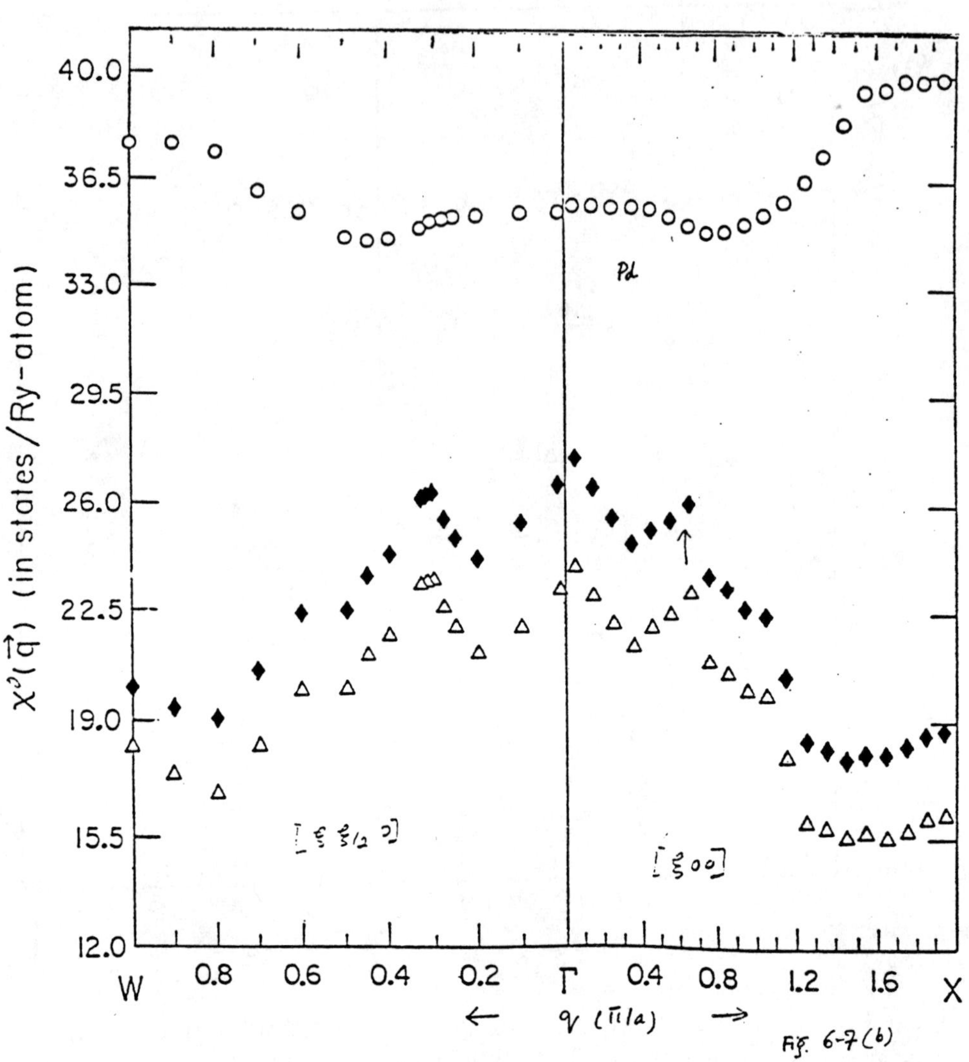

Figure 6.7(b)

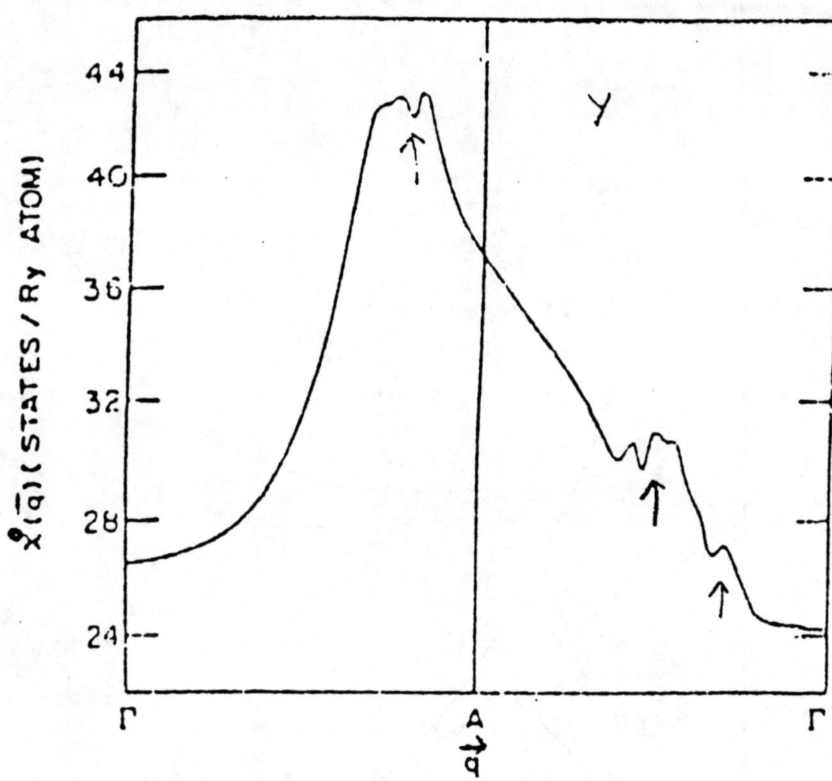

Figure 6.8: $\chi^0(\vec{q})$ versus $\vec{q}$ for yttrium (Y) along $[00\xi]$ direction in the double zone scheme [73]. There are three peaks at $\vec{q} = (2\pi/c)[00\xi]$ with $\xi = 0.375, 0.583$ and 0.75.

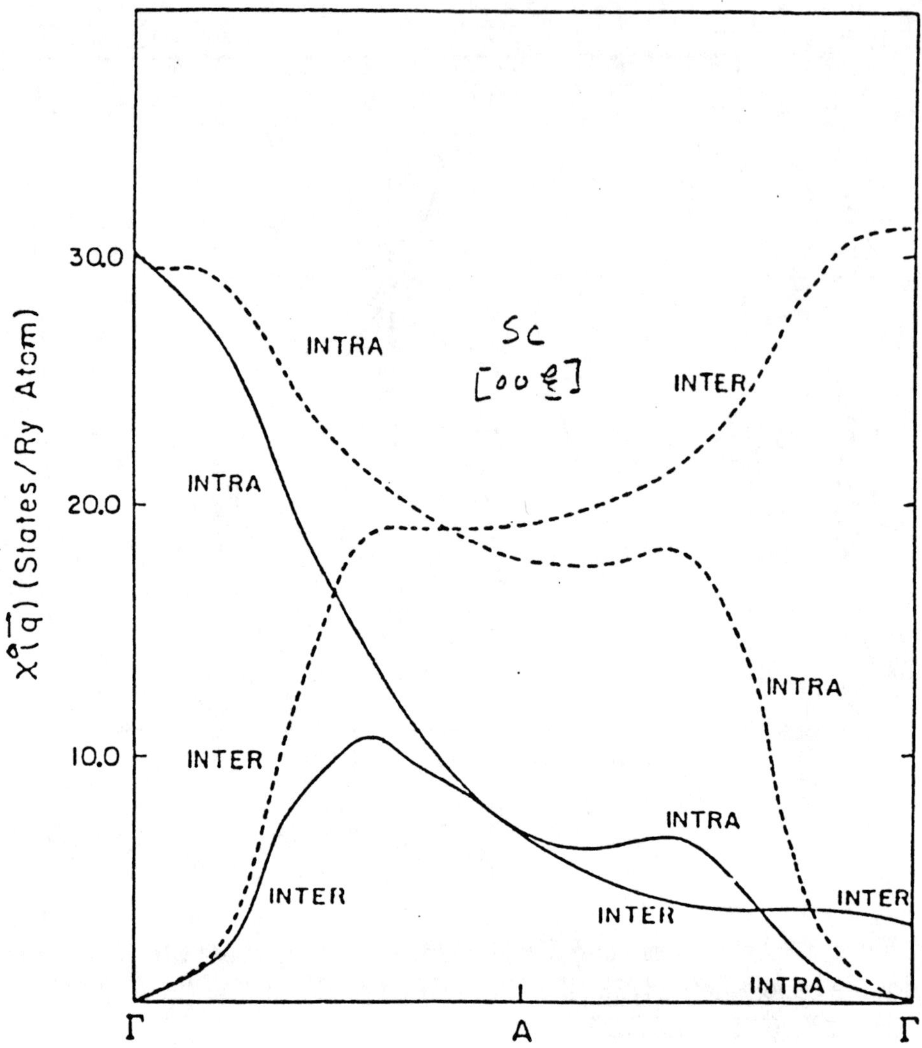

Figure 6.9: Intra-band and inter-band contributions of susceptibility function $\chi^0(\vec{q})$ versus $\vec{q}$ for scandium (Sc) along $[00\xi]$ direction in the double zone scheme [75]. The solid lines show the results when APW overlap matrix elements are accounted for while the dashed lines show the results when the matrix elements are taken as constant.

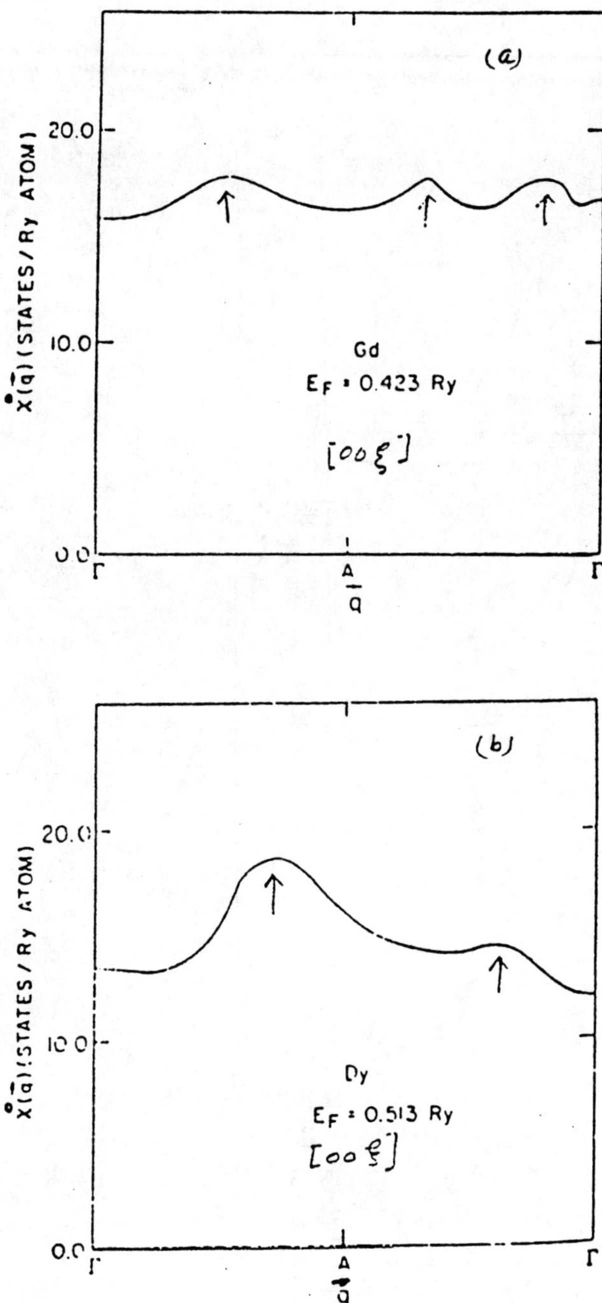

Figure 6.10: $\chi^0(\vec{q})$ versus $\vec{q}$ along $[00\xi]$ direction for (a) Gd and (b) Dy in the double zone scheme. THe broad peaks are marked by arrows.

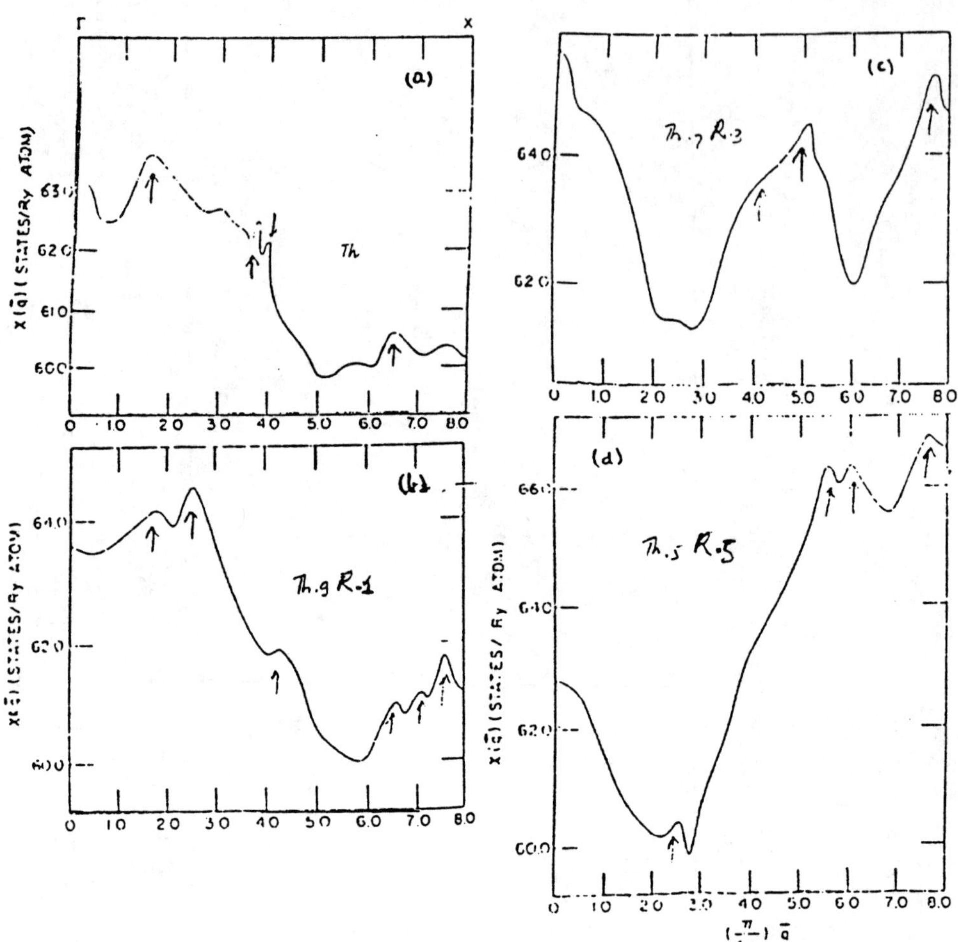

Figure 6.11: $\chi^0(\vec{q})$ versus $\vec{q}$ $(\pi/4a)$ along $[\xi 00]$ direction for (a) Th metal and (b) $Th_{1-x}R_x$ alloys with $x = 0.1$, (c) with $x = 0.3$ and (d) with $x = 0.5$ respectively [73]. The crystal structure is taken as fcc. The broad peaks are denoted by arrows.

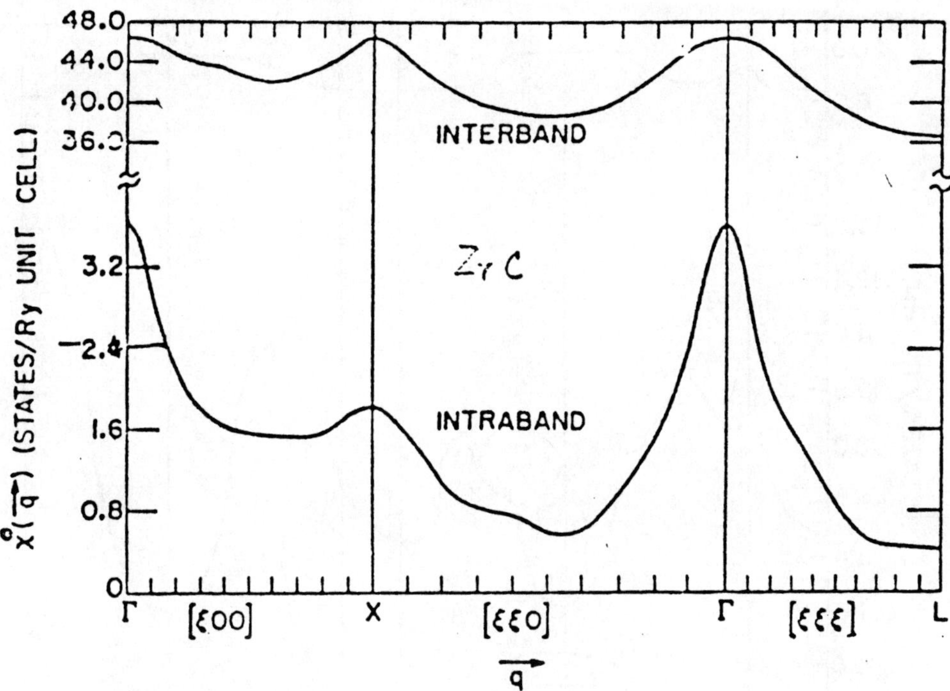

Figure 6.12: $\chi^0(\vec{q})$ versus $\vec{q}$ along $[\xi 00]$, $[\xi\xi 0]$, $[\xi\xi\xi]$, $[\xi, \xi/2, 0]$ and $[1\xi 0]$ directions for NbC. The curves (a), (b) and (c) represent intra-band contributions from the 6th, 5th and 4th bands below Fermi energy. The curve (d) represents the total inter-band contribution [55].

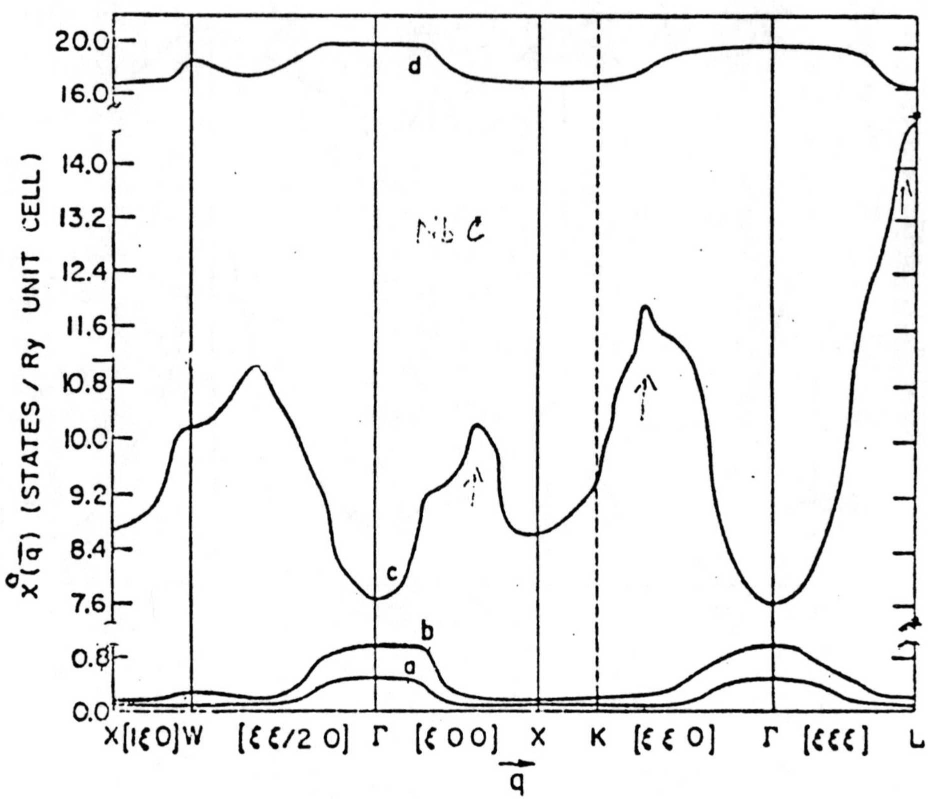

Figure 6.13: $\chi^0(\vec{q})$ versus $\vec{q}$ for ZrC along $[\xi 00]$, $[\xi\xi 0]$ and $[\xi\xi\xi]$ directions [55]. The total intra-band and inter-band contributions are shown separately.

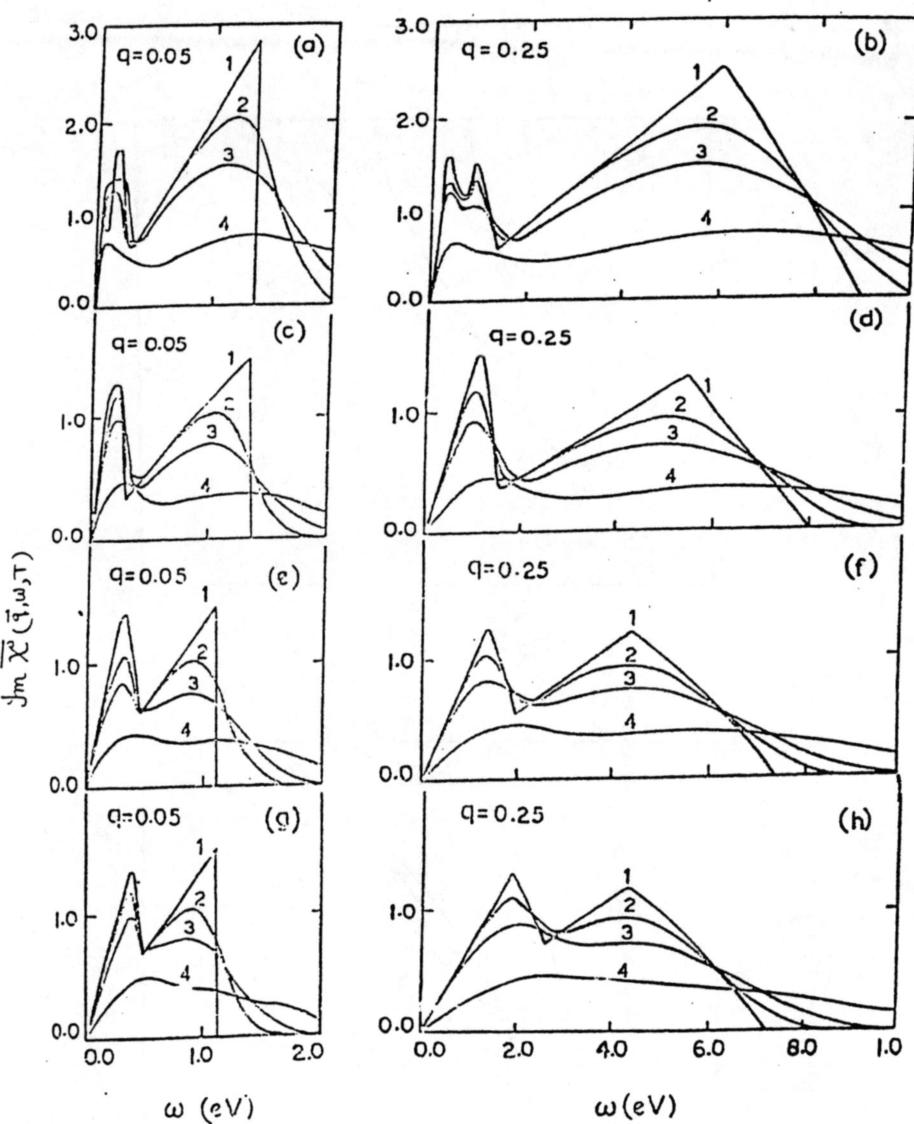

Figure 6.14: $\mathrm{Im}\bar{\chi}^0(\vec{q},\omega,T)$ versus ω for $|\vec{q}| = 0.05$ and 0.25 (a.u.) calculated in the non-interacting band scheme [53]. The curves 1, 2, 3 and 4 correspond to $T/T_F = 0.0, 0.2, 0.4$ and 1.2 respectively. Figures (a) and (b) show the results for ferromagnetic nickel, (c) and (d) for paramagnetic nickel, (e) and (f) for paladium and (g) and (h) for platinum.

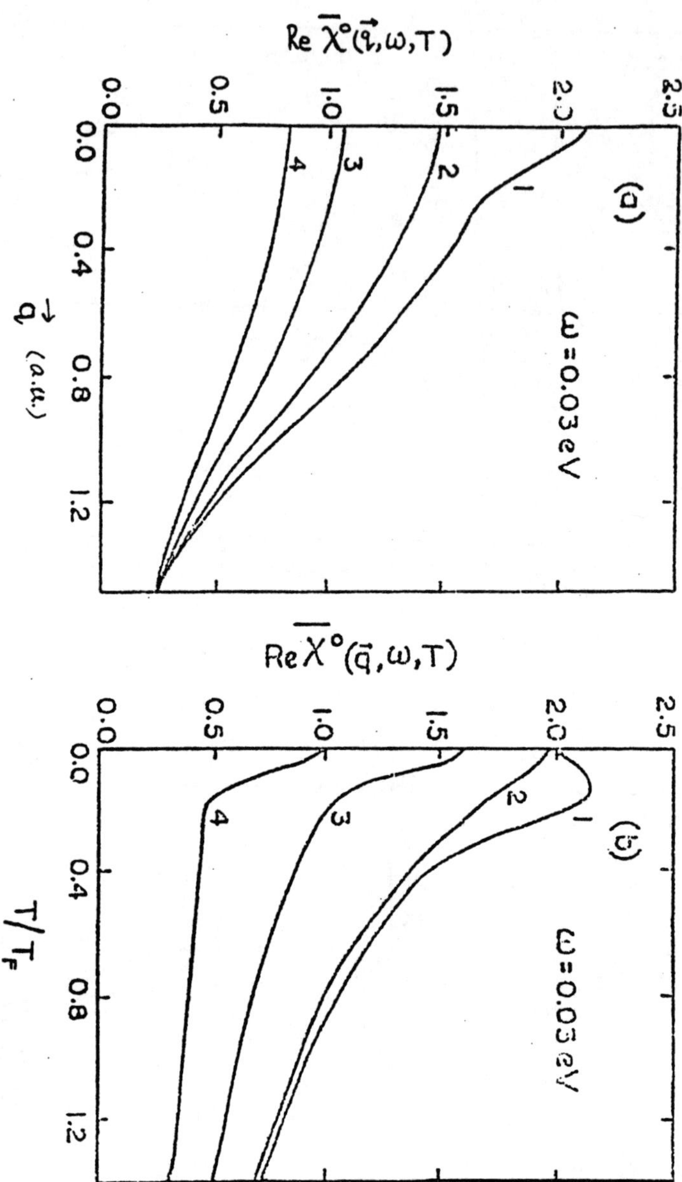

Figure 6.15: $\mathrm{Re}\bar{\chi}^0(\vec{q}, \omega, T)$ as function of $\vec{q}$ and T for $\omega = 0.03$ eV for paladium. In the figure (a) the curves 1, 2, 3 and 4 correspond to T/T_F = 0.2, 0.4, 0.8 and 1.2 respectively and in figure (b) the curves 1, 2, 3 and 4 correspond to $|\vec{q}| = 0.05$, 0.25, 0.85 and 1.25 respectively.

Bibliography

[1] S.Prakash and S.K.Joshi, Phys. Rev. B2, 915 (1970).

[2] W.Hanke, Phys. Rev.B8, 4585 (1973); Phys. Rev. B8, 4591(1973).

[3] N.Singh, J.Singh and S.Prakash, Phys. Rev. B12, 1076 (1975); Phys. Rev. B8, 5415 (1975).

[4] J.Singh, N.Singh and S.Prakash, Phys. Rev. B12, 3159(1975); Phys. Rev. B8, 3166 (1975); Phys. Rev. B18, 2954 (1978).

[5] J. Singh and S. Prakash, Nuovo Cimento 37B, 131 (1977).

[6] W.Kohn, Phys. Rev. Lett. 2, 393 (1959); E.J.Woll and W.Kohn, Phys. Rev. 126, 16 (1962).

[7] L.M.Roth, H.J.Zeiger and T.A. Kaplan, Phys. Rev.149, 519 (1966).

[8] J. Rath and J. Callaway, Phys. Rev. B8, 5398 (1973)

[9] W.M. Lomer, Proc. Phys. Soc. 80, 489 (1962).

[10] R.P. Gupta and S.K. Sinha, Phys. Rev. B3, 2401 (1971).

[11] R.P. Gupta and S.K. Sinha, J.Appl. Phys. 41, 915 (1970).

[12] T.L. Loucks, Augmented Plane Wave Method (Benjamin Inc., New York,1967) and references there in.

[13] J. Korringa. Physica 13, 392 (1947).

[14] W.Kohn and N.Rostoker, Phys. Rev. 94, 111 (1954).

[15] B. Seigal and F.S. Ham, Methods in Computational Physics (Acad. Press, New York, 1968) Vol 8.

[16] G.C. Fletcher, Proc. Phys. Soc. 65, 192 (1952).

[17] J.C. Slater and G.F. Koster, Phys. Rev. 94, 1498 (1954).

[18] F. Stern, Phys. Rev. 116, 1399 (1959).

[19] M. Asdente and J. Friedel, Phys. Rev. 124, 384 (1961).

[20] J.R. Reitz, Solid State Physics (Acad. Press, New York, 1955) Vol 1, p. 1.

[21] J. Callaway, Energy Band Theory (Acad. Press, New York, 1964).

[22] V. Heine, Proc. Phys. Soc. 240A, 340, 361(1957).

[23] T.O. Woodruff, Solid State Physics (Acad. Press, New York, 1957) Vol. 4, p. 367.

[24] L.M. Falicov, Phil. Trans. Roy. Soc. (London) 255, 55(1962).

[25] G.A. Burdick, Phys. Rev. Lett. 7 , 156 (1961).

[26] L. Hodges, H. Ehrenreich and N.D. Lang, Phys. Rev. 152, 505 (1966); H. Ehrenreich and L. Hodges, in " Methods in Computational Physics" (Acad. Press, New York, 1968) Vol. 8.

[27] F.M. Müller, Phys. Rev. 153, 659 (1967).

[28] C.G. Windsor, R.D. Löwde and G. Allan, Phys. Rev. Lett. 22, 849 (1969).

[29] W.E. Evanson, G.S.Fleming and S.H. Liu, Phys. Rev. 178, 783, (1969).

[30] C. Jackson and S. Doniach, Phys. Lett. 30A, 328 (1969).

[31] G. Gilat and G. Dolling, Phys. Lett. 8, 304 (1964).

[32] G. Gilat and L.J. Raubenheimer, Phys. Rev. 144, 390 (1966).

[33] F.M. Müller, J.W. Garland, M.H. Cohen and K.H. Bennemann, Ann. Phys. (N.Y) 67, 19 (1971).

[34] J.F. Cooke and R.F. Wood, Phys. Rev. B5, 1276 (1972).

[35] J.F. Janak, Phys. Lett. 28A, 570 (1969); J.F. Janak, D.E. Eastman and A.R. Williams, Solid State Commun. 8, 271 (1970).

[36] O. Jepsen and O.K. Anderson, Solid State Commun. 9, 1763 (1973).

[37] G. Lehman amd M. Taut, Phys. Stat. Solidi. 54 , 469 (1972).

[38] D. Lipton and R.L. Jacob, J. Phys. C3 , 1388 (1970).

[39] J. Rath and A.J. Freemman, Phys. Rev. B11, 2109 (1975).

[40] G. Lehman, P. Rennert, M. Taut and H. Wonn, Phys. Stat. Solidi 37, K27 (1970).

[41] W.E. Evanson and S.H. Liu, Phys. Rev. Lett. 21, 432 (1968); Phys. Rev. 178, 783 (1969).

[42] W.E. Evanson, G.S. Fleming and S.H. Liu, Phys. Rev.178, 930 (1969).

[43] L.F. Mattheiss, Phys. Rev. 139A, 236 (1965).

[44] G.Shirane and W.J. Takei, J. Phys. Soc. Japan. 17 Suppl. B III, 35 (1962).

[45] J.F. Cooke, H.L. Davis and M. Möstoller, Phys. Rev. B9, 2485 (1974).

[46] W.E. Pickett and P.B. Allen, Phys. Rev. B16, 3127 (1977).

[47] M. Yasui and M Shimizu, J. Phys. F15, 2365 (1985).

[48] E. Stenzil and H. Winter, J. Phys. F15, 1571 (1985); J. Phys. F16, 1789 (1986).

[49] J.E. Hebborn and N.H. March, Adv. Phys. 19, 175 (1970).

[50] S. Doniach, Proc. Phys. Soc. 91, 86 (1967).

[51] R. Singh and S. Prakash, Phys. Rev. B15, 5412 (1977).

[52] R. Singh, J. Singh and S. Prakash, Phys. Rev. B16, 4012 (1977); Phys. Stat. Solidi (b)103, K65 (1981).

[53] R. Singh, S. Prakash and J. Singh, Phys. Rev. B23, 2357 (1981); J. Phys. F10, 1231 (1980).

[54] R. Jullien, M.T. Beal-Monod and B. Coqblin, Phys. Rev. Lett.30, 1057 (1973); Phys. Rev. B9, 1441 (1974).

[55] M. Gupta and A.J. Freeman; Phys. Rev. Lett. 37, 364 (1976); Phys. Rev. B14, 5205 (1976).

[56] F.Y. Fradin, D.D. Koelling, A.J. Freeman and T.J. Watson-Yang, Phys. Rev. B12, 5570 (1975).

[57] T.J. Watson-Yang, B.N. Harmon and A.J. Freeman, J. Mag. and Magn. Materials 2, 334 (1976).

[58] T.J. Watson-Yang, A.J. Freeman and D.D. Koelling, J. Mag and Magn. Materials 5, 277 (1977).

[59] A.J. Freeman, Physica 91B, 103 (1977).

[60] A.J. Freeman, M. Gupta, H.W. Myron, J. Rath and T.J. Watson-Yang, in " Lattice Dynamics" Ed. M. Balkanski (Flammarion, Paris,1977) p.204.

[61] A.J. Freeman, T. J. Watson-Yang and J. Rath, J. Mag. and Magn. Materials 12, 140 (1979).

[62] A.P. Miller, Can. J. Phys. 53, 2491 (1975).

[63] T.L. Loucks, Phys. Rev. 144, 504 (1966); Phys. Rev. 159, 544 (1967).

[64] S.L. Altman and C.J. Bradley, Proc. Phys. Soc. (London) 92,764 (1967).

[65] G.F. Fleming and T.L. Loucks, Phys. Rev 173, 685 (1968).

[66] K.C. Wong, J. Phys. C3, 378 (1970).

[67] S. Wakoh and J. Yamashita, J. Phys. Soc. Japan 28, 1151 (1970).

[68] N. Mori, T. Ukai and S. Kono, J. Phys. Soc. Japan 37, 1278 (1974).

[69] F. Batallan, I. Rosenman and C.B. Sommers, Phys. Rev. B11, 545 (1975).

[70] O. Jepson, O.K. Anderson and A.R. Mackintosh, Phys. Rev. B12, 3084 (1975).

[71] H.R. Child and W.C. Köhler, J. Appl. Phys. 37, 1353 (1966); Phys. Rev. 174, 562 (1968).

[72] W.C. Köhler, H.R. Child, E.O. Wollen and J.W. Cable, J. Appl. Phys. 34, 1335 (1963).

[73] S.H. Liu, R.P. Gupta and S.K. Sinha, Phys. Rev. B4, 1100 (1971).

[74] N. Wakabayashi, S.K. Sinha and F.H. Spedding, Phys. Rev. B4, 2398 (1971).

[75] R.P. Gupta and A.J. Freeman, Phys. Rev. B13, 4376 (1976).

[76] A.H. MacDonald, K.L. Liu and S.H. Vosko, Phys. Rev. B16, 777 (1977).

[77] H.R. Child, W.C. Köhler and A.H. Millhouse, J. Appl. Phys. 39, 1329 (1970).

[78] G.S. Fleming and S.H. Liu, Phys. Rev. B2,164 (1970).

[79] W.C. Köhler, J. Appl. Phys. 36, 1078 (1965).

[80] M.A. Ruderman and C. Kittel, Phys. Rev. 96, 99 (1954).

[81] T. Kasuya, Prog. Theort. Phys. (Kyoto) 16, 45 (1956).

[82] K. Yosida, Phys. Rev. 106, 893 (1957).

[83] S.C. Keeton and T.L. Loucks, Phys. Rev. 168, 672 (1968).

[84] V. Ern and A.C. Switendick, Phys. Rev. 137, 202 (1965).

[85] R.W. Simpson, M.S. Thesis (Uni. of Florida, Gainesville, 1969) unpublished.

[86] K. Schwarz, J. Phys. C10, 195(1977).

[87] C.Y. Fong and M.L. Cohen, Phys. Rev. 136, 3633 (1972).

[88] L.F. Mattheiss, Phys. Rev. B5, 315 (1972).

[89] D.J. Chadi and M.L. Cohen, Phys. Rev. B10, 496 (1974).

[90] A. Neckal, P. Rastl, R. Eibler, P. Weinberger and K. Schwarz, J Phys. C9, 579 (1976).

[91] J.L. Calais, Adv. Phys. 26, 847 (1977).

[92] P. Weinberger, R. Podloucky, C.P. Mallett and A. Neckal, J. Phys C12, 801 (1979).

[93] W. Hanke, J. Hafner and H. Bilz, Phys. Rev. Lett. 37, 1560 (1976).

[94] J. Hafner and W. Hanke, in "Lattice Dynamics" Ed. M. Balkanski (Flammarion, Paris, 1977) p.231.

[95] S.H. Vosko and J.P. Perdew, Cand. J. Phys. 53, 1385 (1975).

[96] J.F. Janak, Phys. Rev. B16, 255 (1977).

[97] J. Singh and S.Prakash, J. Phys. F11, 2409 (1981).

[98] R.W. Shaw and W.A. Harrison, Phys. Rev. 163, 604 (1967).

[99] S. Prakash, Temperature Variation of Dynamical Spin Susceptibility of d-band metals (unpublished).

[100] G. Allan, W.M. Lomer, R.D. Löwde and C.G. Windsor, Phys. Rev. Lett. 20, 933 (1968).

[101] K.S. Singwi, A. Sjölander, M.P. Tosi and R.H. Land, Phys. Rev. B1, 1044 (1970).

Chapter 7

LATTICE DYNAMICS OF METALLIC ALLOYS

Contents

List of Figures

Chapter 7

LATTICE DYNAMICS OF METALLIC ALLOYS

In the earlier chapters we described the lattice dynamics of perfect crystals where the dynamical matrix has the recurrent structure. The calculations of eigenfrequencies and eigenvectors of normal modes of lattice vibrations were rather straight forward. However, the presence of even few impurities, in an otherwise perfect cyrstal, destroys the periodicity of lattice and thus the disorder is introduced. Consequently, the vibrational properties of material are changed [1, 2]. In general, the frequencies of vibrational modes inside the band of allowed frequencies are shifted with respect to the frequencies of an otherwise perfect crystal. Few frequencies which lie near the band edges may even emerge out from the allowed band into the gap region. In these modes, the amplitude of atomic vibrations decreases faster than exponential in the vicinity of impurity atoms. These modes are called local modes. There can also be additional modes whose frequencies lie well within the allowed frequency band of the perfect crystal. These modes are called resonance modes. With the increase of concentration of impurities, the local modes and resonance modes may lead to new phonon bands.

The disorder produced in the defect crystal may be classified in many ways. However, we restrict to substitutional and random disordered alloys which are shown in Fig.(7.1) [3]. The simplest substitutional disorder exist in a crystal with low concentration of impurities.

In these systems, the impurities are separated by large distances and these are heavily screened by conduction electrons of the host metal. Therefore impurity-impurity interaction is almost negligible. Thus the dilute alloy problem reduces to the single impurity problem which can be solved exactly. If the substitutional atom is sufficiently light or tightly bound in a matrix of heavy atoms, it will vibrate in a high frequency local mode which is spatially localized around the light atom. A heavy or a weakly bound impurity atom prefers to vibrate in a low frequency resonance mode or "quasilocalized mode". The lack of exact translational symmetry will give finite width to the localized and resonance modes. Thus, there will be finite life time of the phonons. The width of the phonon peaks will be small if the masses and chemical properties of constituent atoms are nearly the same. If these properties are too far apart, the vibrational spectrum may split into two or more distinct bands.

When the concentration of other constituent atoms increases say more than about 10%. The impurity-impurity interaction becomes important. The concept of host and impurity is lost. The actual positions of impurities which determine the configuration of the alloy become unknown. Then, one has to calculate the ensemble (configurational) averaged values of the physical quantities such as phonon frequencies, phonon width and atomic displacements. Each phonon peak may have the finite width and therefore $\omega(\vec{q})$ versus $\vec{q}$ dispersion relation will also have the finite width. There may also be discontinuities in the phonon dispersion relations at some particular values of phonon wave vectors. With the increase in the concentration of impurities, the localized modes in the forbidden gap may overlap and may form the new band which may eventually merge with the host phonon bands. Thus the analysis of phonon modes in these random alloys becomes more intricate.

The basic problem in studying the lattice dynamics of disordered solids is the incomplete knowledge of microscopic configuration of the solid. This lack of knowledge does not permit to find the exact solution of the eigenvalue problem. Therefore, the concept of ensemble average of observable quantities is introduced. Here an ensemble consists of elements corresponding to all possible atomic configurations. It is assumed that the experiments, in general, provide ensemble averaged values of physical quantities. Therefore, the efforts are made to calculate the ensemble averaged values of the observables. The Green's function method is the most suitable analytical method to study the lattice dynamics of disordered alloys as it is quite compatible with the concept of ensemble average.

The plan of this chapter is as follows: The calculations of normal mode frequencies of perfect and defect crystals by Green's function method are given in sections 7.1 and 7.2. The dynamics of disordered transition metal alloys is described in sections 7.3 and 7.4 in the averaged t-matix approximation (ATA), virtual crystal approximation (VCA) and coherent potential approximattion (CPA). The shell model and double shell model for disordered alloys are discussed in section 7.5 and an epilogue is presented in section 7.6.

7.1 Green's Function for the Perfect Crystals

The equation of motion for the perfect crystal in the harmonic approximation is [1, 2]

$$\sum_{l'\kappa'\beta} \left[M_\kappa \omega^2 \delta_{\alpha\beta} \delta_{ll'} \delta_{\kappa\kappa'} - \Phi_{\alpha\beta}\left(l\kappa, l'\kappa'\right) \right] u_\beta(l'\kappa') \; = \; 0. \qquad (7.1)$$

Equation (7.1) can also be written as

$$\sum_{l'\kappa'\beta} L^0_{\alpha\beta}\left(l\kappa, l'\kappa'; \omega^2\right) u_\beta(l'\kappa') \; = \; 0 \qquad (7.2)$$

where

$$L^0_{\alpha\beta}\left(l\kappa, l'\kappa'; \omega^2\right) \; = \; M_\kappa \omega^2 \delta_{\alpha\beta} \delta_{ll'} \delta_{\kappa\kappa'} - \Phi_{\alpha\beta}(l\kappa, l'\kappa'). \qquad (7.3)$$

In the matrix notations Eq.(7.2) becomes

$$\mathbf{L}^0 \mathbf{u} = 0 \qquad (7.4a)$$

where

$$\mathbf{L}^0 = \mathbf{M}_0 \omega^2 - \mathbf{\Phi}. \qquad (7.4b)$$

$$(7.4)$$

$\mathbf{L}^0(\omega^2)$ is a $3rN \times 3rN$ matrix where r is the number of atoms in a unit cell and N is the number of unit cells in the crystal. $\mathbf{M}_0$ is the atomic mass matrix in the perfect crystal.

The Green's function for the perfect crystal $\mathbf{G}^0$ satisfies the dynamical equation

$$\sum_{l''\kappa''\gamma} \left[M_\kappa \omega^2 \delta_{\alpha\gamma} \delta_{ll''} \delta_{\kappa\kappa''} - \Phi_{\alpha\gamma}\left(l\kappa, l''\kappa''\right) \right]$$

$$\times G^0_{\gamma\beta}\left(l''\kappa'', l'\kappa'; \omega^2\right) = \delta_{\alpha\beta}\delta_{ll'}\delta_{\kappa\kappa'}. \tag{7.5a}$$

In the matrix notations

$$\mathbf{L}^0 \mathbf{G}^0 = \mathbf{I}$$

and hence

$$\mathbf{G}^0 = \mathbf{L}^{0-1} \tag{7.5b}$$

which can also be written as

$$G^0_{\alpha\beta}\left(l\kappa, l'\kappa'; \omega^2\right) = L^{0-1}_{\alpha\beta}\left(l\kappa, l'\kappa'; \omega^2\right). \tag{7.5c}$$

$$(7.5)$$

The eigenvectors of the dynamical matrix $\mathbf{L}^0$, $\left\{ N^{-1/2}\vec{e}(\vec{q}|\kappa)e^{i\vec{q}\cdot\vec{R}(l)} \right\}$ are orthogonal and form a complete set in the crystal space. Therefore

$$\sum_{\vec{q}j} \left\{ N^{-1/2}e_\alpha(\vec{q}j \mid \kappa)e^{i\vec{q}\cdot\vec{R}(l)} \right\}^* \left\{ N^{-1/2}e_\beta(\vec{q}j \mid \kappa')e^{i\vec{q}\cdot\vec{R}(l')} \right\}$$

$$= \delta_{ll'}\delta_{\kappa\kappa'}\delta_{\alpha\beta} \tag{7.6a}$$

and

$$\sum_{l\kappa\alpha} \left\{ N^{-1/2}e_\alpha(\vec{q}j \mid \kappa)e^{i\vec{q}\cdot\vec{R}(l)} \right\}^* \left\{ N^{-1/2}e_\alpha(\vec{q}'j' \mid \kappa)e^{i\vec{q}'\cdot\vec{R}(l)} \right\}$$

$$= \Delta(\vec{q} - \vec{q}')\delta_{jj'}. \tag{7.6b}$$

$$(7.6)$$

Therefore $G^0_{\alpha\beta}(l\kappa, l'\kappa'; \omega^2)$ can be expanded in the double Fourier series as

$$G^0_{\alpha\beta}(l\kappa, l'\kappa'; \omega^2) = N^{-1}(M_\kappa M_{\kappa'})^{1/2}\sum_{\vec{q}j}\sum_{\vec{q}'j'}e_\alpha(\vec{q}j \mid \kappa)$$
$$\times g^0(\vec{q}j, \vec{q}'j'; \omega^2)e_\beta(\vec{q}'j' \mid \kappa')e^{i\vec{q}\cdot\vec{R}(l)+i\vec{q}'\cdot\vec{R}(l')}.$$

$$(7.7)$$

Substituting (7.7) in (7.5a) and using the orthonormality relations given in Eq.(7.6), one finds the expansion coefficients

$$g^0(\vec{q}j, \vec{q}'j'; \omega^2) = \frac{\Delta(\vec{q}+\vec{q}')\delta_{jj'}}{\omega^2 - \omega^2_{0j}(\vec{q})}.$$

$$(7.8)$$

In the derivation of Eq.(7.8), Eqs.(2.54) and (2.67a) are also used. Here $\omega_{0j}(\vec{q})$ are the phonon frequencies of the perfect crystal. Using Eq.(7.8) in (7.7), one gets

$$G^0_{\alpha\beta}(l\kappa, l'\kappa'; \omega^2) = \frac{1}{N(M_\kappa M_{\kappa'})^{1/2}}\sum_{\vec{q}j}\frac{e_\alpha(\vec{q}j \mid \kappa)e^*_\beta(\vec{q}j \mid \kappa')}{\omega^2 - \omega^2_{0j}(\vec{q})}$$
$$\times e^{i\vec{q}\cdot[\vec{R}(l)-\vec{R}(l')]}.$$

$$(7.9)$$

If we define another vector component as

$$w_\alpha(\vec{q}j \mid \kappa) = e^{-i\vec{q}\cdot\vec{R}(\kappa)}e_\alpha(qj \mid \kappa),$$

$$(7.10)$$

Eq.(7.9) can be rewritten as

$$G^0_{\alpha\beta}(l\kappa, l'\kappa'; \omega^2) = \frac{1}{N(M_\kappa M_{\kappa'})^{1/2}}\sum_{\vec{q}j}\frac{w_\alpha(\vec{q}j \mid \kappa)w^*_\beta(\vec{q}j \mid \kappa')}{\omega^2 - \omega^2_{0j}(\vec{q})}$$
$$\times e^{i\vec{q}\cdot[\vec{R}(l\kappa)-\vec{R}(l'\kappa')]}.$$

$$(7.11a)$$

Here $G^0_{\alpha\beta}(l\kappa, l'\kappa'; \omega^2)$ is the function of cell indices l and l' through the relative cell distance $[\vec{R}(l) - \vec{R}(l')]$ which is a consequence of crystal periodicity. From Eq.(7.11a)

$$G^0_{\alpha\beta}(l\kappa, l'\kappa'; \omega^2) = \frac{1}{N(M_\kappa M_{\kappa'})^{1/2}}\sum_{\vec{q}j}\frac{w_\alpha(\vec{q}j \mid \kappa)w^*_\beta(\vec{q}j \mid \kappa')}{\omega^2 - \omega^2_{0j}(\vec{q})}.$$

$$(7.11b)$$

$$(7.11)$$

Thus the eigenfrequencies of perfect crystal are determined by the poles of $\mathbf{G}^0$ or by the zeros of det $|\mathbf{G}^{0-1}|$ given in Eq.(7.11b).

7.2   Phonons in Substitutional Defect Crystals

Let some of the lattice sites of perfect crystal are occupied by substitutional impurities whose masses and charges differ from the mass and charge of host atoms. The Hamiltonian for this defect crystal in the harmonic approximation can be written in the same way as for the perfect crystal [1, 2, 4] i.e.

$$
\begin{aligned}
H^* \;=\; & \frac{1}{2}\sum_{l\kappa\alpha} M_{l\kappa}\ddot{u}_\alpha^2(l\kappa,t) \\
& +\frac{1}{2}\sum_{l\kappa\alpha}\sum_{l'\kappa'\beta} \Phi^*_{\alpha\beta}(l\kappa,l'\kappa')u_\alpha(l\kappa,t)u_\beta(l'\kappa',t)
\end{aligned}
\qquad (7.12)
$$

where $M_{l\kappa}$ and $\vec{u}(l\kappa,t)$ are the atomic masses and displacements of $(l\kappa)$ atoms. The first term of Eq.(7.12) is the kinetic energy and the second term is the total potential energy of the defect crystal. $\Phi^*_{\alpha\beta}(l\kappa,l'\kappa')$ are the force constants and $u_\alpha(l\kappa,t)$ are the components of atomic displacements in the defect crystal.

Following the procedure for the perfect crystal, as described in chapter 2, the equation of motion for the defect crystal becomes

$$
M_{l\kappa}\ddot{u}_\alpha(l\kappa,t) \;=\; -\sum_{l'\kappa'\beta} \Phi^*_{\alpha\beta}(l\kappa,l'\kappa')u_\beta(l'\kappa',t).
\qquad (7.13)
$$

Here $\Phi^*_{\alpha\beta}(l\kappa,l'\kappa')$ depends separately on l and l' because the periodicity of the crystal lattice is destroyed by the presence of impurity atoms. Further $M_{l\kappa}$ will depend upon the cell index l as the atoms in the different cells may have different masses. Assuming that the periodic displacement of atoms $u_\alpha(l\kappa,t) = u_\alpha(l\kappa)e^{-i\omega t}$, Eq.(7.13) becomes

$$
\omega^2 M_{l\kappa}u_\alpha(l\kappa) \;=\; \sum_{l'\kappa'\beta} \Phi^*_{\alpha\beta}(l\kappa,l'\kappa')u_\beta(l'\kappa')
\qquad (7.14)
$$

which can also be written as

$$\sum_{l'\kappa'\beta} L_{\alpha\beta}(l\kappa, l'\kappa'; \omega^2)u_\beta(l'\kappa') \;=\; 0 \qquad (7.15)$$

where

$$L_{\alpha\beta}(l\kappa, l'\kappa'; \omega^2) \;=\; M_{l\kappa}\omega^2\delta_{ll'}\delta_{\kappa\kappa'}\delta_{\alpha\beta} - \Phi^*_{\alpha\beta}(l\kappa, l'\kappa'). \qquad (7.16)$$

In the matrix notations

$$\mathbf{Lu} \;=\; 0 \qquad (7.17)$$

and

$$\mathbf{L} \;=\; \mathbf{M}\omega^2 - \Phi^*. \qquad (7.18)$$

We can also write $\mathbf{L}$ in terms of $\mathbf{L}^0$ as

$$\mathbf{L} \;=\; \mathbf{L}^0 - \mathbf{\Delta L} \qquad (7.19)$$

where $\mathbf{\Delta L}$ is the change in dynamical matrix due to mass defect and change in force constants. Using Eqs.(7.3) and (7.16) in (7.19), one finds

$$\Delta L_{\alpha\beta}\left(l\kappa, l'\kappa'; \omega^2\right) \;=\; \omega^2\epsilon_{l\kappa}M_\kappa\delta_{ll'}\delta_{\kappa\kappa'}\delta_{\alpha\beta} + \Delta\Phi_{\alpha\beta}\left(l\kappa, l'\kappa'\right) \qquad (7.20)$$

where the mass difference parameter

$$\epsilon_{l\kappa} \;=\; (M_\kappa - M_{l\kappa})/M_\kappa \qquad (7.21)$$

and the change in force constants

$$\Delta\Phi_{\alpha\beta}(l\kappa, l'\kappa') \;=\; \Phi^*_{\alpha\beta}(l\kappa, l'\kappa') - \Phi_{\alpha\beta}(l\kappa, l'\kappa'). \qquad (7.22)$$

In matrix notations

$$\mathbf{\Delta L} \;=\; \omega^2\epsilon\mathbf{M} + \mathbf{\Delta\Phi}. \qquad (7.23)$$

In principle, $\mathbf{\Delta L}$ is an infinite dimensional matrix. However, in practice only those elements of $\mathbf{\Delta L}$ are finite where $l\kappa$ and $l'\kappa'$ refer

either to the impurity or to the host atoms which interact with the
impurity atom. Therefore $\Delta\mathbf{L}$ becomes a $3n \times 3n$ matrix where n is
the number of interacting sites including the impurity site. This range
of impurity interaction is called the impurity space of that particular
impurity in the crystal.

In the metallic crystals, the impurity and host ion interaction is
screened by the conduction electrons. Therefore impurity may interact
effectively only with few NN atoms. This may lead to a small impurity
space. In the d- and f-band metals the electrostatic screening may be
large due to quasilocalized nature of conduction electrons. Therefore
$\Delta\mathbf{L}$ may be a low dimensional matrix. This will simplify the lattice
dynamical problem to a great extent.

The potential energy due to interaction between the impurity and
host atoms is unknown. Therefore the elements of $\Delta\mathbf{\Phi}$, in general, are
obtained by comparing the predictions of theory and the experimental
results for some physical property which is affected by the presence of
impurities in the crystal. It is therefore essential that the form of the
matrix $\Delta\mathbf{\Phi}$ should be consistent with the symmetry properties which
apply to the atomic force constants of any perfect or imperfect crystal.
The conditions obtained from the invariance of forces on a given atom
through rigid body translation or rotation etc. in the perfect crystal
are also valid in the defect crystal. Therefore the elements of $\Delta\mathbf{\Phi}$ must
satisfy the relations:

$$\Delta\Phi_{\alpha\beta}(l\kappa, l'\kappa') = \Delta\Phi_{\beta\alpha}(l'\kappa', l\kappa), \tag{7.24}$$

$$\sum_{l'\kappa'} \Delta\Phi_{\alpha\beta}(l\kappa, l'\kappa') = 0 \tag{7.25}$$

and

$$\sum_{l'\kappa'} \Delta\Phi_{\alpha\beta}(l\kappa, l'\kappa')u_{\gamma}(l'\kappa') = \sum_{l'\kappa'} \Delta\Phi_{\alpha\gamma}(l\kappa, l'\kappa')u_{\beta}(l'\kappa').$$

$$\tag{7.26}$$

The Green's function for the defect crystal $G_{\alpha\beta}(l\kappa, l'\kappa'; \omega^2)$ can be defined from Eq.(7.15) as

$$G_{\alpha\beta}(l\kappa, l'\kappa'; \omega^2) = L_{\alpha\beta}^{-1}(l\kappa, l'\kappa'; \omega^2) \tag{7.27}$$

or

$$\mathbf{G} = \mathbf{L}^{-1}$$
$$= (\mathbf{L}^0 - \Delta\mathbf{L})^{-1}. \tag{7.28}$$

The Green's function $\mathbf{G}^0$ for the perfect crystal can be calculated using the symmetry properties of the crystal [5]. The calculation of $\mathbf{G}$ is simplified if it is expressed in terms of $\mathbf{G}^0$ which can be done conveniently by expanding Eq.(7.28) in power series of $\Delta\mathbf{L}$ and using Eq.(7.5(b)) for $\mathbf{G}^0$. These calculations give

$$\mathbf{G} = (\mathbf{I} - \mathbf{G}^0\Delta\mathbf{L})^{-1}\mathbf{G}^0$$
$$= (\mathbf{I} + \mathbf{G}^0\Delta\mathbf{L} + \mathbf{G}^0\Delta\mathbf{L}\mathbf{G}^0\Delta\mathbf{L} + \ldots\ldots)\mathbf{G}^0$$
$$= \mathbf{G}^0 + \mathbf{G}^0\Delta\mathbf{L}\mathbf{G}. \tag{7.29}$$

Equation (7.29) is Dyson equation which includes the effect of multiple scattering due to impurity to all orders. However, Eq.(7.29) is valid only if the det $|\mathbf{G}^0\Delta\mathbf{L}|$ is small which is, in general, guaranteed.

7.2.1 Atomic Displacements

The displacement $\mathbf{u}$ can be calculated from Eq.(7.17) by writing

$$(\mathbf{L}^0 - \Delta\mathbf{L})\mathbf{u} = 0. \tag{7.30}$$

Multiplying Eq.(7.30) by $\mathbf{G}^0$ on the left, we get

$$\mathbf{u} = \mathbf{G}^0\Delta\mathbf{L}\mathbf{u}$$

or

$$u_\alpha(l\kappa) = \sum_{l'\kappa'\beta} \mathbf{G}^0{}_{\alpha\beta}(l\kappa, l'\kappa'; \omega^2)\Delta L_{\alpha\beta}(l\kappa, l'\kappa'; \omega^2)u_\beta(l'\kappa').$$

$$\tag{7.31}$$

Here $\mathbf{u}$ is expressed in terms of $\Delta\mathbf{L}$ which is to be calculated by including the impurity-host interaction upto few NNs. Thus the displacement of any atom in the defect crystal is expressed in the terms of the displacements of small number of atoms affected by the impurity.

The calculation of $\mathbf{u}$ is further simplified if the matrix $\mathbf{\Delta L}$ is partitioned as

$$\mathbf{\Delta L} \;=\; \begin{bmatrix} \mathbf{\Delta}l & 0 \\ 0 & 0 \end{bmatrix} \tag{7.32}$$

where $\mathbf{\Delta}l$ is $3n \times 3n$ matrix whose rows and columns are labelled by indices $(l\kappa\alpha)$ and $(l'\kappa'\beta)$ of atoms affected by the impurity atom. This subspace of the crystal is called the space of $\mathbf{\Delta}l$. The matrices $\mathbf{G}^0$ and $\mathbf{u}$ are also partitioned in the same way i.e.

$$\mathbf{G}^0 \;=\; \begin{bmatrix} g^0 & \mathbf{G}^0_{12} \\ \mathbf{G}^0_{21} & \mathbf{G}^0_{22} \end{bmatrix},$$

$$\mathbf{u} \;=\; \begin{bmatrix} \mathbf{u}_1 \\ \mathbf{u}_2 \end{bmatrix} \tag{7.33}$$

where g^0 is a $3n \times 3n$ matrix, $\mathbf{G}^0_{12}$ and $\mathbf{G}^0_{21}$ are $3n \times (3rN - 3n)$ matrices and G^0_{22} is a $(3rN - 3n) \times (3rN - 3n)$ matrix. Again the rows and columns of g^0 are labelled by $(l\kappa\alpha)$ and $(l'\kappa'\beta)$ of atoms affected by the impurity. $\mathbf{u}_1$ and $\mathbf{u}_2$ have $3n \times 1$ and $(3rN - 3n) \times 1$ dimensions respectively.

Using Eqs.(7.32) and (7.33) in Eq.(7.31), one gets

$$\mathbf{u}_1 \;=\; g^0 \mathbf{\Delta}l \mathbf{u}_1 \tag{7.34}$$

and

$$\mathbf{u}_2 = \mathbf{G}^0_{21} \mathbf{\Delta}l \mathbf{u}_1. \tag{7.35}$$

Equation (7.34) consists of 3n simultanious homogenous equations in $3n$ unknown displacement components $u_{1\alpha}(l\kappa)$. These equations can be solved to find the displacement of atoms which are affected by the impurity. Once the values of $u_{1\alpha}(l\kappa)$ for n atoms are known, the displacement components $u_{2\alpha}(l\kappa)$ of the remaining atoms can be determined with the help of Eq.(7.35).

7.2.2 Phonon Frequencies

The solution for all the normal mode frequencies $\omega_j(\vec{q})$ of the defect crystal can be found with the help of determinental equation

$$det|\mathbf{L}| \;= det|\mathbf{L}^0 - \mathbf{\Delta L}| = \; 0. \tag{7.36}$$

Using the fact that the determinent of product of two matrices is equal to the product of determinents of individual matrices, Eq.(7.36) simplifies as

$$det|\mathbf{L}| = det|\mathbf{L}^0|det|\mathbf{I} - \mathbf{G}^0\mathbf{\Delta L}| = 0. \qquad (7.37)$$

If one has to compute only those frequencies which are affected by the impurity atom, one has to solve the equation

$$\begin{aligned} \mathbf{\Delta}(\omega^2) &= det|\mathbf{I} - g^0\mathbf{\Delta L}| \\ &= det|\mathbf{I} - \mathbf{G}^0\mathbf{\Delta L}| \\ &= 0. \end{aligned} \qquad (7.38)$$

The second equality of Eq.(7.38) is obtained with the help of Eqs.(7.32) and (7.33). Using Eqs.(7.34), (7.37) in (7.38), one gets

$$\mathbf{\Delta}(\omega^2) = det|\mathbf{I} - g^0\mathbf{\Delta L}| = det\left|\frac{\mathbf{L}}{\mathbf{L}^0}\right|. \qquad (7.39)$$

Using Eqs.(7.4b) and (7.18) for $\mathbf{L}^0$ and $\mathbf{L}$, one gets

$$\begin{aligned} \mathbf{\Delta}(\omega^2) &= \frac{det|\mathbf{M}\omega^2 - \Phi^*|}{det|\mathbf{M}_0\omega^2 - \Phi|} \\ &= \frac{det|\mathbf{M}^{1/2}(\omega^2\mathbf{I} - \mathbf{D}^*)\mathbf{M}^{1/2}|}{det|\mathbf{M}_0^{1/2}(\omega^2\mathbf{I} - \mathbf{D})\mathbf{M}_0^{1/2}|} \\ &= \frac{det|\mathbf{M}|}{det|\mathbf{M}_0|}\Pi_{\vec{q}j}\frac{\omega^2 - \omega_j^2(\vec{q})}{\omega^2 - \omega_{0j}^2(\vec{q})} \end{aligned} \qquad (7.40)$$

where the dynamical matrices for the defect and perfect crystals are

$$\mathbf{D}^* = \mathbf{M}^{-1/2}\Phi^* M^{-1/2} \quad \text{and} \quad \mathbf{D} = \mathbf{M}_0^{-1/2}\Phi\mathbf{M}_0^{-1/2} \qquad (7.41)$$

respectively. We see from Eq.(7.40) that the solution of Eq. $\mathbf{\Delta}(\omega^2) = 0$ will give only those vibrational modes of the crystal which are perturbed by the presence of impurity. The factors in the product on the right side of Eq.(7.40) which correspond to the remaining normal modes will cancel between the numerator and denominator of the quotient.

The Hamiltonian of the defect crystal can also be expressed as

$$H^* = \sum_{\vec{q}j=1}^{3rN} \hbar\omega_j(\vec{q}) \left(b_{dj}^\dagger b_{dj} + \frac{1}{2} \right) \qquad (7.42)$$

where $b_{dj}^\dagger$ and b_{dj} are phonon creation and annihilation operators respectively for the defect crystal which can be regarded as a system of 3rN independent harmonic oscillators. Thus H^* is the sum of $3rN$ Hamiltonians and each correspond to each degree of freedom of the defect crystal. The eigenvalues of H^* are

$$E_{n_1,n_2,...,n_{3rN}} = \sum_{\vec{q}j} \hbar\omega_j(\vec{q})(n_j + 1/2) \qquad (7.43)$$

with $n_j = 0, 1, 2, \ldots$.

In a substitutional alloy if one neglects the interaction of impurity atoms with the rest of the crystal, only one lattice site will be affected by the impurity i.e. the impurity site itself. As a result only three degrees of freedom of the host crystal will be perturbed by the presence of impurity. In such a system Δl and hence $g^0\Delta l$ are 3×3 matrices. Equation(7.34) will consist of three equations for three components of displacement $\vec{u}_1$ of impurity atom. The zeros of the determinent $\Delta(\omega^2)$ will give the vibrational frequencies of impurity atom. If one includes the interaction of impurity atom wilh its NNs, the problem becomes more involved. For a substitutional impurity which interacts with its 1NNs in a bcc crystal, the space of Δl will contain nine atoms with 27 degrees of freedom. Thus $g^0\Delta l$ becomes a 27×27 matrix. The numerical solutions of these matrix equations become time consuming. However, the problem gets simplified with the group theoretic considerations [6].

The vibrational spectrum of dilute alloy of vanadium $(Be_{0.033}V_{0.907})$ show the high frequency impurity band upon alloying light impurity Be. Heavy mass impurity Pt causes only very little change in the vibrational spectrum of $(Pt_{0.05}V_{0.95})$ and the spectrum of $(Ni_{0.05}V_{0.95})$ has intermediate features [7]. The large changes in the vibrational spectra are produced by the long range interatomic interactions which arise from the changes in the electronic properties of Ni and Pt in the alloy phase. The references for other measurements are given in chapter 3.

7.2.3 Calculation of Green's Function

The vibrational properties of a defect crystal can be studied if the Green's function for the defect cyrstal $\mathbf{G}$ is known. In the calculation of $\mathbf{G}$, one has to work in the impurity space. Therefore $\mathbf{G}$ has to be written in the partitioned form as

$$\mathbf{G} \;=\; \begin{bmatrix} \mathbf{G}_{11} & \mathbf{G}_{12} \\ \mathbf{G}_{21} & \mathbf{G}_{22} \end{bmatrix} \tag{7.44}$$

which is similar to the form of $\mathbf{G}^0$ given in Eq.(7.33). $\mathbf{G}_{11}$ is a $3n \times 3n$ matrix in the space of Δl. Using Eqs.(7.32), (7.33) and (7.44) in Eq.(7.29), one gets

$$\mathbf{G}_{11} = g^0 + g^0 \Delta l \mathbf{G}_{11} \tag{7.45a}$$

$$\mathbf{G}_{12} = \mathbf{G}_{12}^0 + g^0 \Delta l \mathbf{G}_{12} \tag{7.45b}$$

$$\mathbf{G}_{21} = \mathbf{G}_{21}^0 + \mathbf{G}_{21}^0 \Delta l \mathbf{G}_{11}^0 \tag{7.45c}$$

and

$$\mathbf{G}_{22} = \mathbf{G}_{22}^0 + \mathbf{G}_{21}^0 \Delta l \mathbf{G}_{12}. \tag{7.45d}$$

$$\tag{7.45}$$

Evidently $\mathbf{G}_{11}$ can be obtained by solving Eq.(7.45a). The symmetry properties further reduce the order of these matrix equations.

If the normalized eigenvectors of the defect crystal

$$B_\alpha(\vec{q}j|l\kappa) \;=\; w_\alpha(\vec{q}j|\kappa)e^{i\vec{q}\cdot\vec{R}(l\kappa)} \tag{7.46}$$

are known, Green's function $\mathbf{G}$ can be expanded in terms of $B_\alpha(\vec{q}j|l\kappa)$ as

$$G_{\alpha\beta}(l\kappa, l'\kappa'; \omega^2) \;=\; \frac{1}{N\,(M_{l\kappa}M_{l'\kappa'})^{1/2}} \sum_{\vec{q}j} \frac{B_\alpha(\vec{q}j|l\kappa)B_\beta^*(\vec{q}j|l'\kappa')}{\omega^2 - \omega_j^2(\vec{q}j)}. \tag{7.47}$$

The poles of $G_{\alpha\beta}(l\kappa, l'\kappa'; \omega^2)$ are normal mode frequencies $\omega_j(\vec{q})$ of the defect crystal. However, it is not always possible to determine $B_\alpha(\vec{q}j|l\kappa)$. Therefore the calculation of $\mathbf{G}$ is faciliated by expressing it in terms of eigenvectors of the perfect crystal. Since the defect crystal does not have the translational symmetry, $\mathbf{G}$ depends on the cell indices l and l' separately. Therefore $\mathbf{G}$ is expanded in the double Fourier series as

$$
G_{\alpha\beta}(l\kappa, l'\kappa'; \omega^2) = \frac{1}{N\,(M_\kappa M_{\kappa'})^{1/2}} \sum_{\vec{q}j} \sum_{\vec{q}'j'} w_\alpha(\vec{q}j|\kappa) g(\vec{q}j, \vec{q}'j'; \omega^2)
$$
$$
\times w_\beta(\vec{q}'j'|\kappa') e^{[i\vec{q}\cdot\vec{R}(l\kappa)+i\vec{q}'\cdot\vec{R}(l'\kappa')]}.
$$
$$(7.48)$$

Using Eq.(7.48) for $\mathbf{G}$ and (7.11) for $\mathbf{G}^0$ in Eq.(7.29), we obtain the Fourier coefficient

$$
g(\vec{q}j, \vec{q}'j'; \omega^2) = \frac{\Delta(\vec{q}+\vec{q}')\delta_{jj'}}{\omega^2 - \omega_j^2(\vec{q})} + \frac{1}{\omega^2 - \omega_j^2(\vec{q})}
$$
$$
\times \sum_{\vec{q}_1 j_1} v(\vec{q}j, \vec{q}_1 j_1; \omega^2) g(\vec{q}_1 j_1, \vec{q}'j'; \omega^2) \quad (7.49)
$$

where

$$
v(\vec{q}j, \vec{q}'j'; \omega^2) = \frac{1}{N} \sum_{l\kappa\alpha} \sum_{l'\kappa'\beta} \frac{w_\alpha(\vec{q}j|\kappa)}{M_\kappa^{1/2}} \Delta L_{\alpha\beta}(l\kappa, l'\kappa'; \omega^2)
$$
$$
\times \frac{w_\beta^*(\vec{q}'j'|\kappa')}{M_{\kappa'}^{1/2}} e^{[-i\vec{q}\cdot\vec{R}(l\kappa)+i\vec{q}'\cdot\vec{R}(l'\kappa')]}. \quad (7.50)
$$

If the impurity-host interaction $v(\vec{q}j, \vec{q}'j')$ can be written in the separable form as

$$
v(\vec{q}j, \vec{q}'j') = \sum_{nn'} \sqrt{v_n(\vec{q}j)}\sqrt{v_{n'}(\vec{q}'j')}, \quad (7.51)
$$

Eq.(7.49) can be simplified as

$$
g(\vec{q}j, \vec{q}'j'; \omega^2) = \frac{\Delta(\vec{q}+\vec{q}')\delta_{jj'}}{\omega^2 - \omega_j^2(\vec{q})} + \frac{1}{\left[\omega^2 - \omega_j^2(\vec{q})\right]} \frac{1}{\left[\omega^2 - \omega_{j'}^2(\vec{q}')\right]}
$$
$$
\times \sum_{nn'} \sqrt{v_n(\vec{q}j)} \left[\mathbf{I} - \mathbf{M}(\omega^2)\right]_{nn'}^{-1} \sqrt{v_{n'}(-\vec{q}'j')}
$$
$$(7.52)$$

where

$$M_{nn'}(\omega^2) \;=\; \sum_{\vec{q}j} \frac{\sqrt{v_n(\vec{q}j)}\sqrt{v_{n'}(\vec{q}j)}}{\omega^2 - \omega_j^2(\vec{q})}. \tag{7.53}$$

If $v(\vec{q}j, \vec{q}'j')$ is not in the separable form, Eq.(7.49) is solved by Fredholm method. The expression for $g(\vec{q}j, \vec{q}'j'; \omega^2)$ obtained by this method is

$$g(\vec{q}j, \vec{q}'j'; \omega^2) \;=\; g^0(\vec{q}j, \vec{q}'j'; \omega^2) + \frac{1}{D(\omega^2)}$$
$$\times \sum_{\vec{q}_1 j_1} D(\vec{q}j, \vec{q}_1 j_1; \omega^2) g^0(\vec{q}_1 j_1, \vec{q}'j'; \omega^2) \tag{7.54}$$

where $g^0(\vec{q}_1 j_1, \vec{q}'j'; \omega^2)$, defined in Eq.(7.8), is the Fourier coefficient of perfect crystal Green's function,

$$D(\vec{q}j, \vec{q}'j'; \omega^2) \;=\; \sum_{n=1}^{\infty} d_n(\vec{q}j, \vec{q}'j'; \omega^2), \tag{7.55}$$

$$D(\omega^2) \;=\; 1 + \sum_{n=1}^{\infty} d_n(\omega^2), \tag{7.56}$$

$$d_1(\vec{q}j, \vec{q}'j'; \omega^2) \;=\; \frac{v(\vec{q}j, \vec{q}'j'; \omega^2)}{\omega^2 - \omega_j^2(\vec{q})}, \tag{7.57}$$

$$d_n(\vec{q}j, \vec{q}'j'; \omega^2) \;=\; d_1(\vec{q}j, \vec{q}'j'; \omega^2) d_{n-1}(\omega)$$
$$+ \sum_{\vec{q}_1 j_1} d_1(\vec{q}j, \vec{q}_1 j_1; \omega^2) d_{n-1}(\vec{q}_1 j_1, \vec{q}'j'; \omega^2) \tag{7.58}$$

and

$$d_n(\omega^2) \;=\; -\frac{1}{n} \sum_{\vec{q}j} d_n(\vec{q}j, \vec{q}'j'; \omega^2). \tag{7.59}$$

Here $d_n(\vec{q}j, \vec{q}'j'; \omega^2)$ and $d_n(\omega^2)$ are obtained by recursion formula and these series can be made to converge faster in the numerical calculations.

From Eq.(7.54), the roots of the equation

$$D(\omega^2) \;=\; 0 \tag{7.60}$$

will give the frequencies of the normal modes of defect crystal including the localized modes. The roots of the equation

$$\lim_{\delta \to 0+} D(\omega^2 - i\delta) \;=\; 0 \tag{7.61}$$

will give the frequencies of resonance modes. In practice an attempt is always made to separate $v(\vec{q}j, \vec{q}'j')$ into two parts: The strong perturbation which arises due to mass difference and the weak perturbation which arises due to change in the force constants. For the strong perturbation, Eq.(7.49) is solved by iterative procedure and for the weak perturbation, Eq.(7.54) converges rapidly.

7.2.4 Phonon Density of States

Phonon density of states of the defect crystal is given as

$$g(\omega) = \frac{1}{3rN} \sum_{\vec{q}j} \delta\left(\omega - \omega_j(\vec{q})\right); \qquad w > 0 \tag{7.62}$$

where the normalization condition demands that

$$\int_0^{\omega_m} d\omega\, g(\omega) = 1.$$

Here ω_m is the maximum frequency of the defect crystal. It is often more convenient to calculate the distribution function $g(\omega^2)$ of the square of frequencies. Here $g(\omega^2)d\omega^2$ is the fraction of frequencies in the interval ω^2 and $\omega^2 + d\omega^2$. It is given as

$$g(\omega^2) = \frac{1}{3rN} \sum_{\vec{q}j} \delta\left(\omega^2 - \omega_j^2(\vec{q})\right); \qquad w > 0 \tag{7.63}$$

with the normalization condition

$$\int_0^{\omega_m^2} d\omega^2\, g(\omega^2) = 1.$$

From Eqs.(7.62) and (7.63)

$$\begin{aligned}
g(\omega) &= 2\omega g(\omega^2) \\
&= \frac{2\omega}{3rN} \sum_{\vec{q}j} \delta\left(\omega^2 - \omega_j^2(\vec{q})\right); \qquad w > 0. \tag{7.64}
\end{aligned}$$

If one replaces ω by $\omega + i\delta$ in Eq.(7.47) and take the limit $\delta \to 0^+$, the real and imaginary parts of Green's function can be separated with the help of identity

$$\lim_{\delta \to 0^+} \frac{1}{\omega \pm i\delta} = P\left(\frac{1}{\omega}\right) \mp i\pi\delta(\omega). \qquad (7.65)$$

The real and imaginary parts of Green's function are related by the Kramer-Kronig relation

$$\begin{aligned} \mathrm{Re}G(\omega) &= \pm\frac{1}{\pi}P\int_{-\infty}^{\infty} d\omega' \frac{\mathrm{Im}G(\omega')}{\omega' - \omega} \\ &= \pm\frac{2}{\pi}P\int_{0}^{\infty} d\omega' \frac{\omega'\mathrm{Im}G(\omega'^2)}{\omega'^2 - \omega^2}. \end{aligned} \qquad (7.66)$$

The poles of the retarded Green's function lying in the lower half complex frequency plane near the real axis will give the frequencies of normal modes. Thus the phonon density of states of defect crystal can also be expressed as

$$g(\omega^2) = \frac{1}{3rN\pi}\lim_{\delta \to 0^+}\mathrm{Im}\sum_{l\kappa\alpha} M_{l\kappa}G_{\alpha\alpha}(l\kappa, l\kappa; \omega^2 + i\delta) \qquad (7.67)$$

and

$$g(\omega) = \frac{2\omega}{3rN\pi}\lim_{\delta \to 0^+}\mathrm{Im}\sum_{l\kappa\alpha} M_{l\kappa}G_{\alpha\alpha}(l\kappa, l\kappa; \omega^2 + i\delta). \qquad (7.68)$$

In obtaining Eqs.(7.67) and (7.68), the completeness of eigenvectors is used. In the defect crystal, the mass varies from site to site, therefore, the phonon density of states is given by the imaginary part of mass weighted Green's function.

7.2.5 Other Experimental Results

The ensemble average of displacement square $u^2(l\kappa)$ at temperature T is given by the relation [5]

$$<u^2(l\kappa)> = -\frac{\hbar}{\pi}\int_{0}^{\infty} d\omega \coth\left(\frac{\hbar\omega}{2k_BT}\right) \mathrm{Im}\sum_{\alpha} G_{\alpha\alpha}(l\kappa, l\kappa; \omega). \qquad (7.69)$$

One phonon Raman scattering cross section $R(\omega)$, infrared absorption coefficient $\alpha(\omega)$ and the neutron scattering cross section $d^2\sigma/d\omega d\Omega$ are proportional to the sum [2, 4]

$$\lim_{\delta\to 0+} \sum_{l\kappa\alpha} \sum_{l'\kappa'\beta} A_\alpha(l\kappa) A_\beta(l'\kappa') \mathrm{Im} G_{\alpha\beta}(l\kappa, l'\kappa'; \omega + i\delta). \qquad (7.70)$$

Here the coefficients A_α and A_β are the first order expansion coefficients of the polarizability function for Raman scattering cross section $R(\omega)$, the dipolemoments for infrared absorption coefficient $\alpha(\omega)$ and the components of scattering vectors weighted by the atomic scattering lengths for neutron scattering cross section $(d^2\sigma/d\omega d\Omega)$. Thus the knowledge of Green's function of the defect crystal enables one to compare the calculated results with the Raman scattering, infrared absorption and neutron scattering data. A detailed description of these aspects is given in references [2] and [8]. However, the space limitation does not allow us to go in more details here.

7.3   Phonons in Disordered Alloys

We continue our discussion for the homogeneous substitutional alloys which can be classified as (a) ordered and (b) disordered alloys. Consider an AB alloy. In an ordered alloy, A and B atoms form a regular periodic pattern with respect to each other. This happens when the unlike atoms have the tendency to become the 1NN of each other. However, in a disordered alloy, the distribution of A and B type of atoms is random. There may be a long range order in the disordered alloy at the low temperature. As the temperature increases, the long range order diminishes and at a critical transition temperature the long range order disappears. But at the same time some kind of short range order continue to persist even above the transition temperature [3]. The problem of inhomogeneous disordered alloys needs separate attention [3].

The X-ray diffraction and neutron scattering methods are used to investigate the degree of order. The diffraction lines for the disordered structure are observed in a pattern as if the lattice points are occupied by only one type of atoms. This happens because the scattering power of each atomic plane becomes equal to the average scattering power of A and B type of atoms. On the other hand, in the diffraction spectra of an ordered alloy some extra lines are also observed which were abscent in the diffraction spectra of an otherwise perfect or disordered solid. These extra lines are called superstructure lines [8].

With the above experimental facts, the disordered alloys may be considered to have the perfect crystal lattice but with some averaged properties. On the basis of observed phonon dispersion curves, one may devide disordered alloys broadly into two categories: (i) The alloys which have only one set of phonon frequencies and which are similar to those of a component metal. This characteristic is found in the phonon spectra of alloys such as $NiFe, NbMo$ and TaW [9]-[14]. The local phonon modes have also been observed in some of these alloys. (ii) The alloys which have two sets of phonon frequencies depending upon the nature of constituent atoms present. There is a gap between the two sets of phonon frequencies in the phonon dispersion relations.

We consider a binary alloy $A_{1-x}B_x$ where $(1-x)$ and x are the concentrations of A and B type of atoms respectively. If the potential v takes on the values v_A and v_B at the sites occupied by A and B atoms respectively, the probability distribution for the potential at the site i is

$$P(v_i) \;=\; (1-x)\delta(v_i - v_A) + x\delta(v_i - v_B). \tag{7.71}$$

Equation (7.71) can also be extended for multicomponent alloys. If the distribution of atoms among the lattice sites is random, the total probability distribution becomes

$$P(v_1, v_2 \ldots) \;=\; \Pi_i P(v_i). \tag{7.72}$$

Now the configurational averaged Green's function for the calculation of phonon frequencies, frequency distribution function and phonon width for a disordered alloy can be written as

$$< \mathbf{G} > \;=\; \Pi_i \int dv_i P(v_i)\mathbf{G}. \tag{7.73}$$

The calculation of $< \mathbf{G} >$ is not always trivial. Therefore attempts have been made to obtain the approximate solutions for $< \mathbf{G} >$. In the following, we discuss various assumptions which have been used to calculate $< \mathbf{G} >$ and hence the phonon properties of these alloys [2, 4].

7.3.1   Average t-matrix Approximation (ATA)

The Green's function for defect crystal, given by Dyson Eq.(7.29), can be rewritten as [15, 16]

$$\mathbf{G} \;=\; \mathbf{G}^0 + \mathbf{G}^0 \mathbf{T} \mathbf{G}^0 \tag{7.74}$$

where

$$\begin{aligned}
\mathbf{T} \;&=\; \Delta\mathbf{L} + \Delta\mathbf{L}\mathbf{G}^0\Delta\mathbf{L} + \Delta\mathbf{L}\mathbf{G}^0\Delta\mathbf{L}\mathbf{G}^0\Delta\mathbf{L} + \ldots \\
&=\; \Delta\mathbf{L}\left[1 - \mathbf{G}^0\Delta\mathbf{L}\right]^{-1}.
\end{aligned} \tag{7.75}$$

Here $\mathbf{T}$ is called the t-matrix and contains the information of all orders of scattering of phonons due to impurities. The matrix $\mathbf{T}$ describes the scattering of an undistorted lattice wave represented by $\mathbf{G}^0$ into a distorted lattice wave represented by $\mathbf{G}$. The scattering is due to perturbation $\Delta\mathbf{L}$ produced by the impuritirs.

Comparing Eqs.(7.29) and (7.74), one finds

$$\mathbf{T} \;=\; \Delta\mathbf{L}\mathbf{G}\mathbf{G}^{0-1}. \tag{7.76}$$

In Eqs.(7.74) and (7.76), $\mathbf{G}$ is expressed in terms of $\mathbf{T}$, therefore a careful calculation of $\mathbf{T}$ is needed through Eq.(7.75). If $\Delta\mathbf{L}$ is a weak perturbation, Eq.(7.75) converges rapidly. However for a strong perturbation, $\mathbf{T}$ has to be calculated by iterative procedure with the help of equation

$$\begin{aligned}
\mathbf{T} \;&=\; \Delta\mathbf{L}\left[1 + \mathbf{G}^0\mathbf{T}\right] \\
&=\; \left[1 + \mathbf{T}\mathbf{G}^0\right]\Delta\mathbf{L}
\end{aligned} \tag{7.77}$$

which is obtained by using Eq.(7.74) in (7.76). Once $\mathbf{T}$ is known, calculation of $\mathbf{G}$ is straight forward.

For the vibrational properties of disordered alloys, one has to calculate the configurational average of $\mathbf{G}$ i.e. $< \mathbf{G} >$. This is equivalent to the calculation of average t-matrix $< \mathbf{T} >$ because by definition $\mathbf{G}^0$ is unaffected by the averaging process. Thus taking the configurational average of Eq.(7.74), one finds

$$< \mathbf{G} > \;=\; \mathbf{G}^0 + \mathbf{G}^0 < \mathbf{T} > \mathbf{G}^0. \tag{7.78}$$

Here the averaging of t-matrix leads to the averaging of scattering due to impurities on different lattice sites. Hence the problem of calculation of $< \mathbf{G} >$ has been shifted to the calculation of $< \mathbf{T} >$. In this procedure, the approximations which are used in the calculation of $< \mathbf{T} >$, may be analysed more carefully.

It is now convenient to define a self-energy operator Σ for averaged Green's function through the relation

$$< \mathbf{G} > \; = \; \mathbf{G}^0 + \mathbf{G}^0 \Sigma < \mathbf{G} > \tag{7.79}$$

where Σ is defined as

$$\Sigma \; = \; \mathbf{G}^{0-1} - < \mathbf{G} >^{-1} . \tag{7.80}$$

Here Σ depends upon the configurational averaged Green's function $< \mathbf{G} >$, therefore Σ can have the translational symmetry. However Σ is frequency dependent complex number. Use of Eqs.(7.78) and (7.79) relates Σ with $< \mathbf{T} >$ through the relation

$$\Sigma \; = \; \left[1 + < \mathbf{T} > \mathbf{G}^0 \right]^{-1} < \mathbf{T} > . \tag{7.81}$$

A comparison of Eq.(7.79) with configurational averaged Eq.(7.29) gives $\Sigma = < \Delta L >$. Thus Σ represents the configurational averaged perturbation due to the presence of impurities. Once Σ is known, $< \mathbf{G} >$ can be calculated with the help of Eq.(7.79).

There may be two procedures to calculate Σ. In the first procedure Σ can be calculated directly with the help of Eq.(7.81) if $< \mathbf{T} >$ is known exactly. This is called Average t-Matrix Approximation. However if one assumes that the average atomic scattering in a disordered alloy vanishes i.e. $< \mathbf{T} > = 0$, Eq.(7.81) will be no more valid and then Σ has to be calculated self-consistently through Eqs.(7.79) and (7.80). This will leads to Coherent Potential Approximation (CPA) which will be discussed later.

The general solution of Eqs.(7.74) and (7.78) are difficult. Therefore these equations are solved using different approximations.

(a) On-Diagonal Disorder Approximation

Comparison of Eqs.(7.29) and (7.75) shows that the poles of $\mathbf{T}$ determine those eigenfrequencies which are affected by the disorder due to

$\Delta \mathbf{L}$. Thus $\mathbf{T}$ has to be calculated using multiple scattering theory [17]. If we assume that the impurity concentration is very low such that the impurity spaces of different impurities do not overlap with each other, we can split the total perturbation $\Delta \mathbf{L}$ into the sum of independent perturbations produced by each impurity at site $(l_i \kappa_i)$ i.e.

$$\Delta \mathbf{L} = \sum_i \Delta L_i. \tag{7.82}$$

Here the matrices ΔL_i and hence the matrix $\Delta \mathbf{L}$ are diagonal matrices in the mass defect approximation as we shall see later. Therefore the above approximation is called "On-Diagonal Disorder Approximation".

Equation (7.82) is quite appropriate if the forces are short-ranged and the changes in the force constants are negligible. Thus one gets from Eqs.(7.20) and (7.25)

$$\Delta L_i(l\kappa, l'\kappa'; \omega^2) = \omega^2 \epsilon_{l\kappa} M_\kappa \delta_{ll'} \delta_{\kappa\kappa'} \delta_{ll_i} \tag{7.83}$$

for the ith impurity at the site l_i on the sublattice κ. Equation (7.83) takes into account only the mass difference between the impurity and host atoms through the mass defect parameter $\epsilon_{l\kappa}$. Therefore, it is also called " Mass Defect Approximation".

Using Eq.(7.82) in (7.77), we get

$$\begin{aligned}
\mathbf{T} &= \sum_i \Delta L_i \left[1 + \mathbf{G}^0 \mathbf{T} \right] \\
&= \sum_i \left[1 + \mathbf{T} \mathbf{G}^0 \right] \Delta L_i \\
&= \sum_i T_i. \tag{7.84}
\end{aligned}$$

In this case the scattering matrix $\mathbf{T}$ becomes the sum of scattering matrices T_i which represent the contribution of individual impurities. Using back the last equality of Eq.(7.84) in the first equality of the same equation and separating the terms $i = j$ and $i \neq j$, one gets

$$T_i = \left[1 + \sum_{j(\neq i)} \mathbf{G}^0 T_j \right] t_i \tag{7.85}$$

where

$$t_i \;=\; \Delta L_i \left[1 - \Delta L_i \mathbf{G}^0\right]^{-1}. \tag{7.86}$$

The expression in the square bracket of Eq.(7.85) describes the effect of all the impurities, except the ith impurity, on the lattice wave propagating to the ith impurity. The atomic t-matrix t_i describes the complete scattering of lattice wave by the ith impurity.

In the average t-matrix approximation, following the same arguments as for Eqs.(7.82) and (7.84), one can write the configurational averaged $< \mathbf{T} >$ as

$$< \mathbf{T} > \;=\; \sum_i < T_i > \tag{7.87}$$

where $< T_i >$ is calculated with the help of Eq.(7.85) as

$$
\begin{aligned}
< T_i > \;=\;& < t_i > + < \sum_{j(\neq i)} [T_j - < T_j > + < T_j >] \\
& \times \mathbf{G}^0 \left[t_i - < t_i > + < t_i >\right] > \\
=\;& < t_i > + < \sum_{j(\neq i)} < T_j > \mathbf{G}^0 < t_i >> \\
& + < \sum_{j(\neq i)} [T_j - < T_j >] \, \mathbf{G}^0 \left[t_i - < t_i >\right] > \\
& + < \sum_{j(\neq i)} < T_j > \mathbf{G}^0 \left[t_i - < t_i >\right] > \\
& + < \sum_{j(\neq i)} [T_j - < T_j >] \, \mathbf{G}^0 < t_i >> .
\end{aligned}
\tag{7.88}
$$

The configurational average of the last two terms of Eq.(7.88) vanieshes, therefore

$$
\begin{aligned}
< T_i > \;=\;& \left[1 + \sum_{j(\neq i)} < T_j > \mathbf{G}^0\right] < t_i > \\
& + < \sum_{j(\neq i)} [T_j - < T_j >] \, \mathbf{G}^0 \left[t_i - < t_i >\right] > .
\end{aligned}
\tag{7.89}
$$

Here the first term describes the scattering of an averaged incident wave by an impurity with an averaged atomic t-matrix $< t_i >$. The second term involves the matrices $[T_j - < T_j >]$ and $[t_i - < t_i >]$ which are the fluctuations in the incident wave T_i and atomic t-matrices t_i respectively. Thus the second term describes the correlations between the fluctuations in the incident wave and the atomic t-matrices. The computation of this term is not always trivial. Therefore in the calculation of $< T_i >$, if the second term is neglected, one finds

$$< T_i > \; = \; \left[1 + \sum_{j(\neq i)} < T_j > \mathbf{G}^0 \right] < t_i > . \qquad (7.90)$$

This is called "Single Site Approximation (SSA)". This amounts to averaging over different impurity configurations around a given impurity before creating the dynamics of that impurity. Hence any specific effects arising from the cluster of impurities are averaged out. Therefore one cannot find the fine structures of phonon spectrum in this approximation [18].

Using (7.90) in (7.87), one finds

$$\sum_i < T_i > \left[1 + \mathbf{G}^0 < t_i > \right] = \sum_i \left[1 + < \mathbf{T} > \mathbf{G}^0 \right] < t_i > .$$

Equating the terms for each i on both the sides

$$< T_i > = \left[1 + < \mathbf{T} > \mathbf{G}^0 \right] < t_i > \left[1 + \mathbf{G}^0 < t_i > \right]^{-1} .$$

Summing over i, one gets

$$< \mathbf{T} > \; = \; \left[1 + < \mathbf{T} > \mathbf{G}^0 \right] \sum_i < t_i > \left[1 + \mathbf{G}^0 < t_i > \right]^{-1} .$$

$$(7.91)$$

Using Eq.(7.91) in (7.81), the self-energy in the single site approximation becomes

$$\Sigma \; = \; \sum_i < t_i > \left[1 + \mathbf{G}^0 < t_i > \right]^{-1}$$

$$= \sum_i (\Sigma_i) \qquad (7.92)$$

where the self-energy of ith impurity

$$\Sigma_i \;=\; <t_i>\left[1+\mathbf{G}^0<t_i>\right]^{-1}. \qquad (7.93)$$

At very low concentration of impurities, the interaction between the impurities is negligible. Therefore the SSA given in Eq.(7.90) is fairly valid. We note from Eq.(7.90) that $<T_i>$ is proportional to $<t_i>$. The configurational average of t_i, given in Eq.(7.86) will depend upon the site occupancy by impurities i.e.

$$<t_i> \;=\; \sum_j x_{\kappa j} t_j \delta_{ji} \qquad (7.94)$$

where $x_{\kappa j}$ denotes the fractional occupancy on the κth sublattice by jth impurity. Thus the use of Eqs.(7.87), (7.90) and (7.94) shows that the average t-matrix $<\mathbf{T}>$ in the single site approximation is proportional to the concentration of impurities i.e.

$$<\mathbf{T}> = x\sum_i t_i.$$

Thus $<\mathbf{T}>$ depends linearly on impurity concentration in the single site approximation. Similarly the self-energy given in Eq.(7.92) also becomes proportional to impurity concentration x and hence

$$\Sigma \;=\; x\sum_i t_i. \qquad (7.95)$$

(b) Off-Diagonal Disorder

When the force constants affected by the impurity are also taken into account, $\mathbf{\Delta L}$ becomes a non-diagonal matrix. This type of disorder is called the off-diagonal disorder [18]. The force constants are invariant under translational symmetry operation. Therefore

$$\sum_{l'\kappa'}\Phi_{\alpha\beta}(l\kappa, l'\kappa') \;=\; 0. \qquad (7.96)$$

Equation (7.96) requires that in the perturbation matrix $\mathbf{\Delta L}$, the changes in the off-diagonal coupling constants must be accompanied by the changes in the diagonal coupling constants at all the atomic sites including the impurity site, in the impurity space. This is illustrated by considering a monatomic AB alloy with an average force constant matrix

$$\Phi^{AB}(l,l') \;=\; \frac{1}{2}\left[\Phi^{AA}(l,l') + \Phi^{BB}(l,l')\right]. \tag{7.97}$$

Here $\Phi^{AB}(l,l')$ is the force constant matrix between the atoms A and B at the sites l and l' respectively. The changes in the force constants in the alloy with respect to reference crystal B are

$$\Delta\Phi(l,l') \;=\; \frac{1}{2}\left[\Phi^{AA}(l,l') - \Phi^{BB}(l,l')\right]\chi_A \tag{7.98}$$

where $\chi_A = 1$ if A atom is either on the site l or l' and $\chi_A = 2$ if A atoms are on both the sites l and l'. The corresponding site diagonal force constant changes the results given in Eqs.(7.96) and (7.98). For example, we consider a single A atom embedded in an environment of B atoms and the interactions are limited only upto the first nearest neighbours. This will give

$$\begin{aligned}
\Delta\Phi(l,l') &= -\Delta\Phi(l,l') \\
&= -\frac{1}{2}\left[\Phi^{AA}(l,l') - \Phi^{BB}(l,l')\right]
\end{aligned} \tag{7.99}$$

and

$$\Delta\Phi(l,l) \;=\; -\sum_{l'}\Delta\Phi(l,l') \tag{7.100}$$

where l is the site of an A atom and l' are the NNs of l. Here the site diagonal changes in force constants, given in Eq.(7.100), are at all the sites in the impurity space including impurity site.

Thus including the mass disorder and off-diagonal disorder for an AB alloy, the impurity matrix ΔL_i can be written as

$$\Delta L_i = M_B\omega^2\epsilon_A\delta_{ll'}\delta_{ll_i} + \frac{1}{2}\left[\Phi^{AA}(l,l') - \Phi^{BB}(l,l')\right]\chi_A$$

or

$$\Delta L_{i\alpha\beta} = M_B \omega^2 \epsilon_A \delta_{ll'} \delta_{ll_i} \delta_{\alpha\beta}$$
$$+ \frac{1}{2} \left[\Phi_{\alpha\beta}^{AA}(l, l') - \Phi_{\alpha\beta}^{BB}(l, l') \right] \chi_A. \tag{7.101}$$

This new definition of ΔL_i is used in all the equations following Eq.(7.84), to obtain the expressions for $< \mathbf{T} >$ and $\mathbf{\Sigma}$. Obviously the computational efforts are greatly enhanced. For further calculations the reader is referred to references[19, 20].

(c) Limiting Cases

Local Modes

In the on-diagonal disorder approximation, given in Eq.(7.82), Green's function for the defect crystal becomes

$$\mathbf{G} = \mathbf{G}^0 + \mathbf{G}^0 \sum_i \Delta L_i \mathbf{G}. \tag{7.102}$$

The local vibrational modes arise due to single impurity scattering process. At the low concentration of impurities, ΔL_i are independent of each other. Therefore all the impurities can be treated on the same footings. Separating the effect of the ith impurity in Eq.(7.102), we can write

$$\mathbf{G} = \mathbf{G}_i + \mathbf{G}^0 \Delta L_i \mathbf{G} \tag{7.103}$$

where

$$\mathbf{G}_i = \mathbf{G}^0 + \mathbf{G}^0 \sum_{j(\neq i)} \Delta L_j \mathbf{G}. \tag{7.104}$$

Here $\mathbf{G}_i$ is an effective field as seen by the atoms in the neighbourhood of an impurity at the ith site. The lattice wave is allowed to scatter off all other impurities in the crystal before it scatters off the impurity at the ith site.

Equation (7.103) can be solved for $\mathbf{G}$ in terms of $\mathbf{G}_i$ by substituting $\mathbf{G}$ repeatedly in the right side of Eq.(7.103). This will give

$$\mathbf{G} = \left[1 - \mathbf{G}^0 \Delta L_i \right]^{-1} \mathbf{G}_i. \tag{7.105}$$

Using Eq.(7.105) again in Eq.(7.103), we get

$$\begin{aligned}
\mathbf{G} &= \mathbf{G}^0 + \mathbf{G}^0 \sum_i t_i \mathbf{G}_i \\
&= \mathbf{G}^0 + \mathbf{G}^0 t_i \mathbf{G}_i + \mathbf{G}^0 \sum_{j(\neq i)} t_j \mathbf{G}_j
\end{aligned} \tag{7.106}$$

where

$$t_i = \Delta L_i \left[1 - \mathbf{G}^0 \Delta L_i \right]^{-1}. \tag{7.107}$$

For a single impurity at the ith site $\mathbf{G}_i = \mathbf{G}^0$ and the third term of Eq.(7.106) vanishes. This gives

$$\mathbf{G} = \mathbf{G}^0 + \mathbf{G}^0 t_i \mathbf{G}^0. \tag{7.108}$$

If the impurity is taken at the origin i.e.$t_i = t$ and $\Delta L_i = \Delta L_0$, Eqs.(7.108) and (7.107) simplify as

$$\mathbf{G} = \mathbf{G}^0 + \mathbf{G}^0 t \mathbf{G}^0 \tag{7.109}$$

and

$$t = \Delta L_0 \left[1 - \mathbf{G}^0 \Delta L_0 \right]^{-1}. \tag{7.110}$$

All the factors on the right side of Eq.(7.110) are known. The poles of t will determine vibrational frequencies affected by the impurity. The real and imaginary parts of $\mathbf{G}$ will determine other vibrational properies of a single impurity.

Dilute Alloys

In the dilute alloys, the problem reduces to the single impurity problem because the impurity-impurity interaction is assumed negligible. The total effect is obtained by multiplying the single impurity effect by impurity concentration. The configurational averaged of the first equality of Eq.(7.106) is

$$< \mathbf{G} > = \mathbf{G}^0 + \mathbf{G}^0 < \sum_i t_i \mathbf{G}_i > . \tag{7.111}$$

For non-interacting impurities, we can write

$$< \sum_i t_i \mathbf{G}_i > \; = \; \sum_i < t_i \mathbf{G}_i > . \qquad (7.112)$$

If we introduce a conditional average

$$< t_i \mathbf{G}_i > \; = \; x_0 t_i < \mathbf{G}_i >_{l_i} \qquad (7.113)$$

where x_0 is the fractional occupancy of impurity site and $<>_{l_i}$ denotes the average over all the configurations with the condition that there is an impurity in the cell l_i, in each impurity space. Using Eqs.(7.112) and (7.113) in (7.111), one gets

$$< \mathbf{G} > \; = \; \mathbf{G}^0 + x_0 \mathbf{G}^0 \sum_i t_i < \mathbf{G}_i >_{l_i} . \qquad (7.114)$$

At very low concentration of impurities, it is sufficient to calculate $< \mathbf{G} >$ upto linear terms in x_0. If we set $< \mathbf{G}_i >_{l_i} = \mathbf{G}^0$, the lowest order approximation, we get

$$< \mathbf{G} > \; = \; \mathbf{G}^0 + x_0 \mathbf{G}^0 \left(\sum_i t_i \right) \mathbf{G}^0. \qquad (7.115)$$

Thus Eq.(7.115) is an extension of (7.108). If the impurity scattering matrices t_i for different impurities and Green's function $\mathbf{G}^0$ for reference crystal are known, $< \mathbf{G} >$ can be calculated for dilute alloy. The same procedure can be adopted to calculate $< \mathbf{T} >$ for the dilute alloy too.

7.3.2 Virtual Crystal Approximation (VCA)

The simplest approximation which can be adopted in the calculation of configurational average of $\mathbf{G}$, given in Eq.(7.29), is to ignore the correlation between $\mathbf{\Delta L}$ and $\mathbf{G}$ i.e.

$$< \mathbf{\Delta L G} > = < \mathbf{\Delta L} > < \mathbf{G} >$$

and hence

$$< \mathbf{G} > \; = \; \mathbf{G}^0 + \mathbf{G}^0 < \mathbf{\Delta L} > < \mathbf{G} > . \qquad (7.116)$$

Here the disorder induced purturbation $< \Delta \mathbf{L} >$ simply changes the Green's function of the perfect crystal $\mathbf{G}^0$ into another averaged perfect crystal Green's function $< \mathbf{G} >$ [18]. For a binary alloy $A_{1-x}B_x$, we can write $< \Delta \mathbf{L} >$ as

$$< \Delta \mathbf{L} > \ = \ [(1-x)\Delta \mathbf{L}_A + x\Delta \mathbf{L}_B] I \qquad (7.117)$$

where $\Delta \mathbf{L}_A$ and $\Delta \mathbf{L}_B$ are the perturbations produced by the atoms of A and B type respectively and I is an unit matrix. For a single impurity, the average perturbation $< \Delta \mathbf{L} >$ adds simply a constant to all the energies. However in a many impurity system, the effect of $< \Delta \mathbf{L} >$ is to replace the actual perturbation at each lattice site by an averaged perturbation which is the weighted average of $\Delta \mathbf{L}$ at different lattice sites occupied by different type of atoms. In this approximation, the effective lattice potential is periodic and therefore the crystal states can be assigned the crystal wave vectors. This approximation is called Virtual Crystal Approximation (VCA) and it is very often used in the calculations of phonon frequencies of those disordered alloys which do not have a gap in the phonon spectrum.

We can also say that in the virtual crystal approximation, the disordered alloy is replaced by an averaged ordered crystal with the identical effective atoms at the lattice sites. Each effective atom has concentration weighted average mass and average force constants [21] i.e.

$$\bar{M} \ = \ (1-x)M_A + xM_B \qquad (7.118)$$

and

$$\bar{\Phi}_{\alpha\beta} \ = \ (1-x)\Phi^A_{\alpha\beta} + x\Phi^B_{\alpha\beta}. \qquad (7.119)$$

Here M_A and M_B are the masses of A and B type of atoms. $\Phi^A_{\alpha\beta}$ and $\Phi^B_{\alpha\beta}$ are the force constants of perfect A and B metals and these are calculated for the original crystal structure of A and B metals. With these simplifications, the formalism of perfect crystals (chapter 2) can be used to find the phonon frequencies and force constants of $A_{1-x}B_x$ alloys provided the crystal potentials and atomic masses of the perfect crystals A and B are known.

The VCA gives the exact phonon frequencies of the metals A and B in the extreme limits. Therefore this method provides an interpolation of phonon frequencies between the phonon frequencies of perfect A and B metals and this interpolation will depend upon the impurity concentration. Although the phonon frequencies change continuously with the change in the concentration, the phonons have infinite life time in VCA. Therefore the local modes or the resonance modes are not predicted in this approximation. Thus VCA does not provide the complete information about the dynamics of a disordered alloy. It can be regarded only as the lowest order calculation. However, the general information about the nature of force constants in the alloy is obtained.

The equation of motion in the Green's function formalism is

$$\omega^2 M_{l\kappa} \mathbf{G}(l\kappa, l'\kappa'; \omega^2) = \delta_{ll'}\delta_{\kappa\kappa'} + \sum_{l''\kappa''(\neq l\kappa)} \mathbf{\Phi}(l\kappa, l''\kappa'')$$
$$\times \left[\mathbf{G}(l''\kappa'', l'\kappa'; \omega^2) - \mathbf{G}(l\kappa, l'\kappa'; \omega^2)\right].$$
$$(7.120)$$

In the VCA, $M_{l\kappa}$ and $\mathbf{\Phi}$ are replaced by their averaged values $\bar{M}$ and $\bar{\mathbf{\Phi}}$. The Green's functions are also averaged with due weightage to concentration. Thus the mass and concentration weighted averaged dynamical Eq. (7.120) is written as

$$\omega^2 \bar{M}_{l\kappa} g(l\kappa, l'\kappa'; \omega^2) = \delta_{ll'}\delta_{\kappa\kappa'} + \sum_{l''\kappa''(\neq l\kappa)} \bar{\mathbf{\Phi}}(l\kappa, l''\kappa'')$$
$$\times \left[g(l''\kappa'', l'\kappa'; \omega^2) - g(l\kappa, l'\kappa'; \omega^2)\right]$$
$$(7.121)$$

where

$$g(l\kappa, l'\kappa'; \omega^2) = \ <\mathbf{G}(l\kappa, l'\kappa'; \omega^2)>_x, \qquad (7.122)$$
$$\bar{M}_{l\kappa} = \sum_{j=A,B} x_{\kappa j} M_{j(l\kappa)}, \qquad (7.123)$$
$$\bar{\mathbf{\Phi}}(l\kappa, l'\kappa') = \sum_{jj'(AB)} x_{\kappa j} x_{\kappa' j'} \mathbf{\Phi}^{jj'}(l\kappa, l'\kappa'), \qquad (7.124)$$

and $x_{\kappa j}$ is the fractional occupancy of the κth site by jth type of atom. Introducing the Fourier series

$$g(l\kappa, l'\kappa'; \omega^2) = \frac{1}{N} \sum_{\vec{q}} g(\kappa, \kappa'; \vec{q}; \omega^2) \exp[i\vec{q} \cdot \{\vec{R}(l\kappa) - \vec{R}(l'\kappa')\}]$$

$$(7.125)$$

in Eq.(7.121), we get

$$
\sum_{\kappa''}\left\{\left[\omega^2 \bar{M}_{l\kappa} + \sum_{l'''\kappa'''}(1 - \delta_{ll'''}\delta_{\kappa\kappa'''})\bar{\bar{\Phi}}(l\kappa, l'''\kappa''')\delta_{\kappa\kappa'''}\right.\right.
$$

$$
\left.\left. - \sum_{l''}(1 - \delta_{ll''}\delta_{\kappa\kappa''})\bar{\bar{\Phi}}(l\kappa, l''\kappa'')\exp[i\vec{q}\cdot\{\vec{R}(l''\kappa'') - \vec{R}(l\kappa)\}]\right]\right\}
$$

$$
\times g(\kappa'', \kappa'; \vec{q}; \omega^2) = \delta_{\kappa\kappa'}. \qquad (7.126)
$$

The zeros of $det|g^{-1}|$ give eigenfrequencies. For a diatomic three dimensional lattice, there will be three acoustic and three optical modes.

The averaged force constants $\bar{\bar{\Phi}}(l\kappa, l'\kappa')$ are determined either with the help of a central potential or with the help of a pseudopotential which can be defined for an alloy as

$$
w(\vec{r}) = \sum_{j(A)} w_A(\vec{r} - \vec{R}_j^A) + \sum_{j(B)} w_B(\vec{r} - \vec{R}_j^B) \qquad (7.127)
$$

where w_A and w_B are the pseudopotentials for the perfect metals A and B respectively. Other description remains the same as that of a perfect crystal. Some times few measured phonon frequencies are also used to fit the force constants. This method has been used to explain the phonon dispersion relations of $Cr_{1-x}W_x$ [22], $Ni_{55}Pd_{45}$ and $Ni_{50}Fe_{50}$ [23]. The interatomic interactions are found long ranged in these alloys. The phonon dispersion relations are also used to calculate the elastic constants, Debye temperature and lattice specific heat. The VCA is also used to explain the phonons in invar alloys $Fe_{0.7}Mn_{0.3}, Fe_{0.65}Mn_{0.35}$ and $Fe_{0.72}Pt_{0.28}$ [24]. However, the microscopic analysis of the force constants in these materials is yet to be worked out.

Simon and Varma [25] used non-orthogonal tight-binding scheme, as discussed in chapter 4, for the force constants $\bar{\bar{\Phi}}$. The concentration weighted force constants and atomic masses are used to fit the phonon dispersion relations of $Nb_{0.8}Zr_{0.2}$ and $Nb_{0.5}Zr_{0.5}$ in the bcc phase. It is shown that bcc to ω-phase transition in $Nb_{1-x}Zr_x$ alloys is electronically deriven and is related to the nesting of Fermi surface. Singh [26] used transition metal pseudopotential for constituent elements of $Pd_{0.72}Fe_{0.28}$ to obtain the force constants. The general features of phonon dispersion relations are explained and the presence of long range forces is indicated. However the anomalous features which arise due to quasilocalized d-electrons could not be explained.

7.3.3 Coherent Potential Approximation (CPA)

In the average t-matrix approximation in Eq.(7.90), the single site approximation was introduced. In this approximation it was assumed that the individual impurities are embedded in an effective medium which arises from the cluster of impurities. In the average t-matrix approximation, the effective medium is determined by using the condition that the average t-matrix, given in Eq.(7.91), is exact at each atomic site. However this effective medium can be chosen arbitrarily.

The coherent potential approximation (CPA) takes one step further. In this approximation, the medium around the impurity atom is taken that of the defect crystal and the medium and the dynamics of impurity atom are calculated self-consistently [27, 28]. This is achieved by making the t-matrix for scattering from each scatterer on average equal to zero i.e.

$$< \mathbf{T} > = 0. \tag{7.128}$$

Therefore in the calculation of $< \mathbf{G} >$, one has to calculate Σ given in Eq.(7.80) self-consistently. It is done conveniently by introducing an unknown diagonal matrix $\Xi = \sigma I$ in Eq. (7.28) which is rewritten for a cubic crystal as

$$\begin{aligned} \mathbf{G}^{-1} &= \mathbf{G}^{0-1} - \mathbf{\Delta L} \\ &= \mathbf{G}^{m-1} - \mathbf{\Delta L}^m \end{aligned} \tag{7.129}$$

where $\mathbf{G}^{m-1}$ and $\mathbf{\Delta L}^m$ are obtained by adding and subtracting Ξ on the right side of first equality of Eq.(7.129) i.e.

$$\mathbf{G}^{m-1} = \mathbf{G}^{0-1} - \Xi \tag{7.130}$$

and

$$\mathbf{\Delta L}^m = \mathbf{\Delta L} - \Xi. \tag{7.131}$$

Here $\mathbf{G}^m$ is new reference crystal Green's function and $\mathbf{\Delta L}^m$ is the corresponding perturbation.

Following Eq.(7.79), the self-energy $\mathbf{\Sigma}^m$ in the new reference crystal is defined as

$$< \mathbf{G} > \;=\; \mathbf{G}^m + \mathbf{G}^m \mathbf{\Sigma}^m < \mathbf{G} > \tag{7.132}$$

where

$$\mathbf{\Sigma}^m \;=\; \mathbf{\Sigma} - \Xi. \tag{7.133}$$

In this new reference crystal the atomic t-matrix is obtained in the same way as in Eq.(7.86) and it is given as

$$t_i^m(\omega) \;=\; \frac{(\Delta L_i - \Xi)}{1 - (\Delta L_i - \Xi)\mathbf{G}^m(\omega)} \tag{7.134}$$

where $\mathbf{G}^m(\omega) = G_{\alpha\alpha}^m(0,0,\omega)$ is the Green's function at the origin.

The self-energy $\mathbf{\Sigma}^m$ has the property

$$\mathbf{\Sigma}^m \;=\; \sum_l \mathbf{\Sigma}_l^m \tag{7.135}$$

where the sum is over all the lattice sites. This is a situation where the same perturbation $\mathbf{\Sigma}_l^m$ is applied to each site and the reference medium has the translational symmetry of the perfect host lattice. The host and impurity atoms can scatter lattice excitations with respect to this reference medium.

Using Eq.(7.81) in (7.133) we get

$$\begin{aligned}
\mathbf{\Sigma} \;&=\; \Xi + \mathbf{\Sigma}^m \\
&=\; \Xi + \frac{< \mathbf{T}^m >}{1 + \mathbf{G}^m(\omega) < \mathbf{T}^m >}.
\end{aligned} \tag{7.136}$$

For the low concentration of impurities, the average t-matrix for the reference crystal, in the SSA, is given as

$$< \mathbf{T}^m > \;=\; \sum_i x_i t_i^m(\omega). \tag{7.137}$$

Using Eq.(7.137) in (7.136), the self-energy becomes

$$\Sigma \;=\; \Xi + \frac{\sum_i x_i t_i^m(\omega)}{1 + \mathbf{G}^m(\omega)\sum_i x_i t_i^m(\omega)}. \qquad (7.138)$$

Equation (7.138) can be evaluated in two ways depending upon the choice of Ξ.

(i) Average t-matrix Approximation (ATA)

Choose some value of Ξ and determine $\mathbf{G}^m$ using Eq.(7.130) and t_i^m using Eqs.(7.134) and hence $< \mathbf{T}^m >$ with the help of Eq.(7.137). These values can be used to calculate self-energy given in Eq.(7.138). This is the average t-matrix approximation as the second term on the right side of Eq.(7.136) represents an effective perturbation on each site and this effective perturbation is determined by the averaged t-matrix. This approach is not self-consistent as the self-energy depends upon the choice of Ξ.

For simplicity if we put $\Xi = 0$ and choose the perfect A crystal as the reference crystal, we will have

$$t_i^m(\omega) \;=\; \frac{\Delta L_i}{1 - \Delta L_i \mathbf{G}^m(\omega)} \qquad (7.139)$$

and

$$\Sigma \;=\; \frac{< \mathbf{T}^m >}{1 + \mathbf{G}^m(\omega) < \mathbf{T}^m >}. \qquad (7.140)$$

For a binary alloy $A_{1-x}B_x$, one can estimate $< \mathbf{T}^m >$ by taking the concentration weighted average of atomic t-matrices for two type of atoms in the alloy, i.e.

$$\begin{aligned}
< T^m > \;&=\; \sum_i x_i t_i^m(\omega) \\
&=\; (1 - x)t_A^m(\omega) + x t_B^m(\omega).
\end{aligned} \qquad (7.141)$$

Using Eq.(7.141) in (7.140), we get the self-energy

$$\Sigma \;=\; \frac{(1 - x)t_A^m(\omega) + x t_B^m(\omega)}{1 + \mathbf{G}^0(\omega)[(1 - x)t_A^m(\omega) + x t_B^m(\omega)]}. \qquad (7.142)$$

Since the reference crystal is taken as the perfect crystal A, therefore $\mathbf{G}^m(\omega) = \mathbf{G}^0(\omega)$.

The atomic t-matrix can be calculated with the help of Eq.(7.139) if the perturbation ΔL_i is knwon. Let us choose the perturbation ΔL_i in the mass defect approximation and keep the reference crystal as the perfect crystal. Thus the use of Eq.(7.83) in (7.139) will lead to

$$t_i^m(\omega) \;=\; \frac{(M - M_i)\omega^2}{1 - (M - M_i)\omega^2 \mathbf{G}^0(\omega)} \tag{7.143}$$

where M is the mass of perfect host atom and M_i is the mass of ith type of impurity atom. In the binary alloy $A_{1-x}B_x$, $M = M_A$ and $M_i = M_B$, therefore

$$< \mathbf{T}^m > \;=\; \sum_i x_i t_i^m(\omega)$$

$$=\; \frac{x(M_A - M_B)\omega^2}{1 - (M_A - M_B)\omega^2 \mathbf{G}^0(\omega)} \tag{7.144}$$

and

$$\mathbf{\Sigma}^m \;=\; \frac{x(M_A - M_B)\omega^2}{1 + (1 - x)(M_A - M_B)\omega^2 \mathbf{G}^0(\omega)}. \tag{7.145}$$

The self-energy given in Eq.(7.145) and hence the resulting Green's function given in Eq.(7.132) are not symmetric in A and B type of atoms. As x increases, the resonant behaviour decreases and as $x \to 1$

$$\mathbf{\Sigma}^m = (M_A - M_B)\omega^2$$

or

$$\Sigma_{\alpha\beta}^m(l, l') \;=\; (M_A - M_B)\omega^2 \delta_{\alpha\beta}\delta_{ll'}. \tag{7.146}$$

Thus the resonance behaviour of self-energy vanishes and this leads to the virtual crystal approximation [29].

If the reference crystal is taken as the virtual crystal and we put $\Xi = \bar{\Delta \mathbf{L}}$ which is an averaged perturbation, t-matrix becomes

$$< \mathbf{T}^v > \;=\; \frac{(1 - x)(\mathbf{\Delta L}_A - \bar{\mathbf{\Delta L}})}{1 - (\mathbf{\Delta L}_A - \bar{\mathbf{\Delta L}})\mathbf{G}^v(0)}$$

$$+ \frac{x(\mathbf{\Delta L}_B - \bar{\mathbf{\Delta L}})}{1 - (\mathbf{\Delta L}_B - \bar{\mathbf{\Delta L}})\mathbf{G}^v(0)} \tag{7.147}$$

where $\mathbf{G}^v(0)$ is Green's function for the virtual crystal. Here $< \mathbf{T}^v >$ is symmetric in A and B type of atoms. For $x = 0$, the average t-matrix $< \mathbf{T}^v >$ is for a crystal of A type of atoms and for $x = 1$, it is for a crystal of B type of atoms.

The self-energy for the virtual crystal reference medium becomes

$$\Sigma^v = \bar{\Delta\mathbf{L}} + \frac{< \mathbf{T}^v >}{1 + \mathbf{G}^v(0) < \mathbf{T}^v >}. \tag{7.148}$$

For a particular case of $\Delta\mathbf{L}_A = \delta/2, \Delta\mathbf{L}_B = -\delta/2$ and

$$\bar{\Delta\mathbf{L}} = (1 - x)\Delta\mathbf{L}_A + x\Delta\mathbf{L}_B, \tag{7.149}$$

$$< \mathbf{T}^v > = \frac{x(1 - x)\delta^2\mathbf{G}^v(0)}{[1 - x\delta\mathbf{G}^v(0)][1 + (1 - x)\delta\mathbf{G}^v(0)]} \tag{7.150}$$

and

$$\Sigma^v = \bar{\Delta\mathbf{L}} + \frac{x(1 - x)\delta^2\mathbf{G}^v(0)}{1 + \bar{\Delta\mathbf{L}}\mathbf{G}^v(0)}. \tag{7.151}$$

Thus $< \mathbf{T}^v >$ and Σ^v are completely determined in ATA. If the second term of Eq.(7.151) is expanded in powers of $\bar{\Delta\mathbf{L}}$ and terms upto second order are retained, one finds [30]

$$\Sigma^v_{\alpha\beta}(l, l') = [\bar{\Delta\mathbf{L}} + x(1 - x)\delta^2\mathbf{G}^v(0)]_{\alpha\beta}\delta_{ll'}. \tag{7.152}$$

The imaginary part of Eq.(7.152) gives the phonon lifetime in a disordered alloy which is the same as obtained by Ng and Brockhouse [31] using mean lattice as the effective medium.

(ii) Coherent Potential Approximation (CPA)

In ATA, the medium around an impurity is replaced by an average reference medium. This is justified for the small concentration of impurities. However the change in the force constants which connect any two atoms, depends upon the nature of these atoms. Both the atoms may be impurity atoms, host atoms or one is an impurity atom and other is an host atom. Therefore, the change in the force constants will depend upon the local environment of these two atoms. If the forces are long ranged, as may be the case for transition metals and their alloys, it is more appropriate to calculate the effect of local environment of an impurity atom in a self-consistant way. This self-consistant procedure is "Coherent Potential Approximation (CPA)".

In this approximation Ξ is identified as the exact self-energy Σ, then $\Sigma^m = 0$ and $\mathbf{G}^m$ becomes equal to exact $< \mathbf{G} >$ i.e.

$$\Sigma \; = \; \Xi \tag{7.153}$$

and

$$< \mathbf{G} > \; = \; \mathbf{G}^m. \tag{7.154}$$

This is achieved by making the scattering term on the right side of Eq.(7.136) equal to zero i.e.

$$< \mathbf{T}^m > \; = \; 0. \tag{7.155}$$

Here Eqs.(7.153-7.155) are solved self-consistently along with the Eqs. (7.129) and (7.130) until the self-energy Σ is found invariant with respect to the choice of Ξ which is the self-energy of reference crystal. To make the procedure tractable, various approximations are used in the self-consistent solution of these equations.

In the SSA, using Eq.(7.155) in (7.141), one gets

$$(1 - x)t_A^m(\omega) + x t_B^m(\omega) \; = \; 0. \tag{7.156}$$

Further use of Eqs.(7.134), (7.153), (7.154) in Eq.(7.155) gives

$$\frac{(1 - x)(\Delta \mathbf{L}_A - \Sigma)}{1 - (\Delta \mathbf{L}_A - \Sigma)\mathbf{G}^m(\omega)} + \frac{x(\Delta \mathbf{L}_B - \Sigma)}{1 - (\Delta \mathbf{L}_B - \Sigma)\mathbf{G}^m(\omega)} \; = \; 0. \tag{7.157}$$

From Eq.(7.130)

$$\mathbf{G}^m(\omega) \; = \; \left[\mathbf{G}^{0-1} - \Sigma\right]^{-1}. \tag{7.158}$$

Equations (7.157) and (7.158) must be solved self-consistently and the final result $\mathbf{G}^m =< \mathbf{G} >$ should be invariant with respect to the choice of Green's function $\mathbf{G}^0$. This invariance can be proved in the mass defect approximation by putting $\mathbf{G}'^{0-1} = \mathbf{G}^{0-1} - X$, $\mathbf{\Sigma}' = \mathbf{\Sigma} - X$ where $X = (M - M')\omega^2$ in Eq.(7.157).

Equation(7.157) which is obtained in CPA, is symmetric in A and B type of atoms. For $x = 0$ and 1, Eq.(7.157) simplifies exactly for the host metals of A and B type of atoms respectively. Thus Eq.(7.157) can be regarded as an appropriate interpolation formula. The solution of Eq.(7.157) gives the self-energy

$$\Sigma \ = \ \frac{(1 - x)\mathbf{\Delta L}_A + x\mathbf{\Delta L}_B - \mathbf{\Delta L}_A\mathbf{\Delta L}_B\mathbf{G}^m(\omega)}{1 - (\mathbf{\Delta L}_A + \mathbf{\Delta L}_B - \Sigma)\mathbf{G}^m(\omega)}. \tag{7.159}$$

For a particular case given in Eq.(7.149)

$$\Sigma \ = \ \frac{\bar{\mathbf{\Delta L}} + (\delta^2/4)\mathbf{G}^m(\omega)}{1 + \Sigma\mathbf{G}^m(\omega)}. \tag{7.160}$$

It has been found that CPA does provide a better qualitative description of the effect of disorder and is considered to be the best SSA. However, CPA is more difficult to implement than ATA from the computational point of view.

Here it is to be mentioned that the following two approximations have been involved throughout in the above description of ATA and CPA. (i) The random part of the effective Green's function has been neglected. (ii) At large concentration x, there are large number of other impurities in the local environment of an impurity. Therefore additional fluctuations are produced in the lattice excitations. These fluctuations are neglected. Although these fluctuations are weak in strength but large in number. Further investigations are needed in this direction [2].

7.4 Calculations and Results

7.4.1 CPA calculations

$Cu_{1-x}Au_x$, $Al_{0.1}Cu_{0.9}$

Taylor [18] used CPA in the mass defect approximation to calculate the phonon density of states of $Cu_{1-x}Au_x$ alloy [32]. Eqations (7.157) and (7.158) are solved self-consistently for a cubic structure. The self-energy $\Sigma(\omega)$ is a complex quantity in ω-space and its real and imaginary parts are written as

$$\lim_{\delta \to 0} \Sigma(\omega + i\delta) \; = \; \omega^2 \mathbf{\Delta}(\omega) + i\omega^2 \mathbf{\Gamma}(\omega) \qquad (7.161)$$

where $\mathbf{\Delta}(\omega)$ is the frequency shift and $\mathbf{\Gamma}(\omega)$ is the phonon width introduced due to disorder. In the $(\vec{q}j)$ representation, Green's function given in Eqs.(7.154) and (7.158) becomes

$$< \mathbf{G}_j(\vec{q},\omega) > \; = \; \left[\omega^2 - \omega_j^2(\vec{q}) - \mathbf{\Sigma}(\omega)\right]^{-1} \qquad (7.162)$$
$$= \; \mathrm{Re} < \mathbf{G}_j(\vec{q},\omega) > + \mathrm{Im} < \mathbf{G}_j(\vec{q},\omega) > . \qquad (7.163)$$

Using Eq.(7.161) in (7.162) and separating the real and imaginary parts, we get

$$\mathrm{Re} < \mathbf{G}_j(\vec{q},\omega) > \; = \; \frac{\omega^2 \left(1 - \mathbf{\Delta}(\omega)\right) - \omega_j^2(\vec{q})}{\left[\omega^2(1 - \mathbf{\Delta}(\omega)) - \omega_j^2(\vec{q})\right]^2 + [\omega^2\mathbf{\Gamma}(\omega)]^2} \qquad (7.164)$$

and

$$\mathrm{Im} < \mathbf{G}_j(\vec{q},\omega) > \; = \; -\frac{\omega^2\mathbf{\Gamma}(\omega)}{\left[\omega^2(1 - \mathbf{\Delta}(\omega)) - \omega_j^2(\vec{q})\right]^2 + [\omega^2\mathbf{\Gamma}(\omega)]^2} . \qquad (7.165)$$

In the single site approximation, the self-energy is diagonal i.e.

$$\Sigma_{jj'}(\omega) \; = \; \delta_{jj'}\mathbf{\Sigma}(\omega) \qquad (7.166)$$

where $\mathbf{\Sigma}(\omega)$, in the mass defect approximation, for cubic crystal simplifies as

$$\mathbf{\Sigma}(\omega) \; = \; \frac{x\epsilon\omega^2}{1 - \mathbf{M}_0\epsilon\omega^2(1 - x)\mathbf{G}^0(\omega)} . \qquad (7.167)$$

Here ϵ is the mass defect parameter and

$$\mathbf{G}^0(\omega) \;=\; \int \frac{g_0(\omega')d\omega'}{\omega^2(1 - \mathbf{\Delta}(\omega)) - \omega'^2}. \tag{7.168}$$

Equations (7.167) and (7.168) are solved self-consistently using the input information of phonon density of states $g_0(\omega)$ for the perfect crystal of copper from the calculations of Sinha [33]. The results for $\Sigma(\omega)$ as a function of $\bar{x} = (\omega/\omega_m)$ for the mass defect parameter $\epsilon = -2.1$ and for the concentration $x = 0.25, 0.75$ and 0.95 are shown in Fig.7.2. Here ω_m is the maximum frequency of the host lattice. The results for $\mathrm{Re}\Sigma(\bar{x})$ and $\mathrm{Im}\Sigma(\bar{x})$ are shown separately. As the concentration increases, the resonance characteristic enhances and resonance peak moves towards the higher frequencies. A gap in the $\mathrm{Im}\Sigma(\bar{x})$ at $\bar{x} = 0.5$ and $x = 0.75$ leads to a divergence in $\mathrm{Re}\Sigma(\bar{x})$. For further higher concentration, the divergence seems to disappear and the strength of singularity decreases.

In Eqs.(7.67) and (7.68), the phonon density of states of the disordered crystal is calculated using $\mathrm{Im}G_{\alpha\alpha}(l\kappa, l\kappa; \omega^2)$. For a monatomic lattice and for the degenerate phonon modes, the density of states $g(\omega)$ in the CPA simplifies as

$$g(\omega) \;=\; \frac{2}{\pi\omega} \int d\omega' g_m(\omega')\mathrm{Im}\left(\frac{\omega'^2}{\omega^2(1 - \mathbf{\Delta}(\omega)) - \omega'^2}\right) \tag{7.169}$$

where $g_m(\omega')$ is the density of states of the reference crystal. The results for $g(\bar{x})$ for $Cu_{1-x}Au_x$ for different values of x and for a particular mass defect parameter $\epsilon = -2.1$ are shown in Fig.(7.3). Since the density of states is proportional to $\mathrm{Im}G^m(\bar{x})$, the results for $\mathrm{Im}G^m(\bar{x})$ are also shown there. With the increase in x, the height of resonant peak increases and two peaks start getting separated. At $x = 0.75$, there is a gap in $g(\bar{x})$ between $\bar{x} = 0.5$ and 0.6. This gap corresponds to the divergence in $\mathrm{Re}\Sigma(\bar{x})$. At the low concentration, the characteristics of phonon density of states of the reference crystal dominate. With the increase in x, the position of low frequency peak is almost unaltered while the high frequency peaks shift towards low ω. This may be due to the fact that the resonance modes in the host phonon band do not alter their positions with the change in concentration. However, the positions of the resonance modes in the impurity phonon band shift towards lower frequency side. It has not been possible to understand the effect of d-electrons as such in the phonon density of states of $Cu_{1-x}Au_x$ alloy.

The quantity

$$< F_j(\vec{q}, \omega) > \; = \; -\frac{1}{\pi}\mathrm{Im} < \mathbf{G_j}(\vec{q}, \omega) > \qquad (7.170)$$

is known as the configurational averaged spectral density function. The Green's function $< \mathbf{G}_j(\vec{q}, \omega) >$ defined in the complex ω-plane with a branch cut along the real axis can be expressed by the Cauchy integral formula as

$$< \mathbf{G}_j(\vec{q}, \omega) > \; = \; \int_{-\infty}^{+\infty} d\omega' \frac{< F_j(\vec{q}, \omega') >}{\omega - \omega'} \qquad (7.171)$$

$$= \; \int_0^{\infty} d\omega'^2 \frac{< F_j(\vec{q}, \omega') >}{\omega^2 - \omega'^2} \qquad (7.172)$$

because $\mathrm{Im} < \mathbf{G}_j(\vec{q}, \omega) >$ is an odd function of ω. We can write the retarded Green's function as

$$< \mathbf{G}_j(\vec{q}, \omega) >^{ret} \; = \; P \int_{-\infty}^{\infty} d\omega' \frac{< F_j(\vec{q}, \omega') >}{\omega - \omega'} d\omega' - i\pi < F_j(\vec{q}, \omega) > .$$

$$(7.173)$$

Here $< \mathbf{G}_j(\vec{q}, \omega) >^{ret}$ is an analytic function in the complex ω upper-half plane and has the poles in the lower half plane. The real part of retarted Green's function describes the phonon dispersion relation and the imaginary part describes the damping of phonons. Since the first term on the right side of Eq.(7.173) is an analytic function, therefore the poles of $< \mathbf{G}_j(\vec{q}, \omega) >^{ret}$ in the complex ω lower-half plane must coincide with the poles of $< F_j(\vec{q}, \omega) >$. Therefore to study the phonons in the disordered alloy, the poles of $< F_j(\vec{q}, \omega) >$ must be investigated [5]. The phonons will be defined when $< F_j(\vec{q}, \omega) >$ for real ω is Lorentzian with small half width.

The neutron scattering cross section is proportional to $\mathrm{Im} < \mathbf{G}_j(\vec{q}, \omega) >$. From Eq.(7.165), for non-zero $\Gamma(\omega)$, the resonances will be (i) at $\omega = \omega_j(\vec{q})$ when $\mathbf{\Delta}(\omega)$ is negligibly small and at (ii) $\omega = \omega_j(\vec{q})/(1 - \mathbf{\Delta}(\omega))$ when $\mathbf{\Delta}(\omega)$ is appreciable. Since $\mathbf{\Delta}(\omega)$ will depend upon the concentration of impurities, therefore the spectral function may show the response for two different frequencies. For the small concentration, one of them may be identified due to heavy impurity resonance and other in the vicinity of $\omega_j(\vec{q})$. Figure (7.4) shows the positions of maxima in these response peaks as a function of x for different values of $\omega_j(\vec{q})$. In case of two peaks the position of the stronger peak is shown by the solid circles and the position of weaker peaks by open circles. The error bars indicate that the half-maximum points of the peaks are quite asymmetric due to ω-dependence of $\boldsymbol{\Delta}(\omega)$. As x increases, the peak associated with $\omega_j(\vec{q})$ reduces in strength and broadens into an impurity band peak. However, the strength of broad resonant peak increases and finally becomes $\omega_j(\vec{q})$ peak for the perfect crystal with $x = 1$. When the mass ratio is small such that there can be no local mode, there is only one main peak which moves smoothly with x. Thus the frequency of any particular mode does not change smoothly with the increase of x. Möstoller and Kaplan [34] extended the mass defect CPA for polyatomic cubic crystals and calculated $g(\omega)$ for $Al_{0.1}Cu_{0.9}$. The local modes are found broad.

$Cu_{75}Pt_{25}$, $Ni_{95}Pt_5$

The phonon dispersion relations of $Cu_{1-x}Pt_x$ are measured by Kunitomi et al. [35] for the $[00\xi]T$ branch. The phonon band gaps are found in the vicinity of $\xi = 0.5(\pi/a)$. The phonon line profiles near the band gap for different wave vectors and for two values of x are shown in Fig.(7.5). The solid lines show the results of mass defect CPA calculations carried out with the help of Eq.(7.70). The dotted lines through the experimental points are drawn for guidence. The width of the peak increases with the increase of concentration and finally two peaks are separated. The mass-defect CPA calculations also predict two peaks. However, the relative height of peaks do not agree with the experimental results. The experimental results show resonance like behaviour rather that of the localized phonon modes. For the detailed understanding of this behaviour, the frequency shifts $\delta(\omega_j)$ in $Cu_{95}Pt_5$ from perfect copper and in $Ni_{95}Pt_5$ from perfect Ni are shown in Fig.(7.6). The mass defect CPA calculations and the low concentration results for $\delta(\omega_j) = (\omega_j/2)\Delta(\omega)$ are also shown there by solid and dashed lines respectively. A comparison with the experimental results shows that the mass defect CPA is insufficient to explain $\delta(\omega_j)$. The change in the force constants should also be included [36, 37] because the force constants of Pt-Pt, Pt-Cu, and Pt-Ni pairs are larger than the force constants of Cu-Cu and Ni-Ni pairs.

CPA theory is also extended to investigate the characteristics of phonons in the mixed diatomic crystals [38, 39]. It is found that CPA can explain the switching from one mode to two mode type behaviour in III-V group mixed crystals. Such calculations for multicomponent transition metal alloys are difficult.

7.4.2 CPA and ATA Comparative Calculations

$$Ni_{1-x}Pd_x, Cu_{96}Al_4, Cu_{90}Al_{10}, Ge_{91}Si_9, Cu_{1-x}Au_x, Cr_{1-x}W_x$$

Kamitakahara and Taylor [40] compared the mass defect CPA and ATA calculations for phonons in the concentrated $Ni_{1-x}Pd_x$ alloys. ATA does not involve the iterative calculations while CPA is self-consistant. The density of states $g(\omega)$ given in Eq.(7.169) is shown in Fig.(7.7) for different values of x for a fixed ϵ and for different values of ϵ for a fixed x. The solid and dashed lines show the results for ATA and CPA calculations respectively.

The force constants of perfect Ni and Pd are used to obtain the averaged force constants appropriate to $Ni_{55}Pd_{45}$ alloy. These force constants are used to obtain the starting frequency distribution function for a perfect monotomic fcc lattice. The ATA spectra has more structure than CPA spectra. These features are associated with the features in the density of states of Ni and Pd. The shaded region of ATA spectra corresponds to the frequency range where spectral function $< F_j(\vec{q},\omega) >$ is δ-function. As compared to ATA, CPA results are found nearer to the computer experiments [41].

In the frequency distribution, the information about crystal dynamics is averaged. Therefore many fine structure details of crystal dynamics are are lost. However the self-energy $\Sigma(\omega)$ consists of more information about the crystal dynamics. The real part of self-energy determines the frequency shift $\Delta(\omega)$ and the imaginary part gives the phonon width $\Gamma(\omega)$. Experimentally it is difficult to separate the impurity induced frequency shifts from the shift due to change in volume[42]. Kamitakahara and Taylor [40] calculated the phonon width function $\Gamma(\omega)$ in CPA and ATA. The width functions are characterized by two peaks. It is found that the sharpness of peaks increases with the increase of concentration. The ATA width function is found sharper and stronger than CPA width function in the high frequency range. The exception is found for $\epsilon = -1.0$ and $x = 0.9$, when according to CPA, an impurity band almost shifts off the main frequency band. The second difference which is found in the calculations is that ATA width function $\Gamma(\omega) = 0$ for $\omega > \omega_m$ although ATA phonon density of states is finite for $\omega > \omega_m$. This behaviour of results in ATA predicts undamped phonons of frequency $\omega > \omega_m$ for a disordered system. Consequently, the spectral function $< F_j(\vec{q},\omega) >$ becomes δ-function and this is the reason for the sharp structure at high frequencies in ATA density of states as shown in Fig.(7.7). Owing to the self-consistent nature of CPA, its width function $\Gamma(\omega)$ is finite wherever its density of state is finite. This gives finite life time for the phonons.

The width functions $\Gamma(\omega)$ calculated in the ATA and CPA are compared with the experimental results of $Ni_{55}Pd_{45}$ for three phonon groups in Fig.(7.8). It is too difficult to get the shift due to disorder experimentally, therefore calculated $\Gamma(\omega)$ are centered on the experimental results given in Fig.(7.8) by adjusting the value of $\omega_j(\vec{q})$. For a particular case shown in Fig.(7.8a), ATA calculations predict a double peak. The experimental results do not show any doubling and CPA calculations agree quite well with the experimental results. A situation where ATA calculations give little doubling is shown in Fig.(7.8b). Here the experimental results and ATA and CPA calculations are in good agreement although ATA still predicts a weak secondary peak. To show that CPA is by no means perfect, a particular case has been shown in Fig.(7.8c) where the results of CPA and ATA calculations agree but these calculations do not agree with the experimental results. A detailed comparison of CPA calculations and experimental results is given by Kamitakahara et al [13] where it is shown that CPA gives an adequate description of the phonons in the $[\xi 00]$T and $[\xi\xi 0]$T branches but a poor description in the $[\xi\xi\xi]$T and all the longitudinal branches. The ATA has been found quite appropriate to explain the experimentals results of $Cu_{96}Al_4$, $Cu_{90}Al_{10}$ and $Ge_{91}Si_9$ [9, 12].

The experimental results in Fig.(7.8) indicate a considerable $\vec{q}$ dependence of the width function. This pattern indicates the presence of significant changes in the force constants which are not included in the calculations for high concentration alloys. One may argu that the effects of mass difference and the changes in the force constants may cancel each other as pointed out earlier in the virtual crystal approximation [13]. But this may not be the case here. The size of the phonon width is a more direct observation of disorder than the dispersion curves themselves. However, it has been shown in the low concentration calculations for $Cu_{1-x}Au_x$ [43] and $Cr_{1-x}W_x$ [44] that the large changes in force constants do not make particularly large changes in the phonon width. The experimental results further show that the resonance like modes persist even in the concentrated alloys. The calculations indicate that the resonances can occur in an alloy even with the mass ratio 2:1.

The complete phonon dispersion relations for $Ni_{55}Pd_{45}$ [13] are shown in Fig.(7.9). Each phonon is associated with certain width, however the dots denote the mean position of the peaks. The widths are found larger near zone boundaries. The general symmetry of $\omega_j(\vec{q})$ versus $\vec{q}$ is that of the fcc lattice. The anomalous behaviour in the transverse branches is the same as that found in Pd. However, in this case, this behaviour may be due to disorder. The solid lines represent the calculated results of Born-vonKarman model for a virtual crystal of $Ni_{55}Pd_{45}$.

$Ni_{1-x}Pt_x$, $Pd_{1-x}Ag_x$, $Cr_{1-x}V_x$

In the alloys such as $Ni_{1-x}Pt_x$ [35], the frequency square and mass ratio of constituent metals are $(\omega^2_{max}(Pt)/\omega^2_{max}(Ni)) = 0.43$ and $(M_0(Ni)/M_0(Pt)) = 0.3$. Therefore the changes in force constants, in addition to mass difference, must also be accounted for. This disorder was first discussed by Grünewald et al [20, 45] and it is referred as off-diagonal disorder (see section 7.3.1(b)). Möstoller and Kaplan [46] rescaled the atomic force constants and carried out the calculations for one dimensional crystal with equal changes in the masses and the force constants. Katayama and Kanamori [37] did the explicit calculations for the phonon density of states, one phonon neutron scattering cross section and phonon dispersion relations for $Ni_{1-x}Pt_x$, $Pd_{1-x}Ag_x$ and $Cr_{1-x}V_x$ alloys including the changes in both the mass and the force constants. The changes in the force constants $\mathbf{\Delta\Phi}(l, l')$ are determined empirically rather than adopting the sum rule (7.99). The Fourier transformed force constants are written as

$$\begin{aligned}
\Phi^{AA}_{\alpha\beta}(\vec{q}, j) &= \lambda'\Phi_{\alpha\beta}(\vec{q}, j), \\
\Phi^{BB}_{\alpha\beta}(\vec{q}, j) &= \mu'\Phi_{\alpha\beta}(\vec{q}, j)
\end{aligned}$$

and

$$\Phi^{AB}_{\alpha\beta}(\vec{q}, j) = \nu'\Phi_{\alpha\beta}(\vec{q}, j) \tag{7.174}$$

and the force constants $\Phi_{\alpha\beta}(\vec{q}, j)$ are determined with the help of phonon spectra of perfect metal. The parameters λ' and μ' and determined by the relative comparison of phonon spectra of perfect A and B metals. The parameter ν' is determined by obtaining the best fit of experimental data. Equations (7.157) and (7.158) are solved self-consistently to obtain the spectral density function $\mathrm{Im}< \mathbf{G}_j(\vec{q}, \omega) >$. These values are used in Eq.(7.70) to obtain one phonon neutron scattering cross-section. These results along with the experimental data are shown in Fig.(7.10) for the $[00\xi]T$ phonon branch of $Ni_{0.95}Pt_{0.05}$ and $Ni_{0.7}Pt_{0.3}$ alloys. The experimental data on peak positions are shown by the circles. The width of the peak determined by half-peak position is shown by the horizental bars. The results for $Ni_{0.95}Pt_{0.05}$ show a single peak (Fig.7.10a). With the increase in concentration x, the experimental results show two peaks while the calculated results show only one peak (Fig.7.10b). The width of the peaks increases with the increase of frequency. The peaks at low frequency and small wave vector are relatively sharp.

The phonon dispersion relations of $Ni_{0.7}Pt_{0.3}$ are shown in Fig.(7.11). The experimental results due to Katyama et al [37] are shown by dashed line. These results show the phonon band gaps at different energies in the longitudinal and transverse branches. These gaps are not there in the phonon dispersion relations of $Ni_{55}Pd_{45}$. Thus the disorder effects are more prominent in $Ni_{0.7}Pt_{0.3}$ than in $Ni_{55}Pd_{45}$. The calculated peak positions and phonon widths in the CPA are shown by circles and bars respectively. The phonon dispersion relations with finite width are shown by solid lines. A comparison of the experimental and theoretical results show that CPA with off-diagonal disorder provides a reasonable description of phonons in these alloys. For further improvement in the calculations, SSA has to be revoked by including correlations among the atoms in the vicinity of impurities.

The effect of force constants disorder in $Pd_{1-x}Ag_x$ and $Cr_{1-x}V_x$ alloys, where mass disorder is small, is also investigated. The experimental results for phonon density of states $g(\omega)$ for $Cr_{0.25}V_{0.75}$ are explained [37, 47]. Here again the emphasis has been on the disorder effects. The effect of electronic structure of these materials on the phonon frequencies is not analysed.

7.5   Disordered Shell Model and Double Shell Model

In the shell model for the lattice dynamics of transition metals, d-electrons are represented by additional vibrational degrees of freedom. In the breathing shell model, the change in the radius of each shell due to lattice vibrations is taken into account. In the double shell model a certain part of electronic charge density is represented by a supershell which adds one more vibrational degree of freedom.

The above ideas are extended by Grünewald [48] for transition metal alloys. But due to absence of translational symmetry, the lattice dynamics is described in terms of displacement-displacement correlation functions which are averaged over all the possible configurations. The Fourier transform of these correlation functions are used to calculate the vibrational properties of alloys in the CPA formalism.

In a random binary alloy $A_{1-x}B_x$, the atoms having masses M_A and M_B respectively are randomly distributed. The cores are located at the equilibrium positions $\vec{R}_l^0$ and these are coupled only to their own surrounding shells by a configuration-dependent spring constant γ_A or γ_B. The coupling between the shells is described by configuration-independent force constant matrix.

Thus the equation of motion for the simple shell model, adopting adiabatic approximation (i.e. shell mass $m_s(l) = 0$), can be written as

$$\mathbf{M}(l)\ddot{\mathbf{u}}_\alpha(l) \;=\; -\bar{\gamma}(l)\mathbf{u}_\alpha(l) - \sum_{l'\beta} \mathbf{\Phi}_{\alpha\beta}(l - l')\mathbf{u}_\beta(l'). \qquad (7.175)$$

Here $\mathbf{M}, \mathbf{u}, \mathbf{\Phi}$ and $\bar{\gamma}$ are 2×2 matrices in the core-shell space i.e.

$$\mathbf{M} = \begin{bmatrix} M_c & 0 \\ 0 & m_s \end{bmatrix}, \mathbf{\Phi} = \begin{bmatrix} 0 & 0 \\ 0 & \Phi \end{bmatrix}, \mathbf{u} = \begin{bmatrix} u \\ v \end{bmatrix}, \bar{\gamma} = \begin{bmatrix} \gamma & -\gamma \\ -\gamma & \gamma \end{bmatrix}. \qquad (7.176)$$

M_c and m_s are core and shell masses respectively, $\mathbf{\Phi}$ is core and the neighbouring shell force constant matrix, u and v are core and shell displacements respectively and γ are the core and its shell force constants.

The force constants $\mathbf{\Phi}_{\alpha\beta}(l - l')$ will depend upon the occupation of the sites l or l' by A or B type of atoms and $\mathbf{\Phi}_{\alpha\beta}(0) = -\sum_{l',(\neq l)} \mathbf{\Phi}_{\alpha\beta}(l - l')$. The corresponding displacement-displacement Green's function is also written as 2 x 2 matrix for the core and shell displacements i.e.

$$\mathbf{G}_{\alpha\beta}(ll', t) = \begin{bmatrix} G_{\alpha\beta}^{uu}(ll', t) & G_{\alpha\beta}^{uv}(ll', t) \\ G_{\alpha\beta}^{vu}(ll', t) & G_{\alpha\beta}^{vv}(ll', t) \end{bmatrix}. \qquad (7.177)$$

After taking the time Fourier transform of Eqs.(7.175) and (7.176), the equation of motion for $\mathbf{G}$ becomes

$$\left[\mathbf{M}(l)\omega^2 - \bar{\gamma}(l)\right] \mathbf{G}(ll', \omega) \;=\; \delta_{ll'}\mathbf{I} + \sum_{l''} \mathbf{\Phi}(l - l'')\mathbf{G}(l''l'; \omega). \qquad (7.178)$$

If we define the Green's function

$$\begin{aligned}
g(l) &= \left[\mathbf{M}(l)\omega^2 - \bar{\gamma}(l)\right]^{-1} \\
&= \begin{bmatrix} g_{uu} & g_{uv} \\ g_{vu} & g_{vv} \end{bmatrix},
\end{aligned} \qquad (7.179)$$

Eq.(7.178) can be simplified to get

$$\mathbf{G}(ll') \;=\; g(l)\delta_{ll'} + \sum_{l''} g(l)\mathbf{\Phi}(l - l'')\mathbf{G}(l''l'). \qquad (7.180)$$

Here every quantity is a 2×2 matrix.

The configurational average of Eq.(7.180) gives

$$< \mathbf{G}(ll') > \;=\; \Upsilon\delta_{ll'} + \sum_{l''} \Upsilon\mathbf{\Phi}(l - l'') < \mathbf{G}(l''l') > \qquad (7.181)$$

where the coherent locator $\Upsilon =< g(l) >$. By writing Eq.(7.181) in the $3N \times 3N$ matrix representation of the sites and then using the single site approximation one can calculate Υ. The result for the cubic lattice is

$$\begin{aligned}
\Upsilon \;&=\; < g(l)[I- < U(00) > (g(l) - \Upsilon)]^{-1} > \\
&=\; \begin{bmatrix} \Upsilon_{uu} & \Upsilon_{uv} \\ \Upsilon_{vu} & \Upsilon_{vv} \end{bmatrix}
\end{aligned} \qquad (7.182)$$

where the interactor

$$< U(00) > \;=\; \frac{1}{3N} \sum_{\vec{q}j}[\mathbf{I} - \mathbf{\Phi}(\vec{q}j)\Upsilon]^{-1}\mathbf{\Phi}(\vec{q}j). \qquad (7.183)$$

Equations (7.182) and (7.183) are solved self-consistantly. The detailed expression for Υ_{uu} etc. are calculated and it is found that only a single non-linear equation for Υ_{vv} is needed to be solved which is given as

$$\begin{aligned}
\Upsilon_{vv}^{-1} \;=\;& (1 - x)g_{vv}^{-1}(A) + xg_{vv}^{-1}(B) \\
&+ \left[\Upsilon_{vv}^{-1} - g_{vv}^{-1}(A)\right]\left[\Upsilon_{vv}^{-1} - g_{vv}^{-1}(B)\right] f(\omega)
\end{aligned} \qquad (7.184)$$

where

$$g_{vv}^{-1}(A/B) \;=\; M_{A/B}\omega^2 + \frac{M_{A/B}\omega^2}{1 - M_{A/B}\omega^2/\gamma_{A/B}} \qquad (7.185)$$

and

$$f(\omega) \;=\; \frac{1}{3N} \sum_{\vec{q}j} \frac{1}{\Upsilon_{vv}^{-1} - \mathbf{\Phi}(\vec{q}j)}. \qquad (7.186)$$

If one keeps $\gamma_A = \gamma_B = \infty$ i.e. the shells are tightly bound to the core, Eq.(7.184) exactly reduces to CPA formalism. If $M_A = M_B = M$, $\gamma_A = \gamma_B = \gamma$, one has the solution $\Upsilon = g$ for the perfect crystal i.e.

$$\Upsilon_{vv}^{-1} = g_{vv}^{-1}$$
$$= \frac{M\omega^2}{1 - M\omega^2/\gamma}. \tag{7.187}$$

Phonon dispersion relations are obtained by solving the equation

$$det| < \mathbf{G} >^{-1} | = det|\Upsilon^{-1} - \mathbf{\Phi}| = 0. \tag{7.188}$$

This gives the solution

$$M\omega_j^2(\vec{q}) = \Phi(\vec{q}j)/[1 + \Phi(\vec{q}j)/\gamma]. \tag{7.189}$$

In general $\alpha_{vv}^{-1}(\omega)$ becomes a complex quantity for a disordered crystal and the concept of phonon dispersion relation, as such, looses its meaning.

Phonon density of states is calculated using the mass weighted imaginary part of Green's function which gives the final result as

$$g(\omega) = -\frac{2}{\pi\omega}\text{Im}\left[\Upsilon_{vv}^{-1}\Upsilon_{vu}\left\{1 - \Upsilon_{vv}^{-1}f(\omega)\right\}\right]. \tag{7.190}$$

Here $\Upsilon_{vv}^{-1}(\omega)$ is calculated self-consistently and then $\Upsilon_{vu}^{-1}(\omega)$ is obtained with the help of $\Upsilon_{vv}^{-1}(\omega)$. For a simplified model where the shells of atoms A and B are rigidly bound to the core and mass disorder is neglected i.e. $\gamma_A = \infty$ and $M_A = M_B = M$, the calculated phonon density of states is shown in Fig.(7.12) for different values of concentration x and the disorder parameter $s = M\omega_m^2/\gamma_B$. The low frequency part of the spectrum is not changed by disorder in the core-shell force constant γ. However, in the high frequency part, the longitudinal peak shifts towards lower frequency side and the height of the peak is also changed [45].

Grünewald [49] further extended the above formalism for double shell model for transition metal alloys. In addition to core and shell, a supershell for d-electron charge density is also added. Thus **u** becomes a 3×1 coloumn matrix and $\mathbf{M}, \mathbf{\Phi}$ and γ become 3×3 matrices. The phonon width is calculated in the single site CPA approximation for $Nb_{1-x}Mo_x$. It is found that the disorder in the double shell-innershell coupling constant leads to significant broadening of the phonons. The phonon line widths are found appreciable only in the regions of phonon anomalies. Further investigations beyond SSA and diagonal disorder are needed to understand the contribution of d-electrons towards the phonon structure in these alloys.

7.6   Epilogue

In the above description of dynamics of disordered alloys, the main emphasis has been on the effects of mass and force constant disorders. In most of the calculations, single site approximation has been used. There are certain effects such as spiky structure in the phonon density of states which arise due to clustering of atoms, cannot be explained by these methods. The new methods such as molecular coherent potential approximation, homomorphic cluster CPA and computer simulations must be used. In fact, lots of work is needed to understand the vibrational complexities due to clustering in these alloys [50, 51].

The dynamics of disordered systems with finite imurity concentration is itself a complete subject. It is equally important to understand the localization of phonon states in three dimensions to explain the transport properties of structurally disordered alloys. The microscopic understanding of these aspects is yet to be developed [52]-[56].

In the transition metal and rare earth metal alloys, partially localized conduction electrons contribute significantly towards the electronic and elastic properties. The dynamics of these materials can be developed using electronic response function as discussed in chapter 4. However, the abinitio understanding of electronic structure of these alloys is yet only partial. Therefore it seems difficult, at least at this stage, to know the effect of partially localized conduction electrons on the phonon structure of these alloys. An attempt through the disordered shell model and double shell model is just a begining. New methods are yet to come up to understand the dynamics of non-periodic systems.

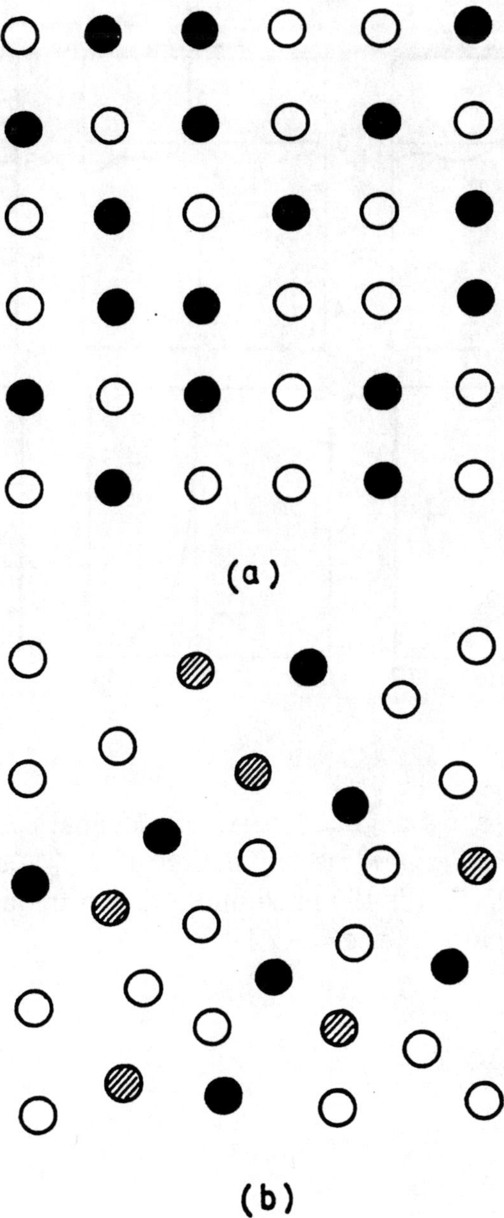

Figure 7.1: (a) Substitutional disorder. (b) Random disorder.

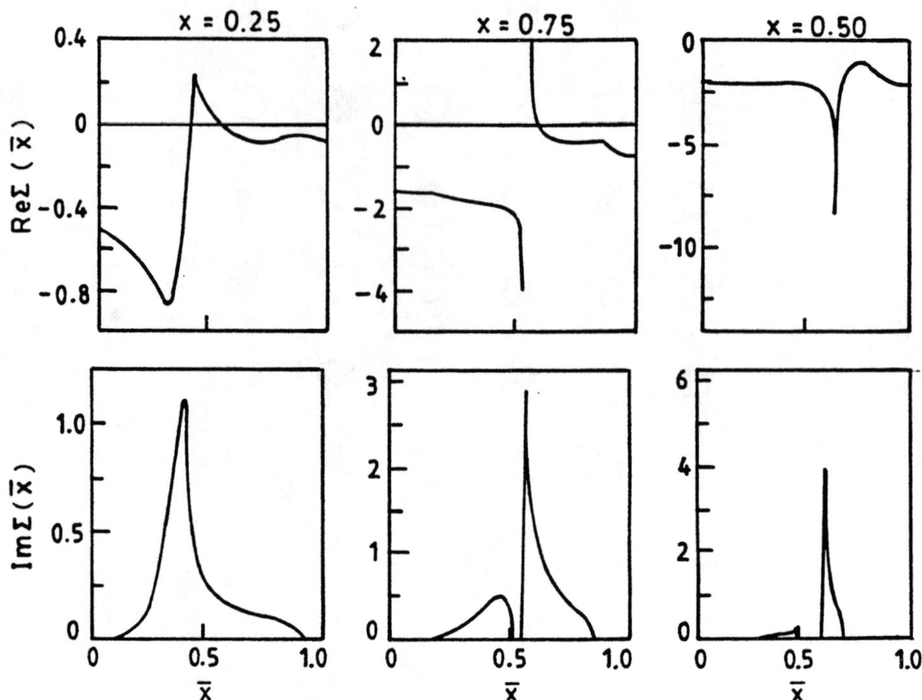

Figure 7.2: Real and imaginary parts of diagonal component of self-energy $\Sigma(\bar{x})$ for $Cu_{1-x}Au_x$ as a function of $\bar{x} = (\omega/\omega_m)$ and $x = 0.25, 0.75$ and 0.95. ω_m is the maximum phonon frequency of Cu-host and mass defect parameter $\epsilon = -2.1$ [18].

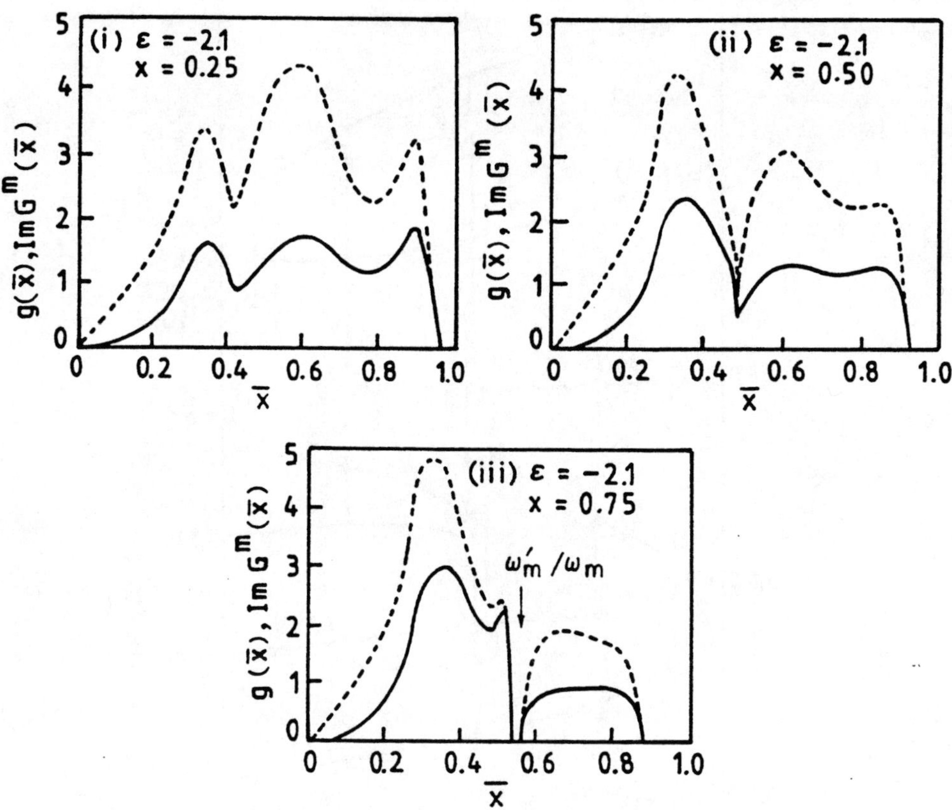

Figure 7.3: Phonon density of states $g(\bar{x})$ (solid line) and $\mathrm{ImG}^m(\bar{x})$ (dashed line) versus $\bar{x}$ for $Cu_{1-x}Au_x$ for mass defect parameter $\epsilon = -2.1$ and $x = 0.25, 0.50$ and 0.75 [18].

 S. *PRAKASH*

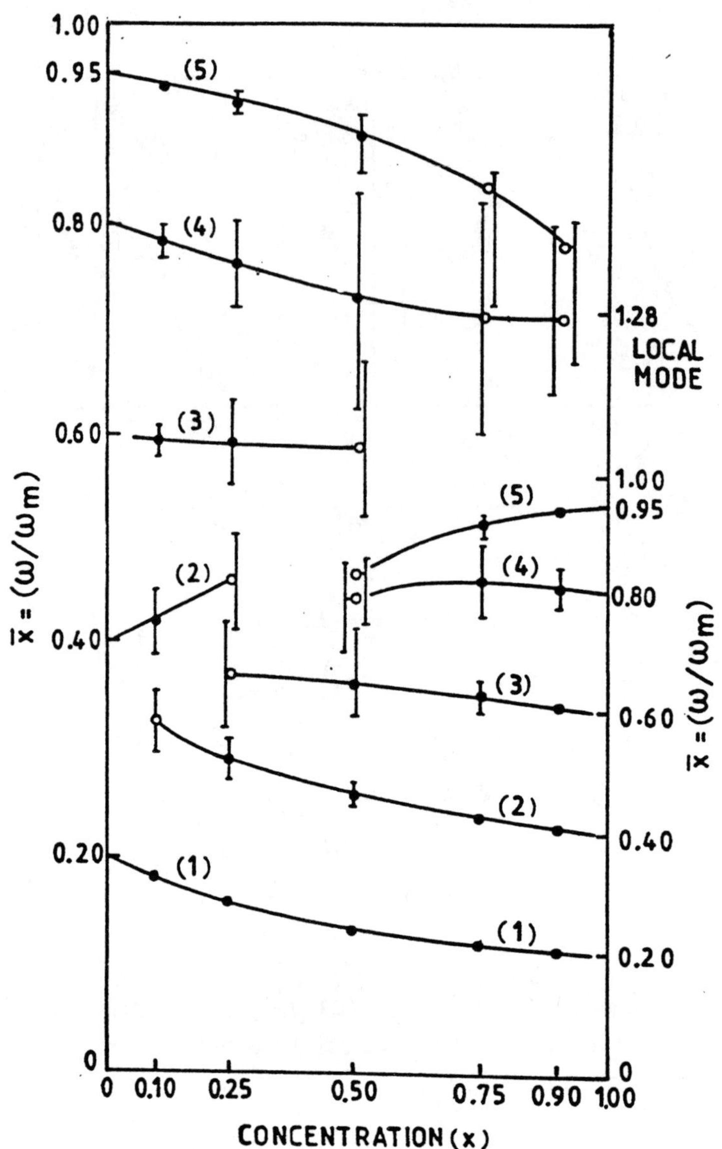

Figure 7.4: The positions and half-maxima points of spectral function of $Cu_{1-x}Au_x$ for $\epsilon = -2.1$ and for different values of concentration x. Different curves are for $\omega_j(\vec{q}) = 0.2\omega_m$ (1), $\omega_j(\vec{q}) = 0.4\omega_m$ (2), $\omega_j(\vec{q}) = 0.6\omega_m$ (3), $\omega_j(\vec{q}) = 0.8\omega_m$ (4) and $\omega_j(\vec{q}) = 0.95\omega_m$ (5). In case of two peaks, stronger peak is shown by solid circles and weaker peak by open circles [18]. The vertical lines are the error bars.

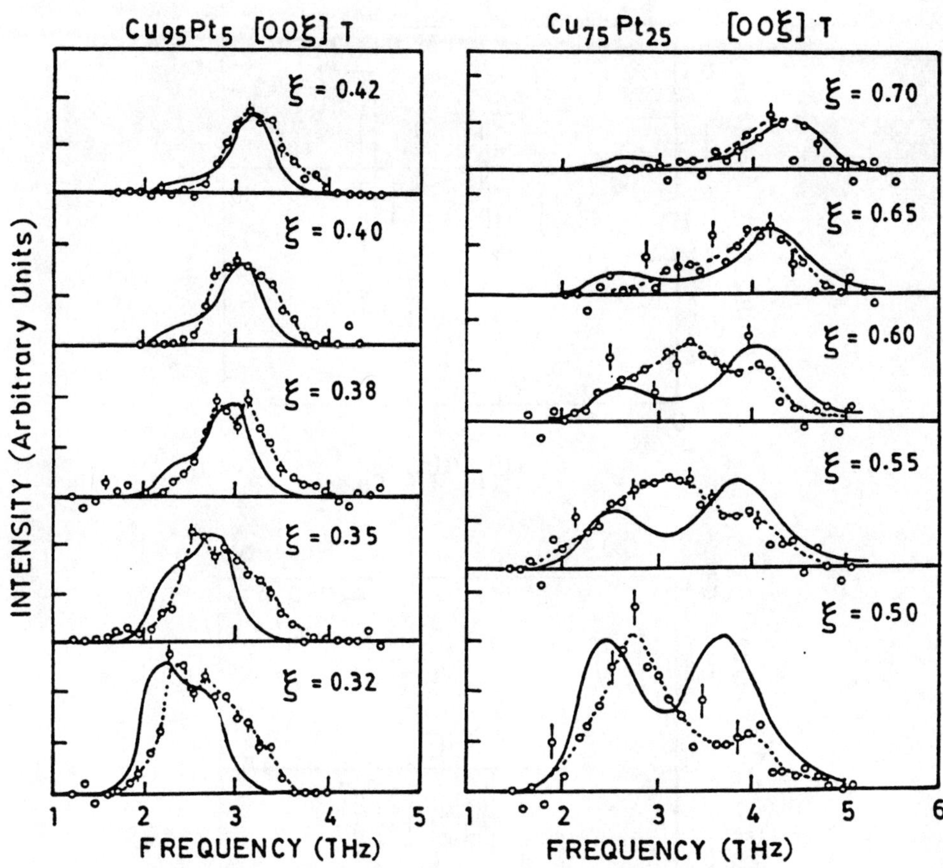

Figure 7.5: Phonon line profiles of the phonon modes in the [00ξ]T branch for different values of phonon wave vectors near the energy gaps in $Cu_{95}Pt_5$ and $Cu_{75}Pt_{25}$. Solid lines show the results of mass defect CPA calculations and dotted lines through the experimental points are just for guidance. The vertical lines are the error bars [35]. ξ is in units of $(2\pi/a)$.

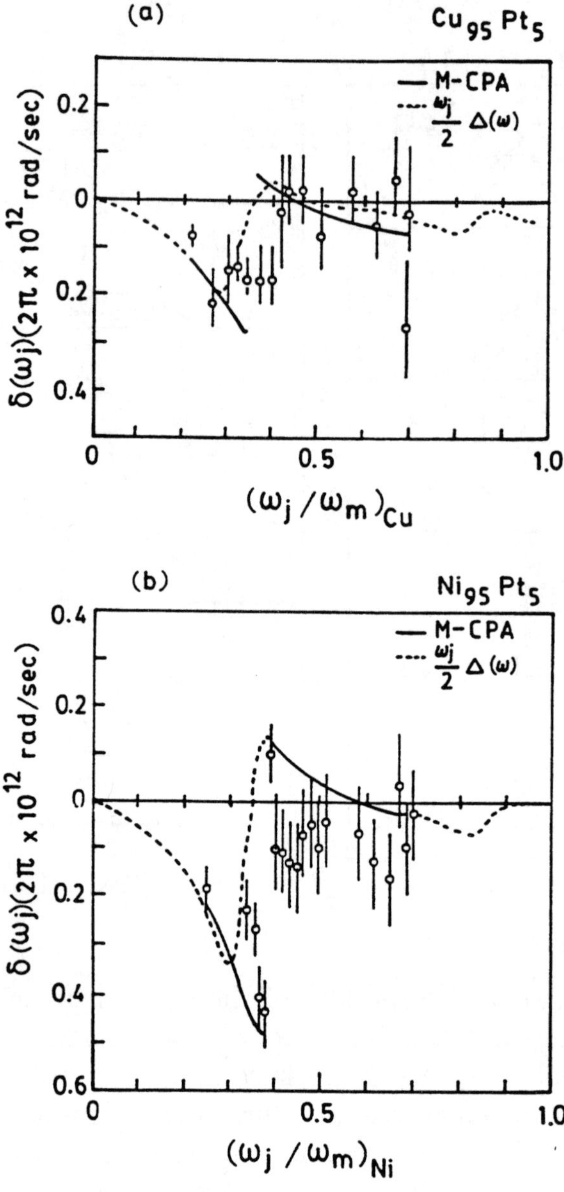

Figure 7.6: Phonon frequency shift $\delta(\omega_j)$ in the $[00\xi]T$ branch for $Cu_{95}Pt_5$(a) and $Ni_{95}Pt_5$(b) from pure Cu and Ni metals respectively. Open circles are the experimental points and vertical lines are the error bars. The solid lines represent the results of mass defect CPA calculations and the dashed lines represent the results for low concentration i.e. $\delta(\omega_j) = (\omega_j/2)\Delta(\omega)$.

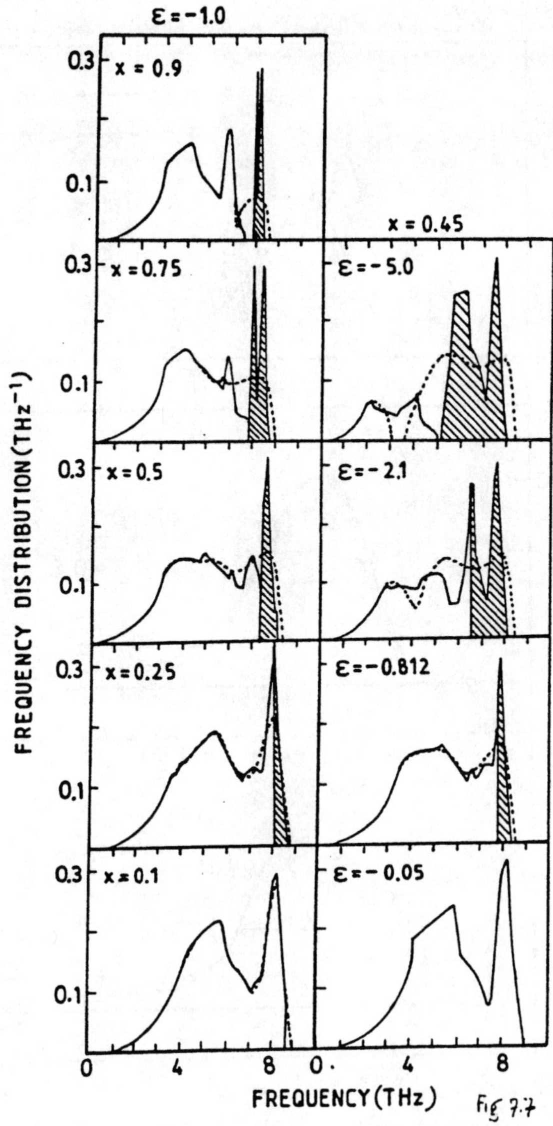

Figure 7.7: Average t-matrix approximation (ATA) and coherent potential approximation (CPA) frequency distributions $g(\omega)$ for fcc random alloy $Ni_{55}Pd_{45}$ with averaged force constants. Solid and dashed lines show the ATA and CPA calculations respectively. The shaded areas show the δ- function contribution. On the left side, the mass ratio $\epsilon = -1$ and concentration x varies from 0.1 to 0.9. On the right side $x = 0.45$ and mass ratio ϵ varies from -0.05 to -5.0[40].

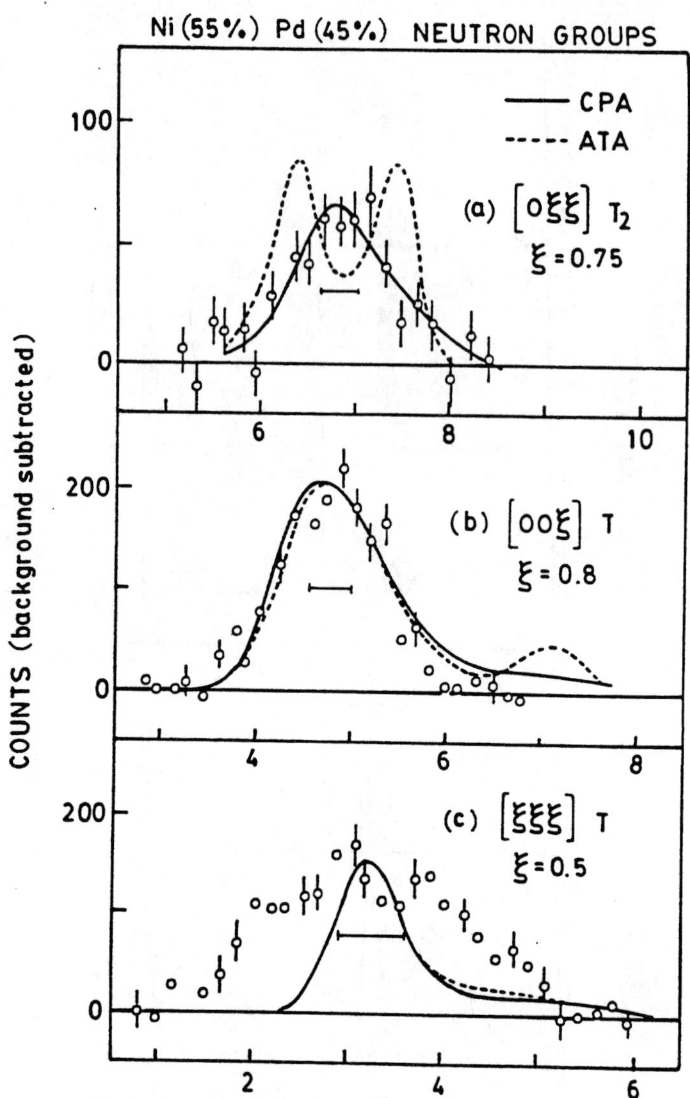

Figure 7.8: Comparison of measured and calculated phonon line width $\Gamma(\omega)$ for $Ni_{55}Pd_{45}$. The dashed and solid lines show ATA and CPA calculations respectively. (a) $[0\xi\xi]T_2$ branch and $\xi = 0.75$. (b) $[00\xi]T$ branch and $\xi = 0.8$. (c) $[\xi\xi\xi]T$ branch and $\xi = 0.5$. ξ is in units of $(2\pi/a)$. The experimental results are shown by circles. Vertical lines are the error bars and horizontal bars indicate the experimental full line width at half-maximum [40].

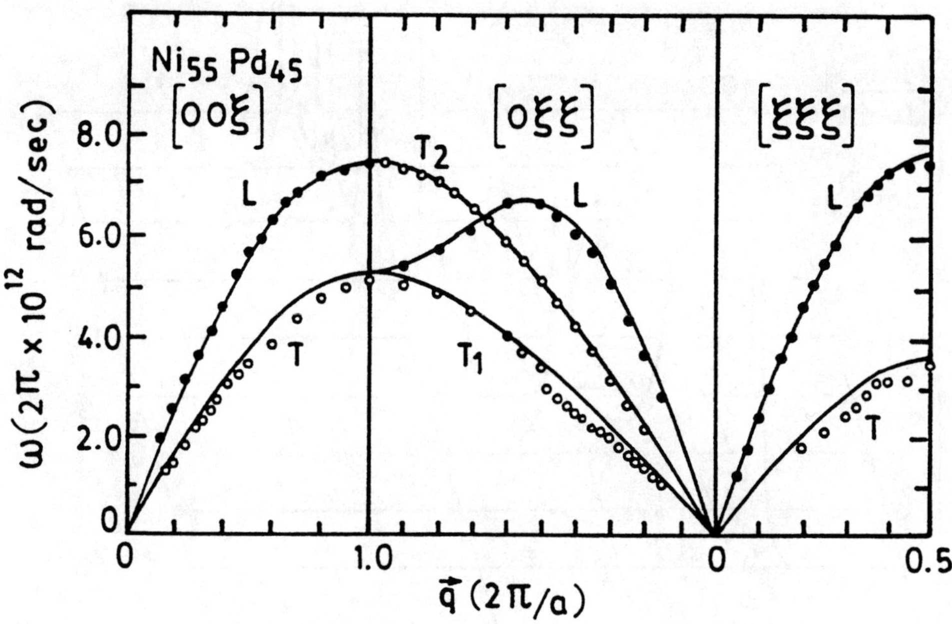

Figure 7.9: Phonon dispersion curves along three principal symmetry directions $[00\xi]$, $[0\xi\xi]$ and $[\xi\xi\xi]$ for $Ni_{55}Pd_{45}$. Dots represent the mean positions of phonon peaks. Solid lines represent the results of Born-von-Karman model calculations for virtual crystal of $Ni_{55}Pd_{45}$ with averaged mass and averaged force constants [13].

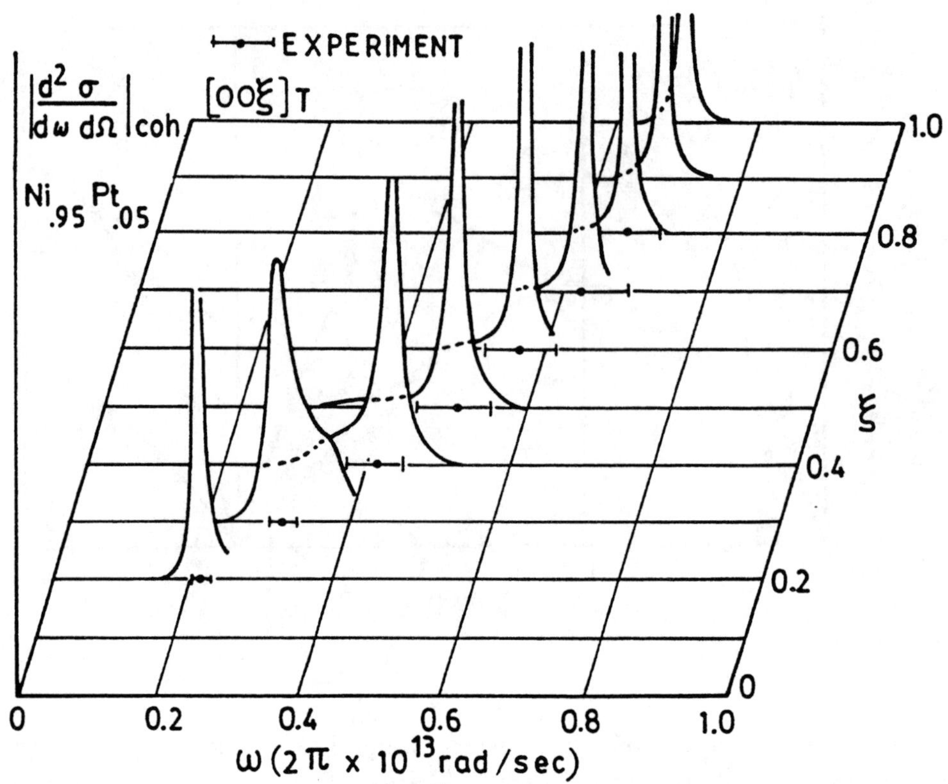

Figure 7.10: (a) One phonon neutron scattering cross-section for transverse phonon mode in the $[00\xi]$ direction for $Ni_{0.95}Pt_{0.05}$. The experimental data on the peak positions are shown by the circles and the width determined by the half peak positions is shown by the horizontal bars. The solid lines show the calculated results [37]. (b) One phonon neutron scattering cross-section for transverse mode in the $[00\xi]$ direction for $Ni_{0.7}Pt_{0.3}$. The description is the same as that of Fig.7.10(a) [37].

Figure 7.10(b) continued

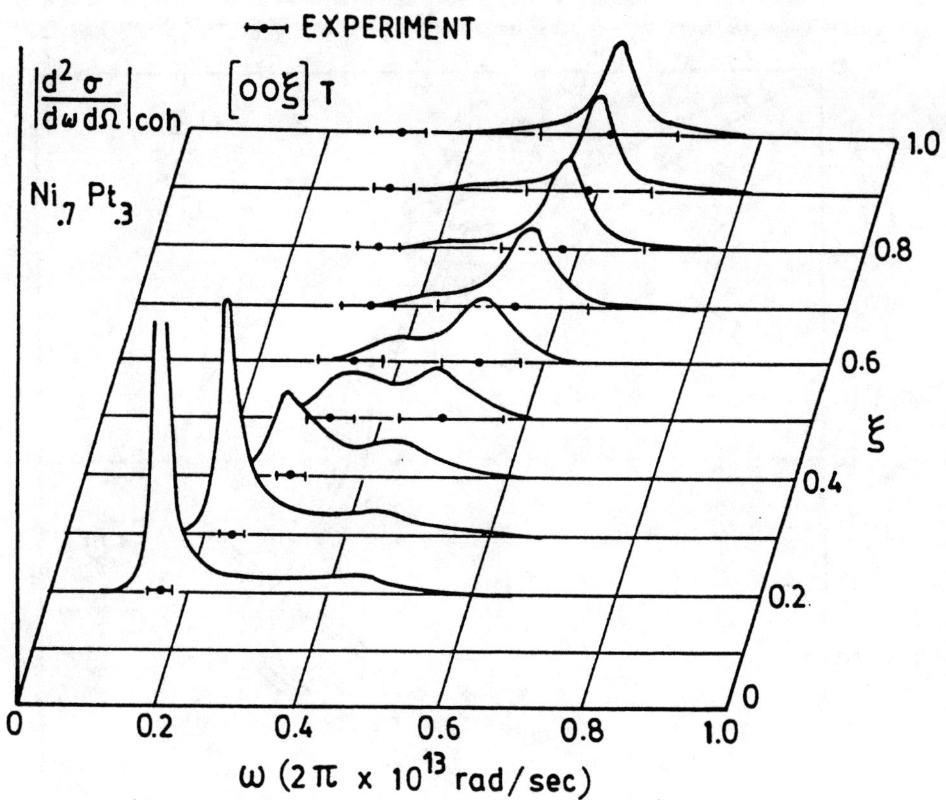

Figure 7.10(b)

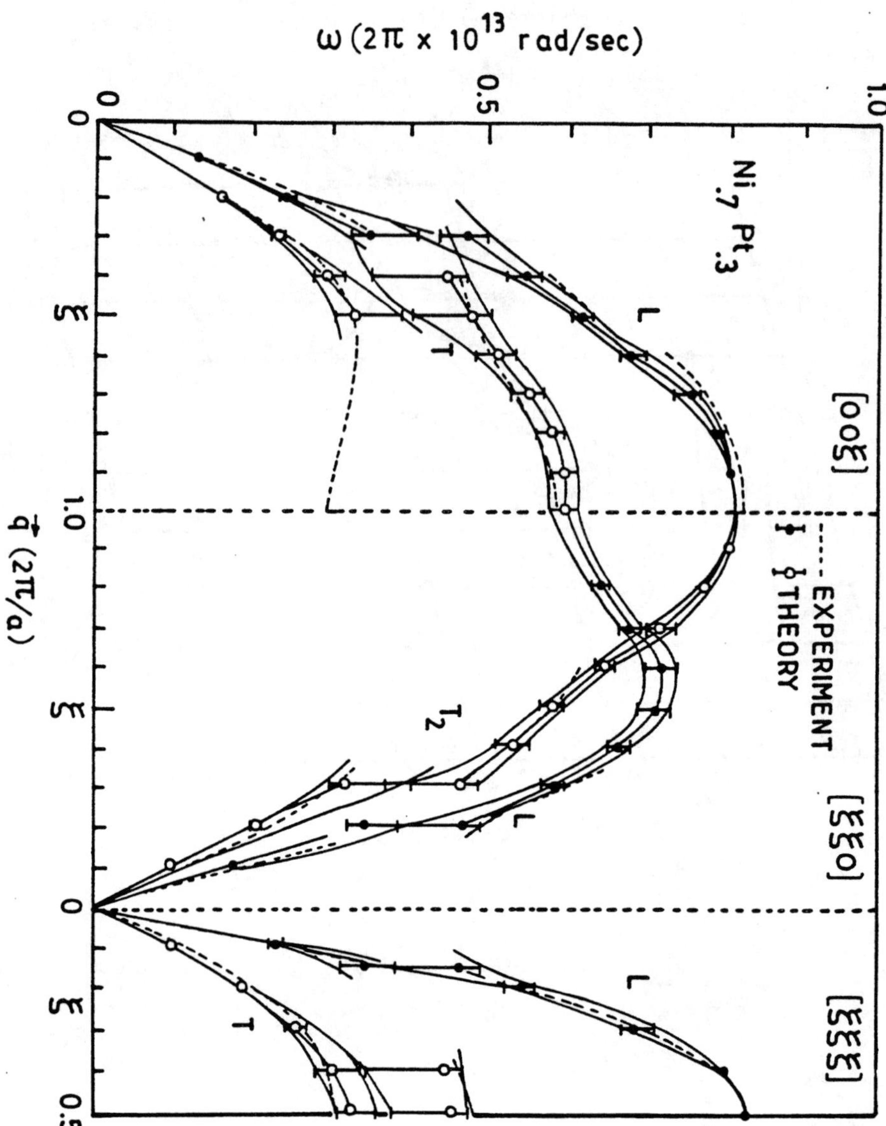

Figure 7.11: Phonon dispersion relations of $Ni_{0.7}Pt_{0.3}$. Dashed lines show the experimental data of peak positions. Upper and lower full lines show the calculated results in CPA for half peak positons and the central full lines show the peak positions. Calculated peak positions and widths are shown by circles and vertical bars respectively [37].

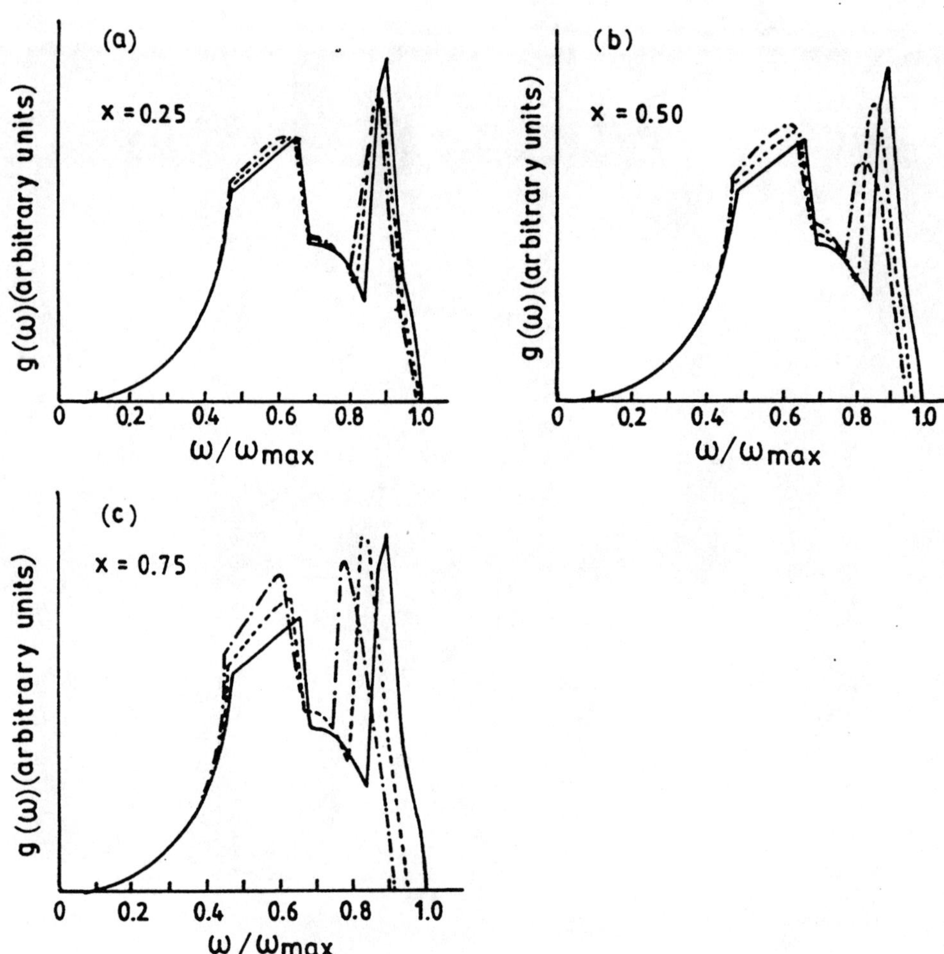

Figure 7.12: Calculated phonon density of states for disordered $Nb_{1-x}Zr_x$ in the disordered shell model. Solid lines show the density of states for an fcc lattice with disorder parameter $s = 0$. Dashed lines are for $s = 0.2$ and dash-dot lines are for $s = 0.4$. The figures (a), (b) and (c) are for $x = 0.25, 0.50$ and 0.75 respectively [48].

Bibliography

[1] A.A. Maradudin, E.W. Montroll, G.H. Weiss and I.P. Ipatova, Theory of Lattice Dynamics in the Harmonic Approximation (Academic Press, New York, 1971 2nd Ed.) p.353.

[2] Harald Böttger, Principles of the Theory of Lattice Dynamics (Physik Verlag, Weinheim, 1980) p.78.

[3] J.M. Ziman, Models of Disorder (Cambridge University Press, Cambridge, London, 1979) p.1.

[4] R.J. Elliot, J.A. Krumhansl and P.L. Leath, Rev. Mod. Phys. 46, 465 (1974) and references therein.

[5] J. Mahanty, The Green's Function Methods in Solid State Physics (Affiliated East-West Press, Pvt. Ltd, New Delhi, 1974) p.23.

[6] V. Heine, Group Theory in Quantum Mechanics (Pergamon Press, Oxford, 1960) p.44.

[7] B. Mozer, K. Otnes and C.Thapar, Phys. Rev. 152, 535 (1966).

[8] U. Bonse and H. Ranch, Neutron Interferometry, (Oxford University Press, 1979).

[9] N. Wakabayashi, R.M. Nicklow and H.G. Smith, Phys. Rev. B4, 2558(1971).

[10] N. Wakabayashi, Phys. Rev. B8, 6015 (1973).

[11] M. Möstoller, T. Kaplant, N. Wakabayashi and R.M. Nicklow, Phys. Rev. B10, 3144(1974).

[12] R.M. Nicklow, P.R. Vijayaraghavan, H.G. Smith and M.K. Wilkinson, Phys. Rev. Letters, 20, 1245 (1968).

[13] W.A. Kamitakahara and B.N. Brockhouse, Phys. Rev. B10, 1200 (1974).

[14] N. Kunitomi, Y. Tsunoda, N. Wakabayashi, R.M. Nicklow and H.G. Smith, Solid State Comm. 25, 921 (1978).

[15] J. Beeby, Proc. Roy. Soc. A279, 82 (1964).

[16] R. N. Aiyer, R. J. Elliot, J. A. Krumhansl and P. L. Leath, Phys. Rev. 181, 1006 (1969).

[17] B. Velicky, S. Kirkpatrick, H. Ehrenreich, Phys. Rev. 175, 747 (1968).

[18] D. W. Taylor, Phys. Rev. 156, 1017 (1967).

[19] M. Möstoller and T. Kaplan, Phys. Rev. B16, 2350 (1977).

[20] G. Grünewald, J. Phys. F6, 999 (1976).

[21] A. P. G. Kutty, Solid State Comm. 14, 213 (1974).

[22] S. Garg, H.C. Gupta, T.K. Bansal and B.B. Tripathi, Physica 125B, 293 (1984).

[23] J. C. Upadhaya and Radheshyam, in "Advances in Statistical Physics of Solids and Liquids" (Eds. S. Prakash and K.N. Pathak, John Wiley, 1990) p.244.

[24] Y. Endoh, Y. Tsunoda and M. Iizumi, J. Phys. Soc. Japan, 50, 469 (1981).

[25] A. L. Simon and C. M. Varma, Solid State Comm. 35, 317 (1980).

[26] N. Singh, Phys. Rev. B42, 8882 (1990).

[27] P. Soven, Phys. Rev. 156, 809 (1967).

[28] P. L. Leath and B. Goodman, Phys. Rev. 181, 1062 (1969).

[29] L. Schwartz, F. Brouers, A.V. Vedyayev and H.Ehrenreich, Phys. Rev. B4, 3338 (1971).

[30] R. J. Elliot and D.W. Taylor, Proc. Roy. Soc. A296, 161 (1967).

[31] S. C. Ng and B.N. Brockhouse, Solid State Comm. 5, 79 (1967).

[32] J. Hafner, in "Hamiltonians to Phase Diagram" (Springer Verlag, Berlin, 1987).

[33] S.K. Sinha, Phys. Rev. 143, 422 (1966).

[34] M. Möstoller and T. Kaplan, Phys. Rev. B9, 353 (1974).

[35] N. Kunitomi, Y. Tsunoda and H. Shirashi, Solid State Comm. 34, 519 (1980).

[36] G. Grünewald and K. Scharnberg, in "Lattice Dynamics" (Ed. M. Balkanski, Flammarion, Paris, 1977) p.443.

[37] H. Katayama and J. Kanamori, J. Phys. Soc. Japan, 45, 1157 (1978).

[38] P. N. Sen and W. M. Hartmann, Phys. Rev. B9, 367 (1974).

[39] D. W. Taylor, J. Phys. C9, 453 (1976).

[40] W. A. Kamitakahara and D. W. Taylor, Phys. Rev. B10, 1190 (1974).

[41] D. N. Payton and W. M. Visscher, Phys. Rev. 175, 120 (1968).

[42] W. A. Kamitakahara and J. R. D. Copley, Phys. Rev. B18, 3772 (1978).

[43] K. M. Kesharwani and B. K. Agrawal, Phys. Rev. B7, 5153 (1973).

[44] K. M. Kesharwani and B. K. Agrawal, Phys. Rev. B6, 2178 (1972).

[45] G. Grünewald and K. Scharnberg, Solid State Comm. 32, 955 (1979).

[46] M. Möstoller and T. Kaplan, Phys. Rev. B19, 3938 (1979).

[47] B. Mozer, Proc. Symp. Copenhagen, IAEA (1968) p.55.

[48] G. Grünewald, J. Phys. C13, 2103 (1980).

[49] G. Grünewald, J. Phys. C14, 595 (1981).

[50] J. E. Hirsch, Phys. Rev. B18, 3976 (1978).

[51] A. Puri and T. Odagaki Phys. Rev. B24, 5541 (1981).

[52] P. W. Anderson, D. J. Thouless, E. Abrahams and D. S. Fisher, Phys. Rev. B22, 3519 (1980); B. Kramer, A. MacKinnor and D. Weaire, Phys. Rev. B23, 6357 (1981); B. Fultz and J. W. Morris, Phys. Rev. B34, 4480 (1986); B. Fultz, A. Hamdet and D.H. Pearson, Acta Metall. 37, 2841 (1989).

[53] J. C. Flores, J. Phys. Condensed Matter 1, 8471 (1989); H. L. Wu and P. Phillips, Phys. Rev. Letters 66, 1966 (1991); R. Darolia, D. Lahrman and R. Field, Scr. Metall. 26, 1007 (1992).

[54] S. E. Burkov, Phys. Rev. B47, 12325 (1993); P. K. Datta, D. Giri and K. Kundu, Phys. Rev. B47 10727 (1993); ibid J. Phys. Condensed Matter 6, 4415 (1994).

[55] C. Djega-Mariadassou, L. Bessais and C. Servant, Phys. Rev. B51, 8830 (1995); P. Kramer, A. Quandt, M. Schlottmann and T. Schneider, Phys. Rev. B51, 8815 (1995); F. D. Adame, E. Diez and A. Sanchez, Phys. Rev. B51 8115 (1995).

[56] R. McCormach, D. deFontaine, C. Wolverton and G. Ceder, Phys. Rev. B51, 15808 (1995).

INDEX